The Madison Symposium
on Complex Analysis

These Proceedings are dedicated to
Walter Rudin
Villas Professor of Mathematics
on the occasion of his retirement

CONTEMPORARY MATHEMATICS

137

The Madison Symposium
on Complex Analysis

Proceedings of the Symposium
on Complex Analysis
held June 2–7, 1991,
at the University of Wisconsin-Madison,
with support from the National Science Foundation
and the William F. Villas Trust Estate

Alexander Nagel
Edgar Lee Stout
Editors

American Mathematical Society
Providence, Rhode Island

The Symposium on Complex Analysis was held at the University of Wisconsin-Madison from June 2–7, 1991, in conjunction with the retirement of Walter Rudin, Villas Professor of Mathematics. This symposium was supported in part by grants from the National Science Foundation and the William F. Villas Trust Estate.

1991 *Mathematics Subject Classification*. Primary 32-xx, 30-xx; Secondary 47-xx, 42-xx.

Library of Congress Cataloging-in-Publication Data

Symposium on Complex Analysis (1991: University of Wisconsin-Madison)
The Madison Symposium on Complex Analysis: proceedings of the Symposium on Complex Analysis held June 2–7, 1991 at the University of Wisconsin-Madison/Alexander Nagel, Edgar Lee Stout, editors.
p. cm.—(Contemporary mathematics; 137)
Includes bibliographical references.
ISBN 0-8218-5147-0
1. Functions of complex variables—Congresses. 2. Mathematical analysis—Congresses. I. Nagel, Alexander, 1945– . II. Stout, Edgar Lee, 1938– . III. Title. IV. Series.
QA331.7.S95 1991
515′.9—dc20

92-23702
CIP

This volume was printed directly from copy typeset by the editors
using $\mathcal{A}_{\mathcal{M}}\mathcal{S}$-TEX, the American Mathematical Society's TEX macro system.

10 9 8 7 6 5 4 3 2 1   97 96 95 94 93 92

C stack
SCIMON 10/19/92

Contents

Preface

A Symposium on Complex Analysis was held on the campus of the University of Wisconsin-Madison from June 2 to June 7, 1991, in conjunction with the retirement of Walter Rudin, Villas Professor of Mathematics. The Symposium was supported in part by grants from the National Science Foundation and from the William F. Villas Trust Estate; for this support we are grateful. We are also grateful to Mary Sheetz for her superb work in preparing these proceedings for publication.

During the week of the conference, a group of about two hundred mathematicians from many nations gathered to discuss recent developments in the broad area of complex analysis and to celebrate the career of Professor Rudin. The papers included in this volume reflect the breadth of the work reported on in the sessions of the Symposium. It is a tribute to Walter Rudin that much of the work contained here makes contact, direct or indirect, with work that he has done over the last forty years.

On behalf of our contributors, as well as on behalf of all those who attended the Symposium, we are pleased to present this volume of papers to Professor Rudin with our heartfelt best wishes.

Alexander Nagel
Edgar Lee Sout

Contemporary Mathematics
Volume **137**, 1992

Hulls of 3-spheres in $\mathbb{C}^3$

PATRICK AHERN & WALTER RUDIN

1. Introduction.

1.1. The *polynomial hull* of a compact set $X \subset \mathbb{C}^n$ is, by definition, the set $\hat{X}$ of all points $p \in \mathbb{C}^n$ at which the inequality

$$(1) \qquad |Q(p)| \leq \max\{|Q(z)| : z \in X\}$$

holds for every polynomial $Q : \mathbb{C}^n \to \mathbb{C}$.

If $\Phi : U \to \mathbb{C}^n$ is a *nonconstant* map whose components are in $H^\infty(U)$ (the space of all bounded holomorphic functions in the open unit disc U), its range $\Phi(U)$ will be called *an H^∞-disc, parametrized by Φ*. If, moreover,

$$(2) \qquad \lim_{r \nearrow 1} \Phi(re^{i\theta}) \in X$$

for almost all $e^{i\theta}$ on the unit circle T, then we say that $\Phi(U)$ is *an H^∞-disc whose boundary lies in X*.

We define the *disc-hull $\mathcal{D}(X)$* to be the *union* of X and all H^∞-discs whose boundaries lie in X. The maximum principle satisfied by every $f \in H^\infty(U)$ shows immediately that

$$(3) \qquad \mathcal{D}(X) \subset \hat{X}.$$

If the components of Φ belong to the disc algebra $A(U)$ then $\Phi(\overline{U})$ will be called *an A-disc*. A perhaps unexpected phenomenon is that there exist very simple sets X (such as the spheres Σ in Section V) which bound very few A-discs even though $\hat{X} \setminus X$ is a large set which is covered by $\mathcal{D}(X)$.

H^∞-discs may be topologically very different from discs, quite apart from possible self-intersections. For example, let M be a finite Riemann surface (such as an annulus, to look at the simplest case) whose boundary lies in X. By the uniformization theorem there is a conformal map Φ from U onto the universal

---

1991 *Mathematics Subject Classification.* Primary 32E20.

This research was partially supported by the National Science Foundation and by the William F. Vilas Trust Estate.

This paper is in final form and no version of it will be submitted for publication elsewhere.

cover $\tilde{M}$ of M, and if $\pi : \tilde{M} \to M$ is the natural projection then $\pi \circ \Phi$ is an H^∞-map, and thus $M = (\pi \circ \Phi)(U)$ is an H^∞-disc.

1.2. In the present paper we confine our attention to compact sets $\Sigma \subset \mathbb{C}^3$ which are homeomorphic to the 3-sphere

$$(1) \qquad S = S^3 = \{(z, w) \in \mathbb{C}^2 : z\overline{z} + w\overline{w} = 1\}$$

via an embedding E of the form

$$(2) \qquad E(z, w) = (z, w, \sigma(z, w)), \quad \Sigma = E(S),$$

where $\sigma : S \to \mathbb{C}$ is, to begin with, an arbitrary continuous function.

The embedded sphere $\Sigma = E(S)$ is thus the graph of σ, over S.

It is known (see [2]; we give another proof in §2.3) that $\hat{\Sigma}$ then covers the closure $\overline{B}$ of the open unit ball B (whose boundary is S) in $\mathbb{C}^2$, and it is therefore natural to ask whether $\hat{\Sigma}$ must be a graph over $\overline{B}$. We shall see that this is true for certain classes of σ's (Examples A and B), but that $\hat{\Sigma}$ can also have nonempty interior (Examples C, F) and can even be the closure of a region in $\mathbb{C}^3$ (Example E).

Another natural question is whether $\hat{\Sigma} = \mathcal{D}(\Sigma)$. The only negative result we have here occurs in Example E.

Here is the reason why we deal only with 3-spheres in $\mathbb{C}^3$ rather than n-spheres in $\mathbb{C}^n$: In [11] and [15] it was shown that $n = 1$ and $n = 3$ are the only dimensions where S^n has a *totally real* embedding in $\mathbb{C}^n$, i.e., one which is of class C^1 and has no complex tangent at any of its points. We gave the first explicit totally real $\mathbb{C}^3$-example in [1], namely the embedding of the form (2) given by

$$(3) \qquad \sigma(z, w) = \overline{zw}(w\overline{w} + iz\overline{z}).$$

This solved a problem posed by Stout and Zame [14; p. 37].

More recently, Forstnerič [9] constructed a totally real 3-sphere in $\mathbb{C}^3$ (by an embedding different from (2)) which contains the boundaries of a one-parameter family of A-discs, and he asked, among other things, whether our above-mentioned example bounds any A-discs at all. We found that it bounds precisely 2 A-discs and also a two-parameter family of H^∞-discs (Theorem 5.2). These were also found, independently, by John Anderson [4], who went further and proved that they fill $\hat{\Sigma}$, so that $\hat{\Sigma} = \mathcal{D}(\Sigma)$. He based his proof on a function-algebraic subharmonicity theorem of Wermer [17; Theorem 20.2]. Our proof of Theorem 5.2 — found after we saw his — looks more elementary to us.

For 2-spheres in $\mathbb{C}^2$ we refer to [5], [6], [2].

Sections II and III contain material that is mostly known. We include it for the sake of completeness, and because some of our proofs are quite a bit simpler than what we could find in the literature.

Throughout this paper, $U, T, B, S, \sigma, \Sigma, \hat{\Sigma}, \mathcal{D}(\Sigma)$ will have the same meanings as in this Introduction. In particular, $\Sigma = E(S)$ will be related to σ as in (2).

2. Some basic facts about $\hat{\Sigma}$.

2.1. We let π denote the projection from $\mathbb{C}^3$ to $\mathbb{C}^2$ given by $\pi(z, w, \zeta) = (z, w)$. If Σ is as in §1.2 and $(z, w) \in \overline{B}$ then the set $\hat{\Sigma} \cap \pi^{-1}(z, w)$ is the *fiber of $\hat{\Sigma}$ over the point* (z, w). The following three facts may be found in [2], [3]. We include our somewhat different proofs because they are short and because the technique used in Proposition 2.4 will be used later, in Examples B and D.

2.2. PROPOSITION. *If $(z_0, w_0) \in S$ and $(z_0, w_0, \zeta_0) \in \hat{\Sigma}$ then $\zeta_0 = \sigma(z_0, w_0)$.*

Thus $\hat{\Sigma}$ has exactly one point over each point of S; each of the corresponding fibers is a singleton.

PROOF. There is a polynomial $P : \mathbb{C}^2 \to \mathbb{C}$ such that $P(z_0, w_0) = 1$ and $|P| < 1$ on the rest of S. Define

$$Q_n(z, w, \zeta) = (\zeta - \sigma(z_0, w_0))P^n(z, w)$$

for $n = 1, 2, 3, \ldots$. Since $(z_0, w_0, \zeta_0) \in \hat{\Sigma}$, and $\zeta = \sigma(z, w)$ on Σ,

$$|\zeta_0 - \sigma(z_0, w_0)| = |Q_n(z_0, w_0, \zeta_0)| \leq \max_{\Sigma} |Q_n|$$

$$= \max_{S} |\sigma(z, w) - \sigma(z_0, w_0)||P^n(z, w)|$$

and this last maximum tends to 0 as $n \to \infty$.

2.3. PROPOSITION. $\pi(\hat{\Sigma}) = \overline{B}$.

PROOF. Put $K = \pi(\hat{\Sigma})$. The inclusions $S \subset K \subset \overline{B}$ are obvious. If $K \neq \overline{B}$ then there is a retraction r of K onto S. Setting $\rho = \sigma \circ r \circ \pi$ we see that ρ retracts $\hat{\Sigma}$ to Σ. Let $i : \Sigma \to \hat{\Sigma}$ be the inclusion map. Then $\rho \circ i$ is the identity map on Σ. If ρ^*, i^* are the induced homomorphisms on the corresponding cohomology groups with coefficients in $\mathbb{C}$, it follows that $i^* \circ \rho^*$ is the identity map on $H^3(\Sigma, \mathbb{C})$. Thus i^* maps

$$\rho^*(H^3(\Sigma, \mathbb{C})) = H^3(\hat{\Sigma}, \mathbb{C})$$

onto $H^3(\Sigma, \mathbb{C})$. But $H^3(\Sigma, \mathbb{C}) = \mathbb{C}$ [13; p. 119] whereas $H^3(\hat{\Sigma}, \mathbb{C}) = 0$, by Browder's theorem [7], [17; p. 94], because $\hat{\Sigma}$ is polynomially convex and is therefore the maximal ideal space of a Banach algebra with 3 generators.

2.4. PROPOSITION. *If $\varphi \in C(\overline{B})$ is pluriharmonic in B, and σ is the restriction of φ to S, then $\hat{\Sigma}$ is the graph of φ.*

PROOF. Fix $(z_0, w_0, \zeta_0) \in \hat{\Sigma}$. By the Riesz representation theorem there is then a probability measure $\tilde{\mu}$ on Σ such that

$$(1) \qquad\qquad P(z_0, w_0, \zeta_0) = \int_{\Sigma} P d\tilde{\mu}$$

for all polynomials $P : \mathbb{C}^3 \to \mathbb{C}$. Since Σ is a graph over S, $\tilde{\mu}$ can be pulled back to a probability measure μ on S which satisfies

$$(2) \qquad\qquad P(z_0, w_0, \zeta_0) = \int_{S} P(z, w, \sigma(z, w)) d\mu(z, w)$$

in place of (1). (μ may be said to "represent" the point (z_0, w_0, ζ_0).) Special cases of (2) are

$$(3) \qquad\qquad \zeta_0 = \int_S \sigma \, d\mu$$

and

$$(4) \qquad\qquad Q(z_0, w_0) = \int_S Q \, d\mu \, ,$$

first for all polynomials $Q : \mathbb{C}^2 \to \mathbb{C}$, then for all Q in the ball algebra $A(B)$ (since the polynomials are dense in $A(B)$), and then also for all Q whose complex conjugate lies in $A(B)$ (since μ is a real-valued measure).

Being pluriharmonic, $\varphi = f + \overline{g}$ in B for some holomorphic f, g. If $r < 1$, the dilation φ_r of φ, defined by $\varphi_r(p) = \varphi(rp)$, $p \in \overline{B}$, decomposes into $\varphi_r = f_r + \overline{g}_r$, and $f_r, g_r \in A(B)$. Thus (4) yields

$$(5) \qquad\qquad \varphi_r(z_0, w_0) = \int_S \varphi_r \, d\mu \, .$$

Since $\varphi_r \to \sigma$, uniformly on S, as $r \nearrow 1$, we conclude from (3) and (5) that

$$(6) \qquad\qquad \zeta_0 = \varphi(z_0, w_0) \, ,$$

as claimed.

2.5. Remark. Let $\varphi = f + \overline{g}$, as above, and put

$$V_c = \{ (z, w, \zeta) : g(z, w) = c, \zeta = f(z, w) + \overline{c} \} \, ,$$

for $(z, w) \in B$, $c \in \mathbb{C}$. Then $\hat{\Sigma} \backslash \Sigma$ is the union of these analytic varieties V_c.

3. Totally Real Embeddings.

3.1. Since we are interested in the disc-hulls $\mathcal{D}(\Sigma)$, it seems appropriate to draw attention to the fact that totally real manifolds of maximal dimension cannot contain the boundaries of "small" H^∞-discs. This follows from the apparently much deeper fact that such manifolds are locally polynomially convex [12; p. 301]. However, Jean-Pierre Rosay showed us an extremely simple and direct proof which we shall now present (for arbitrary $n > 1$).

If M is a totally real manifold in $\mathbb{C}^n$, of real dimension n, then coordinates can be so chosen in $\mathbb{C}^n$ that any preassigned point of M is at the origin, and the tangent space to M at 0 is R^n. Therefore there exist

   (a) convex neighborhoods Ω_0 and Ω_1 of 0 in R^n,
   (b) a C^1-map $y : \Omega_0 \to \Omega_1$, $y(0) = 0$, $y'(0) = 0$,
   (c) a constant $t < 1$,

such that, setting $\Omega = \Omega_0 + i\Omega_1$,

$$(1) \qquad\qquad M \cap \Omega = \{ x + iy(x) : x \in \Omega_0 \}$$

and

(2)
$$|y(\xi) - y(\eta)| \le t|\xi - \eta|$$

for all $\xi, \eta \in \Omega_0$. Here $|\cdot|$ denotes the euclidean norm in R^n.

3.2. Proposition. *In this situation there is no H^∞-disc in $\mathbb{C}^n$ whose boundary lies in $M \cap \Omega$.*

Proof. Suppose $f_k = u_k + iv_k \in H^\infty(U)$, $1 \le k \le n$, put $u = (u_1, \dots, u_n)$, $v = (v_1, \dots, v_n)$, $\Phi = (f_1, \dots, f_n)$, and assume that $\Phi(e^{i\theta}) \in M \cap \Omega$ a.e. (We use the same letter for functions in U and for their boundary values on T.) We will show that Φ is constant (which proves the proposition) by using only the following simple facts:

(i) If $\psi : T \to R^n$ is in L^2 and $\int_T \psi = a$, then

$$\int_T |\psi - a|^2 = \int_T |\psi - b|^2 - |a - b|^2 \le \int_T |\psi - b|^2$$

for every $b \in R^n$. (We write $\int_T f$ for $\int_{-\pi}^{\pi} f(e^{i\theta}) d\theta / 2\pi$.)

(ii) Since the real and imaginary parts of an $f \in H^\infty(U)$ with $f(0) = 0$ have the same L^2-norm on T, we have

$$\int_T |u - u(0)|^2 = \int_T |v - v(0)|^2 \, .$$

Apply (ii), then (i), with $\psi = y \circ u = v$, $a = \int_T v = v(0)$, $b = y(u(0))$, to obtain

$$\int_T |u - u(0)|^2 = \int_T |y \circ u - v(0)|^2 \le \int |y \circ u - y(u(0))|^2$$
$$\le t^2 \int_T |u - u(0)|^2 \, .$$

The last inequality holds because $u(T) \subset \Omega_0$ and $u(0) = \int_T u$ lies in the convex hull of $u(T)$, so that $u(U) \subset \Omega_0$.

Since $t < 1$ it follows that $u = u(0)$ a.e. on T, and this gives $\Phi = \Phi(0)$. $\quad\blacksquare$

As an illustration of the result we just proved we mention that the diameter of every H^∞-disc with boundary in Σ that occurs in Theorem 5.2 is at least 2.

3.3. We now return to our spheres Σ in $\mathbb{C}^3$, given by 1.2(2). There is a simple criterion, in terms of the action of the tangential Cauchy-Riemann operator

(1)
$$L = w\frac{\partial}{\partial \bar{z}} - z\frac{\partial}{\partial \bar{w}}$$

on σ, which decides when Σ is totally real. We proved this in [1], by a calculation with determinants. The following short proof is more geometric and more conceptual.

3.4. PROPOSITION. Σ *is totally real if and only if* $L\sigma \neq 0$ *at every point of* S.

(We assume now that $\sigma \in C^1(S)$.)

PROOF. Fix $(z, w) \in S$. The complex line tangent to S at (z, w) is parametrized by

$$(1) \qquad\qquad \lambda \to (z + \overline{w}\lambda, w - \overline{z}\lambda) \quad (\lambda \in \mathbb{C}).$$

Since $\sigma \in C^1(S)$, it can be extended to a C^1-function on a neighborhood of S, and then it is easy to verify that

$$\sigma(z + \overline{w}\lambda, w - \overline{z}\lambda) = \sigma(z, w) + (\overline{L}\sigma)(z, w)\lambda + (L\sigma)(z, w)\overline{\lambda} + o(|\lambda|)$$

as $\lambda \to 0$, where $\overline{L} = \overline{w}\partial/\partial z - \overline{z}\partial/\partial w$, and L is as in 3.3 (1). The orthogonal projection $\pi : \mathbb{C}^3 \to \mathbb{C}^2$ is $\mathbb{C}$-linear. Therefore the only candidate for a complex line tangent to Σ at $(z, w, \sigma(z, w))$ is parametrized by

$$\lambda \to (z + \overline{w}\lambda, w - \overline{z}\lambda, \sigma(z, w) + (\overline{L}\sigma)(z, w)\lambda + (L\sigma)(z, w)\overline{\lambda})$$

and this is obviously a complex line if and only if $(L\sigma)(z, w) = 0$.

We shall now use Proposition 3.4 to describe a fairly large class of embeddings which produce totally real spheres Σ.

3.5. PROPOSITION. *If* $\Gamma : [0, 1] \to \mathbb{C}$ *is a curve such that*

(a) $\Gamma(t) = t(1 - t)q(t)$ *for some* $q \in C^1([0, 1])$,
(b) *for every* $t \in [0, 1]$, $\Gamma'(t) \neq 0$, *and if* $\Sigma = E(S)$ *where*

$$(1) \qquad\qquad E(z, w) = (z, w, \Gamma(z\overline{z})/zw),$$

then Σ *is totally real.*

PROOF. This E is an embedding like 1.2 (2) in which

$$(2) \qquad\qquad \sigma(z, w) = \Gamma(z\overline{z})/zw = \overline{zw}q(z\overline{z}).$$

(Recall that $z\overline{z} + w\overline{w} = 1$ on S.) Setting $z\overline{z} = t$, a simple calculation gives

$$(L\sigma)(z, w) = (1 - 2t)q(t) + t(1 - t)q'(t) = \Gamma'(t) \neq 0$$

for all $(z, w) \in S$. Now appeal to Proposition 3.4.

3.6. Remarks. (a) We prefer to write the preceding σ in the form $\Gamma(z\overline{z})/zw$ rather than $\overline{z}\,\overline{w}q(z\overline{z})$ in 3.5(1), even though $\Gamma(z\overline{z})/zw$ involves a division, possibly by 0, and even though $\overline{z}\,\overline{w}q(z\overline{z})$ is what we first tried to fit to Proposition 3.4. The reason is that topological properties of the closed curves Γ (they start and end at 0) turn out to have a decisive influence on the shape and size of $\hat{\Sigma}$ in Examples A, B, and C.

(b) Our original example 1.2(3) was of this type, with

$$(1) \qquad\qquad \Gamma(t) = t(1 - t)(1 - t + it).$$

Perhaps the simplest one to satisfy Proposition 3.5 is

$$(2) \qquad \Gamma(t) = 1 - e^{2\pi it} .$$

Both of these are simple closed curves.

(c) Every Σ given by Proposition 3.5 is invariant under the transformations

$$(3) \qquad (z, w, \zeta) \rightarrow (e^{i\alpha}z, e^{i\beta}w, e^{-i(\alpha+\beta)}\zeta) .$$

The same is therefore also true of $\hat{\Sigma}$ and $\mathcal{D}(\Sigma)$.

4. Pairs of Harmonic Functions.

The first conclusion of the following lemma will be used in Examples A and B to show that $\hat{\Sigma}$ is a graph over $\overline{B}$. We thank Jean-Pierre Rosay for suggesting that properly chosen linear combinations of u and v could lead to a proof.

We let H be the open upper half-plane in $\mathbb{C}$, R the real axis, $\overline{H} = H \cup R$.

4.1. LEMMA. *Suppose that u and v are real-valued C^1-functions on $\overline{H}$ which, in H, are the Poisson integrals of their restrictions to R, and that*

    (a) $u' > 0$ *on R,*
    (b) $e^u + e^v = 1$ *on R.*

Then

    (i) *the pair (u, v) separates points on $\overline{H}$,*
    (ii) *its Jacobian has no zero in H, and*
    (iii) *the harmonic conjugate $\tilde{u}$ of u tends to $+\infty$ along every level curve of u.*

Of course, assumption (b) is irrelevant to (iii).

PROOF. The Poisson integral representation of u is

$$(1) \qquad u(x + iy) = \tfrac{1}{\pi} \int_{-\infty}^{\infty} \frac{u(x - \xi)y\,d\xi}{\xi^2 + y^2} \qquad (y > 0) .$$

Combined with (a), it shows that

$$(2) \qquad \partial u / \partial x > 0 \quad \text{in} \quad H .$$

Therefore u separates points on every horizontal line in H.

Now let L be any half-line in $\overline{H}$, starting at a point $c \in R$. There is then a parametrization

$$(3) \qquad L(t) = c + (\alpha + i\beta)t \qquad (0 \leq t < \infty)$$

where $\alpha \in R$, $\beta > 0$. Define

$$(4) \qquad \varphi(t) = e^{u(c)}u(L(t)) + e^{v(c)}v(L(t)) \qquad (0 \leq t < \infty) .$$

We claim that

$$(5) \qquad \varphi'(t) < 0 \qquad (0 < t < \infty) .$$

This will prove that (u, v) separates points on L, for *every* such L, and therefore that (i) holds.

The change of variables $\xi = t\alpha - t\eta$ in (1) (and its analogue with v in place of u) leads to

$$(6) \qquad \varphi(t) = \frac{1}{\pi} \int_{-\infty}^{\infty} \psi(t\eta) \frac{\beta d\eta}{(\alpha - \eta)^2 + \beta^2}$$

where

$$(7) \qquad \psi(t\eta) = e^{u(c)} u(c + t\eta) + e^{v(c)} v(c + t\eta).$$

By (b), $u'e^u + v'e^v = 0$ on R. Hence

$$\frac{\partial \psi}{\partial t} = \eta \left[e^{u(c)} - e^{u(c+t\eta)} \right] u'(c + t\eta) + \eta \left[e^{v(c)} - e^{v(c+t\eta)} \right] v'(c + t\eta).$$

Considering the cases $\eta > 0$ and $\eta < 0$ separately, the monotonicity of u and v on R shows that $\partial \psi / \partial t < 0$ for all $\eta \neq 0$. This gives (5) and hence proves (i).

If the Jacobian of (u, v) were 0 at some $p \in H$ then the gradients of u and v would be parallel at p, their level curves through p would have a common tangent, and therefore, letting L be that tangent line, the derivative of the corresponding φ (given by (4)) would have a zero, contrary to (5). This proves (ii).

Setting $f = u + i\tilde{u}$, (2) shows that Re $f' > 0$ in H. Hence f is univalent in H. Put

$$(8) \qquad g(\lambda) = f\left(i \frac{1 + \lambda}{1 - \lambda} \right) \qquad (\lambda \in U),$$

put $W = f(H) = g(U)$. Since $u' > 0$ on R, the simply connected region W is bounded below by the graph G of a function with domain $u(R)$. We claim that W has no other boundary point in $\mathbb{C}$. If this were not so, the extra boundary points would give rise to more than one (actually, infinitely many) prime ends of W, all of which would correspond, via g, to the point $\lambda = 1$ (since $g(T \backslash \{1\}) = G$). This contradicts Carathéodory's theorem [8; p. 173].

Thus W is a union of vertical half-lines whose lower end-points lie on G. This proves (iii).

4.2. COROLLARY. *There is a real-analytic function Λ on the open set (u, v) (H) such that $\lambda = \Lambda(u_0, v_0)$ if and only if $u_0 = u(\lambda)$ and $v_0 = v(\lambda)$.*

PROOF. Λ is simply the inverse of the real-analytic map $\lambda \to (u(\lambda), v(\lambda))$ whose Jacobian is nowhere zero on H.

5. Example A.

5.1. **Preliminaries.** The main result of this section is Theorem 5.2. It concerns totally real spheres Σ, produced by curves Γ as in Proposition 3.5 which satisfy an additional assumption, namely

(∗) $\Gamma(t_1) \neq \Gamma(t_2)$ *when* $0 \leq t_1 < t_2 \leq 1$, *except that* $\Gamma(0) = \Gamma(1) = 0$.

There is then a simply connected region $\Omega \subset \mathbb{C}$ whose boundary $\partial\Omega$ is $\Gamma([0,1])$. Let $\Psi : \overline{U} \to \overline{\Omega}$ be a Riemann map (a homeomorphism of $\overline{U}$ onto $\overline{\Omega}$ which is holomorphic in U), so chosen that $\Psi(1) = 0$.

We will show that there exist outer functions F, G, H in U such that

(i) $F\overline{F} = \Gamma^{-1} \circ \Psi$ and $F\overline{F} + G\overline{G} = 1$ on $T\backslash\{1\}$,

(ii) F and G are continuous on $\overline{U}\backslash\{1\}$,

(iii) $FGH = \Psi$ on $\overline{U}$, and

(iv) $H \in A(U)$, $H(1) = 0$.

Since $0 \notin \Psi(U)$, $1/\Psi$ is holomorphic and univalent in U. Hence $1/\Psi \in H^p(U)$ if $p < \frac{1}{2}$ [10; pp. 50, 51] and this implies that Ψ has no inner factor. Thus Ψ is an outer function.

Γ^{-1} is one-to-one on $\partial\Omega\backslash\{0\}$, with range $(0,1)$. Since $\Gamma(0) = \Gamma(1) = 0$ and Γ is differentiable at $t = 0$ and $t = 1$, there is a constant $c > 0$ such that

(1) $$\Gamma^{-1}(p) \geq c|p| \quad \text{and} \quad 1 - \Gamma^{-1}(p) \geq c|p|$$

for all $p \in \partial\Omega\backslash\{0\}$. Hence

(2) $$\int_T \log \Gamma^{-1} \circ \Psi > -\infty, \quad \int_T \log(1 - \Gamma^{-1} \circ \Psi) > -\infty .$$

This gives outer functions F and G that satisfy (i).

Since $\Gamma \in C^1$, every point of $\partial\Omega\backslash\{0\}$ satisfies the local chord-arc condition which is used in Lemma 1 of [16] to prove that every point of $T\backslash\{1\}$ has a neighborhood in which $\Psi|_T$ satisfies a Hölder condition. By (i) the same is true of $|F|$, hence of $\log|F|$, hence of its harmonic conjugate $\arg F$. The same is true with G in place of F. This implies (ii).

As θ increases from 0 to 2π, (i) shows that $|F|$ increases from 0 to 1 and $|G|$ decreases from 1 to 0 (or vice versa, depending on the direction in which $\Gamma(t)$ traverses $\partial\Omega$). If we define $(FG)(1)$ to be 0, *it follows that $FG \in A(U)$.*

Since $|\Psi| = |\Gamma(F\overline{F})| = F\overline{F}G\overline{G}|q(F\overline{F})|$ on T, by 3.5(a), we see that

(3) $$|\Psi| \leq \|q\|_\infty |FG|^2 \quad \text{on} \quad T .$$

The same inequality holds also in U because $(FG)^2$ is outer. Setting $H = \Psi/FG$ we get (iii), and (iv) follow from $|H| \leq \|q\|_\infty |FG|$.

Now define

(4) $$D_1 = \{(z,0,0) : |z| \leq 1\}, \quad D_2 = \{(0,w,0) : |w| \leq 1\}$$

and let $D_{\alpha\beta}$ be the range of the map

(5) $$\lambda \to (e^{i\alpha}F(\lambda), e^{i\beta}G(\lambda), e^{-i(\alpha+\beta)}H(\lambda))$$

with domain $\overline{U}\backslash\{1\}$, for $\alpha, \beta \in R$.

5.2. THEOREM. *Let $\Sigma = E(S)$ be a totally real sphere defined by an embedding*

$$(1) \qquad E(z, w) = (z, w, \Gamma(z\overline{z})/zw)$$

in which Γ is a simple closed curve as in §5.1. Then

(a) *each $D_{\alpha\beta}$ is an H^∞-disc with boundary in Σ, and*

$$\hat{\Sigma} = \mathcal{D}(\Sigma) = D_1 \cup D_2 \cup \bigcup_{\alpha,\beta} D_{\alpha\beta}.$$

(b) *$\hat{\Sigma}$ is the graph of a function $\varphi \in C(\overline{B})$ such that*
   (i) *$\varphi(z, w) = \Gamma(z\overline{z})/zw$ on S,*
   (ii) *φ is real-analytic in $B\backslash\{zw = 0\}$,*
   (iii) *$\varphi(z, w) = 0$ if and only if $zw = 0$.*
(c) *$\overline{D}_{\alpha\beta}\backslash D_{\alpha\beta} = D_1 \cup D_2$ for all real α, β,*
(d) *D_1 and D_2 are the only A-discs with boundary in Σ.*

With regard to (b)(ii), we do not know whether φ can be real-analytic in all of B, even in the simple case given by $\Gamma(t) = 1 - \exp(2\pi it)$.

PROOF. That each $D_{\alpha\beta}$ is an H^∞-disc with boundary in Σ follows from (5), (i), (iii) in §5.1. Since the rest of (a) becomes obvious if $=$ is replaced by $\supset$, we have to prove that

$$(1) \qquad \hat{\Sigma} \subset D_1 \cup D_2 \cup \bigcup_{\alpha,\beta} D_{\alpha\beta}.$$

Fix $(z_0, w_0, \zeta_0) \in \hat{\Sigma}$, put $p_0 = z_0 w_0 \zeta_0$.

We claim that $p_0 \in \overline{\Omega}$. If not, there is a polynomial $Q : \mathbb{C} \to \mathbb{C}$ such that $Q(p_0) = 1$, $|Q| < \frac{1}{2}$ on $\overline{\Omega}$. Put $P(z, w, \zeta) = Q(zw\zeta)$. Then $P(z_0, w_0, \zeta_0) = 1$ but $|P| < \frac{1}{2}$ on Σ, because $\Gamma([0, 1]) = \partial\Omega$, and hence

$$(2) \qquad |P(z, w, \Gamma(z\overline{z})/zw| = |Q(\Gamma(z\overline{z}))| < \frac{1}{2} \text{ on } \Sigma.$$

This is impossible, since $(z_0, w_0, \zeta_0) \in \hat{\Sigma}$.

Thus $p_0 \in \overline{\Omega}$, and there is a unique $\lambda_0 \in \overline{U}$ such that

$$(3) \qquad \Psi(\lambda_0) = p_0 = z_0 w_0 \zeta_0.$$

Note that $\lambda_0 = 1$ if and only if $z_0 w_0 \zeta_0 = 0$.

We claim that

$$(4) \qquad |z_0| \leq |F(\lambda_0)|, \quad |w_0| \leq |G(\lambda_0)| \text{ in case } \lambda_0 \neq 1,$$

$$(5) \qquad |\zeta_0| \leq |H(\lambda_0)| \text{ in any case,}$$

$$(6) \qquad z_0 w_0 = 0 \text{ when } \zeta_0 = 0.$$

(These inequalities occur also in the paper [4] that we mentioned in §1.2.) The point $\lambda_0 = 1$ is omitted in (4) simply because F and G are not defined there. Note that (5) and (6) show that $z_0 w_0 \zeta_0 = 0$ only when both $z_0 w_0$ and ζ_0 are 0.

Suppose we have (4), (5), (6). In case $\lambda_0 \neq 1$ it follows then that

$$0 < |z_0 w_0 \zeta_0| \leq |(FGH)(\lambda_0)| = |\Psi(\lambda_0)| = |z_0 w_0 \zeta_0|$$

by (3) and 5.1 (iii). The inequalities in (4) and (5) are thus equalities. Hence

(7) $$z_0 = e^{i\alpha} F(\lambda_0), \quad w_0 = e^{i\beta} G(\lambda_0)$$

for some α, β, and thus, using (3) again,

(8) $$\zeta_0 = \Psi(\lambda_0)/z_0 w_0 = e^{-i(\alpha+\beta)} H(\lambda_0)$$

since $H = \Psi/FG$.

Thus $(z_0, w_0, \zeta_0) \in \bigcup_{\alpha,\beta} D_{\alpha\beta}$ when $\lambda_0 \neq 1$.

When $\lambda_0 = 1$, (5) shows that $\zeta_0 = 0$, since $H(1) = 0$ by 5.1 (iv), and therefore $z_0 w_0 = 0$ by (6).

Thus $(z_0, w_0, \zeta_0) \in D_1 \cup D_2$ when $\lambda_0 = 1$, and we see that conclusion (a) follows from (4), (5), (6). We shall now prove these.

Assume first that $\lambda_0 \neq 1$. Since F is outer, there are outer functions $F_n \in A(U)$ such that $|F_n| \geq |F|$ on $T\backslash\{1\}$, $|F_n| > 0$ on $\overline{U}$, and $F_n(\lambda_0) \to F(\lambda_0)$ as $n \to \infty$. Define

(9) $$Q_n(z, w, \zeta) = z/F_n(\Psi^{-1}(zw\zeta))$$

on $\{(z, w, \zeta) : zw\zeta \in \overline{\Omega}\}$. By Mergelyan's theorem $1/F_n \circ \Psi^{-1}$ is a uniform limit of polynomials on $\overline{\Omega}$. Since $(z_0, w_0, \zeta_0) \in \hat{\Sigma}$, this gives the first inequality in

(10) $$|z_0/F_n(\lambda_0)|^2 = |Q_n^2(z_0, w_0, \zeta_0)| \leq \max_{\Sigma} |Q_n|^2$$
$$= \max_S z\overline{z}/F_n \overline{F}_n(\Psi^{-1}(\Gamma(z\overline{z}))) \leq 1;$$

the last one holds because $|F_n| \geq |F|$ and

(11) $$(F\overline{F} \circ \Psi^{-1} \circ \Gamma)(z\overline{z}) = z\overline{z} \qquad (0 < |z| < 1)$$

by 5.1 (i). Letting $n \to \infty$ in (10) gives the first half of (4).

The second half of (4), as well as (5), are proved in the same way, using the identities

(12) $$(G\overline{G} \circ \Psi^{-1} \circ \Gamma)(z\overline{z}) = w\overline{w},$$

(13) $$|zw(H \circ \Psi^{-1} \circ \Gamma)(z\overline{z})| = |\Gamma(z\overline{z})|$$

in place of (11); these follow from 5.1 (i), (iii).

To prove (6), let $\chi \in A(\Omega)$ peak at 0. (As usual, $A(\Omega)$ is the class of all $f \in C(\overline{\Omega})$ which are holomorphic in Ω.) The assumptions on χ are, more explicitly: $\chi(0) = 1$, $|\chi| < 1$ on $\overline{\Omega}\backslash\{0\}$.) [For example, $\chi = \frac{1}{2}(1 + \Psi^{-1})$.] Define

(14) $$P_n(z, w, \zeta) = zw\chi^n(zw\zeta).$$

On Σ, $P_n(z, w, \Gamma(z\overline{z})/zw) = zw\chi^n(\Gamma(z\overline{z}))$ tends to 0 uniformly, as $n \to \infty$. Thus $P_n(z_0, w_0, 0) \to 0$, because χ^n is a uniform limit of polynomials on $\overline{\Omega}$. Since $P_n(z_0, w_0, 0) = z_0 w_0$ for all n we have (6).

This completes the proof of (a).

We turn to (b). Suppose (without loss of generality) that $\Gamma(t)$ traverses $\partial\Omega$ in the positive direction as t increases from 0 to 1. Then, as θ increases from 0 to 2π, $|F(e^{i\theta})|$ increases from 0 to 1 and $|G(e^{i\theta})|$ decreases from 1 to 0. Let u, v be the harmonic functions given by $F\overline{F} = e^u$, $G\overline{G} = e^v$; they are the Poisson integrals of $\log F\overline{F}$, $\log G\overline{G}$, respectively, and $e^u + e^v = 1$ on $T\backslash\{1\}$. A Moebius transformation, taking U to the upper half-plane and 1 to ∞, transplants Lemma 4.1 to U, and shows therefore that *the pair* $(|F|, |G|)$ *separates points on* $\overline{U}\backslash\{1\}$.

The equations that describe $D_{\alpha\beta}$, namely

$$(15) \qquad z = e^{i\alpha}F(\lambda), \quad w = e^{i\beta}G(\lambda), \quad \zeta = e^{-i(\alpha+\beta)}H(\lambda),$$

therefore determine λ, hence also $e^{i\alpha}$, $e^{i\beta}$ and ζ, if $(z, w) \in B$ is given and $zw \neq 0$. When $zw = 0$, then $\zeta = 0$ by (5), (6).

Thus $\hat{\Sigma}$ is the graph of $\zeta = \varphi(z, w)$, and φ must be continuous since $\hat{\Sigma}$ is compact. Conclusion (b)(i) follows from Proposition 2.2, (ii) follows from Corollary 4.2, and (iii) from (5) and (6), since $z_0 w_0 = 0$ implies $\lambda_0 = 1$, hence $\zeta_0 = 0$.

This proves (b).

To prove (c) it is enough to show that every point of D_1 lies in the closure of D_{00}. (The same argument will apply to D_2.) So fix z, $0 < |z| < 1$, let $\gamma = \{\lambda \in U : |F(\lambda)| = |z|\}$, and let $\lambda \to 1$ along γ. Since $FG \in A(U)$ and $(FG)(1) = 0$, $G(\lambda) \to 0$. Also, $H(\lambda) \to 0$. If $\tilde{u}$ is the harmonic conjugate of $u = \log F\overline{F}$, Lemma 4.1 (iii) shows that $\tilde{u}(\lambda) \to +\infty$. Thus $F(\lambda)$ traverses the circle of radius $|z|$, center 0, infinitely many times as $\lambda \to 1$ along γ. It follows that there is a sequence $\{\lambda_n\}$ on γ such that

$$(16) \qquad \lim_{n\to\infty}(F(\lambda_n), G(\lambda_n), H(\lambda_n)) = (z, 0, 0).$$

This proves (c).

Since no $D_{\alpha\beta}$ is an A-disc, part (d) is a consequence of the following result which shows that every H^∞-disc with boundary in Σ lies either in D_1 or in D_2 or in some $D_{\alpha\beta}$.

5.3. THEOREM. *With Σ as in Theorem 5.2, suppose that*

$$(1) \qquad\qquad\qquad (f, g, h) : U \to \mathbb{C}^3$$

parametrizes an H^∞-disc with boundary in Σ. Then either

   (a) *f is inner, $g \equiv 0$, $h \equiv 0$, or*
   (b) *g is inner, $f \equiv 0$, $h \equiv 0$, or*
   (c) *there is an inner function u such that*

$$(2) \qquad\qquad f = e^{i\alpha}F \circ u, \quad g = e^{i\beta}G \circ u, \quad h = e^{-(\alpha+\beta)}H \circ u$$

for some α, β, where F, G, H are as in §5.1.

PROOF. Define $k = fgh$. Since (f, g, h) maps almost every point of T into Σ, and $zw\zeta = \Gamma(z\bar{z})$ on Σ, we have

$$(3) \qquad k = \Gamma(f\bar{f}) \quad \text{a.e. on } T.$$

The boundary values of the H^∞-function k lie therefore (a.e.) in $\Gamma([0,1]) = \partial\Omega$. There are thus three possibilities:

(i) $k \equiv$ const. $\neq 0$. (ii) $k \equiv 0$. (iii) $k(U) \subset \Omega$.

Assume (i): $k \equiv c \neq 0$. Then $fgh = c$, so f, g, h are outer. By (3), $|f|$ is constant on T. So f is constant. The same is true of g, since $g\bar{g} = 1 - f\bar{f}$ a.e. on T, and hence h is also constant. But then (f, g, h) does not parametrize an H^∞-disc. Thus (i) cannot happen.

Assume (ii): Now (3) shows that $|f|$ can have at most 2 values on T (a.e.), namely 0 and 1. If $f = 0$ on a set of positive measure then $f \equiv 0$. So either $f \equiv 0$ or f is inner. Also, $f \equiv 0$ implies that g is inner (since $f\bar{f} + g\bar{g} = 1$ a.e. on T) and f inner implies $g \equiv 0$. In either case, $h \equiv 0$, since

$$h = \Gamma(f\bar{f})/fg = \bar{f}\,\bar{g}q(f\bar{f}),$$

using the notation in Proposition 3.5. So (ii) gives either (a) or (b).

Assume (iii), put $u = \Psi^{-1} \circ k$. Then u is inner, by (3). Since Ψ is outer, so is $k = \Psi \circ u$, hence f, g, h are outer. Also, a.e. on T,

$$(4) \qquad f\bar{f} = \Gamma^{-1} \circ k = \Gamma^{-1} \circ \Psi \circ u = (F\,\bar{F}) \circ u.$$

Thus f and $F \circ u$ are outer functions whose absolute values coincide a.e. on T. This gives the first equality in (2). It follows also that $|g| = |G \circ u|$ a.e. on T, which gives the second equality in (2). Finally

$$h = k/fg = (\Psi \circ u) \cdot e^{-i(\alpha+\beta)}/(FG) \circ u = e^{-i(\alpha+\beta)}H \circ u.$$

6. Example B.

6.1. The embeddings described in this section are again as in Proposition 3.5, but this time we take the range of Γ to be a figure eight. A simple example is

$$(1) \qquad \Gamma_0(t) = \sin 2\pi t + i\sin 4\pi t \qquad (0 \le t \le 1),$$

but Theorem 6.2 will deal with the following more general situation.

Pick $c > 0$, $s > 0$, so that $c^2 + s^2 = 1$. Choose Γ so that the hypotheses of Proposition 3.5 hold, and so that $\Gamma([0, c^2])$ and $\Gamma([c^2, 1])$ are, respectively, the boundaries of two simply connected regions Ω_1 and Ω_2 whose closures intereesect only at the origin. Thus $\Gamma(0) = \Gamma(c^2) = \Gamma(1) = 0$, but otherwise Γ is one-to-one.

We need the following three subsets of $\overline{B}$:

$$V_1 = \{(z, w) \in B : 0 < |z| < c, s < |w| < 1\}$$
$$V_2 = \{(z, w) \in B : c < |z| < 1, 0 < |w| < s\}$$
$$\Delta = \{(z, w) : |z| \le c, |w| \le s\}.$$

6.2. THEOREM. *If* Γ *is as in* §*6.1,*

(1) $$E(z,w) = (z,w,\Gamma(z\overline{z})/zw)$$

on S, *and* $\Sigma = E(S)$, *then* Σ *is totally real and*

   (a) $\hat{\Sigma} = \mathcal{D}(\Sigma)$,
   (b) $\hat{\Sigma}$ *is the graph of a function* $\varphi \in C(\overline{B})$ *such that*
       (i) $\varphi(z,w) = \Gamma(z\overline{z})/zw$ *on* S,
       (ii) φ *is real-analytic in* $V_1 \cup V_2$,
       (iii) $\varphi(z,w) = 0$ *if and only if* $zw = 0$ *or* $(z,w) \in \Delta$.

Note that even though $\hat{\Sigma}$ is a graph, (iii) shows that it *cannot* be the graph of a real-analytic function, even when Σ is a real-analytic manifold, such as the sphere generated by the curve Γ_0 of 6.1 (1).

The proof of Theorem 6.2 has a great deal in common with that of Theorem 5.2. To avoid repetition we will concentrate on those aspects that are really different.

PROOF. Fix $(z_0,w_0,\zeta_0) \in \hat{\Sigma}$, put $p_0 = z_0 w_0 \zeta_0$. That $p_0 \in \overline{\Omega}_1 \cup \overline{\Omega}_2$ is proved just as $p_0 \in \overline{\Omega}$ was proved in §5.2. Suppose $p_0 \in \overline{\Omega}_1$. (The argument is the same if $p_0 \in \overline{\Omega}_2$.)

There is a Riemann map $\Psi_1 : \overline{U} \to \overline{\Omega}_1$ with $\Psi_1(1) = 0$, and there is a unique $\lambda_0 \in \overline{U}$ such that $p_0 = \Psi_1(\lambda_0)$. As in §5.1 there are outer functions F, G, H such that

(1) $$F\,\overline{F} = \Gamma^{-1} \circ \Psi, \quad F\,\overline{F} + G\,\overline{G} = 1 \quad \text{on} \ T\backslash\{1\}$$

and

(2) $$FGH = \Psi_1 \quad \text{on} \ \overline{U}.$$

Also $H \in A(U)$, $H(0) = 1$.

Since Γ^{-1} maps $\partial\Omega_1\backslash\{0\}$ onto $(0, c^2)$ it follows that

(3) $$0 < |F| < c, \quad s < |G| < 1 \quad \text{on} \ \overline{U}\backslash\{1\}.$$

Thus (F,G) maps U into V_1. (If $p_0 \in \overline{\Omega}_2$ we get an analogous map to V_2.) We claim that

(4) $|z_0| \le |F(\lambda_0)|$, $|w_0| \le |G(\lambda_0)|$ *in case* $\lambda_0 \neq 1$,

(5) $|\zeta_0| \le |H(\lambda_0)|$ *in any case*,

(6) *if* $\zeta_0 = 0$ *then either* $z_0 w_0 = 0$ *or* $(z_0, w_0) \in \Delta$.

These assertions prove (a) and (b) (iii) as in §5.2.

The proofs of (4) and (5) are quite similar to those of 5.2 (4) and 5.2 (5): As regards (4) we again choose outer functions $F_n \in A(U)$ such that $|F_n| \ge |F|$ on $T\backslash\{1\}$, $|F_n| > 0$ on $\overline{U}$, $F_n(\lambda_0) \to F(\lambda_0)$ as $n \to \infty$, but instead of applying Mergelyan's theorem to $1/F_n \circ \Psi^{-1}$ on $\overline{\Omega}$ we apply it on $\overline{\Omega}_1 \cup \overline{\Omega}_2$ to the function which is $1/F_n \circ \Psi_1^{-1}$ on $\overline{\Omega}_1$ and $1/F_n(1)$ on $\overline{\Omega}_2$.

We turn to the proof of (6). There is a function $\chi \in C(\overline{\Omega}_1 \cup \overline{\Omega}_2)$, holomorphic in $\Omega_1 \cup \Omega_2$, such that $\chi(0) = 1$, $|\chi| < 1$ on the rest of $\overline{\Omega}_1 \cup \overline{\Omega}_2$. Since now $(z_0, w_0, 0) \in \hat{\Sigma}$ there is a probability measure μ on S (as in the proof of Proposition 2.4) such that

$$(7) \qquad Q(z_0, w_0, 0) = \int_S Q(z, w, \Gamma(z\overline{z})/zw) d\mu(z, w)$$

for every polynomial $Q : \mathbb{C}^3 \to \mathbb{C}$. Since χ is a uniform limit of polynomials on $\overline{\Omega}_1 \cup \overline{\Omega}_2$, (7) holds also when Q is replaced by

$$(8) \qquad Q_n(z, w, \zeta) = zwP(z, w)\chi^n(zw\zeta)$$

where $P : \mathbb{C}^2 \to \mathbb{C}$ is any polynomial, and $n = 1, 2, 3, \ldots$. Thus, since $\chi^n(0) = 1$, (7) becomes

$$(9) \qquad z_0 w_0 P(z_0, w_0) = \int_S zwP(z, w)\chi^n(\Gamma(z\overline{z})) d\mu(z, w) .$$

The integrand tends boundedly to 0 at all points of S except those where $zw \neq 0$ and $\Gamma(z\overline{z}) = 0$, i.e., where $|z| = c$, $|w| = s$. Thus, for every polynomial P,

$$(10) \qquad z_0 w_0 P(z_0, w_0) = \int_K zwP(z, w) d\mu(z, w),$$

where $K = \{(z, w) : |z| = c, |w| = s\}$.

If $z_0 w_0 \neq 0$ then, since (10) holds for *all* P, it follows that (z_0, w_0) lies in the polynomial hull Δ of the 2-torus K.

The proofs of (b) (i), (ii) are as in Theorem 5.2.

7. Example C.

7.1. In each of the two preceding examples, $\hat{\Sigma}$ was the graph of some $\varphi \in C(\overline{B})$. Its (real) dimension was therefore 4. We shall now see that if Σ is obtained from a curve Γ as in Proposition 3.5, and *if Γ makes a loop around the origin, then $\mathcal{D}(\Sigma)$ has nonempty interior in $\mathbb{C}^3$, and its dimension is therefore 6.*

Perhaps the simplest example of such a curve is

$$\Gamma(t) = e^{2\pi it} - e^{4\pi it} \qquad (0 \leq t \leq 1);$$

here $\Gamma([1/6, 5/6])$ is the boundary of a simply connected region which contains the origin.

7.2. THEOREM. *Pick positive numbers c_1, c_2, s_1, s_2 so that $c_1 < c_2$, $c_1^2 + s_1^2 = c_2^2 + s_2^2 = 1$.*

Suppose that Γ is a curve as in Proposition 3.5 which is one-to-one on $[c_1^2, c_2^2]$ except that $\Gamma(c_1^2) = \Gamma(c_2^2)$, and that $\Gamma([c_1^2, c_2^2])$ is the boundary of a simply connected region $\Omega \subset \mathbb{C}$ which contains the origin.

Let

$$(1) \qquad E(z, w) = (z, w, \Gamma(z\overline{z})/zw) \qquad ((z, w) \in S) .$$

The disc-hull of the totally real 3-sphere $\Sigma = E(S)$ *contains then the polydisc*

$$(2) \qquad \mathcal{P} = \left\{ (z, w, \zeta) : |z| \le c_1, |w| \le s_2, |\zeta| \le \frac{r}{c_2 s_1} \right\}$$

where $r = \min\{|\Gamma(t)| : c_1^2 \le t \le c_2^2\}$.

PROOF. Fix $(z_0, w_0, \zeta_0) \in \mathcal{P}$. Choose $(z, w, \zeta) \in \mathcal{P}$ with $zw\zeta \ne 0$, in such a way that $z = z_0$ if $z_0 \ne 0$, $w = w_0$ if $w_0 \ne 0$, $\zeta = \zeta_0$ if $\zeta_0 \ne 0$. Let m be the number of coordinates of (z_0, w_0, ζ_0) that are 0. Thus $0 \le m \le 3$. Let $\Psi : \overline{U} \to \overline{\Omega}$ be a Riemann map with $\Psi(0) = 0$. Note that $|zw\zeta| < r$, by (2), since $c_1 < c_2$ and $s_2 < s_1$. Thus $zw\zeta \in \Omega = \Psi(U)$. We now consider two cases.

If $m = 0$, let φ be a *singular inner function* in U such that $\Psi(\varphi(0)) = zw\zeta$ $(= z_0 w_0 \zeta_0)$.

If $m > 0$, let φ be a *singular inner function* in U such that $\varphi(0) = zw\zeta/\Psi'(0)$. Note that this is possible, because $|\Psi'(0)| \ge r$, by the Schwarz lemma, applied to $\Psi^{-1} : rU \to U$, and therefore $|zw\zeta/\Psi'(0)| < 1$.

In either case we let F and G be outer functions such that

$$(3) \qquad (F\overline{F})(\lambda) = \Gamma^{-1}(\Psi(\lambda^m \varphi(\lambda))), \quad F\overline{F} + G\overline{G} = 1$$

a.e. on T. Since Γ^{-1} maps $\Psi(T) = \partial\Omega$ onto $[c_1^2, c_2^2]$ we see that

$$(4) \qquad c_1 < |F| < c_2, \ s_2 < |G| < s_1 \ \text{in } U.$$

Put $\alpha = z/F(0)$, $\beta = w/G(0)$. Then $|\alpha| < 1$, $|\beta| < 1$.

When $m = 0$, another application of the Schwarz lemma gives, by (2) and (4),

$$(5) \qquad |\varphi(0)| = |\Psi^{-1}(zw\zeta)| \le \frac{|zw\zeta|}{r} \le \frac{|\alpha\beta F(0)G(0)|}{c_2 s_1} < |\alpha\beta|.$$

When $m > 0$, then

$$(6) \qquad |\varphi(0)| = \frac{|zw\zeta|}{|\Psi'(0)|} \le \frac{|zw\zeta|}{r} < |\alpha\beta|.$$

Thus $|\varphi(0)| < |\alpha\beta|$ in both cases. In particular, $|\varphi(0)| < |\alpha|$ and $|\varphi(0)| < \beta$. Hence there exist $\epsilon_j > 0$, θ_j, such that

$$(7) \qquad \alpha = e^{i\theta_1}[\varphi(0)]^{\epsilon_1}, \quad \beta = e^{i\theta_2}[\varphi(0)]^{\epsilon_2},$$

and $\epsilon_1 + \epsilon_2 < 1$ because $|\varphi(0)| < |\alpha\beta|$.

Being a singular inner function, φ has no zero in U. Hence φ^ϵ is again inner, for every $\epsilon > 0$, and we can define holomorphic functions f, g, h in U by

$$(8) \qquad f(\lambda) = \lambda^a e^{i\theta_1}(\varphi^{\epsilon_1} F)(\lambda)$$

$$(9) \qquad g(\lambda) = \lambda^b e^{i\theta_2}(\varphi^{\epsilon_2} G)(\lambda)$$

$$(10) \qquad h(\lambda) = \Psi(\lambda^{a+b+c}\varphi(\lambda))/(fg)(\lambda)$$

where $a = 0$ if $z_0 \ne 0$ (i.e., $z = z_0$), $a = 1$ if $a_0 = 0$, and b, c are defined analogously, with respect to w_0, ζ_0. Thus $a + b + c = m$.

We claim that $(f, g, h) : U \to \mathbb{C}^3$ is an H^∞-map that takes T (a.e.) into Σ and takes 0 to the prescribed point $(z_0, w_0, \zeta_0) \in \mathcal{P}$.

It is clear that f and g are bounded in U.

Since $\Psi(0) = 0$, $\Psi(\lambda)/\lambda$ is bounded, and since $|F| > c_1$, $|G| > c_2$, we see that, for some $M < \infty$,

$$(11) \qquad |h(\lambda)| \leq M|\lambda|^c|\varphi|^{1-\epsilon_1-\epsilon_2}(\lambda) \qquad (\lambda \in U).$$

Thus $h \in H^\infty(U)$.

It is clear that $f\overline{f} + g\overline{g} = 1$ a.e. on T, since $|f| = |F|$ and $|g| = |G|$ on T. Also

$$(12) \qquad \Gamma(f\overline{f})(\lambda) = \Gamma(F\overline{F})(\lambda) = \Psi(\lambda^m\varphi(\lambda)) = (fgh)(\lambda)$$

a.e. on T.

Thus (f, g, h) parametrizes an H^∞-disc whose boundary lies in Σ.

Finally, by (7) to (10)

$$(13) \qquad f(0) = \alpha F(0)\lambda^a \Big|_{\lambda=0} = z_0$$

$$(14) \qquad g(0) = \beta G(0)\lambda^b \Big|_{\lambda=0} = w_0$$

and

$$(15) \qquad h(0) = \frac{\Psi(\lambda^m\varphi(\lambda))}{zw\lambda^{a+b}} \Big|_{\lambda=0}.$$

When $m = 0$, then $a = b = 0$ and

$$(16) \qquad h(0) = \Psi(\varphi(0))/zw = \frac{zw\zeta}{zw} = \zeta = \zeta_0.$$

When $m > 0$, then, as $\lambda \to 0$

$$(17) \qquad \frac{\Psi(\lambda^m\varphi(\lambda))}{\lambda^{a+b}} = \Psi'(0)\lambda^c\varphi(\lambda) + O(|\lambda|^{m+c})$$

so that $h(0) = 0$ when $c > 0$ (i.e. when $\zeta_0 = 0$), and

$$(18) \qquad zwh(0) = \Psi'(0)\varphi(0) = zw\zeta = zw\zeta_0$$

when $c = 0$. Thus $h(0) = \zeta_0$ in all cases, and the proof is complete.

8. Example D.

8.1. From now on, Σ will no longer be totally real.

Let $\gamma : [0, 1] \to \mathbb{C}$ be continuous and one-to-one. Thus $\gamma^* = \gamma([0, 1])$ is an arc in $\mathbb{C}$. No smoothness beyond continuity is assumed.

For $0 \le c \le 1$ define

$$(1) \qquad K_c = \{(z, w, \zeta) : |z| \le c, |w| \le s, \zeta = \gamma(c^2)\};$$

here $c^2 + s^2 = 1$. Note that K_c is a bidisc if $0 < c < 1$, but K_0 and K_1 are discs. Define

$$(2) \qquad K = \bigcup_{0 \le c \le 1} K_c.$$

Then K is compact and $\dim K = 5$.

There is another way of describing K: Associate to each $(z, w) \in \overline{B}$ the set

$$(3) \qquad L_{z,w} = \gamma([z\overline{z}, 1 - w\overline{w}]).$$

This is an arc when $(z, w) \in B$, is a point when $(z, w) \in S$, and K *is the set of all* (z, w, ζ) *having* $\zeta \in L_{z,w}$.

For $|\alpha| \le 1$ we define

$$(4) \qquad \varphi_\alpha(\lambda) = \frac{\alpha - \lambda}{1 - \overline{\alpha}\lambda}.$$

This is a Moebius transformation of U when $|\alpha| < 1$, and is the constant α when $|\alpha| = 1$.

8.2. Theorem. *Let γ and K be as in §8.1, let Σ be the sphere given by*

$$(1) \qquad \Sigma = E(S), \quad E(z, w) = (z, w, \gamma(z\overline{z})).$$

Then

    (a) $\hat{\Sigma} = \mathcal{D}(\Sigma) = K$, *and*

    (b) *every H^∞-disc with boundary in Σ is parametrized by a map Φ of the form*

$$(2) \qquad \Phi(\lambda) = (c\psi_1(\lambda), s\psi_2(\lambda), \gamma(c^2)) \quad (\lambda \in U)$$

    where $c \ge 0$, $s \ge 0$, $c^2 + s^2 = 1$, and ψ_1, ψ_2 are inner functions.

Note that (a) and 8.1 (3) show that the fiber of $\hat{\Sigma}$ over $(z, w) \in B$ is the arc $L_{z,w}$, a subarc of $\gamma^* = L_{0,0}$.

Proof. Suppose $(z, w) \in B$, $(z, w, \zeta) \in K_c$. Then $z = c\alpha$, $w = s\beta$, $|\alpha| \le 1$, $|\beta| \le 1$, $\zeta = \gamma(c^2)$. The map

$$(3) \qquad \Phi = (c\varphi_\alpha, s\varphi_\beta, \zeta)$$

(see 8.1 (4)) parametrizes an A-disc whose boundary lies in Σ and which contains the point

(4) $$\Phi(0) = (z, w, \zeta)\,.$$

This proves that $K \subset \mathcal{D}(\Sigma)$.

To complete (a), we show that $\hat{\Sigma} \subset K$. Fix $(z_0, w_0, \zeta_0) \in \hat{\Sigma}$. As in Proposition 2.4, there is a probability measure μ on S such that

(5) $$Q(z_0, w_0, \zeta_0) = \int_S Q(z, w, \gamma(z\bar{z}))d\mu(z, w)$$

for all polynomials Q. In particular,

(6) $$f(\zeta_0) = \int_S f(\gamma(z\bar{z}))d\mu(z, w)$$

for every polynomial $f : \mathbb{C} \to \mathbb{C}$, hence also for every continuous f, since γ^* is an arc. Take $f(\zeta) = |\zeta - \zeta_0|$. Then (6) becomes

(7) $$0 = \int_S |\gamma(z\bar{z}) - \zeta_0|d\mu(z, w)\,.$$

Hence μ is concentrated on the subset of S where $\gamma(z\bar{z}) = \zeta_0$. This means that $\zeta_0 = \gamma(c^2)$ for some $c \in [0, 1]$ and that the support of μ lies on the torus (or circle, when $c = 0$ or $c = 1$) given by $|z| = c$, $|w| = s$, where $c^2 + s^2 = 1$. Since

(8) $$P(z_0, w_0) = \int P d\mu$$

for every polynomial P, (z_0, w_0) lies in the polynomial hull of the above-mentioned torus. Thus $|z_0| \le c$, $|w_0| \le s$, $\zeta_0 = \gamma(c^2)$, i.e., $(z_0, w_0, \zeta_0) \in K_c \subset K$.

This proves (a).

To prove (b), suppose $(f, g, h) : U \to \mathbb{C}^3$ parametrizes an H^∞-disc with boundary in Σ. Then

(9) $$f\bar{f} + g\bar{g} = 1, \quad h = \gamma(f\bar{f}) \quad \text{a.e. on } T.$$

Since γ^* is an arc, it follows that $h(U) \subset \gamma^*$, and the open mapping theorem shows that h is constant. Hence $f\bar{f}$ and $g\bar{g}$ are constant a.e. on T, and this means that f and g are constant multiples of inner functions. This completes the proof.

8.3. Remark. If we take, as a special case, $\gamma(t) = t$ in Theorem 8.2, then

$$\hat{\Sigma} = \{(z, w, \zeta) : z\bar{z} \le \zeta \le 1 - w\bar{w}\}$$

is the convex hull of Σ.

9. Example E.

9.1. The formula that defines Σ in this example is as in Theorem 8.2, but γ is now a simple closed curve; i.e., $\gamma : [0, 1] \to \mathbb{C}$ is continuous and one-to-one, except that $\gamma(0) = \gamma(1) = p$, say. Thus $\gamma([0, 1]) = \partial\Omega$ for some simply connected region $\Omega \subset \mathbb{C}$, and $\gamma^{-1} : \partial\Omega\backslash\{p\} \to (0, 1)$ is one-to-one.

There is a Riemann map $\Psi : \overline{U} \to \overline{\Omega}$, $\Psi(1) = p$, which we use to define a function u in $\overline{\Omega}\backslash\{p\}$ by

$$(1) \qquad u \circ \Psi = P[\log \gamma^{-1} \circ \Psi]$$

where P denotes the Poisson integral of the function in the bracket. Thus u is harmonic in Ω if the integrand in (1) is in $L^1(T)$, and is $\equiv -\infty$ in the other case.

We claim that u is well-defined by (1), in the sense that *we get the same u if Ψ is replaced by another Riemann map $\Psi_1 : \overline{U} \to \overline{\Omega}$ on both sides of (1).* The proof of this statement depends on the well-known identity

$$(2) \qquad P[f] \circ \varphi = P[f \circ \varphi],$$

valid for all $f \in L^1(T)$ and all inner functions φ. (Note that (2) is trivial when $f(e^{i\theta}) = e^{in\theta}$, hence (2) holds for all trigonometric polynomials f, and the general case follows.)

If now Ψ and Ψ_1 are as above, then $\Psi_1 = \Psi \circ \varphi$ for some inner function φ, and if u is given by (1), then

$$(3) \qquad \begin{aligned} u \circ \Psi_1 &= u \circ \Psi \circ \varphi = P[\log \gamma^{-1} \circ \Psi] \circ \varphi \\ &= P[\log \gamma^{-1} \circ \Psi \circ \varphi] = P[\log \gamma^{-1} \circ \Psi_1], \end{aligned}$$

which proves our claim.

In the same way, we define v on $\overline{\Omega}\backslash\{p\}$ by

$$(4) \qquad v \circ \Psi = P[\log(1 - \gamma^{-1} \circ \Psi)].$$

Again, v is either harmonic in Ω, or $v \equiv -\infty$.

Note that $e^u = \gamma^{-1}$ on $\partial\Omega\backslash\{p\}$ if u is harmonic, and $e^v = 1 - \gamma^{-1}$ on $\partial\Omega\backslash\{p\}$ if v is harmonic. Thus

$$(5) \qquad e^u + e^v = 1 \quad \text{on} \quad \partial\Omega\backslash\{p\}$$

if both u and v are harmonic.

We define K_c and K as in §8.1:

$$(6) \qquad K_c = \{(z, w, \zeta) : |z| \leq c, |w| \leq s, \zeta = \gamma(c^2)\}$$

where $c^2 + s^2 = 1$, and

$$(7) \qquad K = \bigcup_{0 \leq c \leq 1} K_c.$$

So K is compact, $\dim K = 5$, K_c is a bidisc if $0 < c < 1$, K_0 and K_1 are discs.

Next, we define

(8) $$Y = \{(z, w, \xi) : z\overline{z} \le e^{u(\xi)}, \ w\overline{w} \le e^{v(\xi)}, \ \xi \in \Omega\}.$$

with the understanding the $e^u = 0$ when $u = -\infty$, and $e^v = 0$ when $v = -\infty$. Note that therefore

(i) $\dim Y = 6$ when both u and v are harmonic,
(ii) $\dim Y = 4$ when one of them is harmonic, the other is $-\infty$,
(iii) $\dim Y = 2$ when $u = v = -\infty$.

9.2. THEOREM. *If γ, K, Y are as in §9.1 and $\Sigma = E(S)$, where*

(1) $$E(z, w) = (z, w, \gamma(z\overline{z}))$$

then

(a) $\hat{\Sigma} = K \cup Y$ *in all cases,*
(b) $\mathcal{D}(\Sigma) = K \cup Y$ *if both u and v are harmonic,*
(c) $\mathcal{D}(\Sigma) = K$ *if at least one of u, v is $-\infty$,*
(d) $Y \subset \mathcal{D}(K)$ *in all cases.*

(The phrase "in all cases" means simply that it is irrelevant whether u and/or v are harmonic or $-\infty$.)

Conclusion (a) shows that $\hat{\Sigma}$ is the closure of a region $W \subset \mathbb{C}^3$ and that the 3-sphere Σ is the Silov boundary of the algebra $A(W)$.

Note that $\hat{\Sigma} = \mathcal{D}(\Sigma)$ if and only if both u and v are harmonic, and that $\dim \hat{\Sigma} = 6$ in precisely the same circumstances. The shape and size of $\hat{\Sigma}$ and $\mathcal{D}(\Sigma)$ are thus strongly influenced by the convergence or divergence of the integrals 9.1 (1) and 9.1 (4).

Note also that it follows from the theorem that

(2) $$\hat{\Sigma} = \mathcal{D}(\mathcal{D}(\Sigma))$$

in all cases.

We shall break the proof into 4 steps.

Step 1. $K \subset \mathcal{D}(\Sigma)$.

Step 2. $Y \subset \mathcal{D}(\Sigma)$ if both u and v are harmonic, and $Y \subset \mathcal{D}(K)$ in all cases.

Step 3. $\hat{\Sigma} \subset K \cup Y$.

Step 4. If Y intersects $\mathcal{D}(\Sigma)$ then both u and v are harmonic.

Steps 1 and 2 together prove (d) and show that $K \cup Y \subset \hat{\Sigma}$ in all cases, $K \cup Y \subset \mathcal{D}(\Sigma)$ if both u and v are harmonic. Hence Step 3 completes the proofs of (a) and (b), and then (c) follows from Step 4.

PROOF OF STEP 1. Fix $(z, w, \zeta) \in K_c$ for some $c \in [0, 1]$. There exist $\alpha, \beta, |\alpha| \le 1, |\beta| \le 1$, such that $z = c\alpha$, $w = s\beta$, and $\zeta = \gamma(z\overline{z}) = \gamma(c^2)$. The map (see 8.1 (4) for notation)

(3) $$\Phi = (c\varphi_\alpha, s\varphi_\beta, \zeta)$$

takes T into Σ and 0 to the prescribed point (z, w, ξ).

PROOF OF STEP 2. Fix $(z, w, \zeta) \in Y$. Then $\zeta \in \Omega$ and we can choose our Riemann map Ψ so that $\Psi(0) = \zeta$. There exist $F, G \in H^\infty(U)$ such that, on $T\backslash\{1\}$,

$$(4) \qquad F\,\overline{F} = e^u \circ \Psi, \quad G\,\overline{G} = e^v \circ \Psi,$$

because when u is harmonic, it has a harmonic conjugate $\tilde{u}$ in Ω, and we can define

$$(5) \qquad F = \exp\left(\frac{u + i\tilde{u}}{2}\right) \circ \Psi\,;$$

when $u = -\infty$, put $F = 0$. Ditto for G.

Since $(z, w, \zeta) \in Y$ and $F\overline{F}(0) = \exp u(\zeta)$, we have $|z| \le |F(0)|$. Likewise, $|w| \le |G(0)|$. So $z = \alpha F(0)$, $w = \beta G(0)$, $|\alpha| \le 1$, $|\beta| \le 1$, so that the H^∞-map

$$(6) \qquad \Phi = (F\varphi_\alpha, G\varphi_\beta, \Psi)$$

takes 0 to the prescribed point (z, w, ζ).

When both u and v are harmonic then (see 9.1 (5)) $e^u + e^v = 1$ on $\partial\Omega\backslash\{p\}$, hence $F\overline{F} + G\overline{G} = 1$ on $T\backslash\{1\}$. Also, since $e^u = \gamma^{-1}$ on $\partial\Omega\backslash\{p\}$ in this case,

$$(7) \qquad \gamma(F\overline{F}) = \gamma(e^u \circ \Psi) = \Psi$$

on $T\backslash\{1\}$. Thus (6) maps $T\backslash\{1\}$ into Σ, and therefore $(z, w, \zeta) \in \mathcal{D}(\Sigma)$.

In any case, whether u, v are harmonic or $-\infty$, the definitions of F and G show that

$$(8) \qquad F\overline{F} \le \gamma^{-1} \circ \Psi \le 1 - G\overline{G}$$

on $T\backslash\{1\}$, so that (6) maps $T\backslash\{1\}$ into K. Thus $Y \subset \mathcal{D}(K)$.

PROOF OF STEP 3. Fix $(z_0, w_0, \zeta_0) \in \hat{\Sigma}$. It is clear that $\zeta_0 \in \overline{\Omega}$. (The proof is the same as the one that gave $p_0 \in \overline{\Omega}$ in Theorem 5.2.)

We first look at the case $\zeta_0 = p$. Pick some $\chi \in A(\Omega)$ that peaks at p; $\chi = (1 + \Psi^{-1})/2$ will do nicely. Define

$$(9) \qquad P_n(z, w, \zeta) = zw\chi^n(\zeta)\,.$$

Since $(z_0, w_0, p) \in \hat{\Sigma}$, we have, for all $n = 1, 2, 3, \ldots,$

$$(10) \qquad |z_0 w_0| = |P_n(z_0, w_0, p)| \le \max_\Sigma |P_n|$$
$$= \max_S |P_n(z, w, \gamma(z\overline{z}))| = \max |zw||\chi^n(\gamma(z\overline{z}))|$$

and this last max tends to 0 as $n \to \infty$ because $\gamma(z\overline{z}) = p$ only when $|z| = 0$ or $|z| = 1$, $w = 0$.

Thus $z_0 w_0 = 0$, which shows that $(z_0, w_0, p) \in K_0 \cup K_1 \subset K$.

Next, we suppose $\zeta_0 \in \overline{\Omega}\backslash\{p\}$.

Since Re $A(\Omega)$ is dense in $C_R(\partial\Omega)$ there exist q_n and $Q_n = \exp q_n$ in $A(\Omega)$ such that first

$$(11) \qquad \operatorname{Re} q_n < -\log \gamma^{-1} \quad \text{on } \partial\Omega\backslash\{p\},$$

hence

$$(12) \qquad |Q_n| < 1/\gamma^{-1} \quad \text{on } \partial\Omega\backslash\{p\}$$

and secondly

$$(13) \qquad \lim_{n\to\infty} Q_n(\zeta_0) = \begin{cases} 1/\gamma^{-1}(\zeta_0) & \text{in case } \zeta_0 \in \partial\Omega\backslash\{p\}, \\ \exp(-u(\zeta_0)) & \text{in case } \zeta_0 \in \Omega. \end{cases}$$

(To see (13), look at the Poisson integrals of $\log Q_n \circ \Psi$.)

Define

$$(14) \qquad f_n(z, w, \zeta) = z^2 Q_n(\zeta).$$

Then, if $0 < z\bar{z} < 1$ and $(z, w, \zeta) \in \Sigma$, $\zeta = \gamma(z\bar{z})$, (12) shows

$$(15) \qquad |f_n(z, w, \gamma(z\bar{z}))| = |z^2 Q_n(\gamma(z\bar{z}))| < 1,$$

whereas

$$(16) \qquad \lim_{n\to\infty} f_n(z_0, w_0, \zeta_0) = \lim_{n\to\infty} z_0^2 Q_n(\zeta_0) = \begin{cases} z_0^2/\gamma^{-1}(\zeta_0) \\ z_0^2 \exp(-u(\zeta_0)). \end{cases}$$

Since $(z_0, w_0, \zeta_0) \in \hat{\Sigma}$, it follows from (15) and (16) that

$$(17) \qquad |z_0|^2 \leq \begin{cases} \gamma^{-1}(\zeta_0) & \text{in case } \zeta_0 \in \partial\Omega\backslash\{p\} \\ \exp u(\zeta_0) & \text{in case } \zeta_0 \in \Omega. \end{cases}$$

The analogous inequalities

$$(18) \qquad |w_0|^2 \leq \begin{cases} 1 - \gamma^{-1}(\zeta_0) & \text{in case } \zeta_0 \in \partial\Omega\backslash\{p\} \\ \exp v(\zeta_0) & \text{in case } \zeta_0 \in \Omega \end{cases}$$

are proved in the same way.

Combining (17) and (18) shows that $\hat{\Sigma} \subset K \cup Y$.

PROOF OF STEP 4. Suppose there is a point $(z, w, \zeta) \in Y \cap \mathcal{D}(\Sigma)$. There is then an H^∞-map $(f, g, h) : U \to \mathbb{C}^3$ that sends T (a.e.) into Σ and sends 0 to (z, w, ζ).

Since $h = \gamma(f\bar{f})$ a.e. on T, we have $h(T) \subset \partial\Omega$. Since $(z, w, \zeta) \in Y$, $h(0) = \zeta \in \Omega$. Thus h is not constant, hence neither is $f\bar{f}$ nor $g\bar{g} = 1 - f\bar{f}$ on T. Thus neither f nor g are constant in U.

Choose a Riemann map $\Psi : \overline{U} \to \overline{\Omega}$ with $\Psi(0) = \zeta$. Put $\varphi = \Psi^{-1} \circ h$. Then φ is inner, $\varphi(0) = 0$. Since f is not constant, Jensen's theorem gives the inequality in

$$(19) \qquad -\infty < \int_T \log f\overline{f} = \int_T \log \gamma^{-1} \circ h$$
$$= \int_T \log \gamma^{-1} \circ \Psi \circ \varphi = \int_T \log \gamma^{-1} \circ \Psi \, .$$

(The last equality is a special case of 9.1 (2), since $\varphi(0) = 0$.) Thus $\log \gamma^{-1} \circ \Psi \in L^1(T)$, which shows that u is harmonic. The same argument proves the harmonicity of v.

This completes the proof.

9.3. We end this section by showing explicitly that it is not the shape of Ω or $\partial\Omega$ which determines whether u and/or v are harmonic or $-\infty$ in Theorem 9.3. The important difference lies in the *parametrization* γ of $\partial\Omega$.

Let $\Omega = U$, $\Psi(\lambda) = \lambda$, and consider two parametrizations of $\partial\Omega = T$:

$$(1) \qquad \gamma_1(t) = e^{2\pi i t} \qquad (0 \leq t \leq 1),$$

$$(2) \qquad \gamma_2(t) = \exp\left(\frac{2\pi i}{1 - \log t}\right) \qquad (0 \leq t \leq 1),$$

which gives

$$(3) \qquad \gamma_1^{-1}(e^{i\theta}) = \theta/2\pi \qquad (0 < \theta < 2\pi),$$

$$(4) \qquad \gamma_2^{-1}(e^{i\theta}) = \exp\left(1 - \frac{2\pi}{\theta}\right) \qquad (0 < \theta < 2\pi),$$

Thus $(\log \gamma_1^{-1} \circ \Psi)(e^{i\theta}) = \log \frac{\theta}{2\pi}$ is in L^1, whereas

$$(5) \qquad \int_0^{2\pi} (\log \gamma_2^{-1} \circ \Psi)(e^{i\theta}) d\theta = \int_0^{2\pi} \left(1 - \frac{2\pi}{\theta}\right) d\theta = -\infty \, .$$

Therefore u, given by 9.1 (1), is harmonic in the first case and is $-\infty$ in the second.

10. Example F.

Proposition 2.4 showed that $\hat{\Sigma}$ is a graph over $\overline{B}$ when σ has a pluriharmonic extension, for instance when $\sigma(z, w) = zw$ or $\sigma(z, w) = \overline{zw}$. The situation turns out to be quite different when $\sigma(z, w) = z\overline{w}$.

10.1. THEOREM. *Let $\Sigma = E(S)$, where*

$$(1) \qquad E(z, w) = (z, w, z\overline{w}) \, .$$

Then $\hat{\Sigma}$ contains an open subset V of $\mathbb{C}^3$ such that

    (a) *V contains the disc $\{(0, 0, \zeta) : |\zeta| < 1/2\}$,*

    (b) *V contains the graph $G = \{(z, w, z\overline{w}) : (z, w) \in B\}$,*

    (c) *$G \subset \mathcal{D}(\Sigma)$,*

(d) $V \subset \mathcal{D}(G)$.

PROOF. We define V to be the set of all $(z, w, \zeta) \in \mathbb{C}^3$ with $(z, w) \in B$ to which corresponds a constant t such that $|z| < t$ and

$$
(2) \qquad t^2 + \left| \frac{tw + \overline{\zeta}\lambda}{t + \overline{z}\lambda} \right|^2 < 1 \quad \text{for all } \lambda \in \overline{U}.
$$

It is clear that V is open.

If $(z, w) = (0, 0)$, (2) becomes $t^2 + t^{-2}|\zeta|^2 < 1$, which holds for all ζ with $|\zeta| < 1/2$ when $t = 1/\sqrt{2}$. This proves (a).

If $(z, w) \in B$ and $z\overline{z} < t^2 < 1 - w\overline{w}$, $\zeta = z\overline{w}$, then the left side of (2) is $t^2 + w\overline{w}$. This proves (b).

Next, fix $(z, w, z\overline{w}) \in G$, put $t = \sqrt{1 - w\overline{w}}$, $\alpha = z/t$, note that $|\alpha| < 1$, and define

$$
(3) \qquad \Phi = (\Phi_1, \Phi_2, \Phi_3) = (t\varphi_\alpha, w, t\overline{w}\varphi_\alpha)
$$

where $\varphi_\alpha(\lambda) = (\alpha - \lambda)/(1 - \overline{\alpha}\lambda)$. Then $|\Phi_1|^2 + |\Phi_2|^2 = 1$ and $\Phi_3 = \Phi_1\overline{\Phi}_2$ on T. Thus $\Phi(T) \subset \Sigma$. Since $\Phi(0) = (z, w, \xi)$, we have proved (c).

Finally, fix $(z, w, \xi) \in V$, choose t so that (2) holds, and define

$$
(4) \qquad \Phi(\lambda) = \left(\frac{t(t\lambda + z)}{t + \overline{z}\lambda}, \frac{tw + \overline{\zeta}\lambda}{t + \overline{z}\lambda}, \frac{t(\zeta + t\overline{w}\lambda)}{t + \overline{z}\lambda} \right).
$$

The components Φ_1, Φ_2, Φ_3 of Φ satisfy $|\Phi_1|^2 + |\Phi_2|^2 < 1$ on T, by (2), and $\Phi_3 = \Phi_1\overline{\Phi}_2$ on T, by inspection. This proves (d).

It follows from (d) and (c) that

$$
(5) \qquad V \subset \mathcal{D}(\mathcal{D}(\Sigma)) \subset \hat{\Sigma}.
$$

Thus $\hat{\Sigma}$ has nonempty interior, $\dim \hat{\Sigma} = 6$.

10.2. We know that V is not dense in $\hat{\Sigma}$. It may well be that $\hat{\Sigma}$ is the closure of its interior, but we have not been able to prove this, nor have we found any other really precise information about $\hat{\Sigma}$ in the case of Theorem 10.1.

11. Three Questions.

One can of course formulate any number of questions that are left unanswered by the preceding work. The following three seem particularly intersting to us.

1. Is the disc-hull $\mathcal{D}(X)$ always compact, for every compact $X \subset \mathbb{C}^n$?
2. A propos Proposition 2.3, is $\pi(\mathcal{D}(\Sigma)) = \overline{B}$ for all spheres Σ described in §1.2?
3. For the same spheres, are all fibers of $\hat{\Sigma}$ (see §2.1) always connected?

POSTSCRIPT. Herbert Alexander has shown us how Rosay's simple argument which proved Proposition 3.2 can also be used to prove that the manifold

M is locally polynomially convex. The idea is simply to replace $d\theta/2\pi$ by a representing measure on M.

We let M, Ω, y, and t be as in §3.1.

PROPOSITION. *Every compact subset of $M \cap \Omega$ is polynomially convex.*

PROOF. Let $K \subset M \cap \Omega$ be compact, pick $p \in \hat{K}$. There is then a probability measure μ on K such that $f(p) = \int_K f d\mu$ for every holomorphic polynomial f. With $(f - f(p))^2$ in place of f, the real part of the integral shows that

$$(1) \qquad \int_K \{\text{Re}(f - f(p))\}^2 d\mu = \int_K \{\text{Im}(f - f(p))\}^2 d\mu.$$

Apply (1) to the coordinate functions $f_j(w) = w_j$ $(1 \le j \le n)$ and add. The result is

$$(2) \qquad \int_K |u(w) - a|^2 d\mu(w) = \int_K |v(w) - b|^2 d\mu(w)$$

where $p = a + ib$, $w = u + iv$, and $a, b, u, v \in R^n$.

Note that $v = y(u)$ in (2), since $w \in K \subset M \cap \Omega$.

Next, $p = \int_K w d\mu(w)$. Thus $a = \int_K u d\mu \in \Omega_0$, since Ω_0 is convex, and $b = \int_K v d\mu$. Hence, as in Proposition 3.2, (2) leads to

$$\int_K |u - a|^2 d\mu \le \int_K |y(u) - y(a)|^2 d\mu \le t^2 \int_K |u - a|^2 d\mu.$$

Since $t^2 < 1$ it follows that both integrals in (2) are 0, and this says that the support of μ lies in the set where $u = a$ and $v = b$. In other words, supp $\mu = \{p\}$, so that $p \in K$.

References.

1. Patrick Ahern and Walter Rudin, *Totally real embeddings of S^3 in $\mathbb{C}^3$*, Proc. Amer. Math. Soc. **94** (1985), 460–462.

2. H. Alexander, *Polynomial hulls of graphs*, Pacific J. Math. **147** (1991), 201–212.

3. Herbert Alexander and John Wermer, *Polynomial hulls with convex fibers*, Math. Ann. **271** (1985), 99–109.

4. John T. Anderson, On an example of Ahern and Rudin, preprint.

5. Eric Bedford and Bernard Gaveau, *Envelopes of holomorphy of certain 2-spheres in $\mathbb{C}^2$*, Amer. J. Math. **105** (1983), 975–1009.

6. Eric Bedford and Wilhelm Klingenberg, *On the envelope of holomorphy of a 2-sphere in $\mathbb{C}^2$*, preprint.

7. Andrew Browder, *Cohomology of maximal ideal spaces*, Bull. Amer. Math. Soc. **67** (1961), 515–516.

8. E. F. Collingwood and A. J. Lohwater, *The Theory of Cluster Sets*, Cambridge Univ. Press, 1966.

9. Franc Forstnerič, *A totally real three-sphere in $\mathbb{C}^3$ bounding a family of analytic discs*, Proc. Amer. Math. Soc. **108** (1990), 887–892.

10. Peter L. Duren, *Theory of H^p-Spaces*, Academic Press, 1970.

11. M. Gromov, *Convex integration of differential relations*, Math. USSR-Izv. **7** (1973), 329–343.

12. F. Reese Harvey and R. O. Wells, *Holomorphic approximation and hyperfunction theory on a C^1 totally real submanifold of a complex manifold*, Math. Ann. **197** (1972), 287–318.

13. Witold Hurewicz and Henry Wallman, *Dimension Theory*, Princeton Univ. Press, 1948.

14. Edgar Lee Stout and William R. Zame, *Totally real imbeddings and the universal covering space of domains of holomorphy: some examples*, Manuscripta Math. **50** (1985), 29–48.

15. Edgar Lee Stout and William R. Zame, *A Stein manifold topologically but not holomorphically equivalent to a domain in $\mathbb{C}^N$*, Advances in Math. **80** (1986), 154–160.

16. S. E. Warschawski, *On differentiability at the boundary in conformal mapping*, Proc. Amer. Math. Soc. **12** (1961), 614–620.

17. John Wermer, *Banach Algebras and Several Complex Variables*, 2nd Ed., Springer-Verlag, 1976.

DEPARTMENT OF MATHEMATICS, UNIVERSITY OF WISCONSIN, MADISON, WISCONSIN 53706

Contemporary Mathematics
Volume **137**, 1992

On the totally real spheres of
Ahern and Rudin and Weinstein

H. ALEXANDER

Dedicated to Walter Rudin.

1. John Anderson [4] has recently determined the polynomially convex hull of the totally real 3-sphere imbedded in $\mathbf{C}^3$ which was found by Ahern and Rudin [1]. For topological reasons, a 3-sphere in $\mathbf{C}^3$ cannot be polynomially convex by Serre's theorem [5]; moreover, in the case of the Ahern-Rudin sphere, which is a graph of a continuous complex valued function on the boundary of the unit ball in $\mathbf{C}^2$, it follows from [3] that the hull covers the unit ball. Anderson applies methods used by Wermer [8] for a similar kind of problem. Subsequently, by a different approach, Ahern and Rudin [2] have obtained the polynomial hulls of some classes of totally real 3-spheres. These include their original 3-sphere, for which they show that the hull is a graph over $\mathbf{C}^2$. It is known [6] that the only $n > 1$ for which $\mathbf{C}^n$ contains an imbedded totally real n-sphere is $n = 3$. However Weinstein ([7], p. 26) has exhibited an example for every n of an immersed Lagrangian (hence totally real) n-sphere in $\mathbf{C}^n$. With the n-sphere given by $x_1^2 + \ldots + x_n^2 + a^2 = 1$ in $\mathbf{R}^{n+1}$, the immersion is the map

$$j(x_1, \ldots, x_n, a) = \big(x_1(1 + 2ia), \ldots, x_n(1 + 2ia)\big).$$

The image in $\mathbf{C}^n$, which we shall call L, has a single self-intersection at the origin. It follows again from Serre's theorem, that L is not polynomially convex.

We shall compute the polynomial hull $\hat{L}$ of L. This will turn out to be a family of (planar) complex disks. As in [4], we use the work of Wermer [8]. We show that $P(\hat{L})$, the closure in $\mathbf{C}(\hat{L})$ of the polynomials, agrees with the obvious candidate: it consists of the continuous functions on $\hat{L}$ which are holomorphic on each of these disks. We also consider $P(\hat{M})$ where M is the Ahern-Rudin 3-sphere in $\mathbf{C}^3$. We show that also in this case $P(\hat{M})$ agrees with the obvious candidate.

1991 *Mathematics Subject Classification*. Primary 32E20.

Supported in part by a grant from the National Science Foundation.

This paper is in final form and no version of it will be submitted for publication elsewhere.

2. Let $F(z_1, \ldots, z_n) = z_1^2 + \ldots + z_n^2$. Then

$$F \circ j(x_1, \ldots, x_n, a) = (1 - a^2)(1 + 2ia)^2 :\equiv \phi(a).$$

Hence F maps L onto a simple closed curve $\gamma = \{\phi(a) : -1 \leq a \leq 1\}$ in the complex plane. Note that $\phi : [-1, 1] \to \gamma$ is one-to-one except that $\phi(-1) = \phi(1) = 0$. Moreover, as a plane curve, γ is non-singular except at the origin which is a cusp where the two tangent lines make an angle $\theta_0 = 2 \arctan\left(\frac{4}{3}\right)$. Let Ω be the bounded simply connected component of the complement of γ. Consider the fiber L_ζ of $F|L$ over a point $\zeta \in \gamma$. If $\zeta = 0$, $L_\zeta = \{0\}$, a singleton. If $\zeta \neq 0$, then L_ζ is the $n-1$ sphere $j\{(x_1, \ldots, x_n, a) : x_1^2 + \ldots + x_n^2 + a^2 = 1, a = \phi^{-1}(\zeta)\}$. Moreover L_ζ is polynomially convex. Indeed, $\sum_1^n |z_k|^2 = |\sum_1^n z_k^2| = |\zeta|$ for $z \in L_\zeta$ and so $\sum |z_k|^2 \leq |\zeta|$ on $(L_\zeta)^\wedge$. On the other hand, $L_\zeta \subseteq V_\zeta \equiv \{z : \sum_1^n z_k^2 = \zeta\}$ and so $(L_\zeta)^\wedge \subseteq V_\zeta$. Since V_ζ is disjoint from $\{z : \sum |z_k|^2 < |\zeta|\}$,

$$(L_\zeta)^\wedge = L_\zeta = V_\zeta \cap \{z \in \mathbf{C}^n : \|z\|^2 = |\zeta|\}.$$

(For the last equality, see section 4.) As each point of γ is a peak point for $P(\overline{\Omega})$, we conclude that $(\hat{L})_\zeta = L_\zeta$ for all $\zeta \in \gamma$. Hence $\hat{L} \backslash L \subseteq F^{-1}(\Omega)$.

3. Fix the branch of the function $\sqrt{\lambda}$ on Ω with $\operatorname{Re}(\sqrt{\lambda}) > 0$. For each $x \in S^{n-1} :\equiv \{(x_1, \ldots, x_n) \in \mathbf{R}^n : x_1^2 + \ldots + x_n^2 = 1\}$, define a holomorphic function $\psi_x : \Omega \to \mathbf{C}^n$ by $\psi_x(\lambda) = \sqrt{\lambda} \cdot x$, where we view $\mathbf{R}^n \subseteq \mathbf{C}^n$. Then, for $\zeta \in \gamma = b\Omega$, $\psi_x(\zeta) \in V_\zeta$ and $\|\psi_x(\lambda)\|^2 = |\zeta|$. Hence $\psi_x(\zeta) \in L_\zeta$, ψ_x maps $b\Omega$ into L and $\psi_x(\Omega) \subseteq \hat{L}$ by the maximum principle. Thus $\bigcup_{x \in S^{n-1}} \psi_x(\Omega) \subseteq \hat{L} \backslash L$. Since $\operatorname{Re}(\sqrt{\lambda}) > 0$ on Ω, this is a disjoint union. We next show that this containment is, in fact, equality

4. Define a function ϕ on Ω by

$$\phi(\lambda) = \max_{\substack{z = (z_1, \ldots, z_n) \in \hat{L} \\ F(z) = \lambda}} \left(\log \sum_{k=1}^n |z_k|^2\right).$$

It follows from Wermer's theorem [8] that ϕ is subharmonic on Ω. In fact, for $\alpha \in \mathbf{C}^n$, $\|\alpha\| = 1$, set $\ell_\alpha(z) = \sum_{k=1}^n \alpha_k z_k$ and set $\phi_\alpha(\lambda) = \max_{\substack{z \in \hat{L} \\ F(z) = \lambda}} \log |\ell_\alpha(z)|$. Wermer's theorem says that the ϕ_α are subharmonic on Ω. Then, since ϕ is continuous on Ω (via the local maximum modulus principle) and $\phi = \sup_{\|\alpha\|=1} \phi_\alpha$, we get the subharmonicity of ϕ.

As $\sum |z_k|^2 = |\lambda|$ for all $z \in (\hat{L})_\lambda$ if $\lambda \in b\Omega \backslash \{0\}$, we see that $\phi(\lambda)$ attains continuous boundary values equal to $\log |\lambda|$ on $b\Omega \backslash \{0\}$. The subharmonicity of ϕ on Ω, together with a simple estimate on harmonic measure (cf. Wermer [8], proof of Lemma 4), yields that $\phi(\lambda) \leq \log |\lambda|$ on Ω. On the other hand, if $z \in \hat{L}$ and $F(z) = \lambda$, we have $|\lambda| = |\sum z_k^2| \leq \sum |z_k|^2$ and so $\log |\lambda| \leq \phi(\lambda)$. We conclude that $\phi(\lambda) \equiv \log |\lambda|$ and $|\sum z_k^2| = \sum |z_k|^2 = |\lambda|$ for each $z \in (\hat{L})_\lambda$,

$\lambda \in \Omega$. Equality in the triangle inequality implies that there exist a complex number ζ and n real numbers $r_1, r_2, \ldots, r_n$ such that $z_k = r_k\zeta$, $1 \le k \le n$. Then $\lambda = \sum z_k^2 = (\sum r_k^2)\zeta^2$. Hence $z = \psi_x(\lambda)$ for $x = \pm(\sum r_k^2)^{-\frac{1}{2}}(r_1, \ldots, r_n) \in S^{n-1}$ where the sign is chosen to be the sign in $\zeta = \pm\sqrt{\lambda}/(\sum r_k^2)^{\frac{1}{2}}$.

This proves that $\hat{L}\backslash L = \bigcup_{x \in S^{n-1}} \psi_x(\Omega)$. In particular, $\hat{L}\backslash L$ is homeomorphic to $U \times S^{n-1}$ where U is the open unit disk. It is clear that $\hat{L}$ contains no higher dimensional analytic structure. Indeed, if W were a local subvariety of $\mathbf{C}^n$ of complex dimension > 1 and contained in $\hat{L}$, then, for some $\lambda \in \Omega$, $W \cap F^{-1}(\lambda)$ would be a subvariety of positive dimension contained in $(\hat{L})_\lambda$. But $(\hat{L})_\lambda$ is itself contained in the sphere about the origin of radius $|\lambda|^{\frac{1}{2}}$ which, being strictly pseudoconvex, contains no positive dimensional subvarieties.

When $n = 1$, L is a "figure eight" and $\hat{L}\backslash L$ is the union of two disks (cf. [7] p. 26). The same picture extends to the n-dimensional case: $\hat{L}$ is contained in the union of the complex lines through the origin of the form $\{\lambda\alpha : \lambda \in \mathbf{C}\}$ where $\|\alpha\| = 1$, $\alpha \in \mathbf{R}^n \subseteq \mathbf{C}^n$. Each such line meets L in a figure eight and meets $\hat{L}\backslash L$ in the union of $\psi_\alpha(\Omega)$ and $\psi_{-\alpha}(\Omega)$.

5. Define F_0 on $\hat{L}$ by $F_0 = \sqrt{\lambda} \circ F$. Since $\sqrt{\lambda}$ is in $P(\overline{\Omega})$, $F_0 \in P(\hat{L})$. The following lemma is the main step in determining $P(\hat{L})$. Let g be a continuous complex-valued function on S^{n-1}. Define a corresponding function $\tilde{g}$ on $\hat{L}$ as follows: for $z \ne 0$ in $\hat{L}$, set $\tilde{g}(z) = g(x) \cdot F_0(z) = g(x)\sqrt{\lambda}$, where $z = \psi_x(\lambda)$ uniquely defines x and λ as continuous functions of z. Indeed $\lambda = F(z)$ and then $x = \dfrac{1}{\sqrt{\lambda}} \cdot z$. Setting $\tilde{g}(0) = 0$ gives a continuous function on $\hat{L}$ such that $\tilde{g} \circ \psi_x(\lambda) = g(x) \cdot \sqrt{\lambda}$ for $x \in S^{n-1}$, $\lambda \in \Omega$.

LEMMA 1. *For all* $g \in \mathcal{C}(S^{n-1})$, $\tilde{g} \in P(\hat{L})$.

PROOF. Since $F_0(z) = \sqrt{\lambda} \in P(\hat{L})$ and since the algebra of polynomials in $x_1, \ldots, x_n$ is dense in $\mathcal{C}(S^{n-1})$, it suffices to verify this for g of the form $g(x) = x_1^{\alpha_1} x_2^{\alpha_2} \ldots x_n^{\alpha_n}$ where $|\alpha| = \alpha_1 + \ldots + a_n > 0$. If $|\alpha| = 1$, then $\tilde{g} \equiv z_j$ for some j and so we may assume that $|\alpha| \ge 2$.

For $\delta > 0$, consider the function $\sigma_s(\lambda) = \dfrac{1}{\sqrt{\lambda + \delta}}$; σ_δ is holomorphic on a neighborhood of $\overline{\Omega}$ and so $\sigma_\delta \circ F \in P(\hat{L})$. Define

$$h_\delta(z) = \frac{z^\alpha}{\left(\sqrt{z_1^2 + \ldots + z_n^2 + \delta}\right)^{|\alpha|-1}} = z^\alpha \cdot (\sigma_\delta \circ F)^{|\alpha|-1};$$

$h_\delta \in P(\hat{L})$. For $z \ne 0$ in $\hat{L}$, with $z = \psi_x(\lambda)$, we have

$$h_\delta(z) = x^\alpha(\sqrt{\lambda})^{|\alpha|}\frac{1}{(\sqrt{\lambda + \delta})^{|\alpha|-1}}$$
$$= x^\alpha\sqrt{\lambda}\left(\sqrt{\frac{\lambda}{\lambda + \delta}}\right)^{|\alpha|-1} = \tilde{g}(z) \cdot \tau_\delta(z)$$

where $\tau_\delta(z) :\equiv \left(\sqrt{\frac{\lambda}{\lambda+\delta}}\right)^{|\alpha|-1}$ and so $\tau_\delta \in P(\hat{L})$. From the fact that $|\arg \lambda| \leq \theta_0 < \pi$ on $\overline{\Omega}\backslash\{0\}$, it follows that $\lambda/(\lambda+\delta)$ is bounded on Ω with a bound independent of $\delta > 0$ and so the functions $\{\tau_\delta\}$ are uniformly bounded on $\hat{L}$. Moreover for every neighborhood $\mathcal{U}$ of 0 in $\hat{L}$, τ_δ converges uniformly (and boundedly) to 1 on $\hat{L}\backslash\mathcal{U}$ as $\delta \downarrow 0$. Since $\tilde{g}$ is continuous on $\hat{L}$ and $\tilde{g}(0) = 0$, it follows that h_δ converges uniformly to $\tilde{g}$ on $\hat{L}$ as $\delta \downarrow 0$.

COROLLARY 1. *Let* $g \in \mathcal{C}(S^{n-1})$ *and let* h *be continuous on* $\overline{\Omega}$, *holomorphic on* Ω *and* $h(0) = 0$. *Define a function* h_L *on* $\hat{L}$ *as follows: for* $0 \neq z \in \hat{L}$ *with* $z = \psi_x(\lambda)$, *set* $h_L(z) = g(x) \cdot h(\lambda)$. *Then* $h_L \in P(\hat{L})$.

PROOF. As above, setting $h_L(0) = 0$ gives a continuous function on $\hat{L}$. Since $F_0 = \sqrt{\lambda} \in P(\hat{L})$, Lemma 1 implies that all functions on $\hat{L}$ of the form $z \mapsto g(x)p(\lambda)$, where p is a polynomial in λ with $p(0) = 0$, are in $P(\hat{L})$.

By Mergelyan's theorem there exists a sequence of polynomials $\{p_k(\lambda)\}$ such that p_k converges to h uniformly on $\overline{\Omega}$ and $p_k(0) = 0$. Setting $h_k(z) = g(x)p_k(\lambda)$, we get a sequence of functions $h_k \in P(\hat{L})$ such that h_k converges uniformly to h_L on $\hat{L}$.

6. Suppose that f is continuous on $\hat{L}$ and that $f \circ \psi_x$ is holomorphic on Ω for each $x \in S^{n-1}$. We can now show that $f \in P(\hat{L})$. Subtracting $f(0)$ from f, we may assume that $f(0) = 0$.

Let $\epsilon > 0$. Choose a partition of unity $\{\phi_k\}$ on S^{n-1} such that if x and x' both lie in the support of some ϕ_k in S^{n-1}, then $\sup_{\lambda \in \Omega} |f \circ \psi_x(\lambda) - f \circ \psi_{x'}(\lambda)| < \epsilon$. This is possible by the uniform continuity of f. Fix a point x_k in the support of ϕ_k and set $h_k(\lambda) = f \circ \psi_{x_k}(\lambda)$. By hypothesis, h_k is holomorphic on Ω. As $f(0) = 0$, h_k is continuous on $\overline{\Omega}$ and $h_k(0) = 0$. By the above corollary, setting $\tilde{h}_k(z) = \phi_k(x)h_k(\lambda)$, where $z = \psi_x(\lambda)$, defines a function in $P(\hat{L})$. Let $h = \sum_k \tilde{h}_k \in P(\hat{L})$. Then for $0 \neq z \in \hat{L}$, $z = \psi_x(\lambda)$,

$$f(z) - h(z) = \sum (f(z) - h_k(z))\phi_k(x)$$
$$= \sum (f \circ \psi_x(\lambda) - f \circ \psi_{x_k}(\lambda))\phi_k(x).$$

By the choice of x_k, the modulus of the kth term is $< \epsilon\phi_k(x)$; i.e. $\|f - h\|_{\hat{L}} < \epsilon$. Therefore $f \in P(\hat{L})$.

7. Let M be the 3-sphere of Ahern and Rudin in $\mathbf{C}^3$. We recall the description of $\hat{M}$ given by Anderson [4]. Set $X = \{(z_1, z_2, z_3) \in \mathbf{C}^3 : z_3 = 0, |z_1|^2 + |z_2|^2 < 1, z_1 \cdot z_2 = 0\}$, the two coordinate disks through the origin. Set $F(z_1, z_2, z_3) = z_1 \cdot z_2 \cdot z_3$. Then $F(M)$ is a simple closed plane curve through the origin which bounds a bounded simply connected domain Ω. Let T^2 be the torus $\{\theta : \theta = (e^{i\theta_1}, e^{i\theta_2}), 0 \leq \theta_j \leq 2\pi, j = 1,2\}$. There exist bounded holomorphic functions $\phi_1(\zeta)$ and $\phi_2(\zeta)$ on Ω which are continuous on $\overline{\Omega}\backslash\{0\}$ and non-vanishing on Ω and which parameterize $\hat{M}$ in the following sense. Set $\Phi_\theta : \Omega \to \mathbf{C}^3$ equal to

the map $\zeta \mapsto \left(e^{i\theta_1}\phi_1(\zeta), \; e^{i\theta_2}\phi_2(\zeta), \; \dfrac{e^{-i(\theta_1+\theta_2)}\zeta}{\phi_1(\zeta)\phi_2(\zeta)} \right)$ for each $\theta = (e^{i\theta_1}, e^{i\theta_2}) \in T^2$.
Then

$$\hat{M}\backslash M = X \cup \bigcup_{\theta \in T^2} \Phi_\theta(\Omega),$$

where both unions are disjoint. Φ_θ is a section of the map $F : \hat{M} \to \mathbf{C}$ in that $F \circ \Phi_\theta(\zeta) = \zeta$ for $\zeta \in \Omega$. We shall show that if $f \in \mathcal{C}(\hat{M})$ is such that $f|X$ is analytic and $f \circ \Phi_\theta$ is analytic on Ω for all $\theta \in T^2$, then $f \in P(\hat{M})$.

8. **Lemma 2.** Let $f \in \mathcal{C}(\hat{M})$ with $f = 0$ on X. If $f \cdot F^n \in P(\hat{M})$ for some positive integer n, then $f \in P(\hat{M})$.

PROOF. Let $\delta > 0$. The function $\dfrac{1}{F+\delta} \in P(\hat{M})$ since Ω is contained in the first quadrant. Hence $f_\delta :\equiv fF^n \cdot \left(\dfrac{1}{F+\delta}\right)^n \in P(\hat{M})$. We have $f_\delta = f \cdot \left(\dfrac{F}{F+\delta}\right)^n$.
For every neighborhood $\mathcal{U}$ of X in $\hat{M}$, $F/(F+\delta)$ converges uniformly on $\hat{M}\backslash\mathcal{U}$ to 1. Since $\left|F/(F+\delta)\right| \le 1$ on $\hat{M}$ and since $f = 0$ on X, it follows that f_δ converges uniformly to f on $\hat{M}$ as $\delta \downarrow 0$.

9. Each $z \in \hat{M}\backslash X$ is of the form $z = \Phi_\theta(\zeta)$ for $\theta \in T^2$ and $\zeta \in \overline{\Omega}\backslash\{0\}$ where the θ and the ζ depend uniquely and continuously on z. Indeed, $\zeta = F(z)$ and $e^{i\theta_j} = z_j/\phi_j\big(F(z)\big)$ for $j = 1, 2$. This implies that if g is any continuous function on T^2 and h is any continuous function on $\overline{\Omega}$ with $h(0) = 0$ then a continuous function is defined on $\hat{M}$ as follows: start with $z \mapsto g(\theta)h(\zeta)$ for $z = \Phi_\theta(\zeta) \in \hat{M}\backslash X$ and extend the domain of the function to all of $\hat{M}$ by setting the function $= 0$ on X. In particular a function on $\hat{M}$ of this type is given by $z \mapsto e^{i(\alpha_1\theta_1+\alpha_2\theta_2)}p(\zeta)$ where $(\alpha_1, \alpha_2) \in \mathbf{Z}^2$ and p is a polynomial in ζ with $p(0) = 0$. Fix α_1, α_2, p and denote this function by A.

LEMMA 3. $A \in P(\hat{M})$.

PROOF. We shall abuse notation and write ζ for the function F on $\hat{M}$. Since $\lambda \mapsto \lambda \cdot \phi_1(\lambda)$ is a function in $P(\overline{\Omega})$, $\zeta\phi_1(\zeta) \in P(\hat{M})$. Hence $e^{-i\theta_1}\zeta^2 \equiv (z_2 \cdot z_3) \cdot (\zeta\phi_1(\zeta)) \in P(\hat{M})$. Applying Lemma 2 we get

(a) $$e^{-i\theta_1}\zeta \in P(\hat{M}).$$

By symmetry

(b) $$e^{-i\theta_2}\zeta \in P(\hat{M}).$$

Note that $\lambda \to \lambda/\phi_1(\lambda)$ is continuous on $\overline{\Omega}$, holomorphic on Ω and $= 0$ at 0. In fact, since $z_3 \circ \Phi_0(\lambda) = \dfrac{\lambda}{\phi_1(\lambda)\phi_2(\lambda)}$, we have

$$\left|\lambda/\phi_1(\lambda)\right| \le \left|\phi_2(\lambda) \cdot z_3 \circ \Phi_0(\lambda)\right| \le \left|z_3 \circ \Phi_0(\lambda)\right|.$$

Now $\{z \in \hat{M} : z_3 = 0\} = \overline{X}$ and so $\lambda/\phi_1(\lambda) \to 0$ as $\lambda \in \overline{\Omega}\backslash\{0\} \to 0$.

Therefore $\zeta/\phi_1(\zeta) \in P(\hat{M})$ follows from $\lambda/\phi_1(\lambda) \in P(\overline{\Omega})$. Writing $e^{i\theta_1}\zeta = z_1 \cdot \left(\frac{\zeta}{\phi_1(\zeta)}\right)$ we get

(c) $$e^{i\theta_1}\zeta \in P(\hat{M}).$$

Again by symmetry

(d) $$e^{i\theta_2}\zeta \in P(\hat{M}).$$

By (a)–(d) we get

$$e^{i(\alpha_1\theta_1+\alpha_2\theta_2)}\zeta^{|\alpha|} \in P(\hat{M})$$

for all $\alpha = (\alpha_1, \alpha_2) \in \mathbf{Z}^2$. By Lemma 2 $e^{i(\alpha_1\theta_1+\alpha_2\theta_2)}\zeta \in P(\hat{M})$. Now multiplying by polynomials in ζ gives Lemma 3.

Now let g be continuous on T^2 and let h be holomorphic on Ω, continuous on $\overline{\Omega}$ and with $h(0) = 0$. Let $Q \in \mathcal{C}(\hat{M})$ be the function defined as above satisfying $Q(z) = g(x)h(\zeta)$ for $z \in \hat{M}\backslash X$, $z = \Phi_x(\zeta)$.

COROLLARY 2. $Q \in P(\hat{M})$.

PROOF. We can approximate g by polynomials in $e^{i\theta_1}$, $e^{-i\theta_1}$, $e^{i\theta_2}$ and $e^{-i\theta_2}$ and h by polynomials p in λ with $p(0) = 0$, uniformly on T^2 and $\overline{\Omega}$ respectively. The corollary then follows from Lemma 3.

10. Let $f \in \mathcal{C}(\hat{M})$ with $f|X$ and $f \circ \Phi_\theta$ analytic for $\theta \in T^2$. We show that $f \in P(\hat{M})$. First assume that $f\big|_X \equiv 0$. For $\theta \in T^2$ define $f_\theta(\zeta) := f \circ \Phi_\theta(\zeta)$ on $\overline{\Omega}$; then, since $f|X \equiv 0$, $f_\theta \in P(\overline{\Omega})$ and $f_\theta(0) = 0$. Fix $\epsilon > 0$. Choose a partition of unity $\{\phi_j\}$ on T^2 and θ_j in spt ϕ_j such that

$$\sup_{\zeta \in \Omega} \big|f_\theta(\zeta) - f_{\theta_j}(\zeta)\big| < \epsilon \text{ if } \theta \in \text{ spt } \phi_j.$$

Then, by Corollary 2, $h_\epsilon(z) :\equiv \sum \phi_j(\theta)f_{\theta_j}(\zeta)$ is in $P(\hat{M})$ and $||f - h_\epsilon||_{\hat{M}} < \epsilon$. Hence $f \in P(\hat{M})$.

To remove the assumption that $f\big|_X \equiv 0$ consider, in the general case, f_1, f_2, f_3, three functions on $\hat{M}$ defined as follows:

$$f_1(z_1, z_2, z_3) = f(z_1, 0, 0) - f(0, 0, 0),$$
$$f_2(z_1, z_2, z_3) = f(0, z_2, 0)$$

and $f_3 = f - f_1 - f_2$. Clearly f_1 and f_2 are in $P(\hat{M})$. On X, $f = f_1 + f_2$ and so $f_3\big|_X \equiv 0$. By the previous case, $f_3 \in P(\hat{M})$. Hence $f = f_1 + f_2 + f_3 \in P(\hat{M})$.

References.

1. P. Ahern and W. Rudin, *Totally real imbeddings of S^3 in $\mathbf{C}^3$*, Proc. Amer. Math. Soc. **94** (1985), 460-462.
2. _____, *Hulls of 3-spheres in $\mathbf{C}^3$*, these proceedings.
3. H. Alexander, *Polynomial hulls of graphs*, Pacific J. Math. **147** (1991), 201-212.

4. J. Anderson, *On an example of Ahern and Rudin*, preprint, 1991.

5. J.P. Serre, *Une propriété topologique des domains de Runge*, Bull. Amer. Math. Soc. **6** (1955), 133-134.

6. E.L. Stout and W. Zame, *A Stein manifold topologically but not holomorphically equivalent to a domain in C^n*, Advances Math. **60** (1986), 154–160.

7. A. Weinstein, *Lectures on Symplectic Manifolds*, CBMS Regional Conferences No. 20, AMS, 1977.

8. J. Wermer, *Subharmonicity and hulls*, Pacific J. Math. **58** (1975), 283-290.

DEPARTMENT OF MATHEMATICS, UNIVERSITY OF ILLINOIS AT CHICAGO, CHICAGO, ILLINOIS 60680

Contemporary Mathematics
Volume **137**, 1992

Dominant sets on the unit sphere of $\mathbb{C}^n$

E. AMAR

Introduction.

Let E be a set of positive Lebesgue's measure on the boundary $\partial\mathbb{B}$ of the unit ball $\mathbb{B}$ of $\mathbb{C}^n$, $n \geq 1$. In [5], given a function B in $H^\infty(\mathbb{B})$ for $n = 1$, A.L. Volberg characterized these sets E such that:

(*) $\|f\|_{L^p(\partial\mathbb{B})} \preceq \|f\|_{L^p(E)}$, for all $f \in H^p(\mathbb{B})$ s.t. $\displaystyle\int \overline{f}\varphi B = 0$, $\forall \varphi \in H^q(\mathbb{B})$

He found the following condition:

(**) $|B(z)| + \omega_E(z) \geq \delta > 0$, $\forall z \in \mathbb{B}$;

where ω_E is the harmonic extension in $\mathbb{B}$ of the indicatrix of E. He uses very precise results on Carleson measures by A. Chang and J. Garnett [4].

The aim of this note is to generalize this result to the unit ball $\mathbb{B}$ of $\mathbb{C}^n$ and the method is to use a Corona Theorem in the ball for $H^2(\mathbb{B})$, based on results of [1].

This work started in Leningrad and it is my pleasure to thank professors Nikolskii, Havin, Volberg and all the analysts of the Stecklov Institute for so interesting discussions.

2. A Corona Theorem for $H^2(\mathbb{B})$

Let f_1 and f_2 be two functions in $H^\infty(\mathbb{B})$ such that:

$$D(z)^2 := \left|f_1(z)\right|^2 + \left|f_2(z)\right|^2 \geq \delta^2 > 0, \ \forall z \in \mathbb{B} \tag{2.1}$$

the claim of the Corona in the ball is to find g_1, g_2 in $H^\infty(\mathbb{B})$ such that:

$$f_1 g_1 + f_2 g_2 \equiv 1 \tag{2.2}$$

in $\mathbb{B}$.

This is true for $n = 1$ but unknown for $n > 1$; nevertheless we have:

1991 *Mathematics Subject Classification*. Primary 32A35.

This paper is in final form and no version of it will be submitted for publication elsewhere.

THEOREM 2.1. *Let* f_1, f_2 *two functions in* $H^\infty(\mathbb{B})$ *such that 2.2 is true; then, for any* φ *in* $H^2(\mathbb{B})$ *there are two other functions* φ_1, φ_2 *in* $H^2(\mathbb{B})$ *such that:*
$$\varphi = f_1\varphi_1 + f_2\varphi_2$$
Moreover we have: $\|\varphi_i\|_2 \preceq 1/\delta^3 \|\varphi\|_2$, $i = 1, 2$.

Of course a positive answer to the Corona problem will give this result as an easy corollary.

In order to prove this result, let us introduce the classical form associated to the f_i's:

$$\omega := \frac{\overline{f_1}.\overline{\partial}f_2 - \overline{f_2}.\overline{\partial}f_1}{D^4} \tag{2.3}$$

this $(0,1)$-form is $\overline{\partial}$-closed and we want to apply to it the following proposition, proved in [1]:

PROPOSITION 2.2. *Let* ω *be a (0,1) form* $\overline{\partial}$-*closed in the unit ball* $\mathbb{B}$ *of* $\mathbb{C}^n$ *such that its Carleson norm is bounded; then there is a bounded linear operator* L *from* $H^2(\mathbb{B})$ *to* $L^2(\partial\mathbb{B})$ *such that:*

$$\forall \varphi \in H^2(\mathbb{B}), \ \overline{\partial}_b(L(\varphi)) = \varphi\omega.$$

Let recall here the Carleson norm of the $(0,1)$-form $\omega := \omega_1 d\bar{z}_1 + ... + \omega_n d\bar{z}_n$[1]:

$$\|\omega\|_C^2 := \|(1 - |z|^2)|\omega|^2\|_C + \|(1 - |z|^2)|\partial\omega|^2\|_C + \||\omega \wedge \overline{\partial}\rho|^2\|_C$$

where $\|\mu\|_C$ is the Carleson norm of the measure μ in the ball $\mathbb{B}$.

Now, in order to apply this proposition, we have to compute this norm for ω defined in 2.3.

We have:

$$(1 - |z|^2)|\omega|^2 \preceq (1 - |z|^2)\frac{|f_1|^2|\partial f_2|^2 + |f_2|^2|\partial f_1|^2}{D^8} \tag{2.4}$$

but, because the f_i's are in $H^\infty(\mathbb{B})$ hence in $BMOA(\partial\mathbb{B})$, we have [1]:

$$\left\|(1 - |z|^2)|\partial f_i|^2\right\|_C \preceq 1; \quad \left\||\partial f_i \wedge \partial\rho|^2\right\|_C \preceq 1 \tag{2.5}$$

hence we get:

$$\|(1 - |z|^2)|\omega|^2\|_C \preceq 1/\delta^6.$$

Let us compute $\partial\omega$:

$$\partial\omega = -\frac{\partial D^4 \wedge (\overline{f_1}\overline{\partial}f_2 - \overline{f_2}\overline{\partial}f_1)}{D^8} = -4\frac{(\overline{f_1}\partial f_1 + \overline{f_2}\partial f_2) \wedge (\overline{f_1}\overline{\partial}f_2 - \overline{f_2}\overline{\partial}f_1)}{D^5} \tag{2.6}$$

hence:

$$\|(1 - |z|^2)\partial\omega\|_C \preceq 1/\delta^6,$$

because of 2.5.

Now for $\||\omega \wedge \overline{\partial}\rho|^2\|_C$:

$$\omega \wedge \overline{\partial}\rho = \frac{\overline{f_1}\overline{\partial}f_2 \wedge \overline{\partial}\rho - \overline{f_2}\overline{\partial}f_1 \wedge \overline{\partial}\rho}{D^4} \tag{2.7}$$

and still using 2.5, we get again:

$$\left\| \left| \omega \wedge \overline{\partial}\rho \right|^2 \right\|_C \preceq 1/\delta^6; \tag{2.8}$$

and, finally:

$$\left\| \omega \right\|_C \preceq 1/\delta^3. \tag{2.9}$$

Applying the proposition, we get that the linear operator L is such that:

$$\left\| L \right\| \preceq 1/\delta^3. \tag{2.10}$$

Now let us put:

$$\varphi_1 := \overline{f}_1 \frac{\varphi}{D^2} - f_2 L(\varphi); \quad \varphi_2 := \overline{f}_2 \frac{\varphi}{D^2} + f_1 L(\varphi)$$

then we get:

$$\overline{\partial}\varphi_i = 0, \ i = 1, 2 \text{ and } \varphi = f_1\varphi_1 + f_2\varphi_2.$$

Moreover we have the norm control of the φ_i's:

$$\left\| \varphi_i \right\|_2 \preceq 1/\delta^3 \left\| \varphi \right\|_2. \tag{2.11}$$

Hence the theorem. $\square$

REMARK. *Because L is linear, the φ_i's are also linear in φ.*

3. Application to the dominant sets

The aim of this part is to prove the:

THEOREM. *Let $E \subset \partial\mathbb{B}$ be a closed set and $B \in H^\infty(\mathbb{B})$, $\|B\|_\infty \leq 1$, such that there is a function k holomorphic in $\mathbb{B}$ with the following properties:*
i) $0 \leq \operatorname{Re}k \leq 1$;
ii) $\operatorname{Re}k + |B| \geq \delta > 0$;
iii) $\operatorname{Re}k^* \leq \epsilon$ on E^c;
iv) $3\delta - \epsilon > 2$.
Then there is a constant C_ϵ such that:

$$\text{for all } f \text{ orthogonal to } BH^2(\mathbb{B}), \ \|f\|_2 \leq C_\epsilon \|f\|_{L^2(E)}.$$

These conditions are stronger than those for the disc: in $\mathbb{D}$, A.L. Volberg proves this result with $\epsilon = 0$, $\delta > 0$ (and $\operatorname{Re}k^* = \chi_E$).

Let: δ' such that $0 < \delta' < \delta$ and:

$$F := e^{-(1-k)} \tag{3.12}$$

if $|B| > \delta - \delta'$ we have:

$$\left| F \right|^n + \left| B \right| \geq \delta - \delta' \tag{3.13}$$

and if $|B| \leq \delta - \delta'$ we have:

$$\operatorname{Re}k \geq \delta' \ \Rightarrow \ \left| F \right|^n = e^{-n(1-\operatorname{Re}k)} \geq e^{-n(1-\delta')} \tag{3.14}$$

hence, in any case, we get:

(*) $|F|^{2n} + |B|^2 \geq \gamma_n^2$ with $\gamma_n = e^{-n(1-\delta')}$, provided that n is big enough.

We are now in position to apply the theorem 2.1:

$$\forall \varphi \in H^2(\mathbb{B}), \; \exists \alpha_n, \beta_n \in H^2(\mathbb{B}) \; s.t. \; \varphi = F^n \alpha_n + B \beta_n \qquad (3.15)$$

with the control:

$$\left\| \alpha_n \right\|_2 \preceq 1/\gamma_n^3; \quad \left\| \beta_n \right\|_2 \preceq 1/\gamma_n^3. \qquad (3.16)$$

Let f a function in $H^2(\mathbb{B})$ orthogonal to $BH^2(\mathbb{B})$, we have:

$$\int_{\partial \mathbb{B}} \varphi \overline{f} d\sigma = \int_{\partial \mathbb{B}} \overline{f}(\alpha_n F^n + \beta_n B) d\sigma = \int_{\partial \mathbb{B}} \overline{f} \alpha_n F^n d\sigma \qquad (3.17)$$

because of the orthogonality between f and $\alpha_n B$.

Let us evaluate the last integral on E^c:

$$\left| \int_{E^c} \overline{f} \alpha_n F^n d\sigma \right| \leq e^{-n(1-\epsilon)} \|f\|_2 \|\alpha_n\|_2 \leq \frac{e^{-n(1-\epsilon)}}{\gamma_n^3} \|f\|_2 \|\varphi\|_2 \qquad (3.18)$$

because $|F| \leq e^{\epsilon - 1}$ on E^c and $\|\alpha_n\|_2 \leq 1/\gamma_n^3 \|\varphi\|_2$ and, because $\gamma_n \geq e^{-n(1-\delta')}$, we get:

$$\left| \int_{E^c} \overline{f} \alpha_n F^n d\sigma \right| \preceq e^{-n(1-\epsilon)+3n(1-\delta')} \|f\|_2 \|\varphi\|_2 \qquad (3.19)$$

Now on E:

$$\left| \int_E \overline{f} \alpha_n F^n d\sigma \right| \leq \|F\|_\infty \|f\|_{L^2(E)} \|\alpha_n\|_{L^2(E)} \leq 1/\gamma_n^3 \|f\|_{L^2(E)} \|\varphi\|_2 \qquad (3.20)$$

Finally:

$$\left| \int_{\partial \mathbb{B}} \overline{f} \varphi d\sigma \right| \preceq e^{-n(3\delta'-\epsilon-2)} \|f\|_2 \|\varphi\|_2 + e^{3n(1-\delta')} \|f\|_{L^2(E)} \|\varphi\|_2 \qquad (3.21)$$

hence if n is big enough ($3\delta - \epsilon - 2 > 0$!) we can conclude that:

$$\left\| f \right\|_2 \leq C(n) \left\| f \right\|_{L^2(E)} \qquad (3.22)$$

and the theorem. $\square$

Using directly the Corona theorem in the disc $\mathbb{D}$, Dyakonoff proved a similar result in $\mathbb{D}$ with the better condition $2\delta > 1$. (Communicated to me by Professor Havin from Leningrad)

Note added in proof:

1) M. Andersson proved an $H^2(\mathbb{B})$ Corona theorem in the ball for any number of functions [3].

2) I proved recently an $H^p(\mathbb{B})$ Corona theorem in the ball for $p < \infty$, [2] and using it, one can generalize the results here to the $H^p(\mathbb{B})$ case.

References.

1. E. Amar, *Solutions of $\overline{\partial}_b u = f$ with L^2 estimates and non-factorization*, To appear in the *Proceedings of the Special Year in Complex Analysis*, 1987, Mittag-Leffler Institut.

2. _____, *On the Corona Problem*, Prépublications d'Analyse, Université de Bordeaux I.

3. M. Andersson, *The $H^2(\mathbb{B})$-Corona theorem and $\overline{\partial}_b$*, Preprint, University of Göteborg.

4. S. Y. Chang & J. Garnett, *Analyticity of functions and subalgebras of L^∞ containing H^∞*, Proc. Amer. Math. Soc. **72** (1978), 41-46.

5. A. L. Volberg, *Thin and thick family of rational functions*, Lect. Notes Math. **864** (1981), 440-480.

UNIVERSITÉ DE BORDEAUX I, TALENCE, FRANCE

Contemporary Mathematics
Volume **137**, 1992

The Cauchy transform, the Szegő projection, the Dirichlet problem, and the Ahlfors map

S. BELL

1. Introduction.

The purpose of this paper is to describe a method for studying the Szegő projection based on the Kerzman-Stein theorem [**11**] which states roughly that the Cauchy transform is nearly self adjoint in smooth planar domains and that the Szegő projection is nearly equal to the Cauchy transform. The method gives rise to a concrete description of the Szegő projection in terms of elementary functions and operators. We shall go on to use this description to solve the classical Dirichlet problem on planar domains by means of explicit formulas which can be understood in very concrete terms. We shall also show how this outlook can be used to give an elementary proof of the existence of the Ahlfors map of a multiply connected planar domain.

This paper is expository, and to make my results as comprehensible as possible to as many people as possible, I shall explain the background material that I need from the theory of Hardy space and the Kerzman-Stein theory of the Szegő projection. I shall also restrict myself to studying the objects in the title of this paper on a bounded finitely connected domain Ω in the plane with *real analytic* boundary. The reasoning I give in this setting can easily be adapted to apply to a domain with C^2 smooth boundary. The adaptations involve standard methods of analysis, and so I feel no guilt in making the real analytic assumption. The details of the more general arguments can be found in my forthcoming book [**1**].

Another goal of this paper is to demonstrate that the Kerzman-Stein theorem can be proved with a minimum of machinery, and that once this has been done, it is possible to solve the Dirichlet problem "from scratch" in a small number of pages using only elementary analysis. Furthermore, this can be done in such a way that regularity properties of the solutions can be read off from explicit formulas.

1991 *Mathematics Subject Classification*. Primary 30E20.
Research supported by NSF grant DMS–8922810.
This paper is in final form and no version of it will be submitted for publication elsewhere.

The main results of this paper appeared in [2] and [3]. However, in these earlier papers, the results were deduced from rather advanced facts about the Dirichlet problem. The novelty of the present work is that the same theorems can be deduced without making any advanced assumptions. In fact I have tried to write this paper so that a first year graduate student can read it.

The results described in this paper from a theoretical perspective give rise to practical methods for explicitly computing the objects under study. See [2,9,12,15] for a description of numerical methods stemming from the Kerzman-Stein approach to the Cauchy transform and Szegő projection.

Other authors have rethought some of the classical theorems of potential theory and conformal mapping in the plane in the light of the Kerzman-Stein theorem. For example, results in the same spirit as the present work can be found in Burbea [5,6], and Shapiro [14].

2. The Cauchy transform and its adjoint.

Throughout this paper we shall assume that Ω is a bounded n-connected domain in the plane with real analytic boundary. Thus Ω is bounded by n non-intersecting real analytic simple closed curves.

Given a continuous function u defined on the boundary $b\Omega$ of Ω, the Cauchy transform of u will be written $\mathcal{C}u$ and is defined to be the holomorphic function on Ω given by

$$(\mathcal{C}u)(z) = \frac{1}{2\pi i} \int_{\zeta \in b\Omega} \frac{u(\zeta)}{\zeta - z} \, d\zeta.$$

Let $C^\omega(b\Omega)$ denote the space of complex valued continuous functions on $b\Omega$ which extend to be real analytic in a neighborhood of $b\Omega$. We shall now show that $C^\omega(b\Omega)$ is equal to the space of continuous functions on $b\Omega$ which extend to be *holomorphic* on a neighborhood of $b\Omega$. It is enough to prove this locally because analytic continuation can be used to piece together global functions from local ones. Let $z_0 \in b\Omega$ and let $z(t)$ denote a real analytic parameterization of $b\Omega$ near z_0 such that $z(0) = z_0$. We may express $z(t)$ by a convergent power series $z(t) = \sum_{k=0}^\infty a_k t^k$, and we may replace the real variable t in this formula by a complex variable τ and thereby consider $z(t)$ to be the restriction of a holomorphic function $z(\tau)$ to the real line. Since $z'(0) \neq 0$, this holomorphic function has a holomorphic inverse $\tau(z)$ defined near $z = z_0$. Given $u \in C^\omega(b\Omega)$, the function $u(z(t))$ is real analytic and we may write $u(z(t)) = \sum_{k=0}^\infty c_k t^k$. Now the function U defined via $U(z) = \sum_{k=0}^\infty c_k \tau(z)^k$ is a holomorphic function of z near $z = z_0$. Since $\tau(z(t)) = t$, it follows that U agrees with u on $b\Omega$ and the proof is complete.

The Cauchy transform is particularly well behaved on the space $C^\omega(b\Omega)$.

THEOREM 1. *The Cauchy transform maps $C^\omega(b\Omega)$ into the space $A(\overline{\Omega})$ of holomorphic functions on Ω which extend to be holomorphic on a neighborhood of $\overline{\Omega}$.*

This theorem will allow us to think of the Cauchy transform as an operator mapping $C^\omega(b\Omega)$ into $C^\omega(b\Omega)$. The theorem will follow as a simple consequence of the inhomogeneous Cauchy integral formula and the fact mentioned above that a function in $C^\omega(b\Omega)$ is the restriction to $b\Omega$ of a function that is holomorphic on a neighborhood of $b\Omega$.

PROOF OF THEOREM 1. Given $u \in C^\omega(b\Omega)$, there is a function U which is holomorphic on a neighborhood of $b\Omega$ and which is equal to u on $b\Omega$. By multiplying U by a C^∞ function which is compactly supported inside the set where U is holomorphic and which is equal to one on a small neighborhood of $b\Omega$, we may think of U as being a function in $C^\infty(\overline{\Omega})$ which is holomorphic near $b\Omega$. Let Ψ denote the $C_0^\infty(\Omega)$ function given as $\Psi = \frac{\partial U}{\partial \bar{z}}$.

If $v \in C^\infty(\overline{\Omega})$ and $z \in \Omega$, the inhomogeneous Cauchy integral formula (see Hörmander [10, **Theorem 1.2.1**]) states that

$$v(z) = \frac{1}{2\pi i} \int_{\zeta \in b\Omega} \frac{v(\zeta)}{\zeta - z} \, d\zeta + \frac{1}{2\pi i} \iint_{\zeta \in \Omega} \frac{\frac{\partial v}{\partial \bar{\zeta}}}{\zeta - z} \, d\zeta \wedge d\bar{\zeta}.$$

Apply this formula using $v = U$ to obtain the identity

$$U(z) = (\mathcal{C}u)(z) + \frac{1}{2\pi i} \iint_{\zeta \in \Omega} \frac{\Psi(\zeta)}{\zeta - z} \, d\zeta \wedge d\bar{\zeta}.$$

Since Ψ has compact support, we deduce from this formula that $\mathcal{C}u$ extends smoothly to the boundary. Furthermore, the boundary values of $\mathcal{C}u$ are given by

$$(2.1) \qquad (\mathcal{C}u)(z) = u(z) - \frac{1}{2\pi i} \iint_{\zeta \in \Omega} \frac{\Psi(\zeta)}{\zeta - z} \, d\zeta \wedge d\bar{\zeta} \quad \text{for } z \in b\Omega.$$

Both of the functions on the right hand side of this equation extend to be holomorphic in a neighborhood of $b\Omega$, and hence the right hand side defines a holomorphic extension of $\mathcal{C}u$ to a neighborhood of $\overline{\Omega}$. The proof of Theorem 1 is complete.

Suppose that $z(t)$ parameterizes one of the boundary curves of Ω in the standard sense. If $z_0 = z(t_0)$ is a point on this curve, we define $T(z_0)$ to be equal to $z'(t_0)/|z'(t_0)|$. Thus, for $z \in b\Omega$, $T(z)$ denotes the complex number of unit modulus representing the unit tangent vector to the boundary at z pointing in the direction of the standard orientation. Notice also that T is in $C^\omega(b\Omega)$.

Define $L^2(b\Omega)$ to be the space of complex valued functions defined on the boundary of Ω which are square integrable with respect to the arc length measure ds and define the inner product of two functions in this space via $\langle u, v \rangle = \int_{b\Omega} u \bar{v} \, ds$. Notice that $dz = T \, ds$ and $ds = \overline{T} \, dz$.

We have shown that $\mathcal{C}$ maps $C^\omega(b\Omega)$ into itself. Since $C^\omega(b\Omega)$ is a dense subspace of $L^2(b\Omega)$, we could deduce that the Cauchy transform extends to be a bounded operator on $L^2(b\Omega)$ by proving an L^2 estimate for $\mathcal{C}$ when it acts on functions in $C^\omega(b\Omega)$. We will do this; however, as a preliminary, we must first determine a *formal adjoint identity* for the Cauchy transform when it acts on functions in $C^\omega(b\Omega)$. Suppose that u and v are in $C^\omega(b\Omega)$. By Theorem 1, we know that $\mathcal{C}u$ is in $C^\omega(b\Omega)$ too. We shall now construct a function in $C^\omega(b\Omega)$ which we shall denote by $\mathcal{C}^*v$ that satisfies the adjoint property,

$$\langle \mathcal{C}u, v \rangle = \langle u, \mathcal{C}^*v \rangle.$$

The key to finding the function $\mathcal{C}^*v$ will be identity (2.1). Suppose that U is a function in $C^\infty(\overline{\Omega})$ (like the one used in the proof of Theorem 1) which is holomorphic near $b\Omega$ and which agrees with u on $b\Omega$ and let $\Psi = \frac{\partial U}{\partial \bar{z}}$. If we plug the expression for $\mathcal{C}u$ given by formula (2.1) into $\langle \mathcal{C}u, v \rangle$ and use Fubini's theorem, we obtain

$$\langle \mathcal{C}u, v \rangle = \langle u, v \rangle - \iint_{\zeta \in \Omega} \Psi(\zeta) \left(\frac{1}{2\pi i} \int_{z \in b\Omega} \frac{\overline{v(z)}}{\zeta - z} \, ds \right) \, d\zeta \wedge d\bar{\zeta}.$$

(Because Ψ has compact support, we can use the most basic Fubini's theorem from freshman calculus in this computation.) The boundary integral in the double integral looks like a Cauchy transform. In fact, since $ds = \overline{T}dz$, the quantity inside the last set of parentheses is the Cauchy transform of $-\overline{vT}$, and so we have

$$(2.2) \qquad\qquad \langle \mathcal{C}u, v \rangle = \langle u, v \rangle + \iint_{\Omega} \Psi \, \mathcal{C}(\overline{vT}) \, d\zeta \wedge d\bar{\zeta}.$$

Now recall that $\Psi = \frac{\partial U}{\partial \bar{z}}$. Hence,

$$\langle \mathcal{C}u, v \rangle = \langle u, v \rangle + \iint \frac{\partial}{\partial \bar{\zeta}}[U \, \mathcal{C}(\overline{vT})] \, d\zeta \wedge d\bar{\zeta}$$

and Stokes' theorem reduces the last integral to

$$-\int_{\zeta \in b\Omega} U \, \mathcal{C}(\overline{vT}) \, d\zeta.$$

But $U = u$ on $b\Omega$, so

$$\langle \mathcal{C}u, v \rangle = \langle u, v \rangle - \int_{\zeta \in b\Omega} u\mathcal{C}(\overline{vT}) \, d\zeta$$

and we have shown that $\langle \mathcal{C}u, v \rangle = \langle u, \mathcal{C}^*v \rangle$ where

$$(2.3) \qquad\qquad \mathcal{C}^*v = v - \overline{T\mathcal{C}(\overline{vT})}.$$

For the time being, we think of $\mathcal{C}^*$ as being a *formal* adjoint of $\mathcal{C}$. Soon, however, we will see that this operator agrees with the genuine L^2 adjoint of $\mathcal{C}$ when restricted to functions in $C^\omega(b\Omega)$.

We define the Kerzman-Stein operator $\mathcal{A}$ to be $\mathcal{A} = \mathcal{C} - \mathcal{C}^*$. For the moment, we are thinking of $\mathcal{A}$ as an operator mapping $C^\omega(b\Omega)$ into itself. Kerzman and Stein [11] discovered that $\mathcal{A}$ is a much better operator than either of its component operators. To see that this is so, we shall need to use the Plemelj formula which describes the boundary values of a Cauchy transform in terms of a principal value integral. To make this paper self contained, I shall deduce the Plemelj formula from (2.1). Given $u \in C^\omega(b\Omega)$, identity (2.1) gives a formula for the boundary values of $\mathcal{C}u$. Given a point $z_0 \in b\Omega$, let $D_\epsilon(z_0)$ denote the disc of radius $\epsilon > 0$ about z_0. If $\epsilon > 0$ is small then $D_\epsilon(z_0)$ will not intersect the support of Ψ (which, you will recall, is a compact subset of Ω). Let $\Omega_\epsilon = \Omega - D_\epsilon(z_0)$. Let γ denote the boundary of Ω parameterized in the standard sense, and let γ_ϵ denote the portion of γ which does not intersect $D_\epsilon(z_0)$. Let C_ϵ denote the circular arc of the boundary of $D_\epsilon(z_0)$ which is contained in Ω and assume that C_ϵ has been parametrized so that $\gamma_\epsilon \cup C_\epsilon$ represents the boundary of Ω_ϵ with the standard sense. We may now write

$$(\mathcal{C}u)(z_0) = u(z_0) - \frac{1}{2\pi i} \iint_{\zeta \in \Omega_\epsilon} \frac{\Psi(\zeta)}{\zeta - z_0} \, d\zeta \wedge d\bar\zeta.$$

Stokes' theorem and the fact that $\Psi = \frac{\partial U}{\partial \bar z}$ yields that

$$(\mathcal{C}u)(z_0) = u(z_0) + \frac{1}{2\pi i} \int_{\gamma_\epsilon \cup C_\epsilon} \frac{U}{\zeta - z_0} \, d\zeta.$$

Since $U = u$ on $b\Omega$, it is easy to check that

$$\frac{1}{2\pi i} \int_{C_\epsilon} \frac{U}{\zeta - z_0} \, d\zeta$$

tends to $-\frac{1}{2}u(z_0)$ as $\epsilon \to 0$. We conclude that, given a point $z_0 \in b\Omega$, the principal value integral defined via

$$\mathbf{P.V.} \; \frac{1}{2\pi i} \int_\gamma \frac{u(\zeta)}{\zeta - z_0} \, d\zeta = \lim_{\epsilon \to 0} \frac{1}{2\pi i} \int_{\gamma_\epsilon} \frac{u(\zeta)}{\zeta - z_0} \, d\zeta$$

exists and that

$$(\mathcal{C}u)(z_0) = \frac{1}{2}u(z_0) + \mathbf{P.V.} \; \frac{1}{2\pi i} \int_\gamma \frac{u(\zeta)}{\zeta - z_0} \, d\zeta.$$

Let $z_0 \in b\Omega$. Formula (2.3) allows us to write

$$\mathcal{A}u = \mathcal{C}u - (u - \overline{T\mathcal{C}(\overline{uT})}) = -u + \mathcal{C}u + \overline{T\mathcal{C}(\overline{uT})}.$$

If we now apply the Plemelj formula to the two Cauchy transforms in this identity, we obtain

$$(\mathcal{A}u)(z_0) = \lim_{\epsilon \to 0} \frac{1}{2\pi i} \left[\int_{\gamma_\epsilon} \frac{u(\zeta)}{\zeta - z_0} \, d\zeta - \overline{T(z_0)} \int_{\gamma_\epsilon} \overline{\frac{u(\zeta)T(\zeta)}{\zeta - z_0}} \, d\bar\zeta \right] =$$

$$\lim_{\epsilon \to 0} \frac{1}{2\pi i} \int_{\gamma_\epsilon} u(\zeta) \left[\frac{T(\zeta)}{\zeta - z_0} - \frac{\overline{T(z_0)}}{\bar\zeta - \bar z_0} \right] \, ds.$$

For z and w on the boundary of Ω, define the Kerzman-Stein kernel via

$$A(z, w) = \frac{1}{2\pi i} \left[\frac{T(w)}{w - z} - \frac{\overline{T(z)}}{\bar w - \bar z} \right]$$

when $z \neq w$. This function appears to be very singular when $z = w$. However, the Kerzman-Stein theorem states that $A(z, w)$ extends to $b\Omega \times b\Omega$ as a C^∞ function. In fact, in our setting, we shall prove that $A(z, w)$ is real analytic on $b\Omega \times b\Omega$. It then will follow that the principal value integral for $\mathcal{A}u$ above is a standard integral and we will have proved the following theorem.

THEOREM 2. *The operator $\mathcal{A}$ is given by*

$$(\mathcal{A}u)(z) = \int_{w \in b\Omega} A(z, w)\, u(w)\, ds$$

where $A(z, w)$ is an explicit real analytic kernel on $b\Omega \times b\Omega$. Consequently, $\mathcal{A}$ maps $L^2(b\Omega)$ into $C^\omega(b\Omega)$.

To simplify the computations in the proof of this theorem, let us suppose that the boundary curves of Ω have been parameterized with respect to *arc length*, and let us write $z_j(t)$ to represent the point along the j-th boundary curve which has moved in the standard sense an arc length of t from a fixed boundary point. To further simplify the notation, let us suppress the writing of the subscripts, so that $z(s)$ and $z(t)$ may represent points on the same or different curves.

We wish to see that $A(z(t), z(s))$ can be extended so as to be real analytic as a function of (t, s) as t and s range over the various parameter intervals. Since nothing special is really happening at the endpoints of the parameter intervals, we may assume that t and s stay in the interior of the intervals. Since $A(z(t), z(s))$ is clearly real analytic when t and s are in different parameter intervals, and when t and s belong to the same interval and $t \neq s$, we need only show that $A(z(t), z(s))$ is real analytic in (t, s) when t and s belong to the interior of the same parameter interval and $|t - s| < \epsilon$ for some small $\epsilon > 0$.

The proof rests on the elementary fact that, if $Y(s, t)$ is real analytic in (s, t) and $Y(t, t) = 0$, then $Y(s, t) = (s - t)W(s, t)$ where $W(s, t)$ is also real analytic in (s, t).

Since $z(s) - z(t)$ is a real analytic function of (s, t) which vanishes at (t, t), we may use the fact above to see that the difference quotient $Q(s, t) = (z(s) - z(t))/(s - t)$ is a real analytic function of (s, t). Note that $Q(t, t) = z'(t)$, and therefore, that $Q(s, t)$ is non-vanishing when t and s are close together. Also, since s represents arc length, it follows that $T(z(s)) = z'(s)$. We may now write

$$2\pi i\, A(z(t), z(s)) = \frac{1}{s - t}\left(\frac{z'(s)}{Q(s, t)} - \frac{\overline{z'(t)}}{\overline{Q(s, t)}} \right).$$

Thus, we see that $2\pi i(s - t)A(z(t), z(s))$ is equal to a real analytic function $R(s, t)$. Furthermore, $R(t, t) = 1 - 1 = 0$; so $R(s, t) = (s - t)X(s, t)$ where $X(s, t)$ is also real analytic. After dividing out by $(s - t)$, we conclude that $2\pi i\, A(z(t), z(s)) = X(s, t)$ is real analytic in (s, t), and this is precisely what it means to say that $A(z, \zeta)$ is real analytic on $b\Omega \times b\Omega$. The proof is complete. (By studying the power series above more carefully, it can be shown that $A(z, z) = 0$, but we shall not need this fact.)

We plan to study the Dirichlet problem by means of the Cauchy transform and the Szegő projection. In order to define the Szegő projection, we must first define the Hardy space $H^2(b\Omega)$. We part with tradition at this point and define $H^2(b\Omega)$ to be the closure in $L^2(b\Omega)$ of the space $A(\overline{\Omega})$ of holomorphic functions on Ω which extend to be holomorphic on a neighborhood of $\overline{\Omega}$. Of course, this definition defines the same space as the traditional definition. However, it is usually a rather difficult theorem to prove that $A(\overline{\Omega})$ is dense in $H^2(b\Omega)$. We have avoided this difficulty by making it part of the definition. We can get away with this because we will not need any of the deeper properties of the Hardy space associated with the traditional definition.

Given a function $h \in H^2(b\Omega)$, we may associate to h a holomorphic function H on Ω as follows. Pick a sequence h_j in $A(\overline{\Omega})$ tending to h in $L^2(b\Omega)$. Notice that $h_j(a) = (\mathcal{C}h_j)(a)$ for all $a \in \Omega$. Writing out the Cauchy integrals reveals that the sequence h_j converges uniformly on compact subsets of Ω to the holomorphic function $H(z)$ on Ω given by the Cauchy integral

$$H(z) = \frac{1}{2\pi i} \int_{b\Omega} \frac{h(\zeta)}{\zeta - z} \, d\zeta.$$

It is customary to let h denote both the element in the Hardy space and the associated holomorphic function on Ω. Notice that the formula for H shows that the functional taking $h \in H^2(b\Omega)$ to the value $h(a) = H(a)$ at a point $a \in \Omega$ is a *continuous* linear functional on $H^2(b\Omega)$. Hence, the Riesz representation theorem states that there is an element S_a in the Hardy space which represents this functional, i.e., such that $h(a) = \langle h, S_a \rangle$ for all $h \in H^2(b\Omega)$. Later, we shall identify S_a as the Szegö kernel function.

The Szegö projection P is the orthogonal projection of $L^2(b\Omega)$ onto the subspace $H^2(b\Omega)$. The work we did above to compute $\mathcal{C}^*$ and to prove the Kerzman-Stein theorem will now pay off handsomely. Let I denote the identity operator and consider the operator defined on $C^\omega(b\Omega)$ given by $I + \mathcal{A}$. Since $\mathcal{A} = \mathcal{C} - \mathcal{C}^*$, and since, by (2.3), $\mathcal{C}^*u = u - \overline{T\mathcal{C}(\overline{uT})}$, we may write

(2.4) $$(I + \mathcal{A})u = \mathcal{C}u + \overline{T\mathcal{C}(\overline{uT})}$$

for $u \in C^\omega(b\Omega)$. This last sum is an *orthogonal* sum. Indeed, $(I+\mathcal{A})u = h + \overline{HT}$ where $h = \mathcal{C}u$ and $H = \mathcal{C}(\overline{uT})$ are both in $A(\overline{\Omega})$. It is clear that $h \in H^2(b\Omega)$ because, in fact, $h \in A(\overline{\Omega})$. I now claim that $\overline{HT}$ is orthogonal to $H^2(b\Omega)$. To see this, it will be enough to check that $\overline{HT}$ is orthogonal to $A(\overline{\Omega})$. Given $g \in A(\overline{\Omega})$, we may compute

$$\langle g, \overline{HT} \rangle = \int_{b\Omega} gH \, T \, ds = \int_{b\Omega} gH \, dz,$$

and this last integral is zero by Cauchy's theorem. This orthogonal decomposition of $(I + \mathcal{A})u$ yields the Kerzman-Stein identity

$$P(I + \mathcal{A}) = \mathcal{C}.$$

We have only proved this identity when it acts on functions in $C^\omega(b\Omega)$. However, we can use this identity to learn several interesting things about $\mathcal{C}$ and P. The most obvious consequence that can read off from the identity is that the Cauchy transform extends uniquely to be a bounded operator from $L^2(b\Omega)$ into $L^2(b\Omega)$. Indeed, P, by its definition as a projection, is a bounded operator on $L^2(b\Omega)$, and

$\mathcal{A}$ is given by integration against a kernel which is in $C^{\infty}(b\Omega \times b\Omega)$ by Theorem 2, and hence $\mathcal{A}$ is also bounded as an operator on $L^2(b\Omega)$. Since we now know that $\mathcal{C}$ is bounded in L^2, identity (2.3) allows us to also uniquely extend the formal adjoint $\mathcal{C}^*$ to be a bounded operator on $L^2(b\Omega)$. Furthermore, by continuity, the extensions also satisfy the adjoint identity $\langle \mathcal{C}u, v \rangle = \langle u, \mathcal{C}^*v \rangle$ for all u and v in $L^2(b\Omega)$. Hence, the L^2 extension of the formal adjoint $\mathcal{C}^*$ agrees with the L^2 adjoint of the extension of $\mathcal{C}$ to L^2. Thus, we no longer need to apologize for using the notation $\mathcal{C}^*$.

We have just used the Kerzman-Stein identity to deduce that the Cauchy transform is bounded in L^2 from the same property of the Szegő projection. We can also deduce that the Szegő projection maps $C^{\omega}(b\Omega)$ into itself from the same property of the Cauchy transform. To do this, we must write down the adjoint of the Kerzman-Stein identity:

$$(I - \mathcal{A})P = \mathcal{C}^*.$$

(Here, we have used the facts that $P^* = P$ and $\mathcal{A}^* = -\mathcal{A}$.) If we now subtract this adjoint identity from the Kerzman-Stein identity, we obtain

$$P\mathcal{A} + \mathcal{A}P = \mathcal{C} - \mathcal{C}^* = \mathcal{A}.$$

This shows that $P\mathcal{A} = \mathcal{A}(I - P)$. The operator on the right hand side of this equation maps $L^2(b\Omega)$ into $C^{\omega}(b\Omega)$, and hence so does $P\mathcal{A}$. We may now solve the Kerzman-Stein identity for P to obtain $P = \mathcal{C} - P\mathcal{A}$, and it can now be read off that P preserves the space $C^{\omega}(b\Omega)$.

Next, we shall need to understand in more detail the orthogonal decomposition of $L^2(b\Omega)$ determined by the Szegő projection. The following theorem is due to Schiffer [13].

THEOREM 3. *The space of functions in $L^2(b\Omega)$ orthogonal to $H^2(b\Omega)$ is equal to the space of functions of the form $\overline{HT}$ where $H \in H^2(b\Omega)$. Consequently, a function $u \in L^2(b\Omega)$ can be expressed uniquely as an orthogonal sum*

$$u = h + \overline{HT}$$

where $h = Pu$ and $H = P(\overline{uT})$. Furthermore, if u is in $C^{\omega}(b\Omega)$, then h and H are in $A(\overline{\Omega})$.

To prove this theorem, notice that if v is orthogonal to $H^2(b\Omega)$, it follows that $\mathcal{C}^*v = 0$, and (2.3) then implies that $v = \overline{T\mathcal{C}(\overline{vT})}$, which is a function of the form $\overline{HT}$ with $H \in H^2(b\Omega)$. Conversely, we proved above that functions of the form $\overline{HT}$ with $H \in A(\overline{\Omega})$ are orthogonal to $H^2(b\Omega)$. Since $A(\overline{\Omega})$ is dense in $H^2(b\Omega)$, the same is true when $H \in H^2(b\Omega)$. Let $P^{\perp} = I - P$. Given $u \in L^2(b\Omega)$, we may now write $u = Pu + P^{\perp}u = h + \overline{HT}$ where $h = Pu$ and $\overline{HT} = P^{\perp}u$. To see that $H = P(\overline{uT})$, we multiply the equation $u = h + \overline{HT}$ by T and take the complex conjugate to obtain $\overline{uT} = \overline{hT} + H$. This is an orthogonal sum from which it follows that $H = P(\overline{uT})$. Finally, it follows from the fact that P preserves $C^{\omega}(b\Omega)$ that h and H belong to $A(\overline{\Omega})$ whenever $u \in C^{\omega}(b\Omega)$.

The orthogonal decomposition theorem allows us to define the Szegő and Garabedian kernels with a minimum of effort. Given a point a in Ω, let $C_a(z)$ denote the *complex conjugate of*

$$\frac{1}{2\pi i} \frac{1}{z - a}.$$

If $h \in H^2(b\Omega)$, we have associated with h a holomorphic function on Ω, also denoted by h, whose value at $a \in \Omega$ is given by the Cauchy integral formula,

$$h(a) = \langle h, C_a \rangle.$$

Thus, we may think of C_a as being the kernel of the Cauchy integral operator; we shall call C_a the Cauchy kernel. The Szegő kernel S_a is defined to be PC_a. It is the holomorphic part of the orthogonal decomposition of the Cauchy kernel C_a. The Garabedian kernel comes from the other part of this decomposition. Before we define the Garabedian kernel, we shall take a moment to show that the Szegő kernel is the the kernel of the Szegő projection. It is standard notation to write $S(a, z) = \overline{S_a(z)}$. Given a point a in Ω and a function u in $L^2(b\Omega)$, the value at a of the holomorphic function associated to Pu is given by

$$(Pu)(a) = \langle Pu, C_a \rangle = \langle u, PC_a \rangle = \int_{z \in b\Omega} S(a, z) u(z)\, ds$$

and this shows that $S(a, z)$ is indeed the kernel for P in the classical sense.

Since $S_a = PC_a$, we may write the orthogonal decomposition of C_a in the form $C_a = S_a + \overline{H_a T}$ where $H_a = P(\overline{C_a T})$. Solving this equation for S_a, writing out C_a, and taking complex conjugates gives

$$\overline{S_a(z)} = -i\left(\frac{1}{2\pi} \frac{1}{z - a} - iH_a(z) \right) T(z).$$

The Garabedian kernel L_a is defined to be equal to the function in parentheses, i.e.,

$$L_a(z) = \frac{1}{2\pi} \frac{1}{z - a} - iH_a(z).$$

Notice that Theorem 3 implies that both S_a and L_a extend holomorphically past the boundary of Ω. In fact, $S_a \in A(\overline{\Omega})$ and L_a is meromorphic on a neighborhood of $\overline{\Omega}$ with a single singularity at a that is a simple pole with residue $1/(2\pi)$. It is possible to interpret the Garabedian kernel as being the kernel for the projection $P^\perp$ of $L^2(b\Omega)$ onto the space of functions in $L^2(b\Omega)$ which are orthogonal to $H^2(b\Omega)$, but we shall not do this here (see [1]).

Notice that if $h \in H^2(b\Omega)$, then $h = Ph$, and so $h(a) = (Ph)(a) = \langle h, S_a \rangle$. Thus, S_a represents the functional $h \mapsto h(a)$ in the sense of the Riesz representation theorem. This property uniquely determines S_a and is sometimes taken as the definition of S_a. Taking $h = 1$ in this formula shows that $1 = \langle 1, S_a \rangle$, and hence, that the L^2 norm $\|S_a\|$ is not zero. It follows that $S_a(a) = \langle S_a, S_a \rangle > 0$. We shall need to know this fact when we study the Ahlfors map.

3. Zeroes of the kernels.

We have shown that

(3.1) $\overline{S_a(z)} = -iL_a(z)T(z)$

for $a \in \Omega$ and $z \in b\Omega$. This identity is at the heart of our method for solving the Dirichlet problem.

Before we can solve the Dirichlet problem, we must study the question of where the Szegő and Garabedian kernels vanish, and to study this question, we shall need to know a generalized version of the argument principle allowing zeroes to occur on the boundary. Suppose that h is a not identically zero meromorphic on a neighborhood of $\overline{\Omega}$. Then the zeroes and poles of h are isolated. Suppose further that h has zeroes, but no poles, on $b\Omega$. Let $\{z_i\}_{i=1}^N$ denote the zeroes of h which lie in Ω, let $\{p_i\}_{i=1}^Q$ denote the poles of h which lie in Ω, and let $\{b_i\}_{i=1}^M$ denote the zeroes of h which lie on $b\Omega$. For small $\epsilon > 0$, let $\gamma_\epsilon = b\Omega - \left(\cup_{i=1}^M D_\epsilon(b_i)\right)$. We assume that ϵ is small enough so that the closures of the discs $D_\epsilon(b_i)$ are mutually disjoint and the set γ_ϵ consists of finitely many connected smooth arcs. On each of these arcs, the increment of the argument of $h(z)$ as z moves along the arc in the positive sense is well defined. We shall prove that as $\epsilon \to 0$, the sum of the increments of the argument of $h(z)$ along all these arcs tends to an angle $\Delta \arg h$, and this angle is related to the number of zeroes and poles of h according to the following formula (we let $m(z; h)$ denote the multiplicity of a zero or pole of h at z).

$$\sum_{i=1}^N m(z_i; h) + \frac{1}{2} \sum_{i=1}^M m(b_i; h) - \sum_{i=1}^Q m(p_i; h) = \frac{1}{2\pi} \Delta \arg h.$$

In words, this generalized argument principle says that zeroes or poles in the interior of Ω contribute to the increment of the argument of h as usual, whereas zeroes on the boundary contribute half the normal amount to the increment. To prove this fact, let C_ϵ^i denote the arc of the circle which bounds $D_\epsilon(b_i)$ lying outside of Ω. We assume that the C_ϵ^i are parameterized so that $\gamma_\epsilon \cup \left(\cup_{i=1}^M C_\epsilon^i\right)$ represents the boundary of

$$\Omega_\epsilon = \Omega \cup \left(\cup_{i=1}^M D_\epsilon(b_i)\right),$$

parameterized in the standard sense. We may now apply the classical argument principle to h on Ω_ϵ. As we let ϵ tend to zero, the increment of $\arg h$ on C_ϵ^i is easily seen to approach $\pi m(b_i; h)$, and this completes the proof. We may now prove the following theorem. Our proof follows Nehari's proof as described in Bergman's book [**4, page 117**].

THEOREM 4. *Suppose Ω is a bounded n-connected domain in the plane with real analytic boundary and let $a \in \Omega$ be given. Then $L_a(z)$ is non-vanishing for $z \in \overline{\Omega} - \{a\}$. The function $S_a(z)$ is non-vanishing on $b\Omega$ and has exactly $n - 1$ zeroes in Ω.*

Notice that, in case Ω is simply connected, this theorem yields that S_a in non-vanishing on $\overline{\Omega}$.

Let $f(z) = S_a(z)/L_a(z)$. It was proved by Garabedian [8] that this f is equal to the Ahlfors map associated to a and Ω. The Ahlfors map is an n-to-one map of Ω onto the unit disc (counting multiplicities). Furthermore, among all holomorphic maps h of Ω into the unit disc, the Ahlfors map maximizes Re $h'(a)$. We shall use f to study the zeroes of S_a and L_a. We shall deduce here only the properties of f that we shall need to solve the Dirichlet problem. Later, in §5, we shall prove the extremal property of f.

At the moment, we know nothing about the special properties of f. All we know is that f is a meromorphic function on a neighborhood of $\overline{\Omega}$ and that f has a simple zero at $z = a$ stemming from the facts that $S_a(a) > 0$ and that L_a has a simple pole at $z = a$. Let λ be a complex number of unit modulus. We want to consider how many times $f(z)$ assumes the value λ on $\overline{\Omega}$. To do this, let $G_\lambda(z) = S_a(z) - \lambda L_a(z)$. We shall first prove that $G_\lambda(z)$ has exactly n zeroes on $\overline{\Omega}$, one on each boundary component of Ω. Formula (3.1) implies that, on $b\Omega$,

$$G_\lambda T = S_a T - \lambda L_a T = i\,\overline{L_a} - \lambda i\,\overline{S_a} = -i\lambda(\overline{S_a} - \overline{\lambda L_a}) = -i\lambda\overline{G_\lambda}.$$

Thus, we have the identity,

$$(3.2) \qquad\qquad G_\lambda^2 = -i\lambda|G_\lambda^2|\overline{T}.$$

Let $\{\gamma_i\}_{i=1}^n$ denote the simple closed real analytic curves which represent the n boundary components of Ω. We shall now show that G_λ has at least one zero on each γ_i. Indeed, if G_λ has no zero on γ_i, identity (3.2) shows that the increment of arg G_λ^2 around γ_i is the same as the increment of arg $\overline{T}$. But the increment of the argument of G_λ^2, the *square* of a continuous function, around γ_i is either zero or an *even* multiple of $\pm 2\pi$, and the increment of arg $\overline{T}$ is $\pm 2\pi$. Hence, equality is out of the question. Thus, G_λ must have at least one zero on γ_i.

Let $\{z_i\}_{i=1}^N$ denote the zeroes of G_λ which lie in Ω and let $\{b_i\}_{i=1}^M$ denote the zeroes of G_λ which lie on $b\Omega$. Observe that G_λ has a single simple pole at a. Thus, the generalized argument principle yields

$$(3.3) \qquad -1 + \sum_{i=1}^N m(z_i; G_\lambda) + \frac{1}{2}\sum_{i=1}^M m(b_i; G_\lambda) = \frac{1}{2\pi}\Delta\text{arg } G_\lambda.$$

Now, identity (3.2) reveals that

$$\frac{1}{2\pi}\Delta\text{arg } G_\lambda^2 = \frac{1}{2\pi}\Delta\text{arg } \overline{T} = -1 + (n-1) = n - 2.$$

Therefore, $\frac{1}{2\pi}\Delta\text{arg } G_\lambda = \frac{1}{2}(n-2)$. Also, $\sum_{i=1}^M m(b_i; G_\lambda)$ is at least n because G_λ has at least one zero on each boundary component of Ω. When these numbers are plugged into (3.3), we are forced to conclude that G_λ has no zeroes in Ω, and precisely n zeroes on $b\Omega$, one on each curve γ_i.

Using what we now know about G_λ, we shall show that f is actually *holomorphic* on a neighborhood of $\overline{\Omega}$, that $|f| < 1$ on Ω, and that $|f| = 1$ on $b\Omega$. Indeed, if f had a pole at some point in $\overline{\Omega}$, then there would exist a point $p \in \Omega$

near the pole such that $|f(p)| > 1$ and $L_a(p) \neq 0$. Consider the function $|f(z)|$ along a curve in Ω joining p to a which does not pass through a zero of L_a. Since $|f(p)| > 1$ and $|f(a)| = 0$, the intermediate value theorem implies that there is a point $z_0 \in \Omega$ (which is not a zero of L_a) such that $f(z_0) = \lambda$, a complex number of unit modulus. However, at such a point z_0, the function G_λ associated to this λ that we studied above would have to vanish, and we have shown that it cannot vanish at an interior point. Hence, f has no poles in $\overline{\Omega}$. Now, if there is a point p in $\overline{\Omega}$ with $|f(p)| > 1$, we may apply the same reasoning to obtain the same contradiction. Hence, $|f| \leq 1$ on Ω. Since $f(a) = 0$, the maximum principle implies that $|f| < 1$ on Ω. Identity (3.1) yields that $|f| = 1$ on the dense subset of $b\Omega$ where L_a is non-vanishing. By continuity, $|f| = 1$ on all of $b\Omega$.

Next, we show that L_a is non-vanishing on $\overline{\Omega} - \{a\}$. We have shown that if $|\lambda| = 1$, then $G_\lambda = S_a - \lambda L_a$ has exactly one zero on each boundary component of Ω and no zeroes inside Ω. Also, because $f = S_a/L_a$ is holomorphic in a neighborhood of $\overline{\Omega}$, it follows that if $L_a(z_0) = 0$ with $z_0 \in \overline{\Omega}$, then $S_a(z_0) = 0$ too. Hence, G_λ and S_a must vanish wherever L_a does. Because G_λ cannot vanish in Ω, this yields that L_a cannot vanish in Ω. To see that L_a cannot vanish on $b\Omega$ either, suppose $L_a(z_0) = 0$, $z_0 \in b\Omega$. Then $S_a(z_0) = 0$ by (3.1). Consequently, $G_\lambda(z_0) = 0$ for *any* λ of unit modulus. We have shown that S_a cannot be identically zero on Ω; thus there is a point $\xi_0 \in b\Omega$ such that $S_a(\xi_0) \neq 0$. (Notice that $\xi_0 \neq z_0$ because $S_a(z_0) = 0 \neq S_a(\xi_0)$.) Formula (3.1) shows that $|S_a(\xi_0)| = |L_a(\xi_0)|$. Hence, there is a λ with $|\lambda| = 1$ such that $G_\lambda(\xi_0) = 0$. We have now shown that, for this particular choice of λ, G_λ has zeroes at two points in the boundary, z_0 and ξ_0. This is a contradiction. Therefore, L_a is non-vanishing on $\overline{\Omega} - \{a\}$. Formula (3.1) now shows that S_a is non-vanishing on $b\Omega$ too. We can also read off from (3.1) that

$$-\frac{1}{2\pi}\Delta\arg S_a = \frac{1}{2\pi}\Delta\arg L_a + \Delta\arg T = (-1) + [1 - (n-1)].$$

Hence, $\Delta\arg S_a = 2\pi(n-1)$, and we deduce that S_a has exactly $n-1$ zeroes in Ω. The proof of Theorem 4 is complete.

It is now easy to see that $f = S_a/L_a$ is an n-to-one map of Ω onto the unit disc. Since S_a has $n-1$ zeroes in Ω, since $S_a(a) \neq 0$, and since L_a has a simple pole at a, it follows that f has exactly n zeroes in Ω. Given w in the unit disc, the number of times that f assumes the value w (counting multiplicities) on Ω is given by the integral

$$M(w) = \frac{1}{2\pi i}\int_{b\Omega} \frac{f'(z)}{f(z) - w}\, dz.$$

(Recall that $|f| = 1$ on $b\Omega$.) We know that $M(0) = n$. Since $M(w)$ is a holomorphic function of w on the unit disc taking values in the integers, M must be constant, and so $M(w) \equiv n$. Also, because G_λ vanishes exactly once on each boundary curve of Ω for each complex number λ of unit modulus, it follows that f maps each boundary curve of Ω one-to-one onto the unit circle. We shall study the extremal property of f in §5.

THE CAUCHY TRANSFORM, THE SZEGŐ PROJECTION

4. The Dirichlet problem.

Given a continuous function φ on $b\Omega$, the classical Dirichlet problem is to find a harmonic function u on Ω which extends continuously to $b\Omega$ and which agrees with φ on $b\Omega$. In this section, we shall relate the solution of this problem to the Szegő projection.

To motivate the method, let us assume for the moment that our domain Ω is simply connected and that we know that the Dirichlet problem can be solved in the $C^\omega(b\Omega)$ category. Hence, given a function $\varphi \in C^\omega(b\Omega)$, we assume that the solution u to the Dirichlet problem with boundary data φ exists and is real analytic on a neighborhood of $\overline{\Omega}$. (This is true, but we don't want to assume it without proof. Soon we shall give an elementary proof of this fact from first principles.) It follows that u can be expressed as $h + \overline{H}$ where h and H are in $A(\overline{\Omega})$. These functions h and H are unique up to additive constants, so we may assume without loss of generality that $H(a) = 0$ for some fixed point $a \in \Omega$. Now, restricting to the boundary and multiplying by S_a yields

$$S_a\varphi = S_a h + S_a\overline{H}.$$

Using (3.1), we may replace the last occurence of S_a in this formula by $i\overline{L_a T}$ to obtain

$$S_a\varphi = S_a h + i\overline{L_a HT}.$$

Notice that $L_a H$ is in $A(\overline{\Omega})$ because the zero of H at a cancels the pole of L_a at a. Hence, this is an orthogonal decomposition of $S_a\varphi$. It follows that $S_a h = P(S_a\varphi)$ and $-iL_a H = P(\overline{S_a\varphi T}) = -iP(L_a\overline{\varphi})$. Hence, we have deduced that the solution u to the Dirichlet problem can be expressed as $h + \overline{H}$ where

$$h = \frac{P(S_a\varphi)}{S_a} \quad \text{and} \quad H = \frac{P(L_a\overline{\varphi})}{L_a}.$$

We shall now solve the Dirichlet problem from scratch by showing, without recourse to the special assumptions made above, that these formulas define a solution to the problem.

THEOREM 5. *Suppose Ω is a bounded simply connected domain with real analytic boundary and suppose φ is a function in $C^\omega(b\Omega)$. Let $a \in \Omega$. Then, the function $u = h + \overline{H}$, where h and H are holomorphic functions in $A(\overline{\Omega})$ given by*

$$h = \frac{P(S_a\,\varphi)}{S_a} \quad \text{and} \quad H = \frac{P(L_a\,\overline{\varphi})}{L_a},$$

solves the Dirichlet problem for φ. Note that it follows that u extends to be real analytic on a neighborhood of $\overline{\Omega}$.

It is easy to deduce from Theorem 5 that the solution to the classical Dirichlet problem exists. Indeed, if φ is merely continuous on $b\Omega$, let φ_j be a sequence of functions in $C^\omega(b\Omega)$ which converge uniformly on $b\Omega$ to φ. The Maximum Principle can now be used to see that the solutions u_j to the Dirichlet problems corresponding to φ_j converge uniformly on $\overline{\Omega}$ to a function u which is harmonic on Ω, continuous on $\overline{\Omega}$, and which assumes φ as its boundary values. This same limiting argument reveals that the formula in Theorem 5 expresses the solution to the Dirichlet problem even when φ is only assumed to be continuous on $b\Omega$. It must be pointed out, however, that in this case, the functions h and H need not extend continuously to the boundary, even though the solution $h + \overline{H}$ does.

PROOF OF THEOREM 5. The function $S_a\varphi$ has an orthogonal decomposition $S_a\varphi = g + \overline{GT}$ where $g = P(S_a\varphi)$ and $\overline{GT} = P^\perp(S_a\varphi)$. Identity (3.1) yields that $\overline{T} = -iS_a/\overline{L_a}$. Hence, $S_a\varphi = g - i\overline{G}\,S_a/\overline{L_a}$, and upon dividing this equation by S_a, we obtain

$$\varphi = \frac{g}{S_a} - i\frac{\overline{G}}{\overline{L_a}}.$$

By Theorem 4, S_a does not vanish on $\overline{\Omega}$, and L_a does not vanish on $\overline{\Omega} - \{a\}$. Hence, the first term in this decomposition is seen to be in $A(\overline{\Omega})$ and the second term is the conjugate of a function in $A(\overline{\Omega})$. It follows that the sum is a harmonic function in $C^\infty(\overline{\Omega})$ which agrees with φ on $b\Omega$ and we have found a solution to the Dirichlet problem. (By the way, the maximum principle shows that it is the *unique* solution.)

To finish the proof of the theorem, we must show that $iG = P(L_a\overline{\varphi})$. Recall that $P^\perp v = \overline{T\,P(\overline{vT})}$. Thus, $G = P(\overline{S_a\varphi T})$. But identity (3.1) implies that $\overline{S_aT} = -iL_a$. Therefore, $G = -iP(L_a\overline{\varphi})$, and the proof is finished.

Next, we shall solve the Dirichlet problem in a multiply connected domain with real analytic boundary. In this setting, we are confronted with the problem that not every harmonic function can be decomposed as a sum $h + \overline{H}$ of the type used in the simply connected case. Our first task shall be to prove the following result which states that, in case a harmonic function can be expressed as a sum of the form $h + \overline{H}$, then h and H must be given by the formulas appearing in Theorem 5.

THEOREM 6. *Suppose Ω is a bounded finitely connected domain in the plane with real analytic boundary and suppose φ is a function in $C^\omega(b\Omega)$. Let $a \in \Omega$ be given. Then, on the boundary, the function φ can be decomposed*

$$\varphi = h + \overline{H},$$

where h is a meromorphic function on Ω which extends holomorphically past $b\Omega$ given by

$$h = \frac{P(S_a\,\varphi)}{S_a}$$

and H is a holomorphic function in $A(\overline{\Omega})$ given by

$$H = \frac{P(L_a\,\overline{\varphi})}{L_a}.$$

Furthermore, if φ is equal to $g + \overline{G}$ for some g and G in $H^2(b\Omega)$, then the function h has no poles and is in $A(\overline{\Omega})$. Thus, in this case, $h + \overline{H}$ defines the harmonic extension of φ to Ω.

PROOF. Theorem 3 implies that the operator which sends a function $\varphi \in C^\omega(b\Omega)$ to $h + \overline{HT}$ where $h = P\varphi$ and $H = P(\overline{\varphi}T)$ is the identity operator. Thus, we may decompose $S_a\,\varphi$ on $b\Omega$ as

$$S_a\,\varphi = h + \overline{HT}$$

where $h = P(S_a\varphi)$ and $H = P(\overline{S_a\varphi T})$. Next, using (3.1), we substitute $-iS_a/\overline{L_a}$ for $\overline{T}$ and divide the identity by S_a. What we get is a decomposition of φ on the boundary of Ω as $h + \overline{H}$ where

$$h = \frac{P(S_a\,\varphi)}{S_a} \qquad \text{and} \qquad H = i\frac{P(\overline{S_a\varphi T})}{L_a}.$$

Using (3.1) again, we may replace $\overline{S_a T}$ in the expression for H by $-iL_a$. That h is meromorphic follows from the fact that $S_a(z)$ has exactly $n-1$ zeroes in Ω when Ω is an n-connected domain. That h extends holomorphically past the boundary follows from the facts that P preserves $C^\omega(b\Omega)$, and that $S_a(z)$ is in $A(\overline{\Omega})$, and that $S_a(z)$ is non-vanishing on $b\Omega$. The function H is in $A(\overline{\Omega})$ because $L_a(z)$ has a single simple pole at a and does not vanish for any $z \in \overline{\Omega} - \{a\}$, and furthermore, L_a extends holomorphically past the boundary. The decomposition is proved.

Now suppose that $\varphi \in C^\omega(b\Omega)$ can be decomposed as $\varphi = g + \overline{G}$ where g and G are in $H^2(b\Omega)$. Multiplying by S_a and using (3.1), we see that $S_a\varphi = S_a g + i\overline{G L_a T}$. By subtracting $G(a)$ from G and adding $\overline{G(a)}$ to g, we may assume that $G(a) = 0$. In this case, it follows that GL_a is in $H^2(b\Omega)$. Hence, $S_a g + i\overline{GL_a T}$ is an orthogonal decomposition of $S_a\varphi$. It follows that $S_a g = P(S_a\varphi)$ and this shows that $g \in A(\overline{\Omega})$. In fact, we have shown that $g = P(S_a\varphi)/S_a$ on the boundary, i.e., that $g = h$ on $b\Omega$. Now $g - h$ is a meromorphic function on a neighborhood of $\overline{\Omega}$ vanishing on $b\Omega$. It follows that $g - h \equiv 0$ on $\overline{\Omega}$. Hence, h has no poles in Ω. Since $g = h$, it follows that $G = H$, and the proof is complete.

If Ω is n-connected, let $\{b_i\}_{i=1}^{n-1}$ be a set of points consisting of one point from each of the *bounded* connected components of the complement of $\overline{\Omega}$ in $\mathbb{C}$, and let γ_i denote the boundary curve of Ω bounding the component of $\mathbb{C} - \Omega$ containing b_i. The function $\psi_i = \log|z - b_i|$ is a harmonic function on a neighborhood of $\overline{\Omega}$ which cannot be expressed as the real part of a single valued holomorphic function on Ω. Furthermore, if $\psi = \sum_{i=1}^{n-1} c_i\psi_i$ is the real part of a holomorphic function on Ω, then $c_i = 0$ for each i.

Suppose that $\{c_i\}_{i=1}^{n-1}$ are constants which are not all zero, and let $\psi = \sum_{i=1}^{n-1} c_i\psi_i$ We know that ψ cannot be written as the real part of a holomorphic function on Ω, and therefore ψ cannot be written on $b\Omega$ as $g + \overline{G}$ where $g, G \in H^2(b\Omega)$. By Theorem 6, we may express ψ_i on $b\Omega$ as $h_i + \overline{H_i}$ where h_i is meromorphic on a neighborhood of $\overline{\Omega}$ and $H_i \in A(\overline{\Omega})$. The statement about the impossibility of expressing ψ as $g + \overline{G}$ means that $\sum c_i h_i$ has at least one pole at a zero of the Szegő kernel.

Let $\{a_j\}_{j=1}^{n-1}$ denote the $n-1$ zeroes of $S_a(z)$ in Ω. For simplicity, let us suppose, for the moment, that each zero has multiplicity one. (It is a fact [see [1] or [3]] that the zeroes of S_a become simple zeroes as a tends to the boundary

of Ω. Hence, it is actually possible to choose a so that this condition is met.)
The fact that $\sum c_i h_i$ has at least one pole at a zero of the Szegő kernel if not all
the c_i's are zero means that if

$$\sum_{i=1}^{n-1} c_i \left(P(\psi_i S_a)\right)(a_j) = 0 \qquad \text{for } j = 1, \ldots, n-1,$$

then $c_i = 0$ for all i. This implies that the determinant of this linear system is
non-zero. Hence, given $\varphi \in C^\infty(b\Omega)$, we may solve the linear system,

$$\sum_{i=1}^{n-1} c_i \left(P(\psi_i S_a)\right)(a_j) = P(\varphi S_a)(a_j),$$

$j = 1, \ldots, n-1$, for c_i, $i = 1, \ldots, n-1$. Having solved the system, we deduce
that, when $\varphi - \sum c_i \psi_i$ is expressed as $h + \overline{H}$ via the formulas in Theorem 6, the
function h has no poles at the zeroes of S_a. Hence, the harmonic extension of φ
to Ω is given by

$$h + \overline{H} + \sum c_i \log |z - b_i|.$$

Now, in the case that $S_a(z)$ has zeroes of multiplicity greater than one, we
must solve a linear system analogous to the one above. However, corresponding
to each zero of multiplicity m, there must be m linear equations, one stemming
from point evaluation at the zero, and $m-1$ additional equations arising from
point evaluation of the first $m-1$ derivatives at the zero. The details are not
hard. We leave them to the reader.

We have given a constructive proof of the following theorem.

THEOREM 7. *Suppose Ω is a bounded domain with real analytic boundary and
suppose φ is a function in $C^\omega(b\Omega)$. Then the solution to the Dirichlet problem
with boundary data φ exists and extends to be real analytic on a neighborhood of
$\overline{\Omega}$.*

As in the simply connected case, this theorem solving the Dirichlet problem
in C^ω can be used to prove that the classical Dirichlet problem is solvable for
continuous functions.

It is also possible to solve the classical Neumann problem using methods
analogous to those above. See [1] and [3] for an account of this subject.

5. The Ahlfors map.

Suppose that Ω is a bounded n connected domain in the plane with real
analytic boundary. In §3, we proved that the map $f = S_a/L_a$ is an n-to-one
covering map of Ω onto the unit disc. In this section, we wish to see that f
satisfies the following extremal property. Among all holomorphic functions on
Ω which map into the unit disc, f has the property that $\operatorname{Re} f'(a)$ is as large as
possible. Furthermore, this property characterizes f. Hence, we may think of f
as being a kind of Riemann mapping function for a multiply connected domain.
(A similar treatment of this result can be found in Bergman's book [4].)

First, note that $f'(a) = 2\pi S_a(a)$ because L_a has a simple pole at $z = a$ with residue $1/2\pi$. Since multiplication by $e^{i\theta}$ preserves the class of functions which map into the unit disc, and since $f'(a) > 0$, we may restrict our attention to the class of functions h which map into the disc such that $h'(a) > 0$. Since this class is a normal family, we know extremal functions exist. Furthermore, if h is extremal, it must be that $h(a) = 0$. Indeed, if this is not the case, by forming the composition $M \circ h$ where $M(z)$ is the Möbius transformation $M(z) = (z - h(a))/(1 - \overline{h(a)}\,z)$, we obtain a map in the class with strictly larger derivative at a, which contradicts the extremal assumption. Hence, we may restrict our attention to the class $\mathcal{F}$ of functions h which are holomorphic on Ω mapping Ω into the unit disc such that $h(a) = 0$ and $h'(a) > 0$.

Before we can proceed, we must show that functions in $\mathcal{F}$ are Cauchy integrals of functions in $L^2(b\Omega)$. Recall that for $z \in b\Omega$, $T(z)$ is a complex number of unit modulus representing the unit tangent vector at z pointing in the direction of the standard orientation of $b\Omega$. Hence, $iT(z)$ represents the unit inward pointing normal vector at $z \in b\Omega$. Let ϵ_j be a sequence of small positive real numbers tending to zero as $j \to \infty$. Let Ω_j denote the domain consisting of points in Ω that are a distance greater that ϵ_j from $b\Omega$. Notice that if $z(t)$ parameterizes the boundary of Ω, then $z_j(t) = z(t) + i\epsilon_j T(z(t))$ parameterizes the boundary of Ω_j. Given $h \in \mathcal{F}$, define $h_j(z) = h(z + i\epsilon_j T(z))$ for $z \in b\Omega$. Since h is bounded, it follows that h_j is uniformly bounded in $L^2(b\Omega)$ as $j \to \infty$. Hence, there is a subsequence of h_j which converges *weakly* in $L^2(b\Omega)$ to an element $u \in L^2(b\Omega)$. Let us rename our sequences so that h_j itself tends weakly to u. We now claim that $h = \mathcal{C}u$. To see this, let $\mathcal{C}_j$ denote the Cauchy transform associated to Ω_j. If $a \in \Omega$, then

$$(\mathcal{C}u)(a) - h(a) = (\mathcal{C}u - \mathcal{C}_j h)(a) = \mathcal{C}(u - h_j)(a) + (\mathcal{C}h_j - \mathcal{C}_j h)(a).$$

The weak convergence of h_j to u implies that $\mathcal{C}(u - h_j)(a)$ tends to zero as $j \to \infty$. The integral defining the other term $(\mathcal{C}h_j - \mathcal{C}_j h)(a)$ can be written

$$\frac{1}{2\pi i} \int h(z_j(t)) \left(\frac{z'(t)}{z(t) - a} - \frac{z'_j(t)}{z_j(t) - a} \right) dt.$$

This quantity tends to zero because $h(z_j(t))$ is bounded and the other term in the integral tends uniformly to zero as $j \to \infty$. Hence, we have proved that $h = \mathcal{C}u$, in the sense that the holomorphic function h on Ω agrees with the holomorphic function $\mathcal{C}u$ on Ω. (We have not shown that h can be thought of as an element of $H^2(b\Omega)$, although this would not be hard to do at this point.)

Now consider the function L_a^2. It is meromorphic on a neighborhood of $\overline{\Omega}$ and has a single pole at $z = a$ of order two. We claim that the residue of L_a^2 at a is zero. Indeed, by (3.1), $L_a^2 T = iL_a \overline{S_a}$. Hence, the residue of L_a^2 at a can be computed via

$$2\pi i \mathrm{Res}_a L_a^2 = \int_{b\Omega} L_a^2 T\, ds = \int_{b\Omega} iL_a \overline{S_a}\, ds = \langle iL_a, S_a \rangle,$$

and this last quantity is zero because $L_a = i\,\overline{S_a T}$ is orthogonal to $H^2(b\Omega)$. Hence,

$$L_a^2 = \frac{1}{4\pi^2} \frac{1}{(z - a)^2} + G_a$$

where G_a is holomorphic on a neighborhood of $\overline{\Omega}$. Now suppose that $h \in \mathcal{F}$ is extremal and let u denote the weak limit associated to h that we constructed above. Using the residue theorem and the same reasoning we used to deduce that $h = \mathcal{C}u$, it follows that

$$(5.1) \qquad h'(a) = 4\pi^2 \mathrm{Res}_a(L_a^2 h) = \frac{2\pi}{i} \int_{b\Omega} L_a^2 u T \, ds \leq 2\pi \int_{b\Omega} |L_a^2| \, ds$$

since $|u| \leq 1$ on $b\Omega$. But, by (3.1), the $L^2(b\Omega)$ norm of L_a is equal to that of S_a, which we know is equal to $S_a(a)^{1/2}$. Hence, we have shown that $h'(a) \leq 2\pi S_a(a) = f'(a)$, and therefore, that f is also an extremal function. To finish the proof, we must show that f is the *unique* extremal function in the class $\mathcal{F}$. This turns out to be an easy consequence of a measure theory exercise which asserts that, if $|V| \leq |U|$ and $\int V = \int |U|$, then $V = |U|$. Suppose that $h \in \mathcal{F}$ is extremal, and let $V = -iL_a^2 u T$ where u is the weak limit associated to h as above. Let $U = |L_a|^2$. Because $|u| \leq 1$, it follows that $|V| \leq |U|$ on $b\Omega$. Now (5.1) shows that $\int V = \int |U|$ and it follows that $V = |U|$, i.e., that $-iL_a^2 u T = |L_a|^2$. Solving this equation for u shows that $u = i \overline{L_a T}/L_a$ and using (3.1) yields that $u = S_a/L_a$. Since $h = \mathcal{C}u$, we see that $h = f$ and the proof is complete.

References.

1. S. Bell, *The Cauchy transform, potential theory, and conformal mapping*, CRC Press, Boca Raton, Florida, in press..

2. _____, *Solving the Dirichlet problem in the plane by means of the Cauchy integral*, Indiana Math. J. **39** (1990), 1355–1371.

3. _____, *The Szegő projection and the classical objects of potential theory in the plane*, Duke Math. J. **64** (1991), 1–26.

4. S. Bergman, *The kernel function and conformal mapping*, Math. Surveys 5, AMS, Providence, 1950.

5. J. Burbea, *The Cauchy and the Szegő kernels on multiply connected regions*, Rendiconti del Circolo Math. di Palermo **31** (1982), 105–118.

6. _____, *The Riesz projection theorem in multiply connected regions*, Bolletino della Unione Matematica Italiano **14** (1977), 143–147.

7. G. Folland, *Introduction to partial differential equations*, Princeton U. Press, Princeton, 1976.

8. P. Garabedian, *Schwartz's lemma and the Szegő kernel function*, Trans. Amer. Math. Soc. **67** (1949), 1–35.

9. P. Henrici, *Applied and computational complex analysis*, Vol. 3, John Wiley, New York, 1986.

10. L. Hörmander, *An introduction to complex analysis in several variables*, North Holland, Amsterdam, 1973.

11. N. Kerzman and E. M. Stein, *The Cauchy kernel, the Szegő kernel, and the Riemann mapping function*, Math. Ann. **236** (1978), 85–93.

12. N. Kerzman and M. Trummer, *Numerical conformal mapping via the Szegő kernel*, Numerical conformal mapping, L. N. Trefethen, ed., North Holland, Amsterdam, 1986, pp. 111–123.

13. M. Schiffer, *Various types of orthogonalization*, Duke Math. J. **17** (1950), 329–366.
14. H. S. Shapiro, *The Schwarz function and its generalization to higher dimensions*, to appear..
15. M. Trummer, *An efficient implementation of a conformal mapping method based on the Szegő kernel*, SIAM J. of Numer. Anal. **23** (1986), 853–872.

MATHEMATICS DEPARTMENT, PURDUE UNIVERSITY, WEST LAFAYETTE, INDIANA 47907

Contemporary Mathematics
Volume **137**, 1992

Variations on the theorem of Morera

CARLOS BERENSTEIN*
DER-CHEN CHANG*
DANIEL PASCUAS
LAWRENCE ZALCMAN

Dedicated to Professor Walter Rudin on the occasion of his retirement.

ABSTRACT. We survey some Morera-type theorems in one and several complex variables. We also obtain a Morera-type theorem for CR functions in the unit ball in $\mathbb{C}^n$.

1. Morera-type theorems for holomorphic functions

Let $\mathbb{D} = \{z \in \mathbb{C} : |z| < 1\}$ be the unit disc in the complex plane $\mathbb{C}$. The classical theorem of Morera may be stated in the following form.

THEOREM 1.1. *Let $f \in C(\mathbb{D})$. Then f is holomorphic in $\mathbb{D}$ if*

$$\int_\gamma f(z)dz = 0$$

for all piecewise C^1 Jordan curves γ in $\mathbb{D}$.

For the proof, set

$$\gamma_n(\zeta) = C_n(\zeta) = \{z \in \mathbb{C} : |z - \zeta| = \frac{1}{n}\} \text{ and } D_n(\zeta) = \{z \in \mathbb{C} : |z - \zeta| \leq \frac{1}{n}\}$$

By regularization, we may assume that $f \in C^1(\mathbb{D})$; and then

$$0 = \frac{1}{2i} \int_{C_n(\zeta)} f(z)dz = \int \int_{D_n(\zeta)} \frac{\partial f}{\partial \bar{z}} dx\, dy,$$

1991 *Mathematics Subject Classification*. Primary 32A10. Secondary 30E99.
*Partially supported by a grant from the National Science Foundation.
This paper is in final form and no version of it will be submitted for publication elsewhere.

by Green's theorem. Dividing by the area of $D_n(\zeta)$ and passing to the limit, we see that $\frac{\partial f}{\partial \bar{z}}(\zeta) = 0$ for every $\zeta \in \mathbb{D}$ since $\frac{\partial f}{\partial \bar{z}}$ is continuous. Thus f is analytic on $\mathbb{D}$.

The argument above admits considerable extension; for its ultimate scope, see [Z1, Theorem 1]. The essential point is that each point of $\mathbb{D}$ can be obtained as a limit of a shrinking sequence of contours (in this case, circles). Agranovsky and Val'skii showed that it is possible to replace the infinite sequence of contours by a single contour at each point on which an infinite sequence of moment conditions are satisfied. Denoting by $M(2)$ the Euclidean motion group of transformations

$$z \mapsto e^{i\theta}z + b,$$

where $\theta \in [0, 2\pi)$ and $b \in \mathbb{C}$, we have the following theorem.

THEOREM 1.2. ([AV, Proposition 1]) *Let $f \in C(\mathbb{R}^2)$ and let γ_0 be a piecewise smooth Jordan curve. Then f is analytic on $\mathbb{C}$ (entire) if and only if*

$$\int_{\sigma(\gamma_0)} z^k f(z)dz = 0 \qquad k = 0, 1, 2, \cdots \tag{1.1}$$

for every $\sigma \in M(2)$.

It is well-known that condition (1.1) holds if and only if the restriction of f to $\sigma(\gamma_0)$ can be extended continuously to an analytic function on the region bounded by $\sigma(\gamma_0)$. Theorem 1.2 asserts that these extensions "agree on overlaps" and thus determine a single-valued analytic function on $\mathbb{C}$.

In the rotationally invariant case, *i.e.* when γ_0 is a circle (of radius r, say), the condition (1.1) is minimally restrictive, as only translations come into play. In this case, Theorem 1.2 has a particularly appealing interpretation. It says that f is analytic on all of $\mathbb{C}$ if and only if all Fourier coefficients of negative index vanish in the Fourier expansion of the restriction of f to each circle of radius r. Actually, much more is true.

THEOREM 1.3. ([Z2, Theorem 6]) *Let $f \in C(\mathbb{R}^2)$ and let $r > 0$ be fixed. Suppose there exists $n \in \mathbb{Z}$, $n \neq \pm 1$, such that for all $\zeta \in \mathbb{C}$*

$$\begin{aligned}
\int_0^{2\pi} f(\zeta + re^{i\theta})e^{i\theta}d\theta &= 0 \\
\int_0^{2\pi} f(\zeta + re^{i\theta})e^{in\theta}d\theta &= 0.
\end{aligned} \tag{1.2}$$

Then f is an entire function.

The conditions (1.2) assert the vanishing of the Fourier coefficients of index -1 and $-n$ of the restriction of f to $C_r(\zeta) = \{z : |z - \zeta| = r\}$. That two integral conditions can be sufficient to imply analyticity was first observed in [Z1] and [Sm] (cf. [BST]), where the following theorem is proved.

THEOREM 1.4. ([Z1, Theorem 2]) *Let $f \in C(\mathbb{R}^2)$ and suppose there exist positive numbers r_1, r_2 such that*

$$\int_{C_r(\zeta)} f(z)dz = 0 \qquad (1.3)$$

for every $\zeta \in \mathbb{C}$ and $r = r_1, r_2$. Then f is an entire function so long as $\frac{r_1}{r_2}$ is not a quotient of zeros of the Bessel function J_1.

Special interest attaches to the exceptional set of excluded ratios. It is quite essential to the result: in case $\frac{r_1}{r_2}$ **is** a quotient of zeros of the Bessel function J_1, f need not be holomorphic **anywhere** ([Z1, page 244]); cf. Example 2.2 below.

Theorems 1.2 and 1.4 have hyperbolic (noneuclidean) analogues, in which $\mathbb{C}$ is replaced by $\mathbb{D}$ and the Euclidean motion group $M(2)$ by the Möbius group $\mathcal{M}$ of conformal automorphisms of $\mathbb{D}$. For instance, we have

THEOREM 1.5. ([A]) *Let $f \in C(\mathbb{D})$ and let γ_0 be a piecewise smooth Jordan curve in $\mathbb{D}$. Then f is analytic in $\mathbb{D}$ if and only if*

$$\int_{\sigma(\gamma_0)} z^k f(z)dz = 0 \qquad k = 0, 1, 2, \ldots$$

for every $\sigma \in \mathcal{M}$.

Theorem 1.5 is a sharpening of a result of in [AV], where continuity up to the boundary was required.

The hyperbolic analogue of Theorem 1.4 is

THEOREM 1.6. ([BZ1, pp. 125-126]; cf. [A, Theorem 4]) *Let $f \in C(\mathbb{D})$ and suppose there exist positive numbers r_1, r_2 such that*

$$\int_{\gamma} f(z)dz = 0 \qquad (1.4)$$

for every circle γ in $\mathbb{D}$ of hyperbolic radius r_1 or r_2. Then f is analytic in $\mathbb{D}$ so long as the equations

$$P_z^{-1}(\cosh r) = 0 \qquad r = r_1, r_2$$

have no common solutions $z \in \mathbb{C}$.

Here P_z^{-1} is the associated Legendre function, and we have normalized the hyperbolic metric on $\mathbb{D}$ to have curvature -1. Again, the exceptional set cannot be dispensed with; cf. [BZ1, pp. 125].

Numerous variations on these results are possible. For instance, "one-radius" analogues of Theorems 1.4 and 1.6 (in which there are **no** exceptional sets) hold in the presence of appropriate growth conditions. Thus, Smith [**Sm**, Corollary 2.4] has shown that if (1.3) holds for all circles of a (single) fixed radius and

$$\left| \frac{\partial f}{\partial \bar{z}} \right| = o(|z|^{-\frac{1}{2}}),$$

then f is analytic. Along similar lines, Agranovsky [A, Theorem 3] proves that if (1.4) holds for all circles of a (single) fixed hyperbolic radius, and $f \in C(\mathbb{D}) \cap L^2(\mathbb{D})$, then f is analytic.

A different sort of variation is obtained by placing appropriate restrictions on the Jordan curve γ_0. For instance, as observed in [W] and [B], the work of Brown-Schreiber-Taylor [BST] implies the following theorem.

THEOREM 1.7. *Let $f \in C(\mathbb{C})$ and let γ_0 be a generic Jordan curve. If*

$$\int_{\sigma(\gamma_0)} f(z)dz = 0$$

for all $\sigma \in M(2)$, then f is an entire function.

Using the general method outlined in [BZ2], one can obtain a similar result in $\mathbb{D}$, though the proof is significantly harder (cf. [BS], [B]).

THEOREM 1.8. *Let $f \in C(\mathbb{D})$ and let γ_0 be a generic Jordan curve contained in $\mathbb{D}$. If*

$$\int_{\sigma(\gamma_0)} f(z)dz = 0$$

for all $\sigma \in \mathcal{M}$, then f is analytic in $\mathbb{D}$.

REMARKS. Some comments on the different meanings of the word "generic" used in Theorems 1.7 and 1.8 are in order.

(1) The notion of genericity used in Theorem 1.7 consists in the nonexistence of positive eigenvalues α for the following overdetermined Neumann problem:

$$\Delta u + \alpha u = 0 \qquad \text{in } \Omega$$
$$\frac{\partial u}{\partial \vec{n}} = 0 \qquad \text{on } \gamma_0$$
$$u = 1 \qquad \text{on } \gamma_0.$$

Here Ω is the interior of the region bounded by γ_0 and $\vec{n}$ is the unit outward normal. A sufficient condition for Theorem 1.7 to hold is that γ_0 be Lipschitz but not real analytic. For example, Theorem 1.7 holds if γ_0 is a triangle, but **not** if γ_0 is a circle.

(2) The notion of genericity used in Theorem 1.8 is formally the same as in (1) except that the (Euclidean) Laplacian is replaced by the Laplace-Beltrami operator on $\mathbb{D}$ (viewed as the hyperbolic plane). It remains the case that a sufficient condition for Theorem 1.8 to hold is that γ_0 be Lipschitz but not real analytic.

(3) The conditions in Remarks (1) and (2) may be restated as saying that if $\alpha > 0$, then the equation

$$\Delta v + \alpha v = -\chi_\Omega$$

has no solution of compact support.

A different form of Morera's theorem is obtained by changing the hypothesis

$$\int_{\sigma(\gamma_0)} f(z)dz = 0, \qquad \sigma \in \mathcal{M},$$

in Theorem 1.8 to the condition

$$\int_{\gamma_0} f(\sigma(z))dz = 0, \qquad \sigma \in \mathcal{M}.$$

In this case, we have the following result.

THEOREM 1.9. ([BP]) *Let $f \in C(\mathbb{D})$ and let γ_0 be a generic Jordan curve contained in $\mathbb{D}$. If*

$$\int_{\gamma_0} f(\sigma(z))dz = 0$$

for all $\sigma \in \mathcal{M}$, then f is holomorphic in $\mathbb{D}$.

What is the meaning of "generic" in Theorem 1.9? If γ_0 is sufficiently smooth but not real-analytic, then the result is valid. In case γ_0 is a circle centered at $z_0 \neq 0$ (but **not** if $z_0 = 0$!) the conclusion of the theorem also holds. Thus the notion of genericity in Theorem 1.9 is different from those in Theorems 1.7 and 1.8.

Local versions of many of the above theorems, in which f is defined only on some proper subset of the ambient space, may be found in [Sm], [BG1], [BG2], [BGY1], and [BGY2]. For instance, we have the following result.

THEOREM 1.10. ([BG2, Corollaire 11]) *Let $f \in C(\mathbb{D})$ and let $\gamma_0 \subset \{z \in \mathbb{C} : |z| < \frac{1}{2}\}$ be a Jordan curve generic in the sense of Theorem 1.7. Suppose*

$$\int_{\gamma} f(z)dz = 0 \tag{1.5}$$

for all $\gamma \subset \mathbb{D}$ congruent to γ_0 (in the Euclidean sense). Then f is holomorphic in $\mathbb{D}$.

It is easy to see that the constant $\frac{1}{2}$ in Theorem 1.10 is sharp. Indeed, suppose that the Jordan region bounded by γ_0 contains a circle of radius $r > \frac{1}{2}$. Then the disc $\Delta = \{z \in \mathbb{C} : |z| < 2r - 1\}$ contains no point of any curve congruent to γ_0 which lies entirely in $\mathbb{D}$. Thus f may be prescribed arbitrarily inside Δ without influencing the validity of (1.5).

Finally, we wish to point out that the sufficiency of a single Jordan curve for determining analyticity in Theorems 1.7 and 1.10 depends strongly on the fact that both translations and rotations are used. If only translations are allowed, it is no longer the case that a single curve can serve to determine analyticity [BST, Theorem 4.3]. On the other hand, a suitably chosen collection of curves may suffice. For instance, let γ_1 be a square, and γ_2 and γ_3 the squares obtained by rotating γ_1 through $\frac{\pi}{6}$ and $\frac{\pi}{4}$, respectively. If $f \in C(\mathbb{C})$ satisfies

$$\int_{z+\gamma_j} f(\zeta)d\zeta = 0, \qquad j = 1, 2, 3,$$

for all $z \in \mathbb{C}$, then f is entire. For further results in this direction, see [BT] and [BGY1].

2. Boundary analogues of Morera theorem

We now turn to a discussion of boundary Morera theorems for functions of several complex variables. Denote by $\mathbb{H}^n$ the Heisenberg group

$$\mathbb{H}^n \cong \mathbb{C}^n \times \mathbb{R} = \{(z,t) = (z_1, \ldots, z_n, t) : z \in \mathbb{C}^n, \ t \in \mathbb{R}\}$$

with the group law

$$(z,t) \cdot (w,s) = (z+w, t+s+2\operatorname{Im} \sum_{j=1}^n z_j \overline{w_j}).$$

Then $\mathbb{H}^n$ can be identified (topologically) with the boundary of the Siegel upper half space

$$\mathfrak{D}_{n+1} = \{(z_1, \ldots, z_n, z_{n+1}) \in \mathbb{C}^{n+1} : \operatorname{Im} z_{n+1} > \sum_{j=1}^n |z_j|^2\}$$

by means of the homeomorphism

$$\Phi : \mathbb{H}^n \to \partial \mathfrak{D}_{n+1}$$
$$(z_1, \ldots, z_n, t) \mapsto (z_1, \ldots, z_n, t + i|z|^2).$$

We endow $\mathbb{H}^n$ with the CR structure obtained by transporting the natural CR structure of $\partial \mathfrak{D}_{n+1}$ to $\mathbb{H}^n$ via Φ (a good reference for CR structures is Folland-Kohn [FK]). Then the left-invariant vector fields Z_j and $\overline{Z}_j$ on $\mathbb{H}^n$, $j = 1, \ldots, n$, defined by

$$Z_j f(z,t) = \frac{\partial f}{\partial z_j}(z,t) + i\overline{z}_j \frac{\partial f}{\partial t}(z,t),$$
$$\overline{Z}_j f(z,t) = \frac{\partial f}{\partial \overline{z}_j}(z,t) - iz_j \frac{\partial f}{\partial t}(z,t) \qquad (f \in C^1(\mathbb{H}^n)),$$

form a basis of the subbundle $T^{(1,0)} \oplus T^{(0,1)}$ of the complex tangent bundle $\mathbb{C}T\mathbb{H}^n$. It is easy to see that

$$[Z_j, Z_k] = [\overline{Z}_j, \overline{Z}_k] = 0 \qquad \text{and} \qquad [Z_j, \overline{Z}_k] = -i\delta_{jk} \frac{\partial}{\partial t}.$$

Hence $\mathbb{H}^n$ is a nilpotent Lie group of step 2.

A function $f \in C^1(\mathbb{H}^n)$ is a CR function on $\mathbb{H}^n$ if and only if it satisfies $\overline{Z}_j f \equiv 0$, for $j = 1, \ldots, n$.

The importance of CR functions on $\mathbb{H}^n$ is due to the following fact (see Folland-Stein [FS]):

$f \in C^1(\mathbb{H}^n)$ is a CR function on $\mathbb{H}^n$ if and only if there exists a function $F \in C^1(\overline{\mathfrak{D}}_{n+1})$ which is holomorphic in $\mathfrak{D}_{n+1}$, and satisfies $f = F \circ \Phi$.

We now proceed to define the Hardy spaces $H^p(\mathfrak{D}_{n+1})$. The space $H^p(\mathfrak{D}_{n+1})$ is the set of functions F holomorphic in $\mathfrak{D}_{n+1}$ which satisfy

$$\sup_{0<\rho<\infty} \int_{\mathbb{H}^n} |F_\rho(z,t)|^p \, dV(z,t) = \sup_{0<\rho<\infty} \int_{\mathbb{C}^n \times \mathbb{R}} |F(z,t+i|z|^2+i\rho)|^p \, dm(z) dt < \infty$$

for $0 < p < \infty$. Here $dV(z,t)$ is the Haar measure defined on $\mathbb{H}^n$ and dm is Lebesgue measure on $\mathbb{C}^n$. It is well known that if $F \in H^p(\mathfrak{D}_{n+1})$ then the limit

$$\lim_{\mathfrak{A}(p) \ni q \to p} F(q) \equiv f(p)$$

exists for almost every $p = (z, t+i|z|^2) \in \partial\mathfrak{D}_{n+1}$ and the limit also exists in the L^p norm. Here the admissible region $\mathfrak{A}(p)$ is defined by

$$\mathfrak{A}(p) = \{q = (w, s+i|w|^2+i\mu) \in \mathfrak{D}_{n+1} : |p-q| < \sqrt{\mu}, |(p-q) \cdot \vec{n}_p| < \mu\}$$

where $\vec{n}_p$ is the unit outward normal of $\partial\mathfrak{D}_{n+1}$ at p. The image of $H^p(\mathfrak{D}_{n+1})$ under the injection $F \mapsto f$ is a closed subspace $H^p(\mathbb{H}^n)$ of $L^p(\mathbb{H}^n)$. The boundary value f of F satisfies $\overline{Z}_j f = 0$, for $j = 1, \ldots, n$ in the sense of distributions. Hence we may extend the notion of CR function to $H^p(\mathbb{H}^n)$.

Before proceeding further, let us return to the complex plane. Consider the Fourier series of f

$$f(e^{i\theta}) = \sum_{n \in \mathbb{Z}} \widehat{f}(n) e^{in\theta}$$

for $f \in L^2(\partial\mathbb{D})$ and $\theta \in [0, 2\pi]$. It is well-known that f is the boundary value of some holomorphic function

$$F \in H^2(\mathbb{D}) = \{F \text{ holomorphic in } \mathbb{D} : \sup_{0<\rho<1} \int_0^{2\pi} |F(\rho e^{i\theta})|^2 d\theta < \infty\}$$

if and only if

$$\widehat{f}(n) = 0, \qquad \text{for all} \quad n < 0.$$

There is a similar result in $\mathbb{H}^n$. For $f \in L^1(\mathbb{H}^n) \cap L^2(\mathbb{H}^n)$ consider the Fourier transform $\widehat{f}$ with respect to the real variable t defined by

$$\begin{aligned}
\widehat{f}^\lambda(z) &= \int_{-\infty}^\infty f(z, t+i|z|^2) e^{-2\pi i\lambda t} dt \\
&= \int_{-\infty}^\infty f(z, s) e^{-2\pi i\lambda(s-i|z|^2)} ds \\
&= \int_{-\infty}^\infty e^{-2\pi\lambda|z|^2} f(z, s) e^{-2\pi i\lambda s} ds.
\end{aligned}$$

By the Plancherel theorem,

$$\int_{-\infty}^\infty \int_{\mathbb{C}^n} |\widehat{f}^\lambda(z)|^2 dm(z) d\lambda = \int_{-\infty}^\infty \int_{\mathbb{C}^n} e^{-4\pi\lambda|z|^2} |f(z,s)|^2 dm(z) ds.$$

In [G], Gindikin shows that $f \in H^2(\mathbb{H}^n)$ if and only if $\widehat{f}^\lambda \equiv 0$ for $\lambda < 0$ and $\widehat{f}^\lambda \in \mathfrak{F}_n^\lambda$ for $\lambda > 0$. Here $\mathfrak{F}_n^\lambda = \{f : f \text{ is entire on } \mathbb{C}^n \text{ and } \int_{\mathbb{C}^n} |f(z)|^2 e^{-4\pi\lambda|z|^2} dm(z) <$

$\infty\}$ is the Fock space (see Folland [F]). Inspired by this theorem, we have the following result (see [ABCP]).

For every $g \in \mathbb{H}^n$ we consider the *left translation by g*

$$\tau_g(h) = g \cdot h, \qquad \text{for } h \in \mathbb{H}^n.$$

Then if f is a function defined on $\mathbb{H}^n$ we can consider the *left-translation of f by* $g \in \mathbb{H}^n$

$$\tau_g f = f \circ \tau_g.$$

For $k = 1, \ldots, n$, consider the following differential forms on $\mathbb{C}^n$

$$\omega_k = \omega_k(z) = dz_1 \wedge \cdots \wedge dz_n \wedge d\bar{z}_1 \wedge \cdots \wedge \widehat{d\bar{z}_k} \wedge \cdots \wedge d\bar{z}_n.$$

THEOREM 2.1. *Let* $\rho_1, \ldots, \rho_n > 0$ *and let* $f \in L^p(\mathbb{H}^n)$ *where* $1 \leq p \leq 2$. *Then* f *is a CR function on* $\mathbb{H}^n$ *if and only if*

$$\int_{|z|=\rho_k} \tau_g f(z, 0) \, \omega_k(z) = 0 \quad \text{for every } g \in \mathbb{H}^n \text{ and } k = 1, \ldots, n. \qquad (2.1)$$

This theorem is a new criterion for CR functions on the Heisenberg group in the spirit of Morera's Theorem. Note that Theorem 2.1 requires only one radius per variable. We can even assume that $\rho_1 = \cdots = \rho_n = \rho$. On the other hand, the integrability condition on the function f is essential. Indeed, we have

EXAMPLE 2.2. Following [Z1, p. 244], we shall show that, given arbitrary $\rho_1, \ldots, \rho_n > 0$, there exists a function (not in $L^p(\mathbb{H}^n)$) which satisfies (2.1) but is not a CR function.

In fact, let $\alpha > 0$, and consider the function

$$f(z, t) = f(z) = e^{-2\pi i \alpha \operatorname{Re} z_n}, \qquad (z, t) \in \mathbb{H}^n.$$

Clearly, $f \in L^\infty(\mathbb{H}^n)$ but $f \notin L^p(\mathbb{H}^n)$ for $1 \leq p < \infty$. Since f does not depend on t and

$$\frac{\partial f}{\partial \bar{z}_n}(z, t) = -\pi i \alpha f(z) \neq 0, \qquad (z, t) \in \mathbb{H}^n,$$

f is not a CR function on $\mathbb{H}^n$.

On the other hand, if $g = (w, s) \in \mathbb{H}^n$,

$$\tau_g f(z, 0) = f(z + w) = f(w) f(z),$$

so that

$$\int_{|z|=\rho_k} \tau_g f(z, 0) \, \omega_k(z) = f(w) \int_{|z|=\rho_k} f(z) \, \omega_k(z).$$

By Stokes' theorem we have

$$\int_{|z|=\rho_k} f(z) \, \omega_k(z) = c_{n,k} \int_{|z|<\rho_k} \frac{\partial f}{\partial \bar{z}_k}(z) \, dm(z) = -i\pi c_{n,k} \delta_{n,k} \alpha \int_{|z|<\rho_k} f(z) \, dm(z).$$

Now by [SW, p. 155] and [E, p. 333(3)]

$$\int_{|z|<\rho_n} f(z)\,dm(z) = \frac{2\pi}{\alpha^{n-1}} \int_0^{\rho_n} J_{n-1}(2\pi\alpha x)x^n\,dx$$

$$= \frac{1}{(2\pi\alpha^2)^n} \int_0^{2\pi\alpha\rho_n} J_{n-1}(x)x^n\,dx$$

$$= \left(\frac{\rho_n}{\alpha}\right)^n J_n(2\pi\alpha\rho_n).$$

Thus we have only to pick $\alpha > 0$ so that $2\pi\alpha\rho_n$ is a zero of the Bessel function J_n to fulfill the condition (2.1) of Theorem 2.1.

We note that the hypothesis of Theorem 2.1 is in some sense as weak as possible since we integrate over spheres, *i.e.* $2n-1$ dimensional sets. Integral conditions on lower dimensional sets, e.g. circles embedded on $\mathbb{H}^n$, are stronger. Recently, Globevnik and Stout [GS] proved the following theorem.

THEOREM 2.3. *Let Ω be a bounded domain in $\mathbb{C}^n$, $n \geq 2$, with C^2 boundary. Let $1 \leq k \leq n-1$. Assume that $f \in C(\partial\Omega)$ satisfies*

$$\int_{\Lambda\cap\partial\Omega} f\beta = 0$$

for all complex k-planes Λ that intersect $\partial\Omega$ transversely and for all $(k, k-1)$ forms β on $\mathbb{C}^n$ with constant coefficients. Then f is a CR-function.

When Ω is the unit ball $\mathbb{B}_n$ of $\mathbb{C}^n$, we only need to check the hypotheses of Theorem 2.3 for a restricted family of k-planes in order to insure holomorphy.

THEOREM 2.4. *Let $n \geq 2$, $1 \leq k \leq n-1$, and $0 < r < 1$. Assume that $f \in C(\partial\mathbb{B}_n)$ satisfies*

$$\int_{\Lambda\cap\partial\mathbb{B}_n} f\beta = 0 \tag{2.2}$$

for every complex k-plane Λ at distance r from the origin and for every $(k, k-1)$ form β with constant coefficients on $\mathbb{C}^n$. Let E be the set of all r's, $0 < r < 1$, such that $\frac{r^2}{1-r^2}$ is a root of one of the following polynomials

$$\mathfrak{P}_{p,q}(x) = \sum_{\ell=\max\{p+1-q,0\}}^{p} \frac{(-1)^\ell x^\ell}{\ell!(p-\ell)!(\ell+q-p-1)!(n+p-\ell-1)!}, \quad p \geq 1, q \geq 1.$$

Suppose that one of the following conditions holds:

(1) $k < n-1$.
(2) $k = n-1$ and $r \notin E$.

Then f is a CR function.

Moreover, if $r \in E$ then there exists $f \in C(\partial\mathbb{B}_n)$ which is not a CR function but satisfies (2.2) for every $(n-1)$-plane Λ and for every $(n-1, n-2)$ form β with constant coefficients on $\mathbb{C}^n$.

PROOF. Let Y be the subspace of all functions $f \in C(\partial\mathbb{B}_n)$ satisfying (2.2). Let $\mathfrak{H}(p, q)$ be the space of all harmonic homogeneous polynomials of total degree

p in the variables $z_1, \ldots, z_n$ and of total degree q in the variables $\overline{z}_1, \ldots, \overline{z}_n$, for every $p, q \geq 0$. Let $\mathfrak{U}$ be the unitary group on $\mathbb{C}^n$. It is clear that Y is a $\mathfrak{U}$-invariant subspace of $C(\partial\mathbb{B}_n)$. By a result of Nagel and Rudin [NR], every function in Y is a CR function if and only if $\mathfrak{H}(p, q) \nsubseteq Y$ for each $p \geq 0$ and $q \geq 1$. In fact, either $\mathfrak{H}(p, q) \subseteq Y$ or $\mathfrak{H}(p, q) \cap Y = \{0\}$.

Case (1): $k < n-1$. Let us check that the function $f(z) = z_n^p \overline{z}_k^q$ does not belong to Y, for $p \geq 0$, $q \geq 1$. Let

$$\beta = dz_1 \wedge \cdots \wedge dz_k \wedge d\overline{z}_1 \wedge \cdots \wedge d\overline{z}_{k-1}.$$

Consider the k-plane $\Lambda = U(\Lambda_0)$ where

$$\Lambda_0 = \{\, \zeta \in \mathbb{C}^n : \zeta_{k+1} = r, \zeta_j = 0 \ (k+1 < j \leq n)\,\},$$

and $U \in \mathfrak{U}$, which will be selected later in order to assure that the integral in (2.2) does not vanish. Then

$$\int_{\Lambda \cap \partial\mathbb{B}_n} f\beta = \int_{\Lambda_0 \cap \partial\mathbb{B}_n} U^*(f\beta) = \int_{\Lambda_0 \cap \partial\mathbb{B}_n} (f \circ U)U^*\beta$$

$$= \int_{\Lambda_0 \cap \partial\mathbb{B}_n} (U_n(\zeta))^p \, (\overline{U_k(\zeta)})^q dU_1(\zeta) \wedge \cdots \wedge dU_k(\zeta) \wedge d\overline{U_1(\zeta)} \wedge \cdots \wedge d\overline{U_{k-1}(\zeta)}.$$

If $(u_{j,\ell})_{j,\ell=1,\ldots,n}$ is the matrix of U in the canonical basis of $\mathbb{C}^n$, then on Λ_0 we have

$$U_j(\zeta) = u_{j,1}\zeta_1 + \cdots + u_{j,k}\zeta_k + u_{j,k+1}r = V_j(\zeta_1, \ldots, \zeta_k),$$

$$dU_j(\zeta) = u_{j,1}d\zeta_1 + \cdots + u_{j,k}d\zeta_k,$$

$$dU_1(\zeta) \wedge \cdots \wedge dU_k(\zeta) = \Delta \, d\zeta_1 \wedge \cdots \wedge d\zeta_k,$$

$$d\overline{U_1(\zeta)} \wedge \cdots \wedge d\overline{U_{k-1}(\zeta)} = \sum_{\ell=1}^{k} \overline{\Delta}_\ell \, d\overline{\zeta}_1 \wedge \cdots \wedge \widehat{d\overline{\zeta}_\ell} \wedge \cdots \wedge d\overline{\zeta}_k,$$

where Δ is the determinant of the matrix $(u_{j,\ell})_{j,\ell=1,\ldots,k}$, and Δ_ℓ is the minor obtained from the above matrix deleting the last row and the ℓth column. Then by Stokes' theorem,

$$\int_{\Lambda \cap \partial\mathbb{B}_n} f\beta = (-q)\Delta \left[\sum_{\ell=1}^{k} (-1)^{k+\ell} \overline{u}_{k,\ell} \overline{\Delta}_\ell \right] \int_{\Lambda_0 \cap \mathbb{B}_n} (U_n(\zeta))^p \, (\overline{U_k(\zeta)})^{q-1} \omega$$

$$= (-q)|\Delta|^2 \int_{\sqrt{1-r^2}\mathbb{B}_k} (V_n(\zeta))^p \, (\overline{V_k(\zeta)})^{q-1} \omega,$$

where $\omega = d\zeta_1 \wedge \cdots \wedge d\zeta_k \wedge d\overline{\zeta}_1 \wedge \cdots \wedge d\overline{\zeta}_k$.

The multinomial expansion shows that the last integral is $p!(q-1)!$ times

$$\sum \frac{r^{p_{k+1}+q_{k+1}}}{p_1!q_1!\cdots p_{k+1}!q_{k+1}!} u_{n,1}^{p_1} \overline{u}_{k,1}^{q_1} \cdots u_{n,k+1}^{p_{k+1}} \overline{u}_{k,k+1}^{q_{k+1}} \int_{\sqrt{1-r^2}\mathbb{B}_k} \zeta_1^{p_1} \overline{\zeta}_1^{q_1} \cdots \zeta_k^{p_k} \overline{\zeta}_k^{q_k} \omega,$$

$$\tag{2.3}$$

where the sum is over $p_1 + \cdots + p_{k+1} = p$ and $q_1 + \cdots + q_{k+1} = q - 1$. Since the integral in (2.3) vanishes when $(p_1, \ldots, p_k) \neq (q_1, \ldots, q_k)$, (2.3) equals

$$\sum \frac{r^{p_{k+1}+q_{k+1}}(u_{n,1}\overline{u}_{k,1})^{p_1}\cdots(u_{n,k}\overline{u}_{k,k})^{p_k} u_{n,k+1}^{p_{k+1}}\overline{u}_{k,k+1}^{q_{k+1}}}{(p_1!)^2\cdots(p_k!)^2 p_{k+1}! q_{k+1}!} \int_{\sqrt{1-r^2}\mathbb{E}_k} |\zeta_1|^{2p_1}\cdots|\zeta_k|^{2p_k}\omega,$$

$$(2.4)$$

where the sum is over $p_1 + \cdots + p_k = p - p_{k+1} = q - 1 - q_{k+1}$. But

$$\int_{\sqrt{1-r^2}\mathbb{E}_k} |\zeta_1|^{2p_1}\cdots|\zeta_k|^{2p_k}\omega = c_k \left(\sqrt{1-r^2}\right)^{2(k+p_1+\cdots+p_k)} \int_{\mathbb{E}_k} |\zeta_1|^{2p_1}\cdots|\zeta_k|^{2p_k} dm_k$$

$$= c_k \left(\sqrt{1-r^2}\right)^{2(k+p_1+\cdots+p_k)} \frac{p_1!\cdots p_k!}{(k+p-p_{k+1})!}$$

$$= c_k \left(\sqrt{1-r^2}\right)^{2k+p-p_{k+1}+q-1-q_{k+1}} \frac{p_1!\cdots p_k!}{(k+p-p_{k+1})!}.$$

Here dm_k is Lebesgue measure on $\mathbb{C}^k$.

We have shown that $\int_{\Lambda\cap\partial\mathbb{E}_n} f\beta$ equals $-c_k|\Delta|^2 p! q! \left(\sqrt{1-r^2}\right)^{2k+p+q-1}$ times

$$\sum \frac{(u_{n,1}\overline{u}_{k,1})^{p_1}\cdots(u_{n,k}\overline{u}_{k,k})^{p_k} u_{n,k+1}^{p_{k+1}}\overline{u}_{k,k+1}^{q_{k+1}}}{p_1!\cdots p_k! p_{k+1}! q_{k+1}!(k+p-p_{k+1})!} \left(\frac{r}{\sqrt{1-r^2}}\right)^{p_{k+1}+q_{k+1}}. \quad (2.5)$$

We may pick U such that

$$\Delta \neq 0 \quad \text{and} \quad u_{n,1} = \cdots = u_{n,k+1} = u_{k,1} = \cdots = u_{k,k+1} = \lambda \in \mathbb{R}\setminus\{0\}.$$

(It is at this point that we use the assumption $k \neq n - 1$.) Then $\int_{\Lambda\cap\partial\mathbb{E}_n} f\beta$ equals $-c_k|\Delta|^2 p! q! \left(\sqrt{1-r^2}\right)^{2k+p+q-1} \lambda^{p+q-1}$ times

$$\sum \frac{1}{p_1!\cdots p_k! p_{k+1}! q_{k+1}!(k+p-p_{k+1})!} \left(\frac{r}{\sqrt{1-r^2}}\right)^{p_{k+1}+q_{k+1}},$$

which obviously is nonzero.

Case (2): $k = n - 1$. Let $f(z) = z_{n-1}^p \overline{z}_n^q$, for $p \geq 0$ and $q \geq 1$. We will check that $f \in Y$ if and only if $\mathfrak{P}_{p,q}\left(\frac{r^2}{1-r^2}\right) = 0$.

Consider a typical generating $(n-1, n-2)$ form with constant coefficients on $\mathbb{C}^n$:

$$\beta = dz_1 \wedge \cdots \wedge \widehat{dz_i} \wedge \cdots \wedge d\overline{z}_1 \wedge \cdots \wedge \widehat{d\overline{z}_{j_1}} \wedge \cdots \wedge \widehat{d\overline{z}_{j_2}} \wedge \cdots \wedge d\overline{z}_n,$$

with $1 \leq i \leq n$ and $1 \leq j_1 < j_2 \leq n$.

Let $\Lambda_0 = \{\zeta \in \mathbb{C}^n : \zeta_n = r\}$. Then a generic $(n-1)$-plane is $\Lambda = U(\Lambda_0)$, where $U \in \mathfrak{U}$. Computations similar to those above show that

$$\int_{\Lambda\cap\partial\mathbb{E}_n} f\beta = -c_n p! q! (\sqrt{1-r^2})^{2n+p+q-3}\Delta_i \overline{\Delta}_J \times$$

$$\times \sum \frac{(u_{n-1,1}\overline{u}_{n,1})^{p_1}\cdots(u_{n-1,n-1}\overline{u}_{n,n-1})^{p_{n-1}} u_{n-1,n}^{p_n}\overline{u}_{n,n}^{q_n}}{p_1!\cdots p_n! q_n!(n+p-p_n-1)!} \left(\frac{r}{\sqrt{1-r^2}}\right)^{p_n+q_n}$$

where the sum is over $p_1 + \cdots + p_{n-1} = p - p_n = q - 1 - q_n$ and

$$
\Delta_i = \begin{vmatrix} u_{1,1} & \cdots & u_{1,n-1} \\ \cdots & & \\ \widehat{u_{i,1}} & \cdots & \widehat{u_{i,n-1}} \\ \cdots & & \\ u_{n,1} & \cdots & u_{n,n-1} \end{vmatrix}, \qquad \Delta_J = \begin{vmatrix} u_{1,1} & \cdots & u_{1,n-1} \\ \cdots & & \\ \widehat{u_{j_1,1}} & \cdots & \widehat{u_{j_1,n-1}} \\ \cdots & & \\ \widehat{u_{j_2,1}} & \cdots & \widehat{u_{j_2,n-1}} \\ \cdots & & \\ u_{n,1} & \cdots & u_{n,n-1} \\ u_{n,1} & \cdots & u_{n,n-1} \end{vmatrix}, \qquad J = (j_1, j_2).
$$

It is clear that $\Delta_J = 0$ when $j_2 \neq n$. Therefore we have only to consider the case $J = (j, n)$ for $1 \leq j < n$, and then $\Delta_J = \Delta_j$.

Let
$$
\mathfrak{V} = \bigcup_{\substack{1 \leq i \leq n \\ 1 \leq j < n}} \mathfrak{V}_{i,j}
$$
where
$$
\mathfrak{V}_{i,j} = \{ U \in \mathfrak{U} : \Delta_i = \Delta_i(U) \neq 0, \qquad \Delta_j = \Delta_j(U) \neq 0 \}.
$$
It is clear that $\mathfrak{V}$ is a non-empty open set of $\mathfrak{U}$.

Consider the following real-analytic function on $\mathfrak{U}$:

$$
F_r(U) =
$$
$$
\sum \frac{(u_{n-1,1}\overline{u}_{n,1})^{p_1} \cdots (u_{n-1,n-1}\overline{u}_{n,n-1})^{p_{n-1}} u_{n-1,n}^{p_n} \overline{u}_{n,n}^{q_n}}{p_1! \cdots p_n! q_n! (n + p - p_n - 1)!} \left(\frac{r}{\sqrt{1-r^2}} \right)^{p_n + q_n},
$$

where the sum is over $p_1 + \cdots + p_{n-1} = p - p_n = q - 1 - q_n$. Then from the above discussion, $f \in Y$ is equivalent to $F_r \equiv 0$ on $\mathfrak{V}$. Since $\mathfrak{U}$ is connected (see Chevalley [C, page 37]), this last condition means $F_r \equiv 0$ on $\mathfrak{U}$, by analytic continuation.

Now for $U \in \mathfrak{U}$, we have

$$
u_{n-1,1}\overline{u}_{n,1} + \cdots + u_{n-1,n-1}\overline{u}_{n,n-1} = -u_{n-1,n}\overline{u}_{n,n}.
$$

Thus, using the multinomial expansion, we obtain

$$
F_r(U)
$$
$$
= \sum_{p - p_n = q - 1 - q_n} \frac{(-1)^{p - p_n}}{p_n! q_n! (n + p - p_n - 1)!(p - p_n)!} \left(\frac{r}{\sqrt{1-r^2}} \right)^{p_n + q_n} u_{n-1,n}^{p} \overline{u}_{n,n}^{q-1}
$$
$$
= (-1)^p \left(\frac{r}{\sqrt{1-r^2}} \right)^{q-1-p} \mathfrak{P}_{p,q}\left(\frac{r^2}{1-r^2} \right) u_{n-1,n}^{p} \overline{u}_{n,n}^{q-1}.
$$

Finally, since there exists $U \in \mathfrak{U}$ such that $u_{n-1,n} = u_{n,n} \neq 0$, the above identity shows that $F_r \equiv 0$ on $\mathfrak{U}$ if and only if $\mathfrak{P}_{p,q}\left(\frac{r^2}{1-r^2} \right) = 0$. $\square$

For $n = 2$, Theorem 2.4 is a result of Globevnik and Stout [GS, Theorem 2.5.1]. Note that in this case only hyperplanes arise ($n = 2$, $k = 1$). Observe also that the cases $k < n - 1$ and $k = n - 1$ exhibit completely different behavior, in the sense that in the first case there are no exceptional r's, while in the second case such exceptional r's do appear. In fact, $E \neq \varnothing$, since it is easy to check that $\mathfrak{P}_{p,q}$ has at least one positive root, for p odd and $p + 1 \leq q$. The simplest example is $p = 1$ and $q = 2$; then the corresponding value of r is $\frac{1}{\sqrt{n+1}}$.

Let us show that when $r = 0$, the conclusion of Theorem 2.4 **never** holds.

EXAMPLE 2.5. Let $f(z) = z_j^p \bar{z}_\ell$, $1 \leq j, \ell \leq n$, $p \geq 1$. It is clear that f is not a CR function on $\partial \mathbb{B}_n$, but it satisfies condition (2.2) for every k-plane Λ passing through the origin and every $(k, k - 1)$ form β with constant coefficients on $\mathbb{C}^n$. In fact, computations similar to those in the proof of Theorem 2.4, show that the integral in (2.2) is a constant times

$$\int_{\mathbb{B}_k} (u_{j,1}\zeta_1 + \cdots + u_{j,k}\zeta_k)^p \, \omega,$$

which obviously vanishes.

The theorems of Nagel-Rudin and Rudin (cf. [R, pp. 253-287]) concerning the classification of Möbius invariant and unitary invariant subalgebras of $C(\overline{\mathbb{B}}_n)$ and $C(\partial \mathbb{B}_n)$ generalize the work of Agranovsky and Val'skii [AV] and lead to some Morera-type characterizations of holomorphic functions. They have been further extended by Stout [St], Agranovsky [A], and Grinberg [Gr1,Gr2], to smooth domains.

Let us consider the relation between Theorems 2.1 and 2.4. Recall that the generalized Cayley transform is defined as follows

$$w_n = \frac{z_n + 1}{i(z_n - 1)}, \qquad w_k = \frac{z_k}{i(z_n - 1)}, \qquad k = 1, \ldots, n - 1.$$

This transformation maps the unit ball in $\mathbb{C}^n$ biholomorphically to the Siegel upper half space $\mathfrak{D}_n$ and the boundary of $\mathbb{B}_n$ to the boundary of $\mathfrak{D}_n$. The images of those spheres in the hypothesis of Theorem 2.4 are spheres or complex hyperplanes in $\partial \mathfrak{D}_n$. Under the homeomorphism Φ, the images are spheres of different radii and complex hyperplanes in the Heisenberg group $\mathbb{H}^n$. In Theorem 2.1, on the other hand, we consider the sphere $\{z \in \mathbb{C}^n : |z| = \rho\}$ translated by the Heisenberg group. Basically, we are working on generalized ellipsoids in $\partial \mathfrak{D}_n$. It would be very interesting to understand the connection between these two theorems.

We conclude with some related results. Let $\mathfrak{B}_n$ be the open unit ball in $\mathbb{R}^n$ and $S_x(r)$ the sphere of radius r centered at x.

THEOREM 2.6. ([BG1, Corollary 14]) *Let* $f \in C(\mathfrak{B}_3)$ *and suppose that* $0 < r_1, r_2$, $r_1 \neq r_2$, *and* $r_1 + r_2 < 1$. *If*

$$\int_{S_x(r_1)} f(y) d\sigma(y) = f(x),$$

for every x such that $|x| < 1 - r_1$ *and*

$$\int_{S_x(r_2)} f(y) d\sigma(y) = f(x),$$

for every x such that $|x| < 1 - r_2$, then f is harmonic in $\mathfrak{B}_3$.

A corresponding result holds in $\mathbb{R}^n$. One of the features of this theorem is that the centers x are never near the boundary $\partial \mathfrak{B}_3$. In fact, if $r_1 + r_2 < \frac{1}{2}$, it is easy to see that x does not even need to lie near the origin. It is natural to ask whether we can restrict the possible centers even further. The following result of Globevnik and Rudin [GR] does precisely that. Denote by Γ a C^∞ surface bounding a compact, strictly-convex set K such that $K \subset \mathfrak{B}_n$ and $0 \in \overset{o}{K}$. Let $d\omega_\Gamma$ represent the harmonic measure of Γ with respect to $x = 0$.

THEOREM 2.7. *A function $f \in C(\mathfrak{B}_n)$ is harmonic in $\mathfrak{B}_n$ if and only if for every C^∞ strictly-convex Γ as above we have*

$$\int_\Gamma f d\omega_\Gamma = f(0).$$

For further results along these lines see [Gl1], [Gl2].

References.

[A] M. Agranovsky, *Tests for holomorphy in symmetric regions*, Siberian Math. J. **22** (1981), 171-179.

[ABCP] M. Agranovsky, C. Berenstein, D. C. Chang & D. Pascuas, *A Morera type theorem for L^2 functions in the Heisenberg group*, J. Analyse Math. **56** (1991), in press.

[AV] M. Agranovsky & R. E. Val'skii, *Maximality of invariant algebras of functions*, Sibirsk. Mat. Zh. **12** (1971), 3-12.

[B] C. Berenstein, *A test for holomorphy in the unit ball of $\mathbb{C}^n$*, Proc. Amer. Math. Soc. **90** (1984), 88-90.

[BG1] C. Berenstein & R. Gay, *A local version of the two-circles theorem*, Israel J. Math. **55** (1986), 267-288.

[BG2] _____, *Le problème de Pompeiu local*, J. Analyse Math. **52** (1988), 133-166.

[BGY1] C. Berenstein, R. Gay & A. Yger, *The three-square theorem, a local version*, Analysis and Partial Differential Equations (C. Sadosky ed.), Marcel Dekker (1990), pp. 35-50.

[BGY2] _____, *Inversion of the local Pompeiu transform*, J. Analyse Math. **54** (1990), 259-287.

[BP] C. Berenstein & D. Pascuas, *Morera and mean value type theorems in the hyperbolic disc*, (in preparation) (1991).

[BS] C. Berenstein & M. Shahshahani, *Harmonic analysis and the Pompeiu problem*, Amer. J. Math. **105** (1983), 1217-1229.

[BST] L. Brown, B. Schreiber & B. A. Taylor, *Spectral synthesis and the Pompeiu problem*, Ann. Inst. Fourier **23** (1973), 125-154.

[BT] C. Berenstein & B. A. Taylor, *The "three squares" theorem for continuous functions*, Arch. Rational Mech. Anal. **63** (1977), 253-259.

[BZ1] C. Berenstein & L. Zalcman, *Pompeiu's problem on spaces of constant curvature*, J. Analyse Math. **30** (1976), 113-130.

[BZ2] _____, *Pompeiu's problem on symmetric spaces*, Comment. Math. Helv. **55** (1980), 593-621.

[C] C. Chevalley, *Theory of Lie Groups*, Princeton University Press, Princeton, New Jersey, 1946.

[E] A. Erdélyi, W.Magnus, F. Oberhettinger & F.G. Tricomi, *Tables of Integral Transforms,*, vol. 2, McGraw-Hill, New York, 1954.

[F] G. B. Folland, *Harmonic Analysis in Phase Space*, Ann. of Math. Studies #122, Princeton University Press, Princeton, New Jersey, 1989.

[FK] G.B. Folland & J.J. Kohn, *The Neumann Problem for the Cauchy- Riemann Complex*, Ann. of Math. Studies #75, Princeton University Press, Princeton, New Jersey, 1972.

[FS] G.B. Folland & E.M. Stein, *Estimates for the $\overline{\partial}_b$ complex and analysis on the Heisenberg group*, Comm. Pure Appl. Math. **27** (1974), 429–522.

[G] S. Gindikin, *Analysis in homogeneous domains*, Russian Math. Surveys **19** (1964), 1-89.

[Gl1] J. Globevnik, *Integrals over circles passing through the origin and a characterization of analytic functions*, J. Analyse Math. **52** (1989), 199-209.

[Gl2] _____, *Zero integrals on circles and characterizations of harmonic and analytic functions*, Trans. Amer. Math. Soc. **317** (1990), 313-330.

[Gr1] E. Grinberg, *A boundary analogue of Morera's theorem in the unit ball of $\mathbb{C}^n$*, Proc. Amer. Math. Soc. **102** (1988), 114-116.

[Gr2] _____, *Boundary values of holomorphic functions and the one dimensional extension property*, preprint.

[GR] J. Globevnik & W. Rudin, *A characterization of harmonic functions*, Indag. Math. **91** (1988), 419-426.

[GS] J. Globevnik & E. L. Stout, *Boundary Morera theorems for holomorphic functions of several complex variables*, Univ. of Ljubljana, Dept. of Math. Preprint series 322 **28** (1990).

[NR] A. Nagel & W. Rudin, *Möbius-invariant function spaces on the balls and spheres*, Duke Math. J. **43** (1976), 841-865.

[OV] R. Ogden & S. Vági, *Harmonic analysis on a nilpotent group and function theory on Siegel domains of type 2*, Adv. Math. **33** (1979), 31-92.

[R] W. Rudin, *Function Theory in the Unit Ball of $\mathbb{C}^n$*, Springer-Verlag, Berlin, New York, 1980.

[Sm] J. D. Smith, *Harmonic analysis of scalar and vector fields in $\mathbb{R}^n$*, Proc. Camb. Phil. Soc. **72** (1972), 403-416.

[St] E. L. Stout, *The boundary values of holomorphic functions of several complex variables*, Duke Math. J. **44** (1977), 105-108.

[SW] E. M. Stein and G. Weiss, *Fourier Analysis on Euclidean Spaces*, Princeton University Press, Princeton, New Jersey, 1971.

[W] S. Williams, *Analyticity of the boundary of Lipschitz domains without the Pompeiu property*, Indiana Univ. Math. J. **30** (1981), 357-369.

[Z1] L. Zalcman, *Analyticity and the Pompeiu problem*, Arch. Rational Mech. Anal. **47** (1972), 237-254.

[Z2] _____, *Mean values and differential equations*, Israel J. Math. **14** (1973), 339-352.

[Z3] _____, *Offbeat integral geometry*, Amer. Math. Monthly **87** (1980), 161-175.

SYSTEM RESEARCH CENTER AND DEPARTMENT OF MATHEMATICS, UNIVERSITY OF MARYLAND, COLLEGE PARK, MARYLAND 20742 USA

DEPARTMENT OF MATHEMATICS, UNIVERSITY OF MARYLAND, COLLEGE PARK, MARYLAND 20742 USA

DEPARTMENT DE MATEMÀTICA APLICADA I ANÀLISI, FACCULTAT DE MATEMÀTIQUES, UNIVERSTAT DE BARCELONA, 08071 BARCELONA SPAIN

DEPARTMENT OF MATHEMATICS, BAR-ILAN UNIVERSITY, 52 900 RAMAT GAN, ISRAEL

Contemporary Mathematics
Volume **137**, 1992

A remark on local continuous extension of proper holomorphic mappings

FRANÇOIS BERTELOOT

Dedicated to Walter Rudin.

1. Introduction.

The continuous extension of proper holomorphic mappings between bounded domains in $\mathbb{C}^N$ ($N > 1$) has been studied by many authors. The first result is due to Henkin [9]:

Let $f : D_1 \to D_2$ be a proper holomorphic map between bounded domains in $\mathbb{C}^N$. If D_1 admits a global plurisubharmonic defining function and if D_2 has a C^2 strictly pseudoconvex boundary then f extends to a Hölder continuous map $\tilde{f} : \bar{D}_1 \to \bar{D}_2$ with exponent $1/2$.

This was generalised by Bedford-Fornaess and Diederich-Fornaess ([2], [6]) to the case where D_1 is bounded pseudoconvex with C^2 boundary and D_2 is bounded pseudoconvex with real analytic boundary ; the map is shown to be Hölder continuous for some exponent α, depending on the domains. (See [7] and [11] for other references).

The proofs of these theorems are based on precise estimates of some kind of infinitesimal metric in the image domain D_2. More recently, Forstneric and Rosay ([8]) proved a sharp localization principle for the Kobayashi metric and obtained the following local version of Henkin's theorem :

Let $f : D_1 \to D_2$ be a proper holomorphic map of a domain $D_1 \subset \mathbb{C}^N$ onto a bounded domain $D_2 \subset \mathbb{C}^N$. If a point z_0 in the boundary of D_1 satisfies condition (P) and if the cluster set $C(f, z_0)$ contains a point w_0 at which the boundary of D_2 is C^2 strictly pseudoconvex, then f extends on a neighborhood of z_0 in $\bar{D}_0$ to a Hölder continuous map with exponent $1/2$. (See § 2, for a definition of condition (P)).

1991 *Mathematics Subject Classification.* Primary 32H35.
Key words and phrases. Proper holomorphic mappings, Hölder continuity, Finite type.
This paper is in final form and no version of it will be submitted for publication elsewhere.

This theorem is also true if D_2 admits "good" local holomorphic peak functions near w_0 (see [8], lemma 2.2).

The aim of this note is to observe that such local extension phenomenons occur if the boundary of D_2 satisfies a simple attraction property for analytic disks in a neighborhood of w_0, even when D_2 is unbounded. The proof is elementary and, since the attraction property directly provides estimates of $\|f'\|$, doesn't require any consideration of infinitesimal metric.

By previous results of Berteloot and Cœuré ([4],[5]), the attraction property holds if $\bar{D}_2$ is contained in a taut domain and if its boundary satisfies certain conditions in a neighborhood of w_0. This provides the following :

THEOREM. *Let* $f : D_1 \rightarrow D_2$ *be a proper holomorphic map of a domain* $D_1 \subset \mathbb{C}^N$ *onto a domain* $D_2 \subset \mathbb{C}^N$. *Assume that* $\bar{D}_2$ *admits a taut neighborhood. Let* z_0 *be a boundary point of* D_1 *which satisfies condition (P) and* w_0 *be a point in* $C(f, z_0) \cap bD_2$. *Then* f *extends on a neighborhood of* z_0 *in* $\bar{D}_1$ *to a Hölder continuous map with exponent* α *in the following cases :*

1) bD_2 *is strictly pseudoconvex at* w_0, $\alpha = \frac{1}{2}$;

2) $D_2 \subset \mathbb{C}^2$ *and* bD_2 *is pseudoconvex and finite type* $2k$ *near* w_0, $\alpha = \frac{1}{2k}$;

3) $D_2 \subset \mathbb{C}^N$, $N > 2$ *and* bD_2 *is convex and finite type* $2k$ *near* w_0, $\alpha = \frac{1}{2k} - 0$.

REMARK. *When* D_2 *is assumed to be bounded, 1) and 3) follow directly from the theorem of Forstneric and Rosay ([8]) since "good" local holomorphic peak functions exist. In case 2), we can use the same result and the peak functions constructed by E. Bedford and J.E. Fornaess ([3]), but this doesn't give a precise value for the exponent* α.

When D_2 *is not bounded, our result seems to be new, even in cases 1) and 3), since the proof of the localization principle for the Kobayashi metric given in [8] requires the boundedness of* D_2.

2. Notations and results.

If D is a domain in $\mathbb{C}^N$, bD is the boundary of D. We denote by $d(\bullet, bD)$ the euclidean distance to the boundary.

If $f : D_1 \rightarrow \mathbb{C}^N$ is a continuous map on a domain $D_1 \subset \mathbb{C}^N$ and $z_0 \in bD_1$ is a boundary point of D_1, we denote by $C(f, z_0)$ the cluster set of f at z_0 :

$$C(f, z_0) = \{w \in \mathbb{C}^N s.t : w = \lim f(z_j), z_j \in D_1 \text{ and } \lim z_j = z_0\} .$$

If $A(x)$ and $B(x)$ are depending on a variable x, $A(x) \lesssim B(x)$ means that there is a constant K, $0 < K < +\infty$, such that $A(x) \leq KB(x)$ for all x.

We shall use the following **Condition (P)** which was introduced in [8] : A point $z_0 \in bD_1$ satisfies Condition (P) if the boundary is of class $\mathcal{C}^{1+\epsilon}$ near z_0 for some $\epsilon > 0$ and if there exist a continuous negative plurisubharmonic (p.s.h.) function ρ on D_1 and a neighborhood U of z_0 such that $\rho(z) \gtrsim -d(z, bD_1)$, $z \in U \cap D_1$.

Next we introduce the **attraction property** for analytic disks. We say that bD_2 satisfies the **attraction property of order** α $(0 < \alpha < 1)$ near $w_0 \in bD_2$ if there exist a function $C : [0,1[\to [0, +\infty[$ and a neighborhood W of w_0 such that the following estimate occurs for any analytic disk $g : \Delta \to D_2$ and any $\eta \in bD_2 \cap W$:

$$|u| \le r < 1 \Rightarrow |g(u) - \eta| \le C(r)|g(0) - \eta|^\alpha .$$

(Here Δ is the open unit disk in $\mathbb{C}$).

We are now in order to stay our key result :

PROPOSITION. *Let $f : D_1 \to D_2$ be a proper holomorphic map of a domain $D_1 \subset \mathbb{C}^N$ onto a domain $D_2 \subset \mathbb{C}^N$. Let z_0 be a boundary point of D_1 which satisfies condition (P) and w_0 be a point in $C(f, z_0) \cap bD_2$. If bD_2 is of class $\mathcal{C}^2$ and satisfies the attraction property of order α near w_0, then f extends to a Hölder continuous map with exponent α on a neighborhood of z_0 in $\bar{D}_1$.*

Let D_2 be a domain in $\mathbb{C}^2$ (resp. $\mathbb{C}^N$, $N > 2$) and $w_0 \in bD_2$. Assume that $\bar{D}_2$ admits a taut neighborhood. It follows from [5] theorem 1 (resp. [4], prop. 3.1) that bD_2 satisfies the attraction property of order $\frac{1}{2k}$ (resp. $\frac{1}{2k} - o$) near $w_0 \in bD_2$ if bD_2 is pseudoconvex (resp. convex) and finite type $2k$ in a neighborhood of w_0 . Using methods of [5], it is not hard to see that bD_2 satisfies the attraction property of order $\frac{1}{2}$ if bD_2 is strictly pseudoconvex near w_0. This allows us to deduce the theorem from the proposition.

3. Proof of the proposition.

Let ρ be the continuous negative p.s.h function on D_1 and U the neighborhood of z_0 which are given by condition (P).

Since f is proper and holomorphic, $\tau(w) = \sup\{\rho(z); z \in D, f(z) = w\}$ is a negative continuous p.s.h. function on D_2. (See [1]). Let V be a small neighborhood of w_0 such that $V \cap bD_2$ is of class $\mathcal{C}^2$. By the Hopf lemma, $\tau(w) \lesssim -d(w, bD_2)$ for all $w \in D_2 \cap V$ and therefore :

(1) $\qquad \forall z \in D_1 : f(z) \in V \Rightarrow \rho(z) \lesssim -d(f(z), bD_2).$

On the other hand :

(2) $\qquad \forall z \in U \cap D_1 : \rho(z) \gtrsim -d(z, bD_1) .$

By combining (1) et (2), we get :

(3) $\qquad \forall z \in U \cap D_1 : f(z) \in V \Rightarrow d(f(z), bD_2) \lesssim d(z, bD_1) .$

Let $z \in D_1 \cap U$. Assume that $f(z) \in V$ and consider the analytic disk $g : \Delta \to D_2$ defined by $g(u) = f(z + u\epsilon\vec{x})$, where $\epsilon = d(z, bD_1)$ and $\vec{x}$ is a fixed unit vector in $\mathbb{C}^N$.

Let $\eta \in bD_2$ such that $|\eta - f(z)| = d(f(z), bD_2)$. By (3), we have :

(4) $\qquad |\eta - f(z)| \lesssim \epsilon .$

We may assume that $V \subset\subset W$ where W is neighborhood of w_0 given by the attraction property. It follows from (4) that $\eta \in W$ if U is small enough and therefore :

$$(5) \qquad\qquad |u| \leq \frac{1}{2} \Rightarrow |g(u) - \eta| \lesssim \epsilon^\alpha .$$

Since $\alpha > 1$, we deduce from (4) and (5) that

$$(6) \qquad\qquad |u| \leq \frac{1}{2} \Rightarrow |g(u) - g(0)| \lesssim \epsilon^\alpha .$$

hence by Cauchy's inequality :

$$(7) \qquad\qquad\qquad |g'(0)| \lesssim \epsilon^\alpha .$$

Let us denote by f'_z the holomorphic tangent map of f. Since $f'_z \cdot \vec{x} = \frac{1}{\epsilon} g'(0)$, we have proved the following crucial estimate :

$$(8) \qquad z \in U \text{ and } f(z) \in V \Rightarrow \|f'_z \cdot \vec{x}\| \lesssim d(z, bD_1)^{\alpha-1} \|\vec{x}\|, \forall \vec{x} \in \mathbb{C}^N .$$

Since this estimate is not automatically satisfied for $z \in U$, we have to use a slight modification of the Henkin's technique in order to end the proof. Suppose that g does not extend continuously to z_0 : there are an open ball $B \subset V$ (with center at w_0) and a neighborhoods basis U_j of z_0 such that $f(D_1 \cap U_j)$ is connected (bD_1 is $\mathcal{C}^{1+\epsilon}$ near z_0) and never contained in B. Then, since $w_0 \in C(f, z_0)$, there is a sequence $(z'_j)_j$, $z'_j \in U_j$, such that $f(z'_j) \in bB$ and $\lim f(z'_j) = w'_0 \in bB \cap bD_2$.

Let $(z_j)_j$ be a sequence such that $z_j \in D_1$, $\lim z_j = z_0$ and $\lim f(z_j) = w_0$.

Let $l_j = |z_j - z'_j|$ and $\gamma_j : [0, 3l_j] \to D_1 \cap U$ be a $\mathcal{C}'$-path such that :

i) $\gamma_j(0) = z_j$ and $\gamma_j(3l_j) = z'_j$.

ii) $d(\gamma_j(t), bD_1) \geq t$ on $[0, l_j]$; $d(\gamma_j(t), bD_1) \geq l_j$ on $[l_j, 2l_j]$; $d(\gamma_j(t), bD_1) \geq 3l_j - t$ on $[2l_j, 3l_j]$.

iii) $\|\frac{d\gamma_j}{dt}(t)\| \lesssim 1, t \in [0, 3l_j]$.

(See [10], page 203, prop. 2).

Choose $t_j \in [0, 3l_j]$ such that $f \circ \gamma_j([0, t_j]) \subset \bar{B}$ and $f \circ \gamma_j(t_j) \in bB$. It follows from (8), ii) and iii) that $|f(z_j) - f \circ \gamma_j(t_j)| \lesssim l_j^\alpha$: a contradiction. Hence f extends continuously to z_0.

We may now assume that $f(U) \subset V$ and apply the technique above for proving that f is Hölder continuous with exponent α on $U \cap \bar{D}_1$. $\qquad\qquad\square$

References.

1. H. Alexander, *Proper holomorphic mappings in* $\mathbb{C}^n$, Indiana Univ. Math. J. **26** (1977), 137-147.

2. E. Bedford and J.E. Fornaess, *Biholomorphic maps of weakly pseudoconvex domains*, Duke Math. J. **45** (1979), 711-749.

3. _____, *A construction of peak functions on weakly pseudoconvex domains*, Annals of Mathematics **107** (1978), 555-568.

4. F. Berteloot, *Hölder continuity of proper holomorphic mappings*, Studia Mathematica **100** no. 3 (1991), 229-235.

5. F. Berteloot and G. Cœuré, *Domaines de $\mathbb{C}^2$, pseudoconvexes et de type fini ayant un groupe non compact d'automorphismes*, Ann. Inst. Fourier **41** no. 1 (1991), 77-86.

6. K. Diederich and J.E. Fornaess, *Proper holomorphic maps onto pseudoconvex domains with real analytic boundary*, Ann. Math. **II**, Série 110, 575–592 (1979).

7. F. Forstneric, *Proper holomorphic mappings : a survey*, Preprint.

8. F. Forstneric and J.P. Rosay, *Localization of the Kobayashi metric and the boundary continuity of proper holomorphic mappings*, Math. Ann. **279** (1987), 239-252.

9. G.M. Henkin, *An analytic polyhedron is not biholomorphicaly equivalent to a strictly pseudoconvex domain*, Math. USSR., Dkl. 14, 858–862 (1973).

10. G.M. Henkin and J. Leiterer, *Theory of functions on complex manifolds*, Monographs in Mathematics, Vol. 79, Birkhäuser Verlag, 1984.

11. S.I. Pincuk, *Holomorphic maps in $\mathbb{C}^n$ and the problem of holomorphic equivalence*, Encyclopædia of Math. Sc., Several Complex Variable, 1989, pp. 173-201.

UNIVERSITÉ DE LILLE FLANDRES ARTOIS, UFR DE MATHÉMATIQUES, URA CNRS D 751, 59655 VILLENEUVE D'ASCQ (FRANCE)

Contemporary Mathematics
Volume **137**, 1992

A Multivariable Version of
the Müntz-Szasz Theorem

THOMAS BLOOM

1. Introduction.

1.1   Let $0 < b_0 < b_1 < \cdots$. The Müntz-Szasz theorem gives necessary and sufficient conditions for the linear span of the function 1 and the monomials $t^{b_0}, t^{b_1}, \ldots$ to be dense in the space of continuous functions on the interval $I = [0, 1]$.

The condition is

$$(1.1.1) \qquad\qquad \sum_{j=0}^{\infty} \frac{1}{b_j} = +\infty.$$

The Müntz-Szasz theorem may be deduced from the Blashke condition on zeroes of bounded analytic functions in a half-plane. An excellent exposition and proof of these questions may be found in Rudin's book [R].

1.2   In several variables the corresponding problem would be: Given $B_0, B_1, \ldots$ a sequence of points in $(\mathbb{R}^+)^n$, find conditions for which the linear span of the monomials $t^{B_0}, t^{B_1}, \ldots$ is dense in $\mathcal{C}_0(I^n)$ where $\mathcal{C}_0(I^n)$ denotes the continuous functions on the cube $I^n = \{y \in \mathbb{R}^n | 0 \le y_j \le 1 \text{ for } j = 1, \ldots, n\}$ which vanish on the union of the coordinates axes. Interesting sufficient conditions have been found by Korevaar-Hellerstein [K-H], Ronkin [Ro] and Berndtsson [B]. However finding necessary and sufficient conditions seems very difficult.

1.3   A different point of view was adopted in [Bl]. The family of monomials $t^{B_0}, t^{B_1}, \ldots$ is canonically enlarged to a family of functions $\mathcal{F}(B)$ (see §4) in $\mathcal{C}_0(I^n)$. The idea is to consider the function of two sets of variable $t^z = (t_1^{z_1}) \ldots (t_n^{z_n})$. The monomials are "values" of t^z at $B_0, B_1, \ldots$ as a function of the z-variables. A canonical multivariable polynomial interpolation procedure (see §2) is used to interpolate these "values". The resulting functions are, in general, rational functions in the coordinates $(t_1, \ldots, t_n)$ and their logarithms. The

1991 *Mathematics Subject Classification*. Primary 32A10.
Supported by a NSERC of Canada grant OGP0007535.
This paper is in final form and no version of it will be submitted for publication elsewhere.

following problem is considered as a multivariable generalization of the Müntz-Szasz theorem: Is the linear span of $\mathcal{F}(B)$ dense in $\mathcal{C}_0(I^n)$? For $n = 1$ the linear span of $\mathcal{F}(B)$ is trivially the same as the original monomials (see [Bl]) so the question is answered positively, in this case, by the Müntz-Szasz theorem. For $n > 1$ it was conjectured in [Bl] (assuming $B_0, B_1, \ldots$ satisfy condition 3.1.2) and proved in [Y] that the necessary and sufficient condition that the linear span of $\mathcal{F}(B)$ be dense in $\mathcal{C}_0(I^n)$ is

$$(1.3.1) \qquad\qquad \sum_{j=0}^{\infty} \frac{1}{|B_j|} = +\infty.$$

This result is given as theorem 4.2 of this paper.

In this paper we will prove a uniqueness theorem for bounded analytic functions on a product of half-planes (theorems 3.2 and 3.3). Analogous to the one variable case this theorem may be used to deduce the multi-variable version of the Müntz-Szasz theorem (see §4). Theorems 3.2 and 3.3 give a positive response to conjecture 4.7 of [Bl].

2. Polynomial Interpolation

2.1   Let g be holomorphic on an open set $V \subset \mathbb{C}$ and let $P_0, \ldots, P_m$ be points of V. We will denote the Lagrange-Hermite polynomial interpolant to g at $P_0, \ldots, P_m$ by $L_m(P_0, \ldots, P_m; g)(z)$. For $P_0, \ldots, P_m$ distinct it is the unique polynomial of degree $\leq m$ which takes the same values as g at $P_0, \ldots, P_m$.

In several variables, there is, in general, no unique polynomial interpolant. There is, however, a canonical one due to Kergin [Ke], [M.-M.]. We recall some of its properties. (For more details see [A.-P.1], [Bl]).

2.2   Let $P_0, \ldots, P_m$ be points in an open convex set $V \subset \mathbb{C}^n$. There is a natural projection $\chi_m : \mathcal{O}(V) \to \mathcal{P}^m(V)$ where $\mathcal{O}(V)$ denotes the holomorphic functions on V and $\mathcal{P}^m(V)$ the (analytic) polynomials of total degree $\leq m$.

The projection satisfies:

$$(2.2.1) \qquad\qquad \chi_m(f)(P_j) = f(P_j) \quad \text{for} \quad j = 0, \ldots, m$$

(2.2.2)   If $P_{j_0}, \ldots, P_{j_s}$ are $(s+1)$ elements $\subset \{P_0, \ldots, P_m\}$ and χ_j denotes the Kergin interpolant for $P_{j_0}, \ldots, P_{j_s}$, then $\chi_j \chi_m = \chi_j$.

(2.2.3)   If $f = g \circ h$ where $h : \mathbb{C}^m \to \mathbb{C}$ is linear and g is analytic on $h(V)$ then

$$\chi_m(f) = L_m(h(P_0), \ldots, h(P_m); g)(h(z))$$

where L_m is the one-variable Lagrange-Hermite interpolant to g at $h(P_0), \ldots, h(P_m)$.

2.3 Remark   Kergin interpolants, for analytic functions can be defined on more general open sets than convex ones. The natural domains for Kergin interpolants are the $\mathbb{C}$-convex ones (see [A.-P.1], [A.-P.2], [A.-P.-S.]).

3. Bounded Analytic Functions on Products of Half Planes

3.1   Let Q denote the quadrant $(\mathbb{R}^+)^n = \{y \in \mathbb{R}^n | y_j > 0 \text{ for } j = 1, \dots, n\}$ and let T_Q denote the tube over Q i.e.

$$(3.1.1) \qquad T_Q = \{z \in \mathbb{C}^n | Im(z) \in Q\} = \mathbb{R}^n + iQ$$

Q is a self-dual cone ([S.-W.], [C.-M.]).

Let $B_0, B_1, \dots$ be a sequence of points in $(\mathbb{R}^+)^n$ and $iB_0, iB_1, \dots$ the corresponding points with purely imaginary coordinates in T_Q. We assume that for some $\epsilon > 0$ we have

$$(3.1.2) \qquad B_{jk} \geq \epsilon \quad \text{for all} \quad j = 0, 1, 2, \cdots \quad \text{and} \quad k = 1, \dots, n.$$

For f holomorphic on T_Q and m an integer ≥ 0, we denote by $\chi_m(f)$ the Kergin interpolant to f at $iB_0, iB_1, \dots, iB_m$. We have:

3.2 THEOREM. *Let f be bounded and holomorphic on T_Q. Suppose* $\sum\limits_{j=0}^{\infty} \frac{1}{|B_j|} = +\infty$ *and* $\chi_m(f) \equiv 0$ *for* $m = 0, 1, \dots$. *Then* $f \equiv 0$.

3.3 THEOREM. *Suppose* $\sum\limits_{j=0}^{\infty} \frac{1}{|B_j|} = < +\infty$. *Then there exists a function f, analytic, bounded on T_Q and not identically zero such that $\chi_m(f) \equiv 0$ for $m = 0, 1, \dots$.*

3.4 Remark: In the case of one variable and all B_j distinct, $\chi_m(f)$ is just the Lagrange interpolant at $iB_0, \dots, iB_m$. Hence $\chi_m(f) = 0$ for all m, if and only if, $f(iB_j) = 0$ for all j. Theorems 3.2 and 3.3 give the classical Blashke conditions on the zeroes, on the imaginary axis, of a bounded analytic of the upper half-plane.

In the n-variable case, $\chi_m(f) \equiv 0$ implies that $f(iB_j) = 0$ for $j = 0, \dots, m$ but also $\binom{m+n}{n} - (m+1)$ other linear functionals (not given by point evaluation) annihilate f.

3.5 PROOF OF THEOREM 3.3. Consider the map $\tau : \mathbb{C}^n \to \mathbb{C}$ given by

$$(3.5.1) \qquad \tau(z_1, \dots, z_n) = (z_1 + \cdots + z_n).$$

We have $\tau(T_Q) = H^+ = \{z | Im(z) > 0\}$ and the points $\tau(iB_j)$ lie on the positive imaginary axis. Furthermore the condition

$$\sum_{j=0}^{\infty} \frac{1}{|B_j|} < +\infty \Rightarrow \sum_{j=0}^{\infty} \frac{1}{|\tau(iB_j)|} < +\infty.$$

Now, by the classical one variable result, there exists a bounded analytic function g on H^+, not identically zero such that $g(\tau(iB_j)) = 0$ for all j. Hence using 2.2.3, we see that $f = g \circ \tau$ has the required properties.

3.6 PROOF OF THEOREM 3.2. We will consider analytic functions on T_Q of the form

$$(3.6.1) \qquad f(z) = \langle S_{(t)}, e^{2\pi i(z \cdot t)} \rangle$$

where $(z \cdot t) = z_1 t_1 + \cdots + z_n t_n$, $S_{(t)}$ is a tempered distribution in the t-variables with supp$(S_{(t)}) \subset \overline{Q}$. The pairing $\langle \ , \ \rangle$ is that of a tempered distribution in the t-variables and a rapidly decreasing function in the Schwartz space (in the t-variables).

Any bounded analytic function on T_Q may be represented in the form (3.6.1) [C.-M., Theorem 6.6.1]. We will, in fact, prove Theorem 3.2 for functions of the form (3.6.1). The proof will be based on 3.7, 3.8 and 3.9. It follows from 3.7, 3.8 and 3.9 that if $\sum_{j=0}^{\infty} \frac{1}{|B_j|} = +\infty$, that $\chi_m(f)(z)$ converges to $f(z)$ for all $z = iy$ for y in an open subset E of Q. Hence if $\chi_m(f) \equiv 0$ for all m, we deduce that $f(z) = 0$ for all $z = iy$, $y \in E$. Since f is analytic we may conclude that $f \equiv 0$. This concludes the proof of 3.2.

3.7 LEMMA. *Let f be holomorphic on T_Q of the form (3.6.1). Then*

$$\chi_m(f) = \langle S_{(t)}, L_m \left((iB_0 \cdot t), \ldots (iB_m \cdot t); \exp 2\pi i \right) (z \cdot t) \rangle.$$

PROOF. (see [Bl, prop. 4.4])   For any multi-index α,

$$(3.7.1) \qquad \frac{\partial^\alpha f}{\partial z^\alpha} = \langle S_{(t)}, \frac{\partial^\alpha}{\partial z^\alpha} \exp(2\pi i(z \cdot t)) \rangle.$$

There is an integral formula for the difference of a function and its interpolant [M.-M.]

$$f(z) - \chi_m(f)$$

$$= \int_{\Delta^{m+1}} d^{m+1} f(v_{m+1} z + \sum_{j=0}^{m} iv_j B_j) dv_1 \ldots dv_{m+1} (z - iB_0, \ldots, z - iB_m)$$

where d^{m+1} is the total $(m+1)^{st}$ derivative and Δ^{m+1} is the unit simplex in $\mathbb{R}^{m+1}$. Applying this formula to $\exp(2\pi i(z \cdot t))$, using (3.7.1), (2.2.3) and the fact that integration over a compact set in T_Q in the z-variables commutes with pairing with $S_{(t)}$, the result follows.

3.8 COROLLARY. *Let $f(z)$ be of the form (3.6.1). Then, (for z fixed) $\lim_{m \to \infty} \chi_m(f)(z) = f(z)$ if, for all multi-indices α, β*

$$\lim_{m \to \infty} \sup_{t \in Q} \left\{ t^\beta \frac{\partial}{\partial t^\alpha} \left(\exp(2\pi i(z \cdot t)) - L_m((iB_0 \cdot t), \ldots, (iB_m \cdot t); \exp 2\pi i)(z \cdot t) \right) \right\}$$

$$= 0.$$

PROOF. supp$(S_{(t)}) = \overline{Q}$, so using the characterization of convergence in the Schwartz space, it follows that

$$(3.8.1) \qquad \lim_{m \to \infty} \langle S_{(t)}, L_m((iB_0 \cdot t), \ldots, (iB_m \cdot t); \exp 2\pi i)(z \cdot t) \rangle = f(z)$$

Then applying lemma 3.7 to the left hand side of (3.8.1) completes the proof.

3.9 LEMMA. *Let $B_0, B_1, \ldots$ be a sequence of points in $(\mathbb{R}^+)^n$ satisfying (3.1.2). There is an open subset E of Q such that for all $z = iy$, $y \in E$ and all multi-indices α, β there are constants $C_i = C_i(\alpha, \beta, y) > 0$ for $i = 1, 2$ independent of m such that*

$$\sup_{t \in Q} t^\beta \frac{\partial}{\partial t^\alpha} \{\exp(2\pi i)(iy \cdot t)) - L_m((iB_0 \cdot t), \ldots, (iB_m \cdot t); \exp 2\pi i)(iy \cdot t))\}$$

$$\leq C_1 \exp\left(-C_2 \left(\sum_{j=0}^m \frac{1}{|B_j|}\right)\right).$$

PROOF. First we will do the case $\alpha = 0$. Using the Hermite remainder formula for the difference of a function and its Lagrange-Hermite interpolant [D] we have
(3.9.1)
$$\exp(2\pi i \xi) - L_m\left((iB_0 \cdot t), \ldots, (iB_m \cdot t); \exp(2\pi i)\right)(\xi)$$

$$= (\xi - (iB_0 \cdot t)) \cdots (\xi - (iB_m \cdot t)) \int_Y \frac{e^{2\pi i w} \, dw}{(w - (iB_0 \cdot t)) \cdots (w - (iB_m \cdot))(w - \xi)}$$

where Y is a contour in the w-plane that surrounds $(iB_0 \cdot t), \ldots, (iB_m \cdot t), \xi$. Now, we set

(3.9.2) $\quad \rho = |t|, \quad t = \rho u, \quad w = \rho w_1, \quad y = \rho \lambda, \quad \xi = (iy \cdot t) = i\rho(\lambda \cdot u).$

Here ρ is a positive real, $u \in (\mathbb{R}^+)^n$ and $|u| = 1, \lambda \in (\mathbb{R}^+)^n$. We obtain for the expression (3.9.1)

(3.9.3) $\quad i^{m+1}((y - B_0) \cdot u) \cdots ((y - B_m) \cdot u)$

$$\int_{Y_1} \frac{e^{2\pi i \rho w_1} \, dw_1}{(w_1 - (iB_0 \cdot u)) \cdots (w_1 - (iB_m \cdot u))(w_1 - (iy \cdot u))}$$

where Y_1 is the image of Y under $w = \rho w_1$ and Y_1 surrounds the points $(iB_0 \cdot u), \ldots, (iB_m \cdot u), (iy \cdot u)$ which all lie on the positive imaginary axis.

Let Y_1 be the contour in the w_1 plane consisting of the straight line from $-R + i\epsilon_1$ to $R + i\epsilon_1$ and then the semicircle of radius R in the upper half plane from $R + i\epsilon_1$ to $-R + i\epsilon_1$. Here R, ϵ_1 are positive numbers to be specified.

Now on substituting $w_1 = w_2 + i\epsilon_1$ into

(3.9.4) $\quad \int_{Y_1} \frac{e^{2\pi i w_1} \, dw_1}{(w_1 - (iB_0 \cdot u)) \cdots (w_1 - (iB_m \cdot u))(w_1 - (iy \cdot u))}$

we obtain

(3.9.5) $\quad e^{-2\pi \rho \epsilon_1} \int_{Y_2} \frac{e^{2\pi i w_2} \, dw_2}{(w_2 - i((B_0 \cdot u) - \epsilon_1)) \cdots (w_2 - i((B_m \cdot u) - \epsilon_1))(w_2 - i((y \cdot u) - \epsilon_1))}$

where Y_2 is the semicircle Y_1 translated so that its diameter coincides with the real axis.

Using (3.1.2) we have $B_j \cdot u \geq \epsilon$ for all j. Let
(3.9.6)
$$E = \left\{ y \in (\mathbb{R}^+)^n \mid |y| < \epsilon \text{ and } y_j > \eta \text{ for } j = 1, \ldots, n \quad \text{where } 0 < \eta < \frac{\epsilon}{\sqrt{n}} \right\}.$$

Then $y \cdot u > \eta$ for all $y \in E$. Take $\epsilon_1 < \eta$. Then estimating the integral (3.9.5) as in [Y] we have that the expression (3.9.5) is bounded above by

(3.9.7)
$$\frac{e^{-2\pi\rho\epsilon_1}C}{(\eta - \epsilon_1)((B_0 \cdot u) - \epsilon_1) \ldots ((B_m \cdot u) - \epsilon_1)}$$

where the constant C depends only on $|B_0|$, $|B_1|$. It follows that (3.9.2) is bounded in absolute value by

(3.9.8)
$$C e^{-2\pi\rho\epsilon_1} \left| \prod_{j=0}^{m} \frac{((y - B_j) \cdot u)}{((B_j \cdot u) - \epsilon_1)} \right|.$$

Since, for $y \in E$, $|y| < \epsilon$ it follows that $y \cdot u < B_j \cdot u$ for $j = 0, \ldots, m$ and all u. Thus (3.9.8) is equal to

(3.9.9)
$$C e^{-2\pi\rho\epsilon_1} \prod_{j=0}^{m} \left(1 - \left(\frac{((y \cdot u) - \epsilon_1)}{((B_j \cdot u) - \epsilon_1)} \right) \right)$$

which is

$$\leq C e^{-2\pi\rho\epsilon_1} \prod_{j=0}^{m} \left(1 - \frac{(\eta - \epsilon_1)}{|B_j|} \right).$$

Since for any multi-index β, $t^\beta \leq \rho^{|\beta|}$, the lemma follows in the case $\alpha = 0$.

Next, we consider the case $\alpha \neq 0$.

Substituting $\xi = iy \cdot t$ in (3.9.1) we obtain

(3.9.10) $\exp(-2\pi(y \cdot t)) - L_m((iB_0 \cdot t), \ldots, (iB_m \cdot t); \exp(2\pi i)((iy \cdot t)))$
$$= i^{m+1}((y - B_0) \cdot t) \cdots ((y - B_m) \cdot t)$$
$$\times \int_Y \frac{e^{2\pi i w} dw}{(w - (iB_0 \cdot t)) \cdots (w - (iB_m \cdot t))(w - (iy \cdot t))}$$

We apply $\frac{\partial}{\partial t^\alpha}$ to the above expression. We assume $m \gg |\alpha|$, It will be a sum of terms of the form

(3.9.11) $\prod_{j \in F_1} (y - B_j)_{s_j} \prod_{\substack{j \notin F_1 \\ j=0}}^{m} (y - B_j \cdot t)$

$$\times \int_Y \frac{e^{2\pi i w} dw \prod_{j \in F_2} (iB_j)_{s_j}}{\prod_{j=0}^{m}(w - (iB_j \cdot t)) \prod_{j \in F_2}(w - (iB_j \cdot t)(w - (iy \cdot t))^{1+p}}$$

where s_j indicates a component of a vector, p is an integer $0 \leq p \leq |\alpha|$, F_1 and F_2 are subsets of $\{0, \ldots, m\}$ and $p + \text{card}(F_1) + \text{card}(F_2) = |\alpha|$. Since the number of terms in the sum is independent of m and each term may be estimated in the same manner as the case $\alpha = 0$ (the expression (3.9.11) differs from (3.9.3) by finitely many factors, the number of factors being independent of m) the lemma is proved in general.

4. Application to the Müntz-Szasz theorem.

4.1   Let $B_0, B_1, \ldots$ be a sequence in $(\mathbb{R}^+)^n$. The family of functions $\mathcal{F}(B)$ is defined as follows (see [Bl] or [Y]). $\mathcal{F}(B)$ will be a countable union of subfamilies

$$(4.1.1) \qquad \mathcal{F}(B) = \bigcup_{s=0}^{\infty} \mathcal{F}_s(B)$$

$\mathcal{F}_0(B)$ consists of the monomials $t^{B_0}, t^{B_1}, \ldots$. The functions in $\mathcal{F}_s(B)$ may be described as follows: Each collection of $(s + 1)$ elements of the sequence $B_0, B_1, \ldots$ say $D_0, \ldots, D_s$ together with an n-multi index ν such that $|\nu| = s$ determine a function in $\mathcal{F}_s(B)$ given by

$$(4.1.2) \qquad (\log t)^\nu \sum_{k=0}^{s} \frac{t^{D_k}}{\prod_{\substack{j \neq k \\ j=0}}^{s} ((D_j - D_k) \cdot \log t)}.$$

Here $\log t = (\log t_1, \ldots, \log t_n)$. Formula (4.1.2) makes sense even in the case there are repeated elements in $D_0, \ldots, D_s$ and (4.1.2) defines a function $\mathcal{C}_0(I^n)$.

4.2 THEOREM. [Y]. *Let $B_0, B_1, \ldots$ be a sequence in $(\mathbb{R}^+)^n$ satisfying (3.1.2). The linear span of $\mathcal{F}(B)$ is dense in $\mathcal{C}_0(I^n)$ if and only if $\sum_{j=0}^{\infty} \frac{1}{|B_j|} = +\infty$.*

PROOF. We will show that $\sum_{j=0}^{\infty} \frac{1}{|B_j|} = +\infty$ implies that the linear span of $\mathcal{F}(B)$ is dense. (For the converse see ([Bl], [Y]).) The proof will be by contradiction. Suppose the linear span of $\mathcal{F}(B)$ is not dense in $\mathcal{C}_0(I^n)$. Let $d\mu$ be a positive, non zero, finite Borel measure on I^n such that

$$(4.2.1) \qquad \int_{I^n} f d\mu = 0 \qquad \text{for all} \quad f \in \mathcal{C}_0(I^n).$$

We may assume $d\mu$ assigns no mass to the union for the coordinate axes. Now let

$$(4.2.2) \qquad F(z) = \int_{I^n} t^{-iz} d\mu(t).$$

Then F is analytic, bounded on T_Q and

$$(4.2.3) \qquad \chi_m(F) = \int_{I^n} L_m((iB_0 \cdot t), \ldots, (iB_n \cdot t); \exp)(-iz \cdot \log t) d\mu(t)$$

Since $L_m((iB_0 \cdot t), \ldots, (iB_m \cdot t); \exp)(-iz \cdot \log t)$ is a finite linear combination of elements of $\mathcal{F}(B)$ (see [Bl]) it follows that $\chi_m(F) \equiv 0$ for all m. Hence, by theorem 3.2 we have $F \equiv 0$ and so we may deduce ([Bl],[R]) that $d\mu \equiv 0$. This contradiction gives the result.

4.3 Remark: A proof of the one variable Müntz-Szasz theorem via polynomial interpolation was given by Feller [F].

References.

[A.-P.1] M. Anderson and M. Passare, *Complex Kergin interpolation*, J. of Approx. Theory **64** no. No. 2 (1991), 214–225.

[A.-P.2] ———, *Complex Kergin interpolation and the Fantappié transform*, Preprint.

[A.-P.-S.] M. Anderson, M. Passare and R. Sigursson, *Analytic Functionals and $\mathbb{C}$-convexity*, Preprint.

[Be] B. Berndtsson, *Zeros of analytic functions of several variables*, Arkiv für Matematik **16** no. No. 2 (1978), 251–262.

[Bl] T. Bloom, *A spanning set for $\mathcal{C}(I^n)$*, Tr. Am. Math. Soc. **321** no. No. 2 (1990), 741–759.

[C.-M.] R. Carmichael and D. Mitrović, *Distributions and Analytic Functions*, Pitman Research Notes # 206, Longman Scientific, Essex, 1989.

[D] P.J. Davis, *Interpolation and Approximation*, Dover, 1974.

[F] W. Feller, *on Müntz' theorem and completely monotone functions*, Am. Math. Monthly **75** (1968), 342–349.

[Ke] P. Kergin, *A natural interpolation of C^k functions*, J. of Approx. Theory **29** (1980), 278–293.

[K.-H.] J. Korevaar and S. Hellerstein, *Discrete sets of uniqueness for bounded holomorphic functions*, Proc. Symp. Pure Math. XI, Am. Math. Soc., Providence, R.I., 1968, pp. 273–284..

[M.-M.] C. Micchelli and P. Milman, *A formula for Kergin interpolation in $\mathbb{R}^k$*, J. of Approx. Theory **29** (1980), 294–296.

[Ro] L.I. Ronkin, *Discrete sets of uniqueness for entire functions of several variables of exponential type*, Soviet Math. Dokl. **15** (1974), 1415–1419.

[Ru] W. Rudin, *Real and Complex Analysis*, McGraw-Hill, 1974.

[S.-W.] E. Stein and G. Weiss, *Introduction to Fourier Analysis on Euclidean Spaces*, Princeton, 1971.

[Y] X. Yang, *Une generalisation à plusieurs variables du théorème de Müntz-Szasz*, C.R. Acad. Sci. Paris **312**, série I (1991) 575–578.

DEPARTMENT OF MATHEMATICS, UNIVERSITY OF TORONTO, TORONTO, ONTARIO, M56 1A1, CANADA

Contemporary Mathematics
Volume **137**, 1992

ON A^∞ INTERPOLATING
SETS LYING ON CURVES

JOAQUIM BRUNA AND JOAQUIM M. ORTEGA

Dedicated to Walter Rudin.

1. Introduction

Let B^n be the unit ball in $\mathbb{C}^n$ and $S = bB^n$ its boundary. We will consider smooth simple curves $\gamma : [0, L] \to S$, parametrized by arc-length i.e. $|\dot{\gamma}(t)| = 1$, for all t. The image of γ will be also denoted by γ; thus, γ is either a simple arc or a simple closed curve, and functions on γ will be identified with functions of $t \in [0, L]$ (L-periodic in case γ is closed).

For a closed subset E of γ we consider two notions of interpolation. We call E an interpolation set for the ball algebra $A(B^n)$, shortly an *A-interpolation set* if for any continuous function φ on γ there is $f \in A(B^n)$ such that $f = \varphi$ on E; we call E an interpolation set of infinite order for $A^\infty(B^n)$, or simply an A^∞-*interpolating set*, if for any $\varphi \in C^\infty(\gamma)$ there is $f \in A^\infty(B^n)$ such that

$$\frac{d^k}{dt^k} f(\gamma(t)) = \frac{d^k}{dt^k} \varphi(t), \text{ all } k, \ \gamma(t) \in E.$$

In dimension $n = 1$, these two kind of sets are respectively characterized by the conditions $|E| = 0$ (Rudin-Carleson theorem) and

(ATW) $$\int_I \log \frac{1}{d_E(t)} \, dt \le C|I| \left(1 + \log \frac{1}{|I|} \right), \text{ all arcs } I \subset T$$

where $d_E(t)$ denotes the distance from e^{it} to E (see [**1**]).

In dimension $n > 1$ there are two extreme classes of curves, according to the position of the tangent to the curve with respect to the distribution of complex-tangent spaces $T_z^C(S) = \{v \in \mathbb{C}^n; \ v \cdot \bar{z} = 0\}$. The curve γ is called *transverse* if $\dot{\gamma}(t) \notin T_{\gamma(t)}^C(S) \ \forall t$, i.e., $\dot{\gamma}(t) \cdot \overline{\gamma(t)} \ne 0 \ \forall t$, and is called *complex-tangential* if $\dot{\gamma}(t) \cdot \overline{\gamma(t)} \equiv 0$. The most typical example of a transverse curve is the boundary

1991 *Mathematics Subject Classification.* Primary 32A40.

Partially supported by CICYT grant n. PB89–0311.

This paper is in final form and no version of it will be submitted for publication elsewhere.

of an analytic disk lying inside B^n. Although not all transverse curves are such boundaries, it is a well-known principle, in fact, that transverse curves behave in many aspects as if they were, and the analysis of each concrete problem becomes similar to the one variable case. For example, any bounded holomorphic function in H^∞ has radial limits a.e. on γ, if γ is transverse ([13], [14]).

On the other hand, complex-tangential curves are very good from the point of view of interpolation, as they are interpolation sets for both $A(B^n)$ and $A^\infty(B^n)$ (see [5], [9], [12], [15] for $A(B^n)$ and [6], [7], [11] for $A^\infty(B^n)$). In fact, something somewhat more general holds: if G is the set of all the complex-tangential points of γ

$$G = \{\gamma(t); \dot\gamma(t) \cdot \overline{\gamma(t)} = 0\}$$

then G is A and A^∞-interpolating. Hence, concerning interpolation problems the transverse case is the worst one and it is natural to expect that the condition on a set E to be interpolating should depend only on $E \backslash G$. Moreover the one-variable condition on $E \backslash G$ should be always sufficient. This is so for the ball algebra: a closed set $E \subset \gamma$ is A-interpolating if and only if $E \backslash G$ has zero length (this follows from results in [10] and [12]).

The purpose of this paper is to prove the corresponding result for $A^\infty(B^n)$:

THEOREM. *If E is a closed subset of γ such that*

(ATW) $$\int_I \log \frac{1}{d_E(\gamma(t))}\, dt \le C|I|\left(1 + \log \frac{1}{|I|}\right) \quad \text{all arcs } I \subset \gamma$$

then $E \cup G$ is A^∞-interpolating.

For the case of a transverse curve, the result had been proved in [4] and [8].

An essential part of the proof consists in constructing holomorphic functions $g(z)$ such that for some $p \in N$,

$$d_\gamma(z)^p \lesssim |g(z)| \lesssim d_E(z).$$

Looking for g in the form $g = \exp(u + i\tilde u)$ one is lead to find pluriharmonic functions u with a prescribed size near γ. The construction of such "outer functions" depends on the pluriharmonic extension kernel introduced in [2].

An auxiliary result on A^∞-zero sets is also proved in section 3.

2. Notations and technical lemmas

Without loss of generality we will assume that γ is closed. Since $\gamma(t) \cdot \bar\gamma(t)$ is pure imaginary we can define *the index of transversality* by

$$\gamma(t) \cdot \overline{\dot\gamma(t)} = iT(t)$$

the transverse points being those for which $T(t) \neq 0$. Then $\gamma(t) \cdot \overline{\ddot\gamma(t)} = i\dot T(t) - 1$, and by Taylor's formula

$$1 - \overline{\gamma(x)}\gamma(t) = -iT(t)(x - t) + \frac{1}{2}(1 - i\dot T(t))(x - t)^2 + O(x - t)^3$$

Hence

$$\mathrm{Re}\ (1 - \overline{\gamma(x)}\gamma(t)) \simeq |x - t|^2$$
$$|1 - \overline{\gamma(x)}\gamma(t)| \simeq |T(x)||t - x| + |t - x|^2$$

For each $z \in B^n$, let $r(z)$ the non-isotropic distance from z to γ

$$r(z) = \min\{|1 - \overline{\gamma(t)} \cdot z|,\ \text{all } t\}$$

and let $s = s(z)$ be a point where this distance is attained.

As $|1 - \bar{\zeta}z|^{1/2}$ satisfies the triangle inequality, it follows that

$$(1) \qquad |1 - \overline{\gamma(t)}z| \simeq r(z) + |T(s)||t - s| + |t - s|^2$$

In the following we will write $d(s, t)$ instead of $|1 - \gamma(s)\overline{\gamma(t)}|$. We use $d_E(z)$ to denote the non-isotropic distance from z to E, and write $d_E(t)$ instead of $d_E(\gamma(t))$. Without loss of generality, we assume $d_E \leq 1$. For $z \in B^n$, we denote by I_z the interval

$$I_z = \{t : d(t, s(z)) \leq r(z)\}$$

so that $|1 - \overline{\gamma(t)} \cdot z| \simeq r(z)$ for $t \in I_z$, by (1).

The estimates that we will obtain depend on whether the point z is close or not to a complex-tangential point of the curve. More precisely, we define the set D as follows

$$D = \{z \in B^n : |T(s(z))| > Cr(z)^{1/2}\}$$

where the constant C will be chosen later, sufficiently big. The set D is a neighbourhood of the transverse points of γ, collapsing at the complex-tangential points. The estimates of several quantities will depend or whether $z \in D$ or not.

LEMMA 2.1. *For $z \in D$ and $t \in I_z$*

$$|T(t)| \simeq |T(s(z))|$$
$$\mathrm{Re}\frac{1}{1 - z\overline{\gamma(t)}} \simeq \frac{1}{r(z)}$$

PROOF. We write s for $s(z)$. Since $|T(t) - T(s)| = O(|s - t|) = O(d(t, s)^{1/2})$, the first estimate is clear choosing the constant C in the definition of D sufficiently big. To prove the second one, note that by the definition of s, the function $|1 - \overline{\gamma(t)}z|^2$ has zero-derivative at s and hence

$$-[\mathrm{Im}\ \bar{z} \cdot \dot{\gamma}(s)][\mathrm{Im}\ z \cdot \overline{\gamma(s)}] = [\mathrm{Re}\ \bar{z} \cdot \dot{\gamma}(s)][\mathrm{Re}\ (1 - z \cdot \overline{\gamma(s)})]$$

Now, choosing C big enough

$$|\mathrm{Im}\ \bar{z} \cdot \dot{\gamma}(s)| \geq |T(s)| - O(|z - \gamma(s)|) = |T(s)| - O(r(z)^{1/2}) \geq\ \text{const}\ |T(s)|$$

and $\mathrm{Re}\ \bar{z} \cdot \dot{\gamma}(s) = O(|z - \gamma(s)|) = O(r(z)^{1/2})$, so that

$$|\mathrm{Im}\ z \cdot \overline{\gamma(s)}| = O(r(z)^{3/2}/|T(s)|)$$

and the estimate follows if C is big enough. $\square$

3. Zeros of infinite order of $A^\infty(B^n)$

In this section we will prove a theorem on zero-sets for $A^\infty(B^n)$, auxiliary in the main result. We recall that $E \subset S$ is called a *zero set* (of infinite order) for $A^\infty(B^n)$ if there exists $g \in A^\infty(B^n)$ such that $g^{-1}(0) = E$ and all derivatives of g vanish on E.

THEOREM 3.1. *If E is a closed subset of γ such that*

$$(2) \qquad \int |T(t)| \log d_E(t) \, dt > -\infty$$

then $E \cup G$ is a zero set for A^∞.

PROOF. We adapt the one-variable construction used in [16]. The first step is to construct a function $\varphi(t)$ with the following properties:

(a) $\int |T(t)| \varphi(t) \, dt < +\infty$, $\varphi(t) \geq 0$ $t \in [0, L]$.

(b) φ is C^∞ out of E, and for each k there exist C_k and $p_k \in N$ such that $|\varphi^{(k)}(t)| \leq C_k d_E(t)^{-p_k}$.

(c) for all $c > 0$, $\varphi(t) + c \log d_E(t) \to +\infty$ as $d_E(t) \to 0$.

Next, one defines

$$\Psi(z) = \int_0^L \frac{|T(t)| \varphi(t)}{1 - z\overline{\gamma(t)}} \, dt, \, z \in \overline{B^n} \backslash \gamma,$$
$$F(z) = \exp -\Psi(z).$$

Note that Ψ has positive real part and that Ψ and all its derivatives have a continuous extension in the neighbourhood of each point not belonging to $E \cup G$ (this is because in this neighbourhood $T(t)$ does not vanish and φ is C^∞ so that an integration by parts can be done indefinitely, see details below). Hence $F \in C^\infty(\bar{B}^n \backslash (E \cup G))$, $|F| \leq 1$. We look now at the behaviour of F and its derivatives in the neighbourhood of $E \cup G$. A derivative of Ψ has the form

$$D^\alpha \Psi(z) = \int_0^L \frac{|T(t)| \varphi(t) X_\alpha(t, z)}{(1 - \overline{\gamma(t)} \cdot z)^{|\alpha|+1}} \, dt$$

for some smooth function $X_\alpha(t, z)$; from this the estimate

$$|D^\alpha \Psi(z)| = O(r(z)^{-|\alpha|-1}), \, z \in \bar{B}^n \backslash \gamma$$

trivially follows.

Now we will see that these estimates can be improved for points z in the region D of Lemma 2.1. For $z \in D$, by Lemma 2.1

$$\mathrm{Re} \ \Psi(z) \geq \int_{I_z} |T(t)| \varphi(t) \ \mathrm{Re} \ \frac{1}{1 - z\overline{\gamma(t)}} \, dt \geq \ \mathrm{const} \ \frac{|T(s)|}{r(z)} \int_{I_z} \varphi(t) \, dt.$$

For $t \in I_z$, $d_E(t) \leq \ \mathrm{const} \ d_E(z)$, and by property (c) of φ, for all $c > 0$

$$\mathrm{Re} \ \Psi(z) \geq \ \mathrm{const} \ |I_z| \frac{|T(s)|}{r(z)} \left\{ c \log \frac{1}{d_E(z)} + \eta(d_E(z)) \right\},$$

with $\eta(x) \rightarrow +\infty$ as $x \rightarrow 0$. Since $|I_z| \simeq r(z)/|T(s)|$ for $z \in D$, we conclude that

$$\text{Re } \Psi(z) + c \log d_E(z) \longrightarrow +\infty \text{ as } d_E(z) \longrightarrow 0, \, z \in D$$

with arbitrary big c, hence $|F(z)| = 0(d_E(z)^N)$ for all N in the region D.

To improve the estimate for the derivatives of Ψ at points $z \in D$ let $m(z) = \min\{|T(s)|^2, d_E(z)\}$, which is $\gtrsim r(z)$. Then in D we claim that

$$|D^\alpha \Psi(z)| = O(m(z)^{-q_\alpha})$$

for some $q_\alpha \in N$. It is clear that the contribution of the arc $\{t : d(t,s) \geq \varepsilon m(z)\}$ in the integral defining $D^\alpha \Psi(z)$ satisfies this. For $d(t,s) \leq \varepsilon m(z)$, choosing ε small enough $d_E(t) \simeq d_E(z)$, and by property (b) of φ, it is enough to prove that if $\omega(t,z)$ is a smooth function such that

$$\left| \frac{\partial^k \omega}{\partial t^k}(t,z) \right| = O(m(z)^{-q_k}) \text{ for all } k,$$

then

$$\int_{d(t,s) \leq \varepsilon m(z)} \frac{\omega(t,z)}{(1 - \overline{\gamma(t)}z)^{l+1}} \, dt = O(m(z)^{-q_l}) \text{ for all } l.$$

But for small enough ε, $|\dot{\overline{\gamma}}(t) \cdot z| > \text{const } |T(s)|$ and the estimate is obtained by repeated integration by parts.

In conclusion, the function F satisfies $|F(z)| \leq 1$, $|D^\alpha F(z)| = O(r(z)^{-q_\alpha})$ everywhere and $|F(z)| = O(d_E(z)^N)$, $|D^\alpha F(z)| = O(d_E(z)^N m(z)^{-q_\alpha}) \, \forall N$, in D.

The last step is to multiply F by a convenient function which is flat at G. Let $h(z)$ be the function introduced by A. Nagel in [12]

$$(3) \qquad h(z) = \int_0^L \frac{dt}{(1 - \overline{\gamma(t)}z)^q}$$

with a fixed q, $1/2 < q < 1$. Then $h(z)$ takes values in a sector and

$$(4) \qquad |h(z)| \simeq \text{Re } h(z) \simeq \int_0^L \frac{dt}{|1 - \overline{\gamma(t)}z|^q} \simeq (r(z) + T^2(s))^{\frac{1}{2}-q},$$

$$(5) \qquad |D^\alpha h(z)| = O((r(z) + T^2(s))^{\frac{1}{2}-q-|\alpha|})$$

using (1). Then $H = e^{-h}$ satisfies

$$|D^\alpha H(z)| = O((r(z) + T^2(s))^N) \text{ for all } N, \text{ all } \alpha.$$

Let now $g = FH$. Then in D,

$$|D^\alpha g(z)| = O(T^2(s)^N d_E(z)^N m(z)^{-q_\alpha}) = O(m(z)^N)$$
$$= O(d_{E \cup G}(z)^N) \text{ for all } N,$$

and in the complement of D, for all N

$$|D^\alpha g(z)| = O(r(z)^N r(z)^{-q_\alpha}) = O(r(z)^N) = O(d_{E \cup G}(z)^N)$$

which finishes the proof. □

Remark 1. When the curve is transverse, i.e. $|T(t)|$ is bounded below, the condition (2) is necessary and sufficient as in the one-variable case, as it is to be expected (see [8]). In general, one can obtain necessary conditions of the type

$$\int \log d_E(t)\omega(|T(t)|)\,dt > -\infty$$

with $\omega(x) = o(x)$ as $x \to 0$ which is very far from (2). It is an open question to find the sharp necessary condition for a general curve.

Remark 2. Let $N_\varepsilon(E)$ be the minimum number of (non-isotropric) balls of radious ε needed to cover E, and let E_ε be the set of points at distance $\leq \varepsilon$ from E. Then

$$\int |T(t)| \log \frac{1}{d_E(t)}\,dt = \int_0^1 \left\{ \int_{E_\varepsilon} |T(t)|\,dt \right\} \frac{d\varepsilon}{\varepsilon} \leq \int_0^1 N_\varepsilon(E)\,d\varepsilon$$

because if I is the trace on γ of a ball of radious ε, $\int_I |T(t)|\,dt = O(\varepsilon)$. In [8] it is proved that the finiteness of the last integral above is a sufficient condition on a *general set* for A^∞-zero sets.

Remark 3. From the proof it follows that the functions $L_\eta(z) = g(z)^\eta$ satisfy

$$D^\alpha L_\eta(z) = 0 \,\forall\alpha,\, z \in E \cup G$$
$$|L_\eta(z)| \leq 1 \text{ and } L_\eta(z) \to 0 \; z \in B^n \text{ as } \eta \to 0$$

(6)
$$|D_\alpha L_\eta(z)| = O(r(z)^{\eta N - q_\alpha}) \text{ outside } D$$
$$|D_\alpha L_\eta(z)| = O(m(z)^{\eta N - q_\alpha}) \text{ in } D$$

for all α, N, uniformly in η. We will need this fact in the proof of the main theorem, in the next section.

4. A^∞-interpolation sets

In this section we prove the theorem stated in the introduction. We need the pluriharmonic kernel introduced in [2] which we now recall. Let $K(t, z)$ be defined by

$$K(t, z) = \frac{1}{2\pi} i \operatorname{sign} T(t) \frac{(1 + i\dot{T}(t))z \cdot \overline{\dot\gamma(t)} + (1 - i\dot{T}(t))\gamma(t) \cdot \overline{\dot\gamma(t)}}{z\overline{\dot\gamma(t)} - 1}.$$

Note that when γ is a slice, $\gamma(t) = e^{it}\zeta$, one gets

$$K = \frac{1}{2\pi} \frac{1 + z \cdot \overline{\gamma(t)}}{1 - z \cdot \overline{\gamma(t)}}$$

i.e. the usual Herzglotz kernel for the slice. In general, $K(t, z)$ has not positive real part.

The following lemma is easily established:

LEMMA 4.1. *The (ATW) condition implies*

$$\int_{d(s,t)\leq l} |T(t)| \log \frac{1}{d_E(t)}\, dt \leq \text{ const } l\left(\log \frac{1}{l} + 1\right).$$

LEMMA 4.2. *If E is a closed set satisfying the condition (ATW) the holomorphic function*

$$f(z) = \exp \int_0^L K(t,z) \log d_E(t)\, dt$$

is of class C^∞ in $\bar{B}^n \setminus E$ and satisfies

(a) $d_E(z)^{C_1} \lesssim |f(z)| \lesssim d_E(z)^{-C_2}$ *for some constants C_1, C_2*

$$|D^\alpha f(z)| = O(r(z)^{-|\alpha|-1}|f(z)|) \text{ for all } \alpha.$$

(b) *In the region D of Lemma 2.1, and for some constant C_3*

$$|f(z)| \lesssim d_E(z)^{C_3}$$
$$|D^\alpha f(z)| = O(m(z)^{-|\alpha|-1}|f(z)|),$$

(here $m(z)$ is as in the proof of Theorem 3.1).

PROOF. The estimates for the derivatives are obtained as in Theorem 3.1. To obtain the bounds for $|f(z)|$ it is better to write first Re $K(t,z)$ in a more convenient way, exhibiting the principal part of Re $K(t,z)$ and the perturbations. By using Taylor developments at $s = s(z)$, it is routinely checked (see details in [2]) that

(7) $\qquad 2\pi \text{ Re } K(t,z) = \dfrac{|T(t)| \text{ Re } (1 - z\overline{\gamma(s)})}{|1 - \overline{\gamma(t)} \cdot z|^2} + O\left(1 + \dfrac{r(z)^{1/2}}{|1 - \overline{\gamma(t)} \cdot z|}\right).$

For part (a) we must check that

(8) $\qquad \displaystyle\int_0^L \frac{|T(t)|r(z)}{|1 - \overline{\gamma(t)}z|^2} \log \frac{1}{d_E(t)}\, dt \leq \text{ const } \left(\log \frac{1}{d_E(z)} + 1\right),$

(9) $\qquad \displaystyle\int_0^L \frac{r(z)^{1/2}}{|1 - \overline{\gamma(t)}z|} \log \frac{1}{d_E(t)}\, dt \leq \text{ const } \left(\log \frac{1}{d_E(z)} + 1\right)$

We first bound (8) in terms of $r(z)$. The integral over I_z is of the order of

$$\frac{1}{r(z)} \int_{I_z} |T(t)| \log \frac{1}{d_E(t)}\, dt$$

and is thus bounded by $1 + |\log r(z)|$, by lemma 4.1. The integral off I_z is bounded by

$$\sum_{k=1}^N \frac{r(z)}{[2^k r(z)]^2} \int_{2^k I_z \setminus 2^{k-1} I_z} |T(t)| \log \frac{1}{d_E(t)}\, dt.$$

Here $2^k I_z = \{t : d(t,s) \leq 2^k r(z)\}$ and $2^N I_z = \gamma$. By Lemma 4.1, this is in turn bounded by const $(|\log r(z)| + 1)$. Hence (8) is proved in case $r(z) \geq \varepsilon d_E(z)$; if $r(z) \leq \varepsilon d_E(z)$ then we apply the same procedure replacing I_z by $J_z = \{t : d(t,s) \leq \varepsilon d_E(z)\}$. In this case the integral over J_z is, since $d_E(t) \simeq d_E(z)$ (with ε small enough) bounded by

$$\log \frac{1}{d_E(z)} \int_0^L \frac{|T(t)|r(z)}{[r(z) + |T(s)||s-t| + |s-t|^2]^2}\, dt.$$

This last integral is uniformly bounded. Off J_z the bound will be, with obvious notations, as before

$$\text{const} \sum_{k=1}^N \frac{r(z)}{(2^k d_E(z))} \log \frac{1}{2^k d_E(z)} \leq \text{const} \log \frac{1}{d_E(z)} + \text{const}.$$

To deal with (9) the full hypothesis is needed and not only its consequence, Lemma 4.1. Let us see for instance that (9) is bounded by const $|\log r(z)| + $ const. Recall that by (1) an interval I of type $d(t,s) \leq l$ has length bounded by $\min(l/|T(s)|, l^{1/2})$ and hence $|I| \log \frac{1}{|I|} \lesssim l^{1/2} \log \frac{1}{l}$.

The integral over I_z in (9) is then bounded by const $|\log r(z)| + $ const and over the complement of I_z by

$$\sum_k \frac{r(z)^{1/2}}{2^k r(z)} (2^k r(z))^{1/2} \left\{ \log \frac{1}{2^k r(z)} + c \right\} = \text{const} + \text{const} \log \frac{1}{r(z)}.$$

A similar argument as before, replacing I_z by J_z, shows that (9) is satisfied.

Let us prove the inequality $|f(z)| \lesssim d_E(z)^{C_3}$ in the region D. Using that the first term on the right of (7) is positive, Lemma 2.1, that $d_E(t) \lesssim d_E(z)$ in I_z and that the length of I_z is $\frac{r(z)}{|T(s)|}$ for $z \in D$, we see that

$$\int \frac{|T(t)| \,\text{Re}\,(1 - \bar{z}\gamma(s))}{|1 - z\overline{\gamma(t)}|^2} \log d_E(t)\, dt \leq \int_{I_z} \text{id} \leq$$

$$\leq \frac{|T(s)|}{r(z)} \int_{I_z} \log d_E(t)\, dt \leq C_3 \log d_E(z) + \text{const}.$$

It must be shown that the other part, given by the second part of (7), can be absorbed. We consider again

$$\int_0^L \frac{r(z)^{1/2}}{r(z) + |T(s)||s-t| + |s-t|^2} \log \frac{1}{d_E(t)}\, dt.$$

The integral over I_z is bounded, using the hypothesis and $|I_z| \leq r(z)/|T(s)|$ once again, by

$$\frac{1}{r(z)^{1/2}} \int_{I_z} \log \frac{1}{d_E(t)}\, dt \leq \frac{|I_z|}{r(z)^{1/2}} \left(\text{const} \log \frac{1}{|I_z|} + \text{const} \right) \leq$$

$$\leq \frac{r(z)^{1/2}}{|T(s)|} \left(\text{const} \log \frac{1}{r(z)} + \text{const} \right) \leq \frac{1}{C} \left(\text{const} \log \frac{1}{r(z)} + \text{const} \right)$$

(here C is the constant in the definition of D). The integral over the complement of I_z is dominated by

$$\sum_k \frac{r(z)^{1/2}}{2^k r(z)} |2^k I_z| \left(\log \frac{1}{|2^k I_z|} + 1\right).$$

For those k for which $|2^k I_z|$ is $\simeq (2^k r(z))^{1/2}$, $k > \log T^2(s)/r(z) > \log C$ and their sum is bounded by

$$\sum_{k > \log C} \frac{1}{2^{k/2}} \left(\log \frac{1}{2^k r(z)} + 1\right) \leq \varepsilon(C) \left(\log \frac{1}{r(z)} + 1\right),$$

with $\varepsilon(C) \to 0$ as $C \to \infty$; those k for which $|2^k I_z|$ is $2^k r(z)/|T(s)|$ sum up less than

$$\sum_{k < \log \frac{T^2(s)}{r(s)}} \frac{r(z)^{1/2}}{|T(s)|} \left(\log \frac{|T(s)|}{2^k r(z)} + 1\right) \leq$$

$$\leq \frac{r(z)^{1/2}}{|T(s)|} \log \frac{T^2(s)}{r(z)} \left(1 + \log \frac{1}{r(z)}\right) + \frac{r(z)^{1/2}}{|T(s)|} \left(\log \frac{T^2(s)}{r(z)}\right)^2$$

$$\leq \varepsilon(C) \left(\log \frac{1}{r(z)} + 1\right).$$

Again, replacing I_z by J_z in case $r(z)$ is much smaller than $d_E(z)$, it can be shown that $d_E(z)$ can replace $r(z)$ in the above estimate. Therefore, choosing C big enough in the definition of the region D, part (b) is completely established. $\square$

Remark. If it were possible to construct a *positive* pluriharmonic kernel with size given by the first term in (7) (plus bounded terms) the above proof shows that the condition in Lemma 4.1 -in a sense more natural- would be a sufficient condition for the existence of f and hence for A^∞-interpolation. The estimate (7) implies that

$$\text{Re } K(t, z) \, dt \longrightarrow \delta_x + L(t, x) \, dt$$

weakly as measures as z approaches a transverse point $\gamma(x)$ of the curve, with $L(t, x)$ bounded, and hence it is well suited for pluriharmonic interpolation, as used in [2]. The kernel $\text{Re } (1 - z \cdot \overline{\gamma(t)})^{-1}$ used in the previous section does not satisfy the estimate (7), and has another bad term of type

$$O\left(\frac{|T(t)||t - s|^2}{|1 - \overline{\gamma(t)}z|^2}\right).$$

LEMMA 4.3. *If E is as in Lemma 4.2 and G is the set of complex-tangential points of γ, for each $p \in N$ there exists $g_p \in A^p(B^n)(\overset{\text{def}}{=} H(B^n) \cap C^p(\bar{B}^n))$ such that*

(a) $|D^\alpha g_p(z)| = O(d_{E \cup G}(z))$, *for* $|\alpha| \leq p$.

(b) $|g_p(z)| \geq C_p r(z)^m$ *for some* $m = m(p)$ *and* $C_p > 0$.

(c) $g_p \in C^\infty(\bar{B}^n \backslash (E \cup G))$ and for all α

(10)
$$|D^\alpha g_p(z)| = O(m(z)^{p-|\alpha|}) \text{ in } D,$$
$$|D^\alpha g_p(z)| = O(r(z)^{p-|\alpha|}) \text{ off } D.$$

PROOF. Let h be the function defined in (3). If f is as in Lemma 4.2, we define $g_p = f^{N_2} h^{-N_1}$ with N_1 and N_2 big enough to be chosen later. Part (a) of 4.2 and (4) imply (b). From (4) and (5) it follows that

$$|D^\alpha h^{-N_1}(z)| = O((r(z) + T^2(s))^{N_1(q-\frac{1}{2})-|\alpha|}).$$

In the region D, by part (b) of Lemma 4.2,

$$|g_p(z)| = O(|T(s)|^{N_1(2q-1)} d_E(z)^{N_2 C_3})$$

$$|D^\alpha g_p(z)| = O\left(\sum_{\alpha_1 + \alpha_2 = \alpha} (T^2(s))^{N_1(q-\frac{1}{2})-|\alpha_1|} m(z)^{-|\alpha_2|-1} d_E(z)^{N_2 C_3} \right)$$

$$= O(|T(s)|^{N_1(2q-1)} m(z)^{-|\alpha|-1} d_E(z)^{N_2 C_3}),$$

and outside D, by part (a) of Lemma 4.2,

$$|g_p(z)| = O(r(z)^{N_1(q-\frac{1}{2})}) r(z)^{-N_2 C_2})$$

$$|D^\alpha g_p(z)| = O(r(z)^{N_1(q-\frac{1}{2})-N_2 C_2 - |\alpha|-1}), \qquad |\alpha| \geq 1.$$

Taking into account that $m(z) = \min\{T^2(s), d_E(z)\}$ we see from these inequalities that taking first N_2 and then N_1 big enough (c) holds, which implies (a). $\square$

Once the functions g_p have been constructed, the method used in [3] for the one variable case can be adapted with sligth modifications. Next lemma is the version of [3, Lemma 5.4] in this setting.

LEMMA 4.4. For each $p \in N$ there exists $p^* \in N$, $p^* \geq p$, with the following property: if $g \in A^{p^*}$ satisfies

(a) $D^\alpha g(z) = 0$ for $|\alpha| \leq p^*$, $z \in E \cup G$,
(b) $g \in C^\infty(\bar{B}^n \backslash (E \cup G))$ and (10) hold for g (and p^*)

and $\varepsilon > 0$ is given, there exists $k \in A^\infty(B^n)$ vanishing together with all its derivatives at $E \cup G$ and such that $\|g - k\|_{A^p} \leq \varepsilon$.

PROOF. We consider the functions L_η of Remark 3 of section 3. It is easily seen using (10) and (6) that if p^* is big enough the A^{p+1}-norm of $L_\eta g$ is uniformly bounded in η. Hence, $k = L_\eta g$, which is in A^∞ again by (6) and (10), will approach g in A^p for small η.

PROOF OF THE THEOREM. Let $\varphi \in C^\infty(\gamma)$; without loss of generality we can assume that $\varphi \in C^\infty(\mathbb{C}^n)$ and that $\bar{\partial}\varphi$ is flat on γ. For each $p \in N$, let g_p be as in Lemma 4.3. By 4.3 (b) (c), $\bar{\partial}\varphi/g_p$ is a $\bar{\partial}$-closed $(0,1)$ form smooth up to the boundary. Therefore there exists a solution v_p of $\bar{\partial}v = \bar{\partial}\varphi/g_p$ which is as well smooth up to the boundary. Then

$$f_p = \varphi - v_p g_p$$

is of class C^p in $\bar{B}^n$, interpolates on $E \cup G$ all derivatives up to order p by 4.3 (a), is C^∞ outside $E \cup G$ and its derivatives satisfy (10).

For each p, if p^* is as in Lemma 4.4, we consider now $f_{p^*+1} - f_{p^*} = v_{p^*}g_{p^*} - v_{p^*+1}g_{p^*+1}$. The hypothesis of Lemma 4.4 hold true for this function, hence $k_p \in A^\infty$ flat at $E \cup G$ can be found so that

$$\|f_{p^*+1} - f_{p^*} - k_p\|_{A^p} < 2^{-p}$$

Then, $f = f_{1^*} + \sum_p (f_{p^*+1} - f_{p^*} - k_p)$ is in $A^\infty(B^n)$ and interpolates all derivatives of φ at $E \cup G$. $\square$

References.

1. H. Alexander, B.A. Taylor and D.L. Williams, *The interpolating sets for A^∞*, J. Math. Anal. Appl. **36** (1971), 556–568.

2. Bo Berndtsson, J. Bruna, *Traces of pluriharmonic functions on curves*, Arkiv für Matematik **28, n+ 2** (1990), 221–230.

3. J. Bruna, *Boundary interpolation sets for holomorphic functions smooth to the boundary and B.M.O.*, Transactions of the A.M.S. **264, n+ 2** (1981), 393–404.

4. J. Bruna, J. Ortega, *Interpolation by holomorphic functions smooth to the boundary in the unit ball of $\mathbb{C}^n$*, Math. Ann. **274** (1986), 527–575.

5. D. Burns, E.L. Stout, *Extending functions from submanifolds of the boundary*, Duke Math. J. **43** (1976), 391–404.

6. J. Chaumat, A.M. Chollet, *Ensembles pics pour $A^\infty(D)$*, Ann. Inst. Fourier **29** (1979), 171–200.

7. J. Chaumat, A.M. Chollet, *Caracterizations et proprietes des ensembles localment pics de $A^\infty(D)$*, Duke Math. J. **47** (1980), 763–787.

8. J. Chaumat, A.M. Chollet, *Ensembles de zeros et d'interpolation a la frontiere de domaines strictament pseudoconvexes*, Arkiv für Matematik **24, n+ 1** (1986), 27–57.

9. E.M.Čirka, G.M. Henkin, *Boundary properties of holomorphic functions of several complex variables*, J. Soviet Math. **5** (1976), 612–687.

10. A.M. Davie. B. Øksendal, *Peak interpolation sets for some algebras of analytic functions*, Pacific J. Math. **41** (1972), 81–87.

11. M. Hakim, N. Sibony, *Ensembles pics dans les domaines strictament pseudoconvexes*, Duke Math. J. **45** (1978), 601–607.

12. A. Nagel, *Smooth zero-sets and interpolation sets for some algebras of holomorphic functions on strictly pseudoconvex domains*, Duke Math. J. **43** (1976), 323–348.

13. A. Nagel. W. Rudin, *Local boundary behavior of bounded holomorphic functions*, Can. J. Math. **30** (1978), 583–592.

14. A. Nagel. S. Wainger, *Limits of bounded holomorphic functions along curves*, Ann. of Math. Studies **100**, Princeton Univ. Press, Princeton N.J. (1981), 327–344.

15. W. Rudin, *Function theory in the unit ball of* $\mathbb{C}^n$, Springer-Verlag. New York Inc., 1980.

16. B.A. Taylor, D.L. Williams, *Ideals in rings of analytic functions with smooth boundary values*, Can. J. Math. **22, 6** (1970), 1266–1283.

DEPARTAMENT DE MATEMÀTIQUES,UNIVERSITAT AUTÒNOMA DE BARCELONA, 08193 BEL-LATERRA (BARCELONA) SPAIN

Contemporary Mathematics
Volume **137**, 1992

Vanishing Cup Products on
Pseudoconvex CR Manifolds

LUTZ BUNGART

ABSTRACT. For the smooth compact boundary M of a pseudoconvex domain on a Stein space of dimension n with only isolated singularities the cup product

$$\otimes_{j=1}^{m} H^{k_j}(M) \to H^{\Sigma k_j}(M)$$

vanishes if $k_j < n - 1$ and $\Sigma k_j > n$. The conditions on k_j can be relaxed to $k_j \leq n - 1$ and $\Sigma k_j \geq n$ if M is the link of an isolated singularity (a known result [DH]), but not in general as shown by examples.

1. Introduction.

Let X be a subvariety of a neighborhood of the origin in $\mathbb{C}^N$ of dimension n with an isolated singularity at 0. The intersection of X with a sphere S_ϵ of small radius ϵ is called the link of the singularity, $M = X \cap S_\epsilon$. The homeomorphism class (and even the diffeomorphism class) of M is independent of ϵ as long as ϵ is small (see e.g. [D]).

Assume that there is a holomorphic circle action on M. This is the case, for instance, if X is a cone. Then there is a metric on M which is given by the Levi form in the holomorphic tangent space and by dt^2 in the direction of the circle action. The cohomology groups $H^k(M)$ are isomorphic to spaces of harmonic forms $\mathcal{H}^k$. For $k \leq n - 1$, it can be shown that these harmonic forms do not involve dt ([T], Theorem 13.1) and, by duality via the $*$ operator, harmonic forms involve dt for $k \geq n$.

Now, if $\alpha_j \epsilon \mathcal{H}^{k_j}$ with $k_j \leq n - 1, j = 1, 2$, and if $k_1 + k_2 \geq n$, then $\alpha_1 \wedge \alpha_2 = 0$.

In fact, $\omega = \alpha_1 \wedge \alpha_2$ decomposes $\omega = \omega_h + d\gamma$ with $\omega_h \epsilon \mathcal{H}^{k_1 + k_2}$. Since ω_h and $*\omega$ both involve dt, we have

$$0 = \omega_h \wedge *\omega = \omega_h \wedge *\omega_h + \omega_h \wedge *d\gamma.$$

1991 *Mathematics Subject Classification.* Primary 32C16. Secondary 32F40, 32C35, 14F25, 14F40.

This paper is in final form and no version of it will be submitted for publication elsewhere.

Integrating and noting $d^*\omega_h = 0$ yields the claim:

$$0 = \int \omega_h \wedge *\omega_h + \int \omega_h \wedge *d\gamma$$

$$=\| \omega_h \|^2 + \int d^*\omega_h \wedge *\gamma = \| \omega_h \|^2 .$$

In [DH], Durfee and Hain prove the above for the link of an arbitrary isolated algebraic singularity by establishing that $H^k(M)$ carries a mixed Hodge structure which assigns weights less than or equal to k if $k < n$ and weights greater than or equal to $k + 1$ if $k \geq n$, and that the cup product preserves weights. (The condition of algebraicity can be dropped by a theorem of Artin [A]). In a verbal communication, Hain raised the question to what extent the theorem on vanishing cup products generalizes to arbitrary strictly pseudoconvex CR manifolds. We will give counter examples when $k_1 + k_2 = n$ or when $k_1 = n - 1$, and then proceed to prove the vanishing of the cup product when $k_j < n - 1$ and $k_1 + k_2 > n$. Since the methods of Durfee and Hain do not generalize, we begin with an alternate proof of their theorem which is easily adapted to the more general situation.

The author likes to thank Professor Richard Hain for many helpful conversations on mixed Hodge structures on links.

In the following,

$$H^*(Y) = H^*(Y, \mathbb{C})$$

denotes cohomology with complex coefficients. For a smooth manifold Y, we identify $H^*(Y)$ with deRham cohomology. Then cup product is induced by the exterior product on differential forms. If $Y_0 \subset Y$ is a closed submanifold and $K \subset Y$ a compact subset, then the induced maps

$$H^*(Y) \to H^*(Y_0), H^*(Y) \to H^*(Y \setminus K)$$

are morphisms of rings (preserve cup product).

2. Vanishing Cup Products on Links.

The following theorem is due to Durfee and Hain. We will give another proof[1] more suited to generalization.

THEOREM [DL]. *Let M be the link of an isolated singularity of an n-dimensional analytic variety. If $s, t < n$ and $s + t \geq n$, then the cup product*

$$H^s(M) \otimes H^t(M) \to H^{s+t}(M)$$

vanishes.

---

[1]It was pointed out by the referee that D. Sullivan used a similar argument in [S] in the surface case ($n = 2$).

PROOF. Let X be an n-dimensional analytic variety in a neighborhood of 0 in some $\mathbb{C}^N$ with an isolated singularity at 0. Set

$$X_\epsilon = \{z \in X : |z| < \epsilon\}.$$

For small $\epsilon > 0$, $X_\epsilon \setminus \{0\}$ is homotopy equivalent to the link of the singularity. If we represent M as

$$M = \{z \in X : |z| = \epsilon\}$$

then

$$H^*(M) \cong H^*(X \setminus \{0\})$$

as rings. Let

$$p : \tilde{X} \to X$$

be a resolution of singularities for X and set

$$\tilde{X}_\epsilon = p^{-1}(X_\epsilon), E = p^{-1}(0).$$

We have a long exact sequence of cohomology

$$\to H^k_c(\tilde{X}_\epsilon) \to H^k(\tilde{X}_\epsilon) \to H^k(M) \to .$$

To prove the theorem, we shall show that the map,

$$j^k : H^k_c(\tilde{X}_\epsilon) \to H^k(\tilde{X}_\epsilon),$$

is injective for $k \leq n$ and, by duality, surjective for $k \geq n$. This implies that the map

$$H^k(\tilde{X}_\epsilon) \to H^k(M),$$

which is a morphism of rings, is surjective for $k < n$ and equal to 0 for $k \geq n$. The theorem follows now by lifting cohomology classes from $H^k(M)$ to $H^k(\tilde{X}_\epsilon)$.

From the long exact sequence,

$$\to H^k_E(\tilde{X}_\epsilon) \to H^k(\tilde{X}_\epsilon) \to H^k(\tilde{X}_\epsilon \setminus E) \to$$

and the isomorphism $H^k(\tilde{X}_\epsilon \setminus E) \cong H^k(M)$ we see that

$$H^k_c(\tilde{X}_\epsilon) \cong H^k_E(\tilde{X}_\epsilon).$$

The maps j^k embed, therefore, in the sequence

$$H^k_E(\tilde{X}) \cong H^k_E(\tilde{X}_\epsilon) \cong H_c(\tilde{X}_\epsilon) \xrightarrow{j^k} H^k(\tilde{X}_\epsilon) \to H^k(E),$$

and injectivity of j^k for $k \leq n$ follows from injectivity of

$$H^k_E(\tilde{X}) \to H^k(E), k \leq n.$$

The injectivity of this map was established by Navarro Aznar [NA, Theorem 5.1] in the algebraic category. By Artin's algebraization theorem [A, Theorem 3.8], the same is true then in the analytic category.

3. Counter Examples.

In $\mathbb{C}^n$, let

$$\rho = 16 \sum_{j=1}^{n} (|z_j| - 1)^2.$$

Then

$$\partial\bar{\partial}\rho = 16 \sum_{j=1}^{n} \left(1 - \frac{1}{2|z_j|}\right) dz_j \wedge d\bar{z}_j$$

and thus ρ is strictly plurisubhamonic in a neighborhood of the closure of the "fat" torus

$$\Omega = \{z : \rho < 1\}.$$

The boundary M of Ω is strictly pseudoconvex and contains the n-torus

$$T = \{z : |z_j| = 1 + \frac{1}{4n^{1/2}}\}.$$

For any $k, 1 \le k \le n-1$, let

$$\alpha = \frac{dz_1}{z_1} \wedge \cdots \wedge \frac{dz_k}{z_k} \quad , \quad \beta = \frac{dz_{k+1}}{z_{k+1}} \wedge \cdots \wedge \frac{dz_n}{z_n}.$$

These differential forms are closed and, therefore, define cohomology classes in $H^k(M)$ and $H^{n-k}(M)$, respectively.

The class of $\omega = \alpha \wedge \beta$ in $H^n(M)$ does not vanish, i.e., ω is not exact, since $\int_T \omega \neq 0$.

For the metric induced on M from $\mathbb{C}^n$ consider $*\omega_h$ where ω_h is the harmonic form in the cohomology class of ω, i.e.,

$$\omega = \omega_h + d\gamma.$$

Note that $*\omega_h$ is closed and thus defines a cohomology class in $H^{n-1}(M)$.

The cohomology class of $*\omega \wedge \alpha$ in $H^{n+k-1}(M)$ does not vanish since

$$\int_M *\omega_h \wedge \alpha \wedge \beta = \int_M *\omega_h \wedge (\omega_h + d\gamma) = \|\omega_h\|^2 \neq 0.$$

The above examples show that the cup products

$$H^k(M) \otimes H^{n-k}(M) \to H^n(M)$$
$$H^{n-1}(M) \otimes H^k(M) \to H^{n+k-1}(M)$$

do not vanish. It will follow from the next section that the cup product

$$H^k(M) \otimes H^s(M) \to H^{k+s}(M)$$

does vanish if we avoid the middle dimensions $n - 1$ and n, i.e., if $k, s < n - 1$ and $k + s > n$.

4. Vanishing Cup Products on
Boundaries of Pseudoconvex Manifolds.

Let X be a n-dimensional Stein space with only a finite number of isolated singularities, and

$$\tilde{X} \to X$$

a resolution of singularities of X. Define

$$H^k_\infty(X) = \varinjlim\{H^k(X \setminus K): K \subset X \text{ compact}\}$$
$$= \varinjlim\{H^k(\tilde{X} \setminus \tilde{K}): \tilde{K} \subset \tilde{X} \text{ compact}\}.$$

Then $H^k_\infty(X)$ inherits a cup product in a natural way and we have a long exact sequence

$$\to H^k_c(\tilde{X}) \to H^k(\tilde{X}) \to H^k_\infty(X) \to$$

where the map $H^*(\tilde{X}) \to H^*_\infty(X)$ is a morphism of rings. For the purpose of our discussion we say X has compact boundary M if there is a compact topological space $Y \supset X$ so that $M = Y \setminus X$, $H^k(M)$ is finite dimensional for all k, and we have an isomorphism

$$H^k_\infty(X) \cong H^k(M)$$

induced by the maps

$$H^k_\infty(X) \leftarrow \varinjlim\{H^k(U): U \supset M \text{ open }\} \to H^k(M).$$

This is, for instance, the situation if Y is a C^1 manifold with (piecewise) smooth boundary near M. If X has boundary M, then $\tilde{X}$ has boundary M also in a natural way.

LEMMA. $H^k(\tilde{X})$ *is finite dimensional if $k > n$ and the maps*

$$j^k: H^k_c(\tilde{X}) \to H^k(\tilde{X}), k > n$$

are surjective. If $H^k_\infty(X)$ is finite dimensional for $k \geq n$ then $H^k(\tilde{X})$ is finite dimensional for $k \neq n$ and the maps j^k are injective for $k < n$.

PROOF. X can be embedded as a closed subvariety of some $\mathbb{C}^N$. There is a strictly plurisubharmonic exhaustion function $\phi \geq 0$ for $\mathbb{C}^N$ so that $\{z: \phi(z) = 0\}$ is precisely the singular locus $\{x_1, \cdots, x_s\}$ of X and so that, for small $\epsilon > 0$, $\{\phi = \epsilon\} \cap X$ is the disjoint union of links of the singularities $x_1, \cdots, x_s$. (For instance, let $h_1, \cdots, h_N$ be entire functions on $\mathbb{C}^n$ which vanish at $\{x_1, \cdots, x_s\}$ and give local coordinates there and let $\{f_1, \cdots, f_K\}$ have $\{x_1, \cdots, x_s\}$ as zero locus. Then a smoothed out version of

$$\sup\left\{\sum_{j=1}^N |h_j|^2, \sum_{j=1}^K |f_j|^2 - \delta, |z|^2 - C\right\}$$

will do for ϕ with suitable choices of $\delta, C > 0$.)

Let $\tilde{\phi}$ be the lift of $\phi \mid X$ to $\tilde{X}$ and

$$\tilde{X}_t = \{z \in \tilde{X} : \tilde{\phi}(z) < t\}.$$

Since the real Hessian of $\tilde{\phi}$ has at least n positive eigenvalues when $\tilde{\phi}(z) > 0$, by Morse theory, $H^k(\tilde{X}_t) = H^k(\tilde{X})$ for $k > n, t > 0$. This implies in particular that $H^k(\tilde{X})$ is finite dimensional for $k > n$. By the theorem in section 1, for small ϵ,

$$H_c^k(\tilde{X}_\epsilon) \to H^k(\tilde{X}_\epsilon)$$

is an epimorphism if $k > n$. The commutative diagram

$$
\begin{array}{ccc}
H_c^k(\tilde{X}_\epsilon) & \to & H^k(\tilde{X}_\epsilon) \\
\downarrow & & \downarrow \cong \\
H_c^k(\tilde{X}) & \overset{j^k}{\to} & H^k(\tilde{X})
\end{array}
$$

implies that j^k is an epimorphism. If the $H_\infty^k(X)$ are finite dimensional for $k \geq n$ then the long exact sequence

$$\to H_c^k(\tilde{X}) \to H^k(\tilde{X}) \to H_\infty^k(X) \to$$

shows that $H_c^k(\tilde{X}), k > n$, is finite dimensional. The remainder of the lemma follows from duality.

As an immediate consequence we have:

THEOREM. *If X is a n-dimensional Stein space with a finite number of isolated singularities such that $H_\infty^k(X)$ is finite dimensional for all k, then the cup products*

$$\otimes_{j=1}^m H_\infty^{k_j}(X) \to H_\infty^{\Sigma k_j}(X)$$

vanish if $k_j < n - 1, 1 \leq j \leq m$, and $\Sigma k_j > n$.

PROOF. Let $\tilde{X}$ be a resolution of singularities for X. From the long exact sequence

$$\to H_c^k(\tilde{X}) \to H^k(\tilde{X}) \to H_\infty^k(X) \to$$

and the lemma, we see that the maps

$$H^k(\tilde{X}) \to H_\infty^k(X)$$

are epimorphisms for $k < n - 1$ and equal to 0 for $k > n$. The theorem follows now by lifting cohomology classes from $H_\infty^k(X)$ to $H^k(\tilde{X})$.

COROLLARY. *If X is a n-dimensional Stein space with compact boundary M and a finite number of isolated singularities, then the cup products*

$$\otimes_{j=1}^m H^{k_j}(M) \to H^{\Sigma k_j}(M)$$

vanish if $k_j < n - 1, 1 \leq j \leq m$, and $\Sigma k_j > n$.

The corollary applies in particular to a compact strictly pseudoconvex CR manifold of dimension $2n - 1 > 3$, which can be embedded in some $\mathbb{C}^N$ and, therefore, is the boundary of some Stein space by results of Harvey and Lawson (see, for instance, [Y], Theorem 5.12).

Since presenting these results at the Symposium in Complex Analysis, the author has come across a paper by Ohsawa [O] in which the epimorphism in our lemma is derived from an analytic Hodge decomposition of $H^k(\tilde{X})$ for $k > n$.

References.

[A] Artin, M., *Algebraic Approximation of Structures over Complete Local Rings*, Inst. Hautes Édudes Sci. Publ. Math. **36** (1969), 23-58.

[D] Durfee, A., *Neighborhoods of Algebraic sets*, Trans. Am. Math. Soc. **276** (1983), 517–530.

[DH] Durfee, A., Hain, R., *Mixed Hodge Structures on the Homotopy of Links*, Math. Ann. **280** (1988), 69–83.

[NA] Navarro Aznar, V., *Sur la théorie de Hodge des variétés algébrique à singularités isolées*, Asterisque **130** (1985), 272–307.

[O] Ohsawa, T., *Reduction Theorem for Cohomology Groups of Very Strongly q-Convex Kähler Manifolds*, Invent. Math. **66** (1982), 391–393.

[S] Sullivan, S., *On the Intersection Ring of Compact Three Manifolds*, Topology **14** (1975), 275–277.

[T] Tanaka, N., *A Differential Geometric Study on Strongly Pseudo-Convex Manifolds*, Lectures in Mathematics **9**, Department of Mathematics, Kyoto University, Kinokuniya Bookstore Co., Ltd..

[Y] Yau, S.S.-T., *Kohn-Rossi Cohomology and Its Application to the Complex Plateau Problem, I*, Ann. of Math. **113** (1981), 67–110.

DEPARTMENT OF MATHEMATICS, UNIVERSITY OF WASHINGTON, SEATTLE, WASHINGTON 98195

Contemporary Mathematics
Volume **137**, 1992

Ext for Hilbert modules

JON F. CARLSON AND DOUGLAS N. CLARK

1. Introduction.

A Hilbert module is a Hilbert space H which is also a module over a function algebra A. That is, there is an associative, bilinear multiplication $A \times H \longrightarrow H$ which is continuous in both variables. The first systematic study of Hilbert modules appears in the monograph of Douglas and Paulsen [2]. In that volume, the authors reinterpreted many fundamental operator theory results as statements about the category $\mathcal{H}(A)$ of all Hilbert modules over the function algebra A. In addition, they also considered some homological properties of the category such as the projectivity and injectivity of Hilbert modules and constructions such as tensor products. These are only a few of the structures from homological algebra which might be applicable to the analysis of Hilbert spaces and operators.

In this paper we report on an investigation into the problem of classifying extensions of Hilbert modules. Specifically we consider the functor $Ext^1_{\mathcal{H}(A)}(-,-)$. Basically, $Ext^1_{\mathcal{H}(A)}(K, H)$ is the set of all equivalence classes of short exact sequences of Hilbert modules over A that begin with H and end with K. Details of the definition are given in the next section. Such extensions are of interest because they provide new examples of Hilbert modules and also because they give similarity invariants for operators.

The definitons of the Ext groups are standard from homological algebra, except that some extra care is required in this case. The difficulty is that the category $\mathcal{H}(A)$ of Hilbert modules over A is not, in general, an abelian category (see for example [1] p. 20). Specifically the so-called "First Isomorphism Theorem" fails as in the case of the dense embedding $H^2 \longrightarrow B^2$ of the Hardy space into the Bergman space. If $A(D)$ is the function algebra on the disc then the embedding is a map of $A(D)$-modules, and the cokernel of the map is zero. But the embedding is not surjective as a map of sets. Nevertheless the category $\mathcal{H}(A)$ is certainly additive. Constructions such as pushouts and pullbacks exist

1991 *Mathematics Subject Classification.* Primary 47D99.

Partially supported by N.S.F. Grants.

The final (detailed) version of this paper will be submitted for publication elsewhere.

and the functor $Ext^1(-,-)$ can be defined by short exact sequences as in [**3**]. However we must insist that, in any short exact sequence $0 \to H \xrightarrow{\alpha} B \xrightarrow{\beta} K \to 0$ of Hilbert modules, the map β is surjective as a map of sets and $\alpha(H) \subseteq B$ is a closed subspace.

For the purposes of this paper we restrict ourselves to the case in which $A = A(D)$ is the function algebra of analytic functions on the disk D, which extend continuously to $\overline{D}$. As an illustration of the methods which we are using, we present an outline of the proof that $Ext^1_{\mathcal{H}(A)}(H^2/BH^2, H^2) \cong BMO \cap (BH^2)^{\perp}$, where H^2 is the Hardy space, $B \in H^{\infty}$ is an inner function, and $(BH^2)^{\perp}$ is the orthogonal complement of BH^2 in H^2. The isomorphism is actually an isomorphism of $A(D)$-modules. The techniques can also be used to show that $Ext^1_{\mathcal{H}(A)}(H^2, H^2) = \{0\}$.

A complete account of these results will appear in a forthcoming manuscript which will include a more thorough description of the algebraic theory as well as other results. The computation of $Ext^1_{\mathcal{H}(A)}(K, H)$, where H and K are finite dimensional Hilbert modules, is routine when the best methods are applied. The same is true of $Ext^1_{\mathcal{H}(A)}(K, H^2)$ where K is finite dimensional. If B is an infinite Blaschke product then the theorem on $Ext^1_{\mathcal{H}(A)}(H^2/BH^2, H^2)$ can be obtained by almost purely algebraic means. We have obtained a characterization of $Ext^1_{\mathcal{H}(A)}(B^2, H^2)$ but it is not so explicit as the results of the present paper. We have no examples of Hilbert modules C such that $Ext^1_{\mathcal{H}(A)}(H^2, C) \neq 0$; thus it is still not known if H^2 is a projective object in $\mathcal{H}(A)$.

2. Extensions.

Let

$$E_i : \qquad 0 \to H \xrightarrow{\alpha_i} J_i \xrightarrow{\beta_i} K \to 0 \qquad i = 1, 2$$

be two exact sequences of Hilbert modules over a function algebra A, where α_i, β_i are Hilbert module maps $i = 1, 2$. We say that E_1 and E_2 are *equivalent* provided there is a Hilbert module map $\gamma : J_1 \to J_2$ such that the diagram

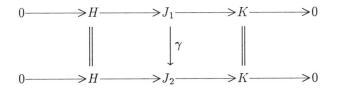

commutes. The set of equivalence classes of such sequences is denoted $Ext^1(K, H)$. The group operation is provided exactly as in [**3**] but it can be obtained as well from what follows.

For a given exact sequence of Hilbert modules

$$E : \qquad 0 \to H \xrightarrow{\alpha} J \xrightarrow{\beta} K \to 0$$

we can write $J = \alpha(H) \oplus \alpha(H)^\perp$ as the (orthogonal) direct sum of (isomorphic copies of) H and K. But $\alpha(H)^\perp$ will not, in general, be a Hilbert submodule of J. Specifically it might not be closed under multiplications by elements of A.

Define a Hilbert space map

$$\gamma : J = \alpha(H) \oplus \alpha(H)^\perp \to H \oplus K$$

by $\gamma(f, g) = (\alpha^{-1}f, \beta g), f \in \alpha(H), g \in \alpha(H)^\perp$. Since $\ker(\alpha) = 0, \gamma$ is well defined and the diagram

$$
\begin{array}{ccccccccc}
0 & \longrightarrow & H & \overset{\alpha}{\longrightarrow} & J & \overset{\beta}{\longrightarrow} & K & \longrightarrow & 0 \\
& & \| & & \downarrow{\scriptstyle\gamma} & & \| & & \\
0 & \longrightarrow & H & \overset{\alpha'}{\longrightarrow} & H \oplus K & \overset{\beta'}{\longrightarrow} & K & \longrightarrow & 0
\end{array}
$$

commutes, where $\alpha'(f) = (f, 0), \beta'(f, g) = g$. The lower row of the diagram is then equivalent to the exact sequence E, provided we equip $H \oplus K$ with a Hilbert module structure which makes γ, α' and β' into Hilbert module maps. We do this by defining

$$a(f, g) = (af + \sigma(a, g), ag), \qquad (f, g) \in H \oplus K, a \in A,$$

where $\sigma(a, g) = \alpha^{-1}[a\delta(a) - \delta(ag)], \delta : K \to J$ a Hilbert space (not necessarily module) map satisfying $\beta\delta = I$. So defined, σ is a bounded bilinear map of $A \times K$ to H.

The above construction allows us to classify members of $Ext^1(K, H)$ (extensions in the category of Hilbert A-modules). We have

THEOREM 1. $Ext^1(K, H) \cong \mathfrak{A}/\mathfrak{B}$, where $\mathfrak{A}$ is the set of all continuous (in both variables) bilinear maps $\sigma : A \times K \to H$ such that

$$a\sigma(b, k) + \sigma(a, bk) = \sigma(ab, k),$$

where $a, b \in A$ and $k \in K$, and $\mathfrak{B}$ is the set of all $\sigma \in \mathfrak{A}$ having the form

$$(1) \qquad\qquad \sigma(a, k) = a\mu(k) - \mu(ak),$$

where $\mu : K \to H$ is a (bounded) linear operator.

We can define addition of elements in $Ext^1(K, H)$ and the action of A in $Ext^1(K, H)$ in the usual way. Under the identification with $\mathfrak{A}/\mathfrak{B}$, these correspond to pointwise addition and multiplication on functions $\sigma(a, k)$. The zero element of $Ext^1(K, H)$ is the split exact sequence

$$0 \to H \underset{\underset{\gamma}{\leftarrow}}{\overset{\alpha}{\to}} J \overset{\beta}{\to} K \to 0$$

where $\beta\gamma = I_K$, the identity on K (this time γ must be a map of Hilbert modules).

In case A is the disk algebra, an element σ of $\mathfrak{A}/\mathfrak{B}$ (or equivalently of $Ext^1(K, H)$) is determined by the operator $\sigma(z, \cdot) : K \to H$. Indeed, it follows from (1) that

$$(2) \qquad \sigma(z^{n+1}, k) = \sum_{j=0}^{n} z^{n-j} \sigma(z, z^j k)$$

for $n \geq 1$ and for $k \in K$. There is at most one σ on $A \times K$ satisfying (2).

3. Submodules of H^2.

In this section, H^2 is the usual Hardy space of function $f(z)$ analytic in $|z| < 1$ and satisfying

$$f(z) = \sum_{n=0}^{\infty} a_n z^n, \qquad \|f\|_2 = \left(\sum_{n=0}^{\infty} |a_n|^2 \right)^{1/2} < \infty.$$

The function algebra A is the set of analytic functions in $D = \{z : |z| < 1\}$ which extend to be continuous in $\overline{D}$, and the action of A on H^2 is pointwise multiplication. The characterization of the H^2 norm as

$$\|f\|_2^2 = \sup_{0 < r < 1} \int_0^{2\pi} |f(re^{i\theta})|^2 d\theta$$

show that $\|af\|_2 \leq \|a\|_\infty \|f\|_2$, where $\|a\|_\infty$ is the sup norm on A.

According to Beurling's theorem, a nonzero A-submodule of H^2 must have the form

$$BH^2 = \{B(z)f(z) : f \in H^2\},$$

where B is an *inner* function, i.e. $|B(z)| \leq 1$ in $|z| < 1$ and $\lim_{r \uparrow 1} |B(re^{i\theta})| = 1$ holds for almost every θ, with respect to Lebesgue measure on $[0, 2\pi]$. The quotient module H^2/BH^2 is given the structure of a Hilbert A-module in the obvious way:

$$\|[f]\|_{H^2/BH^2} = \inf_{a \in H^2} \|f + Bg\|_2 = \|P_{H^2 \ominus BH^2} f\|_2$$

$$a[f] = [af]$$

where $[f]$ is the equivalence class modulo BH^2 of $f \in H^2$.

THEOREM 2. *The group $Ext^1(H^2/BH^2, H^2)$ is isomorphic to $BMO \cap (H^2 \ominus BH^2)$. The correspondence is given by*

$$k_0 \to \sigma(z, f) = \langle f, k_0 \rangle$$

where $k_0 \in BMO \cap (H^2 \ominus BH^2)$ and $\langle , \rangle$ is the inner product in H^2.

Of course BMO is the space of functions of bounded mean oscillation, defined by the finiteness of

$$\sup_I \frac{1}{|I|} \int_I |f(e^{it}) - f_I| dt,$$

where f_I is the average of f over the arc I and where the sup is over all subarcs of $[0, 2\pi]$.

We sketch the proof of Theorem 2. First of all, granting for the moment that σ has the form

$$\sigma(z, f) = \langle f, k_0 \rangle,$$

we show that, for σ to extend to a bounded bilinear map from $A \times H^2/BH^2$ into H^2, it is necessary and sufficient that $k_0 \in BMO$. The proof uses (2). The necessity part shows how BMO enters in: If $k_0 \in H^2 \ominus BH^2$, then by Nehari's Theorem, $k_0 \in BMO$ if and only if the matrix (b_{n+m}) from ℓ^2 to ℓ^2 is bounded (here $k_0(z) = \sum_{n=0}^{\infty} b_n z^n$). This is the case if and only if

$$(3) \qquad \sum_{j=0}^{\infty} |\langle z^j f, k_0 \rangle|^2 \le C\|f\|^2$$

for every $f \in H^2$. Now if we are given that $\sigma(z, f) = \langle f, k_0 \rangle$ extends to a bounded bilinear functional on $A \times H^2$, we have, for $n = 0, 1, \cdots$,

$$C\|f\|^2 \ge \|\sigma(z^{n+1}, f)\|^2 = \|\sum_{j=0}^{n} z^{n-j} \langle z^j f, k_0 \rangle\|^2,$$

by (2). This implies (3).

Next, we show that any $\sigma(z, f)$ arising from an extension of H^2/BH^2 by H^2 has the form $\langle f, k_0 \rangle$. If $\sigma(z, f)$ is a polynomial

$$\sigma(z, f) = \sum_{j=0}^{n} z^j \langle f, k_j \rangle$$

then σ is easily seen to be equivalent to

$$\sigma'(z, f) = \langle f, \sum_{j=0}^{n} \overline{z}^j k_j \rangle = \langle f, P \sum_{j=0}^{n} e^{-ijt} k_j(e^{it}) \rangle$$

where P is the projection of $L^2(0, 2\pi)$ on H^2. The general case is achieved by taking weak limits. Having obtained $k_0 \in H^2 \ominus BH^2$ such that $\sigma(z, f) = \langle f, k_0 \rangle, k_0 \in BMO$ by step 1.

Finally, we show that $\sigma(z, f) = \langle f, k_0 \rangle$ is equivalent to 0 only if $k_0 = 0$ (assuming $k_0 \in H^2 \ominus BH^2$). Define a map $T : H^2 \oplus H^2/BH^2 \to H^2$ by

$$T(f, g) = Bf + (P_B)\varphi g$$

where $\varphi(e^{it}) = e^{-it} B(e^{it}) \overline{k_0(e^{it})}$ and where P_B is the projection of H^2 on $H^2 \ominus BH^2$. It is seen that T is a Hilbert module map if $H^2 \oplus H^2/BH^2$ is equipped with the Hilbert module structure defined by $\sigma(z, f) = \langle f, k_0 \rangle$. Now if

$$0 \to H^2 \to H^2 \oplus H^2/BH^2 \to H^2/BH^2 \to 0$$

splits, say $\gamma \; : \; H^2/BH^2 \; \rightarrow \; H^2 \oplus H^2/BH^2$ satisfies $\gamma g = (\gamma_1 g, g)$, we have that $T\gamma$ is a Hilbert module map of H^2/BH^2 into H^2 and therefore $T\gamma g = B\gamma_1 g + P_B \varphi g = 0$. We conclude at once that $\varphi \in BH^2$ and therefore $k_0 = 0$.

Corollary. $Ext^1(H^2, H^2) = 0$.

This is actually a corollary of the proof, since $B(z) \equiv 0$ is not an inner function. The proof of Theorem 2 shows that every element of $Ext^1(H^2, H^2)$ has the form

$$0 \rightarrow H^2 \xrightarrow{\alpha} H^2 \oplus H^2 \xrightarrow{\beta} H^2 \rightarrow 0$$

where $\alpha f = (f, 0), \beta(f, g) = g$ and where the action of A on $H^2 \oplus H^2$ is given by $z(f, g) = (zf + \langle g, k_0 \rangle, zg)$, with $k_0 \in BMO$. We can therefore choose $y(t) \in L^2(0, 2\pi) \ominus H^2$ such that $k_0 + y \in L^\infty$. Setting $\varphi = e^{-it}(\overline{k}_0 + \overline{y})$ and $\gamma f = (T_\varphi f, f)$, where $T_\varphi f = P\varphi f$ is the Toeplitz operator with symbol φ, we have that γ is a Hilbert module map from H^2 to $H^2 \oplus H^2$, with the above A-action. Since clearly $\beta \gamma f = f$, for $f \in H^2$, the given element from $Ext^1(H^2, H^2)$ splits.

References.

1. H. Bass, *Algebraic K-Theory*, W. A. Benjamin Inc., New York, 1968.
2. R. G. Douglas and V. I. Paulsen, *Hilbert Modules over Function Algebras*, Longman Scientific & Technical, New York, 1989.
3. S. MacLane, *Homology*, Springer Verlag, New York, 1967.

THE UNIVERSITY OF GEORGIA, ATHENS, GEORGIA 30602

Contemporary Mathematics
Volume **137**, 1992

Hardy Spaces and Elliptic Boundary Value Problems

DER-CHEN CHANG

STEVEN G. KRANTZ

ELIAS M. STEIN

Dedicated to Walter Rudin.

0. Introduction

The purpose of this paper is to describe research we have been doing recently in an area of analysis which lies at the confluence of two directions of research: the study of real-variable Hardy spaces, and the theory of elliptic boundary value problems. Here we intend only to give a rough sketch of our results. Detailed proofs will appear elsewhere.

Our results may be thought of as answers to two (vaguely stated) questions.

Question 1. *Let Ω be an (appropriate) domain in $\mathbb{R}^N$. What are the possible (natural) notions of $H^p(\Omega)$ that generalize the usual Hardy spaces $H^p(\mathbb{R}^N)$?*

We shall see that several versions are possible. We concentrate on the two most relevant here, which may be considered to be the "largest" version, and the "smallest" version.

For the second question consider the Dirichlet and Neumann problems for Ω. The former consists in finding the $u = G(f)$ which solves

$$\Delta u = f \quad \text{in } \Omega, \quad \text{and} \quad u|_{\partial\Omega} = 0.$$

The latter consists in finding the $u = \tilde{G}(f)$ which solves

$$\Delta u = f \quad \text{in } \Omega, \quad \text{and} \quad \frac{\partial u}{\partial \vec{n}}\bigg|_{\partial\Omega} = 0.$$

1991 *Mathematics Subject Classification.* Primary 42B30.

All three authors are supported by grants from the National Science Foundation.

The final (detailed) version of this paper will be submitted for publication elsewhere.

Question 2. *In the context of the relevant* $H^p(\Omega)$, *can one obtain the boundedness of* $f \to \frac{\partial^2 G}{\partial x_i \partial x_j}(f)$, *and* $f \to \frac{\partial^2 \tilde{G}}{\partial x_i \partial x_j}(f)$?

This extension is known for L^p, $1 < p < \infty$, and also in the special case when Ω is the whole space $\mathbb{R}^N$ (where no boundary conditions are imposed).

Our paper is organized as follows. Sections 1 and 2 contain a quick review of background material. Our results are stated in Sections 3 and 4. Some ideas of the proofs and further discussions are contained in Sections 5 to 8.

1. Real Variable Hardy Spaces.

On the upper half plane $U \equiv \{z = x + iy \in \mathbb{C} : y > 0\}$ the spaces of G. H. Hardy were originally defined, for $0 < p < \infty$, to be

$$H^p(U) \equiv \left\{ f \text{ holomorphic on } U : \sup_{y>0} \int_{-\infty}^{\infty} |f(x+iy)|^p \, dx < \infty \right\}$$

The classical result is that an $f \in H^p$ has an almost everywhere limit as $y \to 0^+$ and the limit function f^b is p^{th} power integrable. For us it is also important to observe that for every $p > 0$, the limit

$$\lim_{y \to 0} f(x+iy) = f^b$$

exists in the sense of distributions.

The **real** Hardy space $H^p(\mathbb{R})$ is then defined to consist of those distributions f on $\mathbb{R}$, so that

$$f = f_1^b + \overline{f}_2^b$$

where $f_j \in H^p(U)$, $j = 1, 2$; that is, the real parts of boundary values of holomorphic functions in $H^p(U)$.

Let H denote the Hilbert transform (see [**K**] and [**SW**]). Because the Hilbert transform is bounded on L^p when $1 < p < \infty$, it is straightforward to see that for that range of p the space H^p is isomorphic to $L^p(\mathbb{R})$. The Hilbert transform is not bounded on L^1, but we may see that H^1 can be identified with the space of those $f \in L^1(\mathbb{R})$ with the property that $Hf \in L^1(\mathbb{R})$.

It turns out the space $H^p(\mathbb{R})$ has an extension to $\mathbb{R}^N$. It consists of those distributions f on $\mathbb{R}^N$ which can be characterized by any of the equivalent properties guaranteed by Theorem 1 below. We proceed as follows.

Let $\phi \in C_c^\infty(\mathbb{R}^N)$ satisfy $\int \phi \neq 0$. For $t > 0$ define $\phi_t(x) = t^{-N}\phi(x/t)$. Let f be a function or distribution on $\mathbb{R}^N$. We now define two maximal functions:

$$f^*(x) = \sup_{t>0} |f * \phi_t(x)|$$

and

$$f^+(x) = \sup_{y>0} |P_y * f(x)|.$$

Here P_y is the standard Poisson kernel. We also define the "grand maximal function" f^{**} as

$$f^{**}(x) = \sup_{\phi \in \mathfrak{B}} \sup_{t > 0} |f * \phi_t(x)|,$$

where $\mathfrak{B}$ is an appropriate bounded set of elements of $C_c^\infty(\mathbb{R}^N)$.

THEOREM 1. *Let* $0 < p < \infty$. *The following are equivalent:*

(1) $f \in H^p(\mathbb{R}^N)$;
(2) $f^{**} \in L^p(\mathbb{R}^N)$;
(3) $f^* \in L^p(\mathbb{R}^N)$;
(4) $f^+ \in L^p(\mathbb{R}^N)$.

Recall the notion of a Calderón-Zygmund operator S; it is given by

$$S(f) = f * K;$$

here

$$K = \text{P.V.} \frac{\Theta(x)}{|x|^N},$$

where Θ is smooth on the unit sphere and homogeneous of degree 0, with vanishing mean-value.

THEOREM 2. *([FS]) Let S be a Calderón-Zygmund operator. Then S maps* $H^p(\mathbb{R}^N)$ *to* $H^p(\mathbb{R}^N)$, $0 < p < \infty$.

For us the importance of Theorem 2 resides in the fact that if $\Delta u = f$ on $\mathbb{R}^N$ and (say) u has compact support, then

$$\frac{\partial^2 u}{\partial x_i \partial x_j} = S_{ij}(f)$$

where S_{ij} is a Calderón-Zygmund operator plus a multiple of the identity. Thus

$$\left\| \frac{\partial^2 u}{\partial x_i \partial x_j} \right\|_{H^p(\mathbb{R}^N)} \leq A_p \|f\|_{H^p(\mathbb{R}^N)}.$$

It is this result that we intend to extend to the setting of boundary-value problems.

We mention in passing the following properties of the real variable Hardy spaces:

(1) $H^p(\mathbb{R}^N) \approx L^p(\mathbb{R}^N)$ when $1 < p < \infty$;
(2) $H^1(\mathbb{R}^N) \subset L^1(\mathbb{R}^N)$ but $H^1(\mathbb{R}^N) \neq L^1(\mathbb{R}^N)$;
(3) $H^p(\mathbb{R}^N)$ (unlike $L^p(\mathbb{R}^N)$) has many continuous linear functionals, $0 < p < 1$.

We shall now present a final method for looking at real variable Hardy classes that is perhaps the most flexible of all:

Definition 1. *Let $0 < p \leq 1$. A measurable function a on $\mathbb{R}^N$ is called a p-atom if*

(1) *a is supported in a Euclidean ball $B(P, r)$;*
(2) *$\|a\|_{L^2(\mathbb{R}^N)} \leq (m(B(P, r)))^{\frac{1}{2} - \frac{1}{p}}$, where m is Lebesgue measure;*
(3) *$\int a(x) x^\alpha \, dx = 0$ for all multi-indices α of size $|\alpha| \leq [N(\frac{1}{p} - 1)]$, $[\ \ ]$ is the greatest integer function.*

THEOREM 3. *Fix $0 < p \leq 1$. There is a constant C so that if a is a p-atom then $\|a\|_{H^p} \leq C$. Moreover, if $f \in H^p(\mathbb{R}^N)$ then there are atoms a_j and constants α_j such that $f = \sum_{j=1}^{\infty} \alpha_j a_j$ and*

$$\sum_{j=1}^{\infty} |\alpha_j|^p \leq C \cdot \|f\|_{H^p}^p.$$

Finally,

$$\|f\|_{H^p} \approx \inf \left\{ \sum_{j=1}^{\infty} |\alpha|^p : f = \sum_{j=1}^{\infty} \alpha_j a_j \ \text{ with the } a_j \text{'s being atoms} \right\}.$$

One can successfully define atomic H^p spaces in a rather general setting: what is required is a metric and a measure that satisfy certain compatibility conditions. See [**CW1**], [**CW2**] for more on these "spaces of homogeneous type."

2. Local Hardy Spaces.

The real variable Hardy spaces are designed to behave well under the Calderón-Zygmund operators. In particular, they respect translations, rotations, and dilations. It is a fact that order zero pseudo-differential operators (which are not translation invariant) are not bounded on the real variable Hardy spaces. Closely related to this shortcoming is the fact that the Hardy spaces are not closed under composition with diffeomorphisms nor under multiplication by C_c^∞ functions.

It is with these difficulties in mind that the local Hardy spaces were developed in [**G**]. For brevity, we will explain the local Hardy spaces in the language of atoms:

Definition 2. *Let $0 < p \leq 1$. A measurable function a on $\mathbb{R}^N$ is called a local p-atom if*

(1) *a is supported in a Euclidean ball $B(P, r)$;*
(2) *$\|a\|_{L^2(\mathbb{R}^N)} \leq (m(B(P, r)))^{\frac{1}{2} - \frac{1}{p}}$, where m is Lebesgue measure;*
(3) *If $r \leq 1$ then $\int a(x) x^\alpha \, dx = 0$ for all multi-indices of size $|\alpha| \leq [N(\frac{1}{p} - 1)]$; if $r > 1$ then this moment condition does not apply.*

We *define* a distribution f on $\mathbb{R}^N$ to be in the local Hardy space $h^p(\mathbb{R}^N)$ if there is an atomic decomposition

$$f = \sum_j \alpha_j a_j$$

with each a_j a local p- atom and $\sum |\alpha_j|^p < \infty$.

The local h^p spaces enjoy many attractive properties: they are preserved by order zero pseudo-differential operators; they are preserved by composition with a diffeomorphism which is the identity map for sufficiently large x; also if $f \in h^p(\mathbb{R}^N)$ and $\phi \in C_c^\infty(\mathbb{R}^N)$ then $\phi \cdot f \in h^p(\mathbb{R}^N)$. One result is that the local Hardy spaces may be defined on a manifold and are acted on in a natural way by pseudo-differential operators. It is also plausible that (a variant of) these spaces would be suitable for analysis on a domain in space. It is to be noted that the local Hardy spaces can be characterized by a version of Theorem 1, but where the maximal functions that arise involve the supremum only over small values of t.

3. Real Hardy Spaces on a Domain.

Fix a bounded, connected domain $\Omega \subset \mathbb{R}^N$. Assume for now that Ω has Lipschitz boundary. Fix a positive real number p. We then define *two* h^p spaces. We set $h_r^p(\Omega)$ equal to the collection of those distributions f on Ω such that there is an $F \in h^p(\mathbb{R}^N)$ whose restriction to Ω equals f. We define $h_z^p(\Omega)$ to be those f on Ω that arise by restricting to Ω the distributions F in $h^p(\mathbb{R}^N)$ such that $F \equiv 0$ in $^c\overline{\Omega}$. Observe that

(1) $h_r^p(\Omega) = h_z^p(\Omega) = L^p(\Omega), 1 < p < \infty$;
(2) $h_z^p(\Omega) \subset h_r^p(\Omega)$ but $h_z^p(\Omega) \neq h_r^p(\Omega)$ when $0 < p \leq 1$.

Indeed one can show the following. Suppose Ω is a smoothly bounded domain. For $x \in \Omega$, define $d(x) =$ distance of x from the boundary of Ω. Consider the function

$$f(x) = \frac{1}{(d(x))^{\frac{1}{p}-\varepsilon}}, \qquad \varepsilon > 0,$$

which is locally integrable on Ω, and hence defines a distribution on Ω. It can be seen that $f \in h_r^p(\Omega)$, but $f \notin h_z^p(\Omega)$, if $p < 1$, and ε is sufficiently small. Using the function

$$f(x) = \frac{(d(x))^{-1}}{\left(\log \frac{1}{d(x)}\right)^2},$$

one also obtains an $f \in h_r^1(\Omega)$, such that $f \notin h_z^1(\Omega)$.

4. Principal Results.

Let us assume for the moment that $\Omega \subset \mathbb{R}^N$ is bounded and smooth. We will use the notation from our discussion of the Dirichlet and Neumann problems for the Laplacian. Then our principal results are

THEOREM 4. *Let* **G** *be the Green's operator of the Dirichlet problem for the Laplacian on* Ω. *Then the operators*

$$\frac{\partial^2 \mathbf{G}}{\partial x_j \partial x_k}$$

originally defined on $C^\infty(\overline{\Omega})$, *can be extended as bounded operators from* $h^p_r(\Omega)$ *to* $h^p_r(\Omega)$, *for* $1 \leq j, k \leq N$ *and* $\frac{N}{N+1} < p \leq 1$.

It can be shown by examples that the above assertion fails when $p \leq \frac{N}{N+1}$. However if we assume that the data belongs to $h^p_z(\Omega)$, then the result holds for all p. Moreover, for the Neumann problem, for any $p \leq 1$, it does not suffice to assume that the data belongs to $h^p_r(\Omega)$; however, a conclusion analogous to Theorem 4 is valid if we assume the data to belong to $h^p_z(\Omega)$.

THEOREM 5. *Let* **G** *be the Green's operator of the Dirichlet problem for the Laplacian on* Ω. *Then the operators*

$$\frac{\partial^2 \mathbf{G}}{\partial x_j \partial x_k}$$

originally defined on $C^\infty(\overline{\Omega})$, *can be extended as bounded operators from* $h^p_z(\Omega)$ *to* $h^p_r(\Omega)$, *for* $1 \leq j, k \leq N$ *and* $0 < p \leq 1$.

THEOREM 6. *Let* $\tilde{\mathbf{G}}$ *be the Green's operator of the Neumann problem for the Laplacian on* Ω. *Then the operators*

$$\frac{\partial^2 \tilde{\mathbf{G}}}{\partial x_j \partial x_k}$$

originally defined on $C^\infty(\overline{\Omega})$, *can be extended as bounded operators from* $h^p_z(\Omega)$ *to* $h^p_r(\Omega)$, *for* $1 \leq j, k \leq N$ *and* $0 < p \leq 1$.

5. Technical Ideas in the Proofs of the Main Results.

We shall now discuss some of the ingredients that go into the proofs of the main theorems. Some of our constructs are valid on domains with Lipschitz boundaries. The smoothness of the boundary is required later on when we estimate the Green's operators for the boundary value problems. So for now we shall be working on domains with just Lipschitz boundary.

Proposition 7. *Let* $0 < p \leq 1$. *A function* f *is in* $h^p_r(\Omega)$ *if and only if*

$$f = \sum_j \lambda_j a_j,$$

where the a_j *are atoms supported on cubes* $Q_j \subseteq \Omega$, $\sum_j |\lambda_j|^p < \infty$, *and the atoms are of three types:*

(1) Atoms a_j that are supported in a "big" cube Q_j (the diameter of Q_j is greater than 1) and satisfy the standard size but not necessarily the moment conditions.

(2) Atoms a_j that are supported in a "small" cube Q_j (the diameter of Q_j is less than or equal to 1) that is "far" from the boundary (distance at least four times the diameter of Q_j) and that satisfy the standard size and cancellation conditions.

(3) Atoms a_j that are supported in a "small" cube Q_j that is "near" to the boundary (distance less than or equal to four times the diameter of Q_j) satisfy the standard size but not necessarily the moment conditions.

This is essential a result in [**M**]. We call a_j a type (a) atom if it satisfies property (1) or (2) and a type (b) atom if it satisfies property (3). (See Figure 1 below).

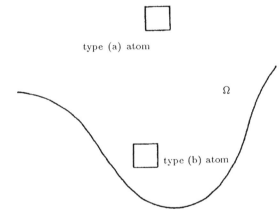

Figure 1

Proposition 8. *Let $0 < p \leq 1$. A function f is in $h_z^p(\Omega)$ if and only if*

$$f = \sum_j \lambda_j a_j,$$

where the a_j are local p- atoms supported on cubes $Q_j \subseteq \Omega$, $\sum_j |\lambda_j|^p < \infty$, and all the atoms except one are classical atoms satisfying both the size and moment conditions. The one exceptional atom has no cancellation and it can be taken to be supported in a unit cube.

Our device for generating an atomic decomposition *with atoms entirely supported in the domain* Ω is a variant of the classical Calderón reproducing formula, and the method of [**CF**] for proving the atomic decomposition in $H^p(\mathbb{R}^N)$, which relies on the area integral.

The reproducing formula is as follows. There exists a pair of functions Φ and Ψ in $C_c^\infty(\mathbb{R}^N)$, so that a fixed number of moments of Φ and Ψ vanish, and with

$$\Phi_t(x) = t^{-N}\Phi\left(\frac{x}{t}\right), \qquad\qquad \Psi_t(x) = t^{-N}\Psi\left(\frac{x}{t}\right)$$

we have

$$f = \int_0^\infty f * \Phi_t(x) * \Psi_t(x)\,\frac{dt}{t}.$$

Indeed, to construct such Φ and Ψ, we need only take a radial Φ, whose Fourier transform vanishes sufficiently rapidly at the origin, but with Φ not identically zero. Then we can choose Ψ to be a constant multiple of Φ.

We define also an associated "area integral" type of expression:

Definition 3. *Let Φ be as in the theorem. Set*

$$S_\Phi(f)(x) = \left(\int_{|x-y|<t} |f * \Phi_t(x-y)|^2\,\frac{dy\,dt}{t^{N+1}}\right)^{\frac{1}{2}}.$$

Theorem 9. *With Φ, S_Φ as above,*

$$\|S_\Phi(f)\|_{L^p} \le A\|f\|_{H^p}, \qquad \text{for } 0 < p < \infty.$$

As already noted, our device is to generalize Calderón's formula to a Lipschitz domain in $\mathbb{R}^N$. For simplicity, we shall restrict attention here to the a *special Lipschitz domain*—that is, the supergraph of a Lipschitz function of $N-1$ variables. Such a domain Ω has the property: there is a cone Γ such that, for any $x \in \Omega$, $x + \Gamma$ lies in Ω. (See Figure 2 below).

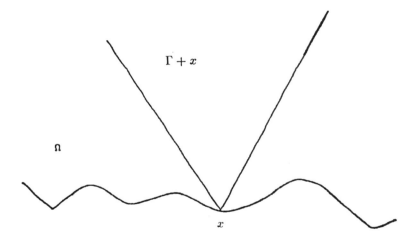

Figure 2. Special Lipschitz domain

The key to our proof of Proposition 8 is the version of the Calderón reproducing formula that is suitable for Lipschitz domains:

THEOREM 10. *Fix any open cone $\Gamma \subset \mathbb{R}^N$ and any positive number M. Then there exists $2N$ functions*

$$\Phi^1, \ldots \Phi^N, \quad and \quad \Psi^1, \ldots, \Psi^N$$

satisfying

$$\int \Phi^\ell \, dx = 0 \quad and \quad \int \Psi^\ell x^\alpha \, dx = 0$$

for $\ell = 1, \ldots, N$ and all multi-indices α such that $0 \leq |\alpha| \leq M$. The functions Φ^ℓ, Ψ^ℓ are all supported in Γ. Most significantly, if $f \in L^1(\mathbb{R}^N)$ then

$$f = \sum_{\ell=1}^{N} \int_0^\infty f * \Phi_t^\ell * \Psi_t^\ell \, \frac{dt}{t}.$$

Observe that if f is supported in Ω then, because we have assumed that Ω is a special Lipschitz domain, we know that $(f * \Phi_t^\ell)(x)$ and $(f * \Phi_t^\ell * \Psi_t^\ell)(x)$ are supported in Ω for all $t > 0$. This allows to decompose the distribution f into an atomic decomposition whose atoms lie in Ω.

6. The Reflection Function.

In the special case that Ω is the upper half space, then the special parity of the Newtonian potential with respect to boundary reflection makes it easy to write down immediately the Green's functions for the Dirichlet and Neumann problems.

In the general case (of a smooth bounded domain Ω) the situation is not as easy, but reflections nevertheless play a role.

Define first the family of reflection functions $x \mapsto r_\zeta(x)$, given for each $\zeta \in \partial\Omega$ and x near the boundary. More precisely, $x \mapsto r_\zeta(x)$ is just the reflection of x in the tangent hyperplane at $\zeta \in \partial\Omega$. (See Figure 3 below).

Moreover for $x \in \Omega$, with x sufficiently close to the boundary, there is also the natural reflection function $x \mapsto r(x)$ which is determines by $r(x) = r_{\zeta_0}(x)$, where ζ_0 is *the* point on the boundary which is nearest to x. Notice that

$$\text{dist}(x, \partial\Omega) = \text{dist}(r(x), \partial\Omega).$$

For us a crucial observation is that

$$|r(x) - r_\zeta(x)| \leq C \cdot |\zeta - x|^2,$$

as $x \to \zeta$.

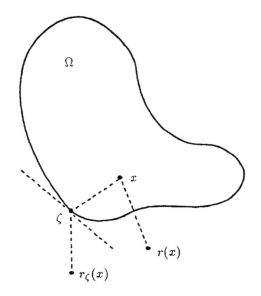

Figure 3. The reflection functions

7. The Calculation for the Dirichlet Problem.

We now sketch the calculation that is necessary to obtain the desired regularity estimate for the Dirichlet problem. In our usual notation, let f be the data function for the Dirichlet problem. We will assume that $f \in h_r^p(\Omega)$, $\frac{N}{N+1} < p \le 1$. Set

$$f_o(x) = \begin{cases} f(x) & \text{if } x \in \Omega \\ -f\big(r^{-1}(x)\big)J_{r^{-1}} & \text{if } x \in^c(\overline{\Omega}) \end{cases}.$$

Here J_g is the Jacobian of g. The solution to our boundary value problem is then

$$G(f) = E * f_o - P\big(R(E * f_o)\big).$$

Here E is the standard Newtonian potential, R is the restriction operator to the boundary of the domain, and P is the standard Poisson operator for the Laplacian.

Then

$$\frac{\partial^2}{\partial x_i \partial x_j} P\big(R(E * f_o)\big) = \int_\Omega K(x,y) f(y) \, dy,$$

where the kernel K is defined by this equation. Non-trivial calculations and the properties of the reflection function described above show that K satisfies the following properties:

$$|K(x,y)| \le A \cdot |x - y|^{-N+1}$$

and

$$\left| \frac{\partial}{\partial y} K(x,y) \right| \le A \cdot |x - y|^{-N}.$$

In order to solve the Dirichlet problem for *all* $0 < p < \infty$ we must restrict attention to data lying in $h_z^p(\Omega)$ (counterexamples show that this is necessary). Again, the solution to our boundary value problem is

$$G(f) = E * f_o - P(R(E * f_o)),$$

but now we have set

$$f_o(x) = \begin{cases} f(x) & \text{if } x \in \Omega \\ 0 & \text{if } x \notin \Omega. \end{cases}$$

Now again the reflection function comes into play, for it can be shown that

$$\frac{\partial^2}{\partial x_i \partial x_j} P(R(E * f)) = \int_\Omega K(x, x - r(y)) f(y) \, dy.$$

The distribution $K(x, z)$ can be seen to have the properties

(1) $|\partial_z^\alpha K(x, z)| \le A_\alpha |z|^{-N - |\alpha|}$;
(2) $\widehat{K}(x, \cdot) \in L^\infty$.

These inequalities hold *uniformly* in x and $\partial_x^\beta K(x, z)$ also satisfies these two inequalities (for all α, β), uniformly in x.

We then have the following lemma:

Lemma 11. *Whenever K satisfies the above properties then the mapping,*

$$f \rightarrow \int_\Omega K(x, x - r(y)) f(y) dy,$$

extends to a bounded mapping of $h_z^p(\Omega)$ to $h_r^p(\Omega)$ for $0 < p < \infty$.

To study the Neumann problem, for all $0 < p < \infty$, we set

$$\tilde{G}(f) = E(f) + P(u_b),$$

where

$$u_b = -(N^+)^{-1} R \frac{\partial E}{\partial \vec{n}}(f).$$

Here N^+ is the "Dirichlet-to-Neumann" operator (see [CNS1],[CNS2]) Then we are led to study

$$\frac{\partial^2 P}{\partial x_i \partial x_j} (N^+)^{-1} R \frac{\partial E}{\partial \vec{n}}(f) = \int_\Omega \tilde{K}(x, x - r(y)) f(y) \, dy.$$

It turns out that the kernel $\tilde{K}$ enjoys the two properties enjoyed by the kernel K that arose in our consideration of the Dirichlet problem. Thus we can show that the operator in $\frac{\partial^2 \tilde{G}}{\partial x_i \partial x_j}$, $i, j = 1, \ldots, N$, originally defined on $C^\infty(\overline{\Omega})$, can be extended as bounded operators from $h_z^p(\Omega)$ to $h_r^p(\Omega)$, $0 < p < \infty$.

8. Smoothness of the Boundary.

The atomic decomposition of h_r^p in fact goes through with rather mild restrictions on the smoothness of the boundary of Ω. In particular, the boundary need not even be Lipschitz (see [M]). We do not know whether a similar assertion holds for our atomic decomposition of h_z^p.

For some applications it would be desirable to find minimal smoothness conditions on $\partial\Omega$ in order for our analysis of the Dirichlet and Neumann problems to remain valid. We do not know whether $C^{1+\epsilon}$ boundary is sufficient in order to obtain h_r^p estimates for the Dirichlet problem when p is near 1. It would seem that, in order to obtain estimates for either the Dirichlet or Neumann problems when p is near to zero, it is necessary for the boundary to be C^k where k has size essentially $\frac{1}{p}$.

The literature for the Dirichlet and Neumann problems for domains with Lipschitz boundary (see [Ke]) teaches us that when the boundary is only Lipschitz then one can expect favorable behavior for a restricted range of p. It would be of interest to explore similar phenomena vis à vis the Hardy spaces introduced here.

References.

[CF] S. Y. A. Chang and R. Fefferman, *A continuous version of duality of H^1 and BMO*, Ann. Math. **112** (1980), 179-201.

[CKS] D. C. Chang, S. G. Krantz, and E. M. Stein, *H^p Theory on a Smooth Domain in $\mathbb{R}^N$ and Applications to Partial Differential Equations*, preprint.

[CNS1] D. C. Chang, A. Nagel, and E. M. Stein, *Estimates for the $\bar{\partial}$-Neumann problem in pseudoconvex domains in $\mathbb{C}^2$ of finite type,*, Proc. Natl. Acad. Sci. USA **85** (1988), 8771-8774.

[CNS2] _____ , *Estimates for the $\bar{\partial}$-Neumann problem in pseudoconvex domains of finite type in $\mathbb{C}^2$*, To appear in Acta Math.

[CW1] R. R. Coifman and G. Weiss, *Analyse Harmonique Non-commutative sur certains espaces homogènes*, Lecture Notes in Math. *242*, Springer-Verlag, Berlin-Heidelberg-Tokyo, 1971.

[CW2] _____ , *Extensions of Hardy spaces and their use in analysis*, Bull. Amer. Math. Soc. **83** (1977), 569-645.

[FS] C. Fefferman and E. M. Stein, *H^p spaces of several variables*, Acta Math. **129** (1972), 137-193.

[G] D. Goldberg, *A local version of real Hardy spaces*, Duke Math. J. **46** (1978), 27-42.

[K] Y. Katznelson, *Introduction to Harmonic Analysis*, John Wiley and Sons, New York, 1971.

[Ke] C. Kenig, *Elliptic boundary value problems on Lipschitz domains*, Beijing Lectures in Harmonic Analysis, Ann of Math. Study #112 (1986), Princeton University Press, Princeton, New Jersey.

[M] A. Miyachi, H^p *spaces over open subsets of* $\mathbb{R}^n$, Studia Math. **XCV, vol 3** (1990), 205-228.

[S] E. M. Stein, *Singular Integrals and Integrability Properties of Functions*, Princeton University Press, Princeton, New Jersey, 1970.

[SW] E. M. Stein and G. Weiss, *On the theory of harmonic functions of several variables, I. The theory of H^p spaces*, Acta Math. **103** (1960), 25-62.

DEPARTMENT OF MATHEMATICS, UNIVERSITY OF MARYLAND, COLLEGE PARK, MARYLAND 20742

DEPARTMENT OF MATHEMATICS, WASHINGTON UNIVERSITY IN ST. LOUIS, ST. LOUIS, MISSOURI 63130

DEPARTMENT OF MATHEMATICS, PRINCETON UNIVERSITY, PRINCETON, NEW JERSEY 08544

Contemporary Mathematics
Volume **137**, 1992

Représentation intégrale de certaines classes de jets de Whitney

JACQUES CHAUMAT ET ANNE-MARIE CHOLLET

dédié à Walter Rudin.

Introduction.

Classiquement, les formules intégrales qui interviennent dans la résolution des équations de Cauchy-Riemann ou la représentation des fonctions holomorphes se construisent de manière parallèle [6].

Ici, on représente par des intégrales les jets de Whitney C^∞ et $\bar{\partial}-$plats sur des compacts irréguliers de $\mathbb{C}^n$, $s-$H convexes, en utilisant un travail précédent des auteurs sur la résolution des équations $\bar{\partial}$ [1], [2]. L'intérêt principal de la formule obtenue réside dans l'holomorphie du noyau intégral. On obtient ainsi très simplement qu'un jet C^∞ et $\bar{\partial}-$plat sur un tel compact K est limite dans $C^\infty(K)$ de fonctions holomorphes au voisinage de K.

Lorsque le compact est $1-$H convexe et pour des classes de jets ultradifférentiables au sens de Carleman ou de Beurling, on généralise des résultats de E. M. Dynkin [5]. Il convient de remarquer que, comme dans le travail de E. M. Dynkin, les hypothèses sur les classes de jets étudiées sont très peu restrictives. Ici encore, comme conséquence, on obtient un théorème d'approximation. Ce dernier résultat a été précédemment obtenu par B. Droste [3], [4] dans le cas où le compact K est contenu dans une variété totalement réelle très régulière.

1. Notations.

Soit $J = (j_1, j_2, \cdots, j_{2n})$, un multi-indice dans $\mathbb{N}^{2n}$. On note $\mid J \mid = j = j_1 + j_2 + \cdots + j_{2n}$ et $J! = j_1! j_2! \cdots j_{2n}!$. Pour tout $x = (x_1, x_2, \cdots, x_{2n})$ de $\mathbb{R}^{2n}$, on pose $x^J = x_1^{j_1} x_2^{j_2} \cdots x_{2n}^{j_{2n}}$ et on note $\mid x \mid$ la norme euclidienne de x.

2. Définitions.

Soit E un sous-ensemble compact de $\mathbb{R}^{2n}$. Un jet F sur E est la donnée d'une suite $\{F^J(\zeta); \zeta \in E, J \in \mathbb{N}^{2n}\}$ de fonctions continues sur E à valeurs dans $\mathbb{C}$.

1991 *Mathematics Subject Classification*. Primary 32A25.

This paper is in final form and no version of it will be submitted for publication elsewhere.

Pour un multi-indice L de longueur ℓ de $\mathbb{N}^{2n}$, on définit la L-ième dérivée du jet F par

$$D^L F = \frac{\partial^\ell F}{\partial x_1^{\ell_1} \cdots \partial x_{2n}^{\ell_{2n}}} = \{F^{J+L}(\zeta); \zeta \in E, J \in \mathbb{N}^{2n}\}$$

Pour un opérateur différentiel P à coefficients continus, on définit de manière analogue le jet PF. On dira que F est P−plat si le jet PF est identiquement nul.

Pour tout jet F, tout ζ de E et tout entier p, on définit le polynôme de Taylor de F, pour tout x de $\mathbb{R}^{2n}$, par

$$(2.1) \qquad T_\zeta^p F(x) = \sum_{J, |J| \leq p} \frac{1}{J!}(x - \zeta)^J F^J(\zeta)$$

et, pour tout entier p, pour tout multi-indice $J, j \leq p$ et tout (ζ, x) de $E \times E$,

$$(2.2) \qquad R_\zeta^{J,p} F(x) = F^J(x) - \sum_{K; |J+K| \leq p} \frac{1}{K!}(x - \zeta)^K F^{J+K}(\zeta).$$

On rappelle qu'un jet F est un jet de Whitney sur E de classe C^∞ et, on note $F \in C^\infty(E)$, si on a, pour tout entier p et pour tout multi-indice $J, j \leq p$,

$$R_\zeta^{J,p} F(x) = o(\mid \zeta - x \mid^{p-j})$$

lorsque $\mid \zeta - x \mid$ tend vers 0 avec (ζ, x) dans $E \times E$.

Si f désigne une fonction de classe C^∞ dans $\mathbb{R}^{2n}$, on note, pour tout multi-indice L de longueur ℓ et tout x de $\mathbb{R}^{2n}$,

$$D^L f(x) = \frac{\partial^\ell f}{\partial x_1^{\ell_1} \cdots \partial x_{2n}^{\ell_{2n}}}(x).$$

On sait, d'après le théorème d'extension de Whitney, que, pour tout jet F de $C^\infty(E)$, il existe une fonction f de classe C^∞ dans $\mathbb{R}^{2n}$, à support compact, telle que, pour tout multi-indice L de $\mathbb{N}^{2n}$ et tout ζ de E, on ait

$$F^L(\zeta) = D^L f(\zeta).$$

3. Suites Régulières.

Soit $\{M_p\}_{p \geq 0}$, une suite de réels positifs vérifiant les propriétés suivantes

$$(3.1) \qquad\qquad M_0 = 1,$$

$$(3.2) \qquad\qquad \{M_p\}_{p \geq 0} \text{ est logarithmiquement convexe,}$$

$$(3.3) \qquad\qquad \lim_{p \to \infty} M_p^{1/p} = \infty.$$

On suppose que la suite $\{M_p\}_{p \geq 0}$ vérifie, de plus, la condition suivante : il existe une constante A, $A > 1$, telle que l'on ait

$$(3.4) \qquad M_{p+1} \leq A^{p+1} M_1 M_p, \text{ pour tout } p.$$

En suivant E. M. Dynkin [5], on dira que la suite $\{p!M_p\}_{p \geq 0}$ est régulière.

4. Propriétés des Suites Régulières et des Fonctions Associées.

L'hypothèse (3.2) assure la croissance des deux suites $M_p^{1/p}$ et $\frac{M_p}{M_{p-1}}$ ainsi que l'inégalité

$$(4.1) \qquad M_p^{1/p} \leq \frac{M_p}{M_{p-1}}.$$

On peut remarquer que (3.2) implique également que ces deux suites ont le même comportement à l'infini.

On note $m_p = \frac{M_p}{M_{p-1}}, p \geq 1$ et on a, d'après (3.3), $\lim_{p \to \infty} m_p = \infty$.

On pose, maintenant, pour tout $r > 0$,

$$(4.2) \qquad N(r) = \min\{p \in \mathbb{N}; \inf_k(M_k r^k) = M_p r^p\}$$

et

$$(4.3) \qquad h(r) = \inf_k M_k r^k = M_{N(r)} r^{N(r)}.$$

On convient de poser $h(0) = 0$.

On a

$$(4.4) \qquad N(r) \text{ est le plus petit entier } p \text{ tel que l'on ait } m_{p+1} \geq 1/r$$

et donc $N(r)$ est décroissante.

5. Classes de Carleman.

Soit $\{p!M_p\}_{p \geq 0}$ une suite régulière. On note $C\{p!M_p\}$, la classe de Carleman des fonctions f indéfiniment dérivables dans $\mathbb{R}^{2n}$ et vérifiant la propriété suivante : il existe des constantes C_1 et C_2, $C_1 \geq 0, C_2 \geq 1$ telles que l'on ait, pour tout x de $\mathbb{R}^{2n}$ et tout multi-indice P de longueur p,

$$\mid D^P f(x) \mid \leq C_1 C_2^p p! M_p.$$

Soit E un sous-ensemble compact de $\mathbb{R}^{2n}$. Un jet F sur E est un jet de Whitney de classe $C_E\{p!M_p\}$ s'il existe des constantes C_1 et $C_2, C_1 \geq 0$ et $C_2 \geq 1$, telles que

a) pour tout entier j, pour tout multi-indice J de longueur j et tout x de E, on ait

$$\mid F^J(x) \mid \leq C_1 C_2^j j! M_j,$$

b) pour tout entier m, pour tout multi-indice J de longueur $j \leq m$ et tout (ζ, x) de $E \times E$, on ait

$$\mid (R_\zeta^{J,m} F)(x) \mid \leq C_1 C_2^{m+1} j! M_{m+1} \mid x - \zeta \mid^{m+1-j} .$$

Remarque. L'hypothèse (3.4) assure que les classes $C\{p!M_p\}$ et $C_E\{p!M_p\}$ sont stables par dérivation.

6.   THEOREME. [5]. *Soit* $\{p!M_p\}_{p\geq 0}$ *une suite régulière et* E *un sous ensemble compact de* $\mathbb{C}^n$. *Soit* F *un jet de Whitney appartenant à* $C_E\{p!M_p\}$, $\bar{\partial}$ *-plat sur* E, *alors il existe une fonction* f *de classe* C^∞ *et à support compact dans* $\mathbb{C}^n$ *et des constantes* D_1 *et* D_2 *positives telles que l'on ait*

(6.1) $\begin{aligned} &f = F \ \ sur \ \ E, au \ sens \ des \ jets, \\ &\mid \bar{\partial} f(z) \mid \leq D_1 h[D_2 d(z,E)], \ pour \ tout \ z \ de \ \mathbb{C}^n \end{aligned}$

où $d(z,E)$ *désigne la distance euclidienne de* z *à* E.

7. Classes de Beurling.

Soit $\{p!M_p\}_{p\geq 0}$ une suite régulière. On note $B\{p!M_p\}$, la classe de Beurling des fonctions f indéfiniment dérivables dans $\mathbb{R}^{2n}$ et vérifiant la propriété suivante: quel que soit $C_2 > 0$, il existe $C_1 \geq 0$ telle que l'on ait, pour tout x de $\mathbb{R}^{2n}$ et tout multi-indice P de longueur p,

$$\mid D^P f(x) \mid \leq C_1 C_2^p p! M_p.$$

Soit E un sous-ensemble compact de $\mathbb{R}^{2n}$. On définit $B_E\{p!M_p\}$ à partir de $B\{p!M_p\}$ comme l'on a défini, dans le paragraphe 5, $C_E\{p!M_p\}$ à partir de $C\{p!M_p\}$.

Remarque. Toujours d'après (3.4), les classes $B\{p!M_p\}$ et $B_E\{p!M_p\}$ sont stables par dérivation.

On se propose de montrer l'analogue du théorème 6 pour des classes de Beurling.

8.   LEMME. [7]. *Soit* $\{M_p\}_{p\geq 0}$ *une suite de réels positifs vérifiant (3.1), (3.2) et (3.3) et* $\{L_p\}_{p\geq 0}$ *une suite de réels positifs telle que, pour tout* $t > 0$, *il existe* $C(t) > 0$ *vérifiant, pour tout* $p \geq 0$,

(8.1) $L_p \leq C(t) t^p M_p.$

Alors, il existe une suite $\{N_p\}_{p\geq 0}$ *de réels positifs satifaisant (3.1), (3.2) et (3.3) telle que,*
(i) il existe une constante $C > 0$ *vérifiant, pour tout* $p \geq 0$,

(8.2) $L_p \leq C N_p$

et que,
(ii) pour tout $t > 0$, *il existe* $C'(t) > 0$ *vérifiant, pour tout* $p \geq 0$,

(8.3) $N_p \leq C'(t) t^p M_p.$

9.   THEOREME. *Soit* $\{p!M_p\}_{p\geq 0}$ *une suite régulière et* E *un sous-ensemble compact de* $\mathbb{C}^n$. *Soit* F *un jet de Whitney appartenant à* $B_E\{p!M_p\}$, $\bar{\partial}$ *-plat sur* E, *alors il existe une fonction* f *de classe* C^∞ *à support compact dans* $\mathbb{C}^n$ *et, quelle que soit* $D_2, D_2 > 0$, *il existe* $D_1, D_1 \geq 0$ *vérifiant*

(9.1) $\begin{aligned} &f = F \ \ sur \ \ E \ au \ sens \ des \ jets, \\ &\mid \bar{\partial} f(z) \mid \leq D_1 h[D_2 d(z,E)], \ pour \ tout \ z \ de \ \mathbb{C}^n. \end{aligned}$

Preuve. Soit F un jet de Whitney sur E appartenant à $B_E\{p!M_p\}$, $\bar{\partial}$−plat sur E. Il existe, d'après le lemme 8, une suite $\{N_p\}_{p \geq 0}$ vérifiant (3.1), (3.2), (3.3) , (8.3) et telle que F appartienne à $C_E\{p!N_p\}$. On peut remarquer ici que $\{N_p\}_{p \geq 0}$ ne vérifie peut être pas (3.4). On utilise maintenant la construction de E. M. Dynkin [5, pages 41-43] et on obtient une fonction f de classe C^∞ à support compact dans $\mathbb{C}^n$ dont le jet sur E coïncide avec F et qui vérifie la propriété suivante :

il existe deux constantes $D_3 > 0$ et $D_4 > 0$ telles que l'on ait, pour tout z de $\mathbb{C}^n \backslash E$,

$$(9.2) \qquad \mid \bar{\partial} f(z) \mid \leq D_3 d(z,E)^{-1} h_N(D_4 d(z,E))$$

où h_N désigne la fonction associée à la suite $\{N_p\}_{p \geq 0}$ par (4.3).

On a, pour tout $r > 0$ et tout $t > 0$,

$$(9.3) \qquad h_N(r) = \inf_p r^p N_p \leq \inf_p C'(t)(tr)^p M_p \leq C'(t)h(tr)$$

et donc, pour tout z de $\mathbb{C}^n \backslash E$,

$$(9.4) \qquad \mid \bar{\partial} f(z) \mid \leq D_3 C'(t)d(z,E)^{-1} h(tD_4 d(z,E)).$$

La suite $\{M_p\}_{p \geq 0}$ vérifie (3.4) ; en conséquence, pour $0 < r < 1/M_1$, on a

$$(9.5) \quad h(r) = \inf_p r^p M_p \leq \inf_p r^{p+1} M_{p+1} \leq A M_1 r \inf_p (Ar)^p M_p \leq A M_1 r h(Ar).$$

On en déduit puisque f est à support compact qu'il existe $D_5 > 0$, $D_6 > 0$ et $t_0 > 0$ telles que l'on ait, pour tout z de $\mathbb{C}^n \backslash E$ et tout $t, 0 < t < t_0$,

$$(9.6) \qquad \mid \bar{\partial} f(z) \mid \leq D_5 C'(t) \, t \, h(tD_6 d(z,E)).$$

Un choix convenable de t achève la preuve du théorème puisque h est croissante.

10. PROPOSITION. *Soit K un sous ensemble compact de $\mathbb{R}^{2n}$, s un réel $s \geq 1$ et A_1 une constante $0 < A_1 < 1$. On note*

$$V(K) = \{\zeta \in \mathbb{R}^{2n}; d(\zeta, K) < 1\}$$

et

$$O = O(A_1) = \{(z, \zeta) \in \mathbb{R}^{2n} \times (V(K) \backslash K); d(z, K) < A_1 d(\zeta, K)^s\}.$$

Soit Λ un espace métrique compact. Soit $f(z, \zeta, \lambda)$ une fonction définie sur $O \times \Lambda$, à valeurs complexes, telle que, pour tout multi-indice J de $\mathbb{N}^{2n}$, $D_z^J(f(z, \zeta, \lambda)$ soit définie et continue sur $O \times \Lambda$.

On suppose qu'il existe des constantes A_2, A_3 et m, $A_2 \geq 0, A_3 \geq 1$ et $m \geq 0$, telles que pour tout multi-indice J de $\mathbb{N}^{2n}$, de longueur j, on ait sur $O \times \Lambda$

$$(10.1) \qquad \mid D_z^J f(z, \zeta, \lambda) \mid \leq A_2 A_3^j j! d(\zeta, K)^{-js-m}.$$

Soit g une fonction continue et à support compact dans $V(K)$ telle que, pour tout entier p, il existe une constante $C(p)$ vérifiant

$$(10.2) \qquad \mid g(\zeta) \mid d(\zeta, K)^{-p} \leq C(p), \text{ pour tout } \zeta \text{ de } V(K).$$

Soit μ une mesure borélienne bornée sur $V(K) \times \Lambda$.

On définit, pour z dans K et J un multi-indice de $\mathbb{N}^{2n}$,

$$H^J(z) = \int_{(V(K) \backslash K) \times \Lambda} D_z^J f(z, \zeta, \lambda) g(\zeta) d\mu(\zeta, \lambda).$$

Alors $H = \{H^J(z); z \in K, J \in N^n\}$ est un jet de Whitney sur K de classe C^∞.

Pour tout $\epsilon > 0$, on note χ_ϵ une fonction continue dans $\mathbb{R}^{2n}$, à valeurs comprises entre 0 et 1, égale à 1 sur $\{\zeta \in \mathbb{R}^{2n}; d(\zeta, K) > \epsilon\}$ et à 0 sur $\{\zeta \in \mathbb{R}^{2n}; d(\zeta, K) < \frac{\epsilon}{2}\}$. On pose

$$(10.3) \qquad H_\epsilon(z) = \int_{(V(K) \backslash K) \times \Lambda} f(z, \zeta, \lambda) \chi_\epsilon(\zeta) g(\zeta) d\mu(\zeta, \lambda).$$

Alors la fonction $H_\epsilon(z)$ est de classe C^∞ dans un voisinage de K et tend vers H dans $C^\infty(K)$ lorsque ϵ tend vers 0.

Preuve. La première partie de la proposition a été établie dans [2, proposition 13]. On en rappelle brièvement les idées pour la commodité du lecteur.

On note, pour z dans K, λ dans Λ et J un multi-indice de $\mathbb{N}^{2n}$,

$$(10.4) \qquad b^J(z, \zeta, \lambda) = \begin{cases} D_z^J f(z, \zeta, \lambda) g(\zeta), & \text{si } \zeta \in V(K) \backslash K, \\ \\ 0, & \text{si } \zeta \in K. \end{cases}$$

Les fonctions b^J sont continues dans $K \times V(K) \times \Lambda$ et à support compact. Si J est un multi-indice de $\mathbb{N}^{2n}$, p un entier, x et y deux points de K et (ζ, λ) dans $V(K) \times \Lambda$, on pose

$$R_x^{J,p} B(y, \zeta, \lambda) = b^J(z, \zeta, \lambda) - \sum_{L; |L+J| \leq p} \frac{b^{L+J}(z, \zeta, \lambda)(x - y)^L}{L!}.$$

On montre que, pour tout ζ, on a
(10.5)
$$\mid R_x^{J,p} B(y, \zeta, \lambda) \mid \leq A_2 (A_n A_1^{-1} A_3)^{p+1} j! \mid x - y \mid^{p+1-j} \mid g(\zeta) \mid d(\zeta, K)^{-(p+1)s-m}$$

où A_n est une constante strictement plus grande que 1 ne dépendant que de la dimension.
On déduit alors de (10.2) et (10.5)

$$(10.6) \quad \mid R_x^{J,p} B(y, \zeta, \lambda) \mid \leq A_2 (A_n A_1^{-1} A_3)^{p+1} j! \mid x - y \mid^{p+1-j} C((p+1)s + m)$$

et, par intégration par rapport à la mesure μ,

$$(10.7) \quad \mid R_x^{J,p} H(y) \mid \leq A_2 \mid \mu \mid (A_n A_1^{-1} A_3)^{p+1} j! \mid x - y \mid^{p+1-j} C((p+1)s + m).$$

Ici $\mid \mu \mid$ désigne la variation totale de μ. Ceci établit la régularité C^∞ du jet H. Clairement, puisque dans l'intégrale (10.3) définissant H_ϵ n'interviennent que les ζ vérifiant $d(\zeta, K) \geq \frac{\epsilon}{2}$, la fonction H_ϵ est de classe C^∞ dans $\{z \in \mathbb{R}^{2n}; d(z, K) < A_1 \left(\frac{\epsilon}{2}\right)^s\}$.

Avec B_ϵ défini comme B en remplaçant $g(\zeta)$ par $\chi_\epsilon(\zeta)g(\zeta)$, on a, pour tout ζ de $V(K)$,

$$| R_x^{J,p}(B - B_\epsilon)(y,\zeta,\lambda) |\leq$$
$$A_2(A_n A_1^{-1} A_3)^{p+1} j! \, | \, x - y \, |^{p+1-j}| \, g(\zeta) \, | \, (1 - \chi_\epsilon)d(\zeta, K)^{-(p+1)s-m}$$

et donc, d'après la définition de χ_ϵ,

$$| R_x^{J,p}(B - B_\epsilon)(y,\zeta,\lambda) |\leq$$
$$A_2(A_n A_1^{-1} A_3)^{p+1} j! \, | \, x - y \, |^{p+1-j}| \, g(\zeta) \, | \, \epsilon \, d(\zeta, K)^{-(p+1)s-m-1}.$$

De là, d'après (10.2), on a

$$| R_x^{J,p}(B - B_\epsilon)(y,\zeta,\lambda) |\leq A_2(A_n A_1^{-1} A_3)^{p+1} j! \, | \, x-y \, |^{p+1-j} \, \epsilon \, C((p+1)s+m+1).$$

Ceci établit la convergence de H_ϵ vers H dans $C^\infty(K)$ lorsque ϵ tend vers 0.

11. COROLLAIRE. *Soit $\{p!M_p\}$ une suite régulière. Sous les hypothèses de la proposition 10 dans lesquelles on se restreint à $s = 1$, s'il existe deux constantes $C_1' \geq 0$ et $C_2' \geq 1$ telles que l'on ait, pour tout entier p,*

$$(11.1) \qquad C(p) \leq C_1' C_2'^p M_p$$

alors le jet H appartient à $C_K\{p!M_p\}$.

De plus, pour tout $\epsilon > 0$, la fonction $H_\epsilon(z)$ est de classe $C\{p!M_p\}$ dans un voisinage de K et tend vers H dans la classe $C_K\{p!M_p\}$ lorsque ϵ tend vers 0.

Preuve. Le corollaire énoncé dans [2] pour les classes de Gevrey se généralise aisément aux classes de Carleman.

La convergence de H_ϵ vers H dans $C_K\{p!M_p\}$ est immédiate.

Remarque: on peut interpréter les conditions (10.2) et (11.1) en écrivant : il existe deux constantes $C_1' \geq 0$ et $C_2' \geq 0$ telles que l'on ait

$$(11.2) \qquad | \, g(\zeta) \, |\leq C_1' h(C_2' d(\zeta, K)), \text{ pour tout } \zeta \text{ de } V(K).$$

12. Définitions. [2].

Soit s un réel, $s \geq 1$.

Un sous ensemble compact K de $\mathbb{C}^n$ est s-H convexe s'il existe une constante $A_0, 0 < A_0 \leq 1$ telle que, pour tout $\epsilon, 0 < \epsilon \leq 1$, il existe un ouvert pseudoconvexe Ω_ϵ vérifiant

$$(12.1) \qquad \{z \in \mathbb{C}^n; d(z, K) < A_0 \epsilon^s\} \subset \Omega_\epsilon \subset \{z \in \mathbb{C}^n; d(z, K) < \epsilon\}.$$

On note toujours

$$V(K) = \{\zeta \in C^n; d(\zeta, K) < 1\}$$

et, pour toute constante $A_1, 0 < A_1 \leq A_0$,

$$O = O(A_1) = \{(z, \zeta) \in \mathbb{C}^n \times (V(K)\backslash K); d(z, K) < A_1 d(\zeta, K)^s\}.$$

De même, pour ζ dans $V(K)\backslash K$, on pose

$$O(A_1, \zeta) = \{z \in \mathbb{C}^n; d(z, K) < A_1 d(\zeta, K)^s\}.$$

13. Dans [2, proposition 12], on construit $w(z, \zeta)$ une section du fibré de Cauchy-Leray.

PROPOSITION. *Soit K, un compact de $\mathbb{C}^n$, $s-H$ convexe. Soit ρ un réel, $\rho > 0$. Il existe des constantes A_4 et A_5, ne dépendant que de K, de s, de ρ et de n, $0 < A_4 < 1 < A_5$ et des fonctions $w_i(z, \zeta)$, $i = 1, \cdots, n$, de classe C^∞ sur $O(A_4)$, holomorphes en z dans $O(A_4, \zeta)$ pour tout ζ dans $V(K) \backslash K$ et vérifiant pour tout (z, ζ) de $O(A_4)$,*

$$(13.1) \qquad \operatorname{Re} \sum_{i=1}^{n} w_i(z, \zeta)(z_i - \zeta_i) \geq \frac{1}{2},$$

(13.2)
   *pour tout multiindice L de longueur ℓ de $\mathbb{N}^{2n}$, vérifiant $\mid \ell \mid \leq 2$,
   pour tout multiindice J de longueur j de $\mathbb{N}^{2n}$ et pour tout $i, i = 1, \cdots, n$,*
$$\mid D_\zeta^L D_z^J w_i(z, \zeta) \mid \leq j! A_5^{j+1} d(\zeta, K)^{-(1+\ell)(ns+\rho)-js}.$$

14. On introduit, comme dans [2] et [6], les formes différentielles suivantes

$$\omega(\zeta) = d\zeta_1 \wedge \cdots \wedge d\zeta_n.$$

On note, si $v(x, y)$ est une fonction définie pour (x, y) dans $X \times Y$, de classe C^1 en (x, y) et à valeurs dans $\mathbb{C}^n$,

$$\omega'(v) = \sum_{j=1}^{n} (-1)^{j+1} v_j \bigwedge_{k \neq j} dv_k$$

et

$$\omega'_x(v) = \sum_{j=1}^{n} (-1)^{j+1} v_j \bigwedge_{k \neq j} d_x v_k$$

où d_x est la différentielle extérieure par rapport à x. On utilisera également la notation $\omega'_{x,y}(v)$ au lieu de $\omega'(v)$.
Ainsi, on a

$$\omega' \left(\frac{\bar\zeta - \bar z}{\mid \zeta - z \mid^2} \right) = \frac{\sum_{j=1}^{n} (-1)^{j+1} (\bar\zeta_j - \bar z_j) \bigwedge_{k \neq j} (d\bar\zeta_k - d\bar z_k)}{\mid \zeta - z \mid^{2n}}$$

On note, pour tout (z, ζ) de $O(A_4)$ et tout λ de $[0, 1]$,

$$\eta(z, \zeta, \lambda) = (1 - \lambda) \frac{w(z, \zeta)}{\sum_{j=1}^{n} w_j(z, \zeta)(\zeta_j - z_j)} + \lambda \frac{\bar\zeta - \bar z}{\mid \zeta - z \mid^2},$$
$$= (\eta_1(z, \zeta, \lambda), \cdots, \eta_n(z, \zeta, \lambda))$$

et

$$\bar\theta(z, \zeta, \lambda) = \frac{(n-1)!}{(2i\pi)^n} \sum_{j=1}^{n} (-1)^{j+1} \eta_j(z, \zeta, \lambda) \bigwedge_{k \neq j} (\bar\partial_\zeta + \bar\partial_z + d_\lambda) \eta_k(z, \zeta, \lambda).$$

Avec les notations précédentes, on a

$$\bar{\theta}(z,\zeta,\lambda) = \frac{(n-1)!}{(2i\pi)^n}\omega'_{\bar{z},\bar{\zeta},\lambda}(\eta(z,\zeta,\lambda))$$

et, si on note

$$\Phi(z,\zeta) = \sum_{j=1}^{n} w_j(z,\zeta)(\zeta_j - z_j),$$

on a

$$\omega'_{\bar{z},\bar{\zeta}}\left(\frac{w(z,\zeta)}{\Phi(z,\zeta)}\right) = \frac{\sum_{j=1}^{n}(-1)^{j+1}w_j(z,\zeta)\bigwedge_{k\neq j}(\bar{\partial}_\zeta + \bar{\partial}_z)w_k(z,\zeta)}{(\Phi(z,\zeta))^n}.$$

Les formes η et $\bar{\theta}$ sont de classe C^∞ dans $O(A_4) \times [0,1]$ et, puisque, dans $O(A_4) \times [0,1]$, on a

$$\sum_{j=1}^{n} \eta_j(z,\zeta,\lambda)(\zeta_j - z_j) = 1,$$

on a, dans $O(A_4) \times [0,1]$,

(14.1) $$(\bar{\partial}_\zeta + \bar{\partial}_z + d_\lambda)(\bar{\theta} \wedge \omega) = 0.$$

On a alors

(14.2) $$\bar{\theta}(z,\zeta,\lambda)\mid_{\lambda=1} = \frac{(n-1)!}{(2i\pi)^n}\omega'\left(\frac{\bar{\zeta}-\bar{z}}{\mid \zeta - z \mid^2}\right)$$

et

(14.3) $$\bar{\theta}(z,\zeta,\lambda)\mid_{\lambda=0} = \frac{(n-1)!}{(2i\pi)^n}\omega'_{\bar{z},\bar{\zeta}}\left(\frac{w(z,\zeta)}{\Phi(z,\zeta)}\right)$$

15. LEMME. *Soit* $\delta(z,\zeta,\lambda)$ *un coefficient d'une des formes* $\bar{\theta}, \bar{\partial}_\zeta\bar{\theta}, \bar{\partial}_z\bar{\theta}, d_\lambda\bar{\theta}$. *Alors il existe une constante* $A_6 \geq 1$ *telle que l'on ait, pour tout* (z,ζ,λ) *de* $O(A_4) \times [0,1]$ *et tout multi-indice* L *de longueur* ℓ *de* $\mathbb{N}^{2n}$

(15.1) $$\mid D_z^L\delta(z,\zeta,\lambda) \mid \leq A_6^{\ell+1}\ell!d(\zeta,K)^{-(ns+\rho)(2n-2)-1-\ell s}.$$

Preuve. Ceci est une conséquence immédiate de la proposition 13 et de la définition de $\bar{\theta}$.

16. PROPOSITION. *Soit* K *un compact de* $\mathbb{C}^n$, *s-H convexe. Soit* f *une fonction continue et à support compact dans* $V(K)$ *telle que* $\bar{\partial}f$ *soit continue dans* $V(K)$ *et vérifie la propriété suivante : pour tout entier* $p, p > 0$, *il existe une constante* $C(p)$ *vérifiant*

(16.1) $$\mid \bar{\partial}f(\zeta) \mid d(\zeta,K)^{-p} \leq C(p), \text{ pour tout } \zeta \text{ de } V(K).$$

Alors, pour tout z *de* K *et tout multi-indice* L *de* $\mathbb{N}^{2n}$, *on a*

$$(16.2) \qquad \int_{\zeta \in V(K)} \bar{\partial}_\zeta f \wedge D_z^L \omega_\zeta' \left(\frac{\bar{\zeta} - \bar{z}}{|\zeta - z|^2} \right) \wedge \omega(\zeta)$$

$$= \int_{\zeta \in V(K)} \bar{\partial}_\zeta f \wedge D_z^L \omega_\zeta' \left(\frac{w(\zeta, z)}{\Phi(\zeta, z)} \right) \wedge \omega(\zeta).$$

Preuve. On pose

$$\gamma = \bar{\partial}_\zeta f \wedge \bar{\theta} \wedge \omega \text{ dans } O(A_4) \times [0, 1].$$

La forme $\bar{\theta} \wedge \omega$ étant C^∞ dans $O(A_4) \times [0, 1]$, on peut prendre la différentielle extérieure de γ en ζ et λ. On a, au sens des distributions, en utilisant (14.1) dans $O(A_4) \times [0, 1]$,

$$(16.3) \qquad \begin{aligned} (d_\zeta + d_\lambda)\gamma &= (d_\zeta + d_\lambda)(\bar{\partial}_\zeta f \wedge \bar{\theta} \wedge \omega) \\ &= (\bar{\partial}_\zeta + d_\lambda)(\bar{\partial}_\zeta f \wedge \bar{\theta} \wedge \omega) \\ &= (-1)\bar{\partial}_\zeta f \wedge (\bar{\partial}_\zeta + d_\lambda)(\bar{\theta} \wedge \omega) \\ &= \bar{\partial}_\zeta f \wedge \bar{\partial}_z(\bar{\theta} \wedge \omega) \\ &= -\bar{\partial}_z(\bar{\partial}_\zeta f \wedge \bar{\theta} \wedge \omega). \end{aligned}$$

Soit L un multi-indice de $\mathbb{N}^{2n}$. On a aussi dans $O(A_4) \times [0, 1]$

$$(16.4) \qquad (d_\zeta + d_\lambda)D_z^L\gamma = \bar{\partial}_\zeta f \wedge \bar{\partial}_z(D_z^L\bar{\theta} \wedge \omega) = -\bar{\partial}_z(\bar{\partial}_\zeta f \wedge D_z^L\bar{\theta} \wedge \omega).$$

De là, pour z fixé dans K, d'après (16.1) et le lemme 15, $D_z^L\gamma$ se prolonge en une forme continue en ζ sur $V(K) \times [0, 1]$ telle que $(d_\zeta + d_\lambda)D_z^L\gamma$ soit continue dans $V(K) \times [0, 1]$ et l'égalité (16.4) s'étend à $V(K) \times [0, 1]$.

On a donc, en utilisant la formule de Stokes et en notant D_0 un ouvert borné à frontière C^∞ contenant le support de f et inclus dans $V(K)$,

$$(16.5) \qquad \int_{(\zeta,\lambda) \in \partial(D_0 \times [0,1])} D_z^L\gamma = \int_{(\zeta,\lambda) \in D_0 \times [0,1]} (d_\zeta + d_\lambda)D_z^L\gamma$$

et, d'après (16.4),

$$\begin{aligned} \int_{(\zeta,\lambda) \in \partial(D_0 \times [0,1])} D_z^L\gamma &= -\int_{(\zeta,\lambda) \in D_0 \times [0,1]} \bar{\partial}_z(\bar{\partial}_\zeta f \wedge D_z^L\bar{\theta} \wedge \omega) \\ &= -\bar{\partial}_z \int_{(\zeta,\lambda) \in D_0 \times [0,1]} \bar{\partial}_\zeta f \wedge D_z^L\bar{\theta} \wedge \omega. \end{aligned}$$

Puisque $\bar{\partial}_\zeta f$ est nul sur K d'après (16.1), on a

$$(16.6) \qquad \int_{(\zeta,\lambda) \in \partial(D_0 \times [0,1])} D_z^L\gamma = -\bar{\partial}_z \int_{(\zeta,\lambda) \in (V(K) \backslash K) \times [0,1]} \bar{\partial}_\zeta f \wedge D_z^L\bar{\theta} \wedge \omega.$$

On calcule maintenant l'intégrale figurant au premier membre de (16.6). On a

$$\partial(D_0 \times [0, 1]) = \partial D_0 \times [0, 1] + D_0 \times \{1\} - D_0 \times \{0\}.$$

On remarque que l'on a

$$(16.7) \qquad \int_{(\zeta,\lambda) \in \partial D_0 \times [0,1]} \bar{\partial}_\zeta f \wedge D_z^L\bar{\theta} \wedge \omega = 0,$$

car f est à support compact dans $V(K)$.

On déduit de (14.2), (14.3) et de la définition de D_0 que l'on a

(16.8)
$$\int_{(\zeta,\lambda)\in D_0\times\{0\}} \bar{\partial}_\zeta f \wedge D_z^L \bar{\theta} \wedge \omega$$
$$= \frac{(n-1)!}{(2i\pi)^n} \int_{\zeta\in V(K)} \bar{\partial}_\zeta f \wedge D_z^L \omega'_{\bar{z},\zeta} \left(\frac{w(z,\zeta)}{\Phi(z,\zeta)}\right) \wedge \omega(\zeta)$$
$$\int_{(\zeta,\lambda)\in D_0\times\{1\}} \bar{\partial}_\zeta f \wedge D_z^L \bar{\theta} \wedge \omega$$
$$= \frac{(n-1)!}{(2i\pi)^n} \int_{\zeta\in V(K)} \bar{\partial}_\zeta f \wedge D_z^L \omega'_{\bar{z},\zeta} \left(\frac{\bar{\zeta}-\bar{z}}{\mid \zeta - z \mid^2}\right) \wedge \omega(\zeta).$$

On remarque maintenant que $\bar{\partial}_\zeta f$ est de bidegré (0,1) en $(\zeta,\bar{\zeta})$ et que $\omega(\zeta)$ est de bidegré (n,0) en $(\zeta,\bar{\zeta})$. On ne doit donc garder dans $\bar{\partial}_\zeta\bar{\theta}\mid_{\lambda=0}$ et $\bar{\partial}_\zeta\bar{\theta}\mid_{\lambda=1}$ aucune dérivation en $\bar{z}$.

On a donc, en reprenant (16.6) et en utilisant (14.2),

(16.9)
$$\frac{(n-1)!}{(2i\pi)^n} \int_{\zeta\in V(K)} \bar{\partial}_\zeta f \wedge D_z^L \omega'_{\bar\zeta} \left(\frac{w(z,\zeta)}{\Phi(z,\zeta)}\right) \wedge \omega(\zeta)$$
$$\frac{(n-1)!}{(2i\pi)^n} \int_{\zeta\in V(K)} \bar{\partial}_\zeta f \wedge D_z^L \omega'_{\bar\zeta} \left(\frac{\bar{\zeta}-\bar{z}}{\mid \zeta - z \mid^2}\right) \wedge \omega(\zeta)$$
$$= -\bar{\partial}_z \int_{(\zeta,\lambda)\in(V(K)\backslash K)\times[0,1]} \bar{\partial}_\zeta f \wedge D_z^L \bar{\theta} \wedge \omega(\zeta).$$

L'intégrale figurant dans le membre de droite de (16.9) est nulle. En effet, puisqu'il faut prendre dans $\bar{\theta}$ un degré en λ pour l'intégration sur $[0,1]$, il ne reste plus qu'un bidegré total (0,n−2) pour les variables $\bar{z}$ et $\bar{\zeta}$.

On déduit donc (16.2) de (16.9).

17. THEOREME. *Soit K un compact arbitraire de $\mathbb{C}^n$. Soit f une fonction continue et à support compact dans $V(K)$ telle que $\bar{\partial}f$ soit continue dans $V(K)$ et vérifie la propriété suivante : pour tout entier $p, p \geq 0$, il existe une constante $C(p)$ vérifiant*

(17.1)
$$\mid \bar{\partial}f(\zeta) \mid d(\zeta,K)^{-p} \leq C(p), \text{ pour tout } \zeta \text{ de } V(K).$$

Alors, le jet défini sur K, pour tout multi-indice L, par

(17.2)
$$f^L(z) = -\frac{(n-1)!}{(2i\pi)^n} \int_{\zeta\in V(K)} \bar{\partial}_\zeta f \wedge D_z^L \omega'_\zeta \left(\frac{\bar{\zeta}-\bar{z}}{\mid \zeta - z \mid^2}\right) \wedge \omega(\zeta)$$

est $\bar{\partial}$-plat sur K et on a

(17.3)
$$f^0 = f \text{ sur } K.$$

Si, de plus, f est de classe C^k, k entier ou $k = \infty$, alors on a

(17.4)
$$f^L = D^L f \text{ sur } K, \text{pour tout } L \text{ de longueur } \ell, \ell \leq k - 1.$$

Preuve. La formule de Martinelli-Bochner [6] donne, puisque f est à support compact dans $V(K)$, que l'on a, pour tout z de $V(K)$,

$$(17.5) \qquad f(z) = -\frac{(n-1)!}{(2i\pi)^n} \int_{\zeta \in V(K)} \bar{\partial}_\zeta f \wedge \omega'_\zeta \left(\frac{\bar{\zeta} - \bar{z}}{|\zeta - z|^2} \right) \wedge \omega(\zeta).$$

Sans hypothèse supplémentaire de régularité sur f, on montre, en utilisant la proposition 10 et (17.1) que le jet défini sur K, pour tout multi-indice L, par

$$f^L(z) = -\frac{(n-1)!}{(2i\pi)^n} \int_{\zeta \in V(K)} \bar{\partial}_\zeta f \wedge D_z^L \omega'_\zeta \left(\frac{\bar{\zeta} - \bar{z}}{|\zeta - z|^2} \right) \wedge \omega(\zeta)$$

est un jet de Whitney de classe C^∞ sur K. Ceci établit (17.2) et (17.3).

On suppose maintenant f de classe C^k, k entier ou $k = \infty$. Alors, on montre comme dans [2], en utilisant une approximation de l'unité et (17.1) que l'on a, pour tout z de K et pour tout multi-indice L de longueur $\ell, \ell \le k - 1$,

$$(17.6) \qquad \begin{aligned} &D_z^L \int_{\zeta \in V(K)} \bar{\partial}_\zeta f \wedge \omega'_\zeta \left(\frac{\bar{\zeta} - \bar{z}}{|\zeta - z|^2} \right) \wedge \omega(\zeta) \\ &= \int_{\zeta \in V(K)} \bar{\partial}_\zeta f \wedge D_z^L \omega'_\zeta \left(\frac{\bar{\zeta} - \bar{z}}{|\zeta - z|^2} \right) \wedge \omega(\zeta). \end{aligned}$$

De là, on a, d'après (17.5) et (17.6), pour tout multi-indice L de longueur $\ell, \ell \le k - 1$ et pour tout z de K,

$$(17.7) \qquad D_z^L f(z) = \frac{(n-1)!}{(2i\pi)^n} \int_{\zeta \in V(K)} \bar{\partial}_z f \wedge D_z^L \omega'_\zeta \left(\frac{\bar{\zeta} - \bar{z}}{|\zeta - z|^2} \right) \wedge \omega(\zeta).$$

ce qui établit (17.4).

Pour montrer que le jet $f^L(z)$ est $\bar{\partial}$-plat sur K, il faut montrer que, si z_0 est un point de K, $\bar{\partial}_z f^L(z_0)$ est nul. Or, un tel point z_0 est trivialement 1-H convexe et d'après la preuve de la proposition 16, on a

$$(17.8) \qquad \begin{aligned} &\int_{\zeta \in V(K)} \bar{\partial}_\zeta f \wedge D_z^L \omega'_\zeta \left(\frac{\bar{\zeta} - \bar{z}_0}{|\zeta - z_0|^2} \right) \wedge \omega(\zeta) \\ &= \int_{\zeta \in V(K)} \bar{\partial}_\zeta f \wedge D_z^L \omega'_\zeta \left(\frac{\tilde{w}(\zeta, z_0)}{\tilde{\Phi}(\zeta, z_0)} \right) \wedge \omega(\zeta) \end{aligned}$$

avec $\tilde{w}(\zeta, z) = (\bar{\zeta} - \bar{z}_0)/|\zeta - z_0|^2$ et $\tilde{\Phi}(\zeta, z) = \sum_{j=1}^n \tilde{w}_j(\zeta, z)(\zeta_j - z_j)$.

La $\bar{\partial}$−platitude du jet se déduit alors de (17.7) et de (17.8) car le noyau figurant dans l'intégrale de droite est holomorphe en z.

Remarque 1. Si K est un compact arbitraire de $\mathbb{C}^n$. Tout jet de Whitney F, de classe C^∞ sur K et $\bar{\partial}$−plat admet une représentation intégrale de la forme

$$(17.9) \qquad F^L(z) = -\frac{(n-1)!}{(2i\pi)^n} \int_{\zeta \in V(K)} \varphi \wedge D_z^L \omega'_\zeta \left(\frac{\bar{\zeta} - \bar{z}}{|\zeta - z|^2} \right) \wedge \omega(\zeta)$$

où φ est une (0,1) forme en $(\zeta, \bar{\zeta})$, à support compact dans $V(K)$, continue dans $\mathbb{C}^n$ telle que, pour tout $p \geq 0$, il existe une constante $C(p) > 0$ vérifiant

$$| \varphi(\zeta) | \, d(\zeta, K)^{-p} \leq C(p), \text{ pour tout } \zeta \text{ de } \mathbb{C}^n.$$

Ici $\varphi = \bar{\partial} f$ où f est une extension C^∞ et à support compact dans $V(K)$ du jet F.

Remarque 2. On suppose dans l'énoncé du théorème 17 que f vérifie de plus pour tout $p \geq 0$, il existe une constante $C'(p) > 0$ telle que l'on ait

$$(17.10) \qquad | f(\zeta) | \, d(\zeta, K)^{-p} \leq C'(p), \text{ pour tout } \zeta \text{ de } V(K).$$

Alors, le jet défini par la formule (17.2) est identiquement nul. Il s'agit là d'une application de la formule de Stokes.

18. THEOREME. *Soit K un compact de $\mathbb{C}^n$, s-H convexe.*
Les propositions suivantes sont équivalentes :
  (a) F est un jet de Whitney sur K de classe C^∞ et $\bar{\partial}$−plat.
  (b) F admet une représentation intégrale, plus précisément, pour tout multi-indice L, on a

$$(18.1) \qquad F^L(z) = \int_{V(K)} \varphi(\zeta) \wedge D_z^L \omega'_{\bar{\zeta}} \left(\frac{w(\zeta, z)}{\Phi(\zeta, z)} \right) \wedge \omega(\zeta), \text{ pour tout } z \text{ de } K$$

où φ est une (0,1) forme en $(\zeta, \bar{\zeta})$, à support compact dans $V(K)$, continue dans $\mathbb{C}^n$ telle que, pour tout $p \geq 0$, il existe une constante $C(p) > 0$ vérifiant

$$(18.2) \qquad | \varphi(\zeta) | \, d(\zeta, K)^{-p} \leq C(p), \text{ pour tout } \zeta \text{ de } \mathbb{C}^n$$

et où $\omega'_{\bar{\zeta}} \left(\frac{w(\zeta,z)}{\Phi(\zeta,z)} \right)$ est la (0,n −1) forme en $(\zeta, \bar{\zeta})$, holomorphe en z définie dans le paragraphe 14.

Preuve. D'après le théorème de Whitney, le jet F s'étend en une fonction f de classe C^∞ dans $\mathbb{C}^n$ à support compact dans $V(K)$ telle que, pour tout $p \geq 0$, il existe une constante $C(p) > 0$ vérifiant :

$$| \bar{\partial} f(\zeta) | \leq C(p) d(\zeta, K)^p, \text{ pour tout } \zeta \text{ de } \mathbb{C}^n.$$

D'après la proposition 16 et le théorème 17, on a donc (18.1).

Réciproquement, si (18.1) est vérifiée, le jet F est un jet de Whitney sur K de classe C^∞, d'après la proposition 10. Il est $\bar{\partial}$−plat puisque le noyau est holomorphe.

19. COROLLAIRE. *Soit K un compact de $\mathbb{C}^n$, s−H convexe et F un jet de Whitney sur K de classe C^∞ et $\bar{\partial}$-plat. Alors F est limite dans la classe $C^\infty(K)$ de fonctions holomorphes au voisinage de K.*

Preuve. D'après le théorème 18, F admet la représentation

$$(19.1) \qquad F^L(z) = \int_{V(K)} \varphi(\zeta) \wedge D_z^L \omega'_{\bar{\zeta}} \left(\frac{w(\zeta, z)}{\Phi(\zeta, z)} \right) \wedge \omega(\zeta), \text{ pour tout } z \text{ de } K$$

où φ vérifie : pour tout p, il existe une constante $C(p)$ telle que l'on ait

$$|\varphi(\zeta)| \le C(p)d(\zeta, K)^p, \text{ pour tout } \zeta \text{ de } \mathbb{C}^n.$$

On considère la fonction définie par

$$(19.2) \qquad F_\epsilon(z) = \int_{V(K)} \chi_\epsilon(\zeta)\varphi(\zeta) \wedge \omega'_\zeta \left(\frac{w(\zeta, z)}{\Phi(\zeta, z)}\right) \wedge \omega(\zeta),$$

où χ_ϵ a été définie dans la proposition 10. D'après la proposition 10, F_ϵ converge vers F dans la classe $C^\infty(K)$ lorsque ϵ tend vers 0. Ici, $\omega'_\zeta\left(\frac{w(\zeta, z)}{\Phi(\zeta, z)}\right)$ est holomorphe en z; la fonction F_ϵ est donc holomorphe au voisinage de K.

Dans le cas des classes de Carleman et pour les compacts 1–H convexes, on obtient les résultats suivants.

20. THEOREME. *Soit K un compact arbitraire de $\mathbb{C}^n$, 1-H convexe . Soit f une fonction continue et à support compact dans $V(K)$ telle que $\bar{\partial}f$ soit continue dans $V(K)$ et vérifie la propriété suivante : il existe des constantes A_7 et $A_8, A_7 \ge 0, A_8 \ge 1$, telles que l'on ait*

$$(20.1) \qquad |\bar{\partial}f(\zeta)| \le A_7 h[A_8 d(\zeta, K)], \text{ pour tout } \zeta \text{ de } V(K)$$

où h a été définie en (4.3).

Alors, le jet défini sur K, pour tout multi-indice L, par

$$(20.2) \qquad f^L(z) = -\frac{(n-1)!}{(2i\pi)^n} \int_{\zeta \in V(K)} \bar{\partial}_\zeta f \wedge D^L_z \omega'_\zeta \left(\frac{w(\zeta, z)}{\Phi(\zeta, z)}\right) \wedge \omega(\zeta)$$

appartient à $C_K\{p!M_p\}$ et il est $\bar{\partial}$-plat sur K.

On a

$$(20.3) \qquad\qquad\qquad f^0 = f \text{ sur } K.$$

Si, de plus, f est de classe C^k, k entier ou $k = \infty$, alors on a

$$(20.4) \qquad f^J = D^J f \text{ sur } K, \text{pour tout } L \text{ de longueur } \ell, \ell \le k-1.$$

Preuve. Il s'agit là d'une consequence simple du théorème 17.

En effet, l'hypothèse (20.1) implique que, pour tout entier p, on a

$$|\bar{\partial}f(\zeta)| d(\zeta, K)^{-p} \le A_7 A_8^p M_p, \text{ pour tout } \zeta \text{ de } V(K).$$

D'après le corollaire 11, le membre de droite de (20.2) définit un jet de classe $C_K\{p!M_p\}$. On a donc établi (20.3) ou (20.4), selon la régularité imposée à f. On peut remarquer que, sous l'hypothèse de 1-H convexité de K, la $\bar{\partial}$−platitude du jet est évidente puisque le noyau est holomorphe en z.

Le théorème suivant généralise un théorème de E. M. Dynkin [5] aux compacts 1-H convexes de $\mathbb{C}^n$.

21. THEOREME. *Soit K un compact de $\mathbb{C}^n$, 1-H convexe et soit F un et de Whitney sur K de classe C^∞ et $\bar{\partial}$-plat.*
Les propositions suivantes sont équivalentes :

   (a) F appartient à $C_K\{p!M_p\}$,

   (b) F s'étend en une fonction f de classe C^∞ dans $\mathbb{C}^n$ telle qu' il existe des constantes A_9 et A_{10}, $A_9 \geq 0$, $A_{10} \geq 1$, telles que l'on ait

$$(21.1) \qquad |\,\bar{\partial} f(\zeta)\,| \leq A_9 h[A_{10} d(\zeta, K)], \; \textit{pour tout } \zeta \textit{ de } \mathbb{C}^n,$$

où h est la fonction associée à la suite $\{M_p\}_{p \geq 0}$ par (4.3),

   (c) F admet une représentation intégrale ; plus précisément, pour tout multi-indice L, on a

$$(21.2) \qquad F^L(z) = \int_{V(K)} \varphi(\zeta) \wedge D_z^L \omega_\zeta'\left(\frac{w(\zeta, z)}{\Phi(\zeta, z)}\right) \wedge \omega(\zeta), \; \textit{pour tout } z \textit{ de } K$$

où φ est une (0,1) forme en $(\zeta, \bar{\zeta})$, à support compact dans $V(K)$, continue dans $\mathbb{C}^n$ telle qu'il existe des constantes A_{11} et A_{12}, $A_{11} \geq 0$, $A_{12} \geq 1$, vérifiant

$$(21.3) \qquad |\,\varphi(\zeta)\,| \leq A_{11} h[A_{12} d(\zeta, K)], \; \textit{pour tout } \zeta \textit{ de } \mathbb{C}^n$$

et où $\omega_\zeta'\left(\frac{w(\zeta, z)}{\Phi(\zeta, z)}\right)$ est la (0,n −1) forme en $(\zeta, \bar{\zeta})$, holomorphe en z définie dans le paragraphe 14.

Preuve. L'affirmation b) se déduit de a), d'après le théorème 6. On a un contrôle précis des constantes A_9 et A_{10} en fonction des constantes C_1 et C_2 du jet F et des constantes de la suite $\{M_p\}_{p \geq 0}$.

Le théorème 20 montre que b) implique c). En effet, d'après (20.4), le jet f^J défini par (20.2) vérifie

$$f^J = D^J f, \; \textit{sur } K$$

et, par hypothèse, on a

$$F^J = D^J f \; \textit{sur } K.$$

Que c) implique a) est une conséquence du corollaire 11.

22. THEOREME. *Soit K, un compact 1-H convexe ayant la propriété (*) suivante:*
deux jets de Whitney de classe C^∞ et $\bar{\partial}$-plat sur K qui coïncident à l'ordre 0 sur K coïncident à tous les ordres.
Soit F un jet de Whitney sur K de classe C^∞ et $\bar{\partial}$-plat. Les propriétés suivantes sont équivalentes :

   (a) F appartient à $C_K\{p!M_p\}$,

   (b) F^0 s'étend en une fonction f continue et à support compact dans $\mathbb{C}^n$ telle que $\bar{\partial} f$ soit continue et telle qu'il existe des constantes A_9 et A_{10}, $A_9 \geq 0$, $A_{10} \geq 1$, vérifiant

$$(22.1) \qquad |\,\bar{\partial} f(\zeta)\,| \leq A_9 h[A_{10} d(\zeta, K)], \; \textit{pour tout } \zeta \textit{ de } \mathbb{C}^n,$$

(c) F admet une représentation intégrale, plus précisément, pour tout multi-indice L, on a

$$(22.2) \qquad F^L(z) = \int_{V(K)} \varphi(\zeta) \wedge D_z^L \omega_\zeta' \left(\frac{w(\zeta,z)}{\Phi(\zeta,z)} \right) \wedge \omega(\zeta), \text{ pour tout } z \text{ de } K$$

où φ est une (0,1) forme en $(\zeta, \bar\zeta)$, à support compact dans $V(K)$, continue dans $\mathbb{C}^n$ telle qu'il existe des constantes A_{11} et A_{12}, $A_{11} \geq 0, A_{12} \geq 1$, vérifiant

$$(22.3) \qquad\qquad | \varphi(\zeta) | \leq A_{11} h[A_{12} d(\zeta, K)], \text{ pour tout } \zeta \text{ de } \mathbb{C}^n$$

et où $\omega_\zeta' \left(\frac{w(\zeta,z)}{\Phi(\zeta,z)} \right)$ est la (0,n −1) forme en $(\zeta, \bar\zeta)$, holomorphe en z définie dans le paragraphe 14.

Preuve. D'après le théorème 21, il suffit de montrer que b) implique c). Sous l'hypothèse b), d'après le théorème 17, f coïncide à l'ordre 0 sur K avec le jet défini par $f^L(z)$ en (17.2). C'est (17.3).
D'après l'hypothèse (*) sur K, on a

$$f^L(z) = F^L(z) \text{ sur } K.$$

Ceci achève la preuve du théorème 22.

On peut mentionner comme exemples typiques de compacts K ayant la propriété (*) :
(i) les sous-ensembles parfaits sur une courbe de $\mathbb{C}$,
(ii) les convexes compacts d'intérieur relatif non vide de $\mathbb{R}^n$ dans $\mathbb{C}^n$,
(iii) plus généralement, les compacts d'intérieur relatif dense sur une sous-variété totalement réelle de dimension n dans $\mathbb{C}^n$.

Pour tous ces ensembles K, un jet C^∞ et $\bar\partial$−plat sur K est complètement déterminé par sa partie tangente à la variété totalement réelle de dimension n qui contient K. Les hypothèses faites sur K relativement à cette variété impliquent que la partie tangente du jet est imposée par la connaissance de F_0.

23. COROLLAIRE. *Soit K un compact de $\mathbb{C}^n$, $1-H$ convexe et F un jet de Whitney de $C_K\{p!M_p\}$ et $\bar\partial$-plat. Alors F est limite dans $C_K\{p!M_p\}$ de fonctions holomorphes au voisinage de K.*

La preuve est analogue à celle du corollaire 19.
24. Dans le cas des classes de Beurling, on a des énoncés semblables à ceux des théorèmes 20, 21 et 22 et de leur corollaire 23 où chaque affirmation de la forme
(a)   il existe des constantes E_1 et E_2, $E_1 \geq 0, E_2 \geq 0$, vérifiant

$$| \psi(\zeta) | \leq E_1 h[E_2 d(\zeta, K)], \text{ pour tout } \zeta \text{ de } \mathbb{C}^n$$

est remplacée par l'affirmation correspondante
(b)   quel que soit $E_2 > 0$, il existe $E_1 \geq 0$ vérifiant

$$| \psi(\zeta) | \leq E_1 h[E_2 d(\zeta, K)], \text{ pour tout } \zeta \text{ de } \mathbb{C}^n.$$

A titre d'exemple, on peut comparer les énoncés des théorèmes 6 et 9.

25. On se propose maintenant d'établir une réciproque à la remarque 2 du paragraphe 17 et de généraliser un théorème d'unicité de E. M. Dynkin [5]. La preuve nécessite le lemme suivant.

26. LEMME. *Pour tout couple* (x, y) *de* $\mathbb{R}^p \times \mathbb{R}^p$, $x \neq y$, *tout entier* $n \geq 1$ *et tout entier* $k \geq 0$, *on a*

$$(26.1) \quad \frac{1}{\mid x - y \mid^{2n}} = \sum_{j=0}^{k} \frac{1}{\mid y \mid^{2(j+1)n}} \sum_{(\alpha,\beta) \in J(j,n)} C_{(\alpha,\beta)}^{j} x^\alpha y^\beta$$
$$+ \frac{1}{\mid x - y \mid^{2n}} \frac{1}{\mid y \mid^{2(k+1)n}} \sum_{(\alpha,\beta) \in I(k+1,n)} C_{(\alpha,\beta)}^{k+1} x^\alpha y^\beta$$

où α *et* β *sont des multi-indices appartenant à* $\mathbb{N}^p$ *vérifiant*

$$(26.2) \quad C_{(\alpha,\beta)}^0 = 1 \ et \sum_{(\alpha,\beta) \in I(k,n)} \mid C_{(\alpha,\beta)}^k \mid \leq (n(4p)^n)^k.$$

Ici, on a posé

$$I(k,n) = \{(\alpha,\beta); k \leq \mid \alpha \mid \leq 2n + k - 1, \mid \alpha \mid + \mid \beta \mid = 2kn\}$$

et

$$J(k,n) = \{(\alpha,\beta); k = \mid \alpha \mid, \mid \alpha \mid + \mid \beta \mid = 2kn\}.$$

Remarque. On note, pour tout $k \geq 0$, $P(n,k)$ et $R(n,k)$ respectivement le premier et le deuxième terme du membre de droite de (26.1) ; on a

$$(26.3) \quad \frac{1}{\mid x - y \mid^{2n}} = P(n,k) + R(n,k)$$

où $P(n,k)$ peut être considéré comme la partie principale du développement de Taylor par rapport à la variable x en 0 de $\frac{1}{\mid x-y \mid^{2n}}$ à l'ordre k et $R(n,k)$ comme le reste au même ordre.

Preuve. La formule (26.1) se démontre par récurrence sur k.
On vérifie aisément en utilisant l'identité classique

$$\mid y \mid^{2n} - \mid x - y \mid^{2n} = (\mid y \mid^2 - \mid x - y \mid^2) \sum_{j=0}^{n-1} \mid y \mid^{2j} \mid x - y \mid^{2(n-1-j)}$$

que l'on a

$$(26.4) \quad \mid y \mid^{2n} - \mid x - y \mid^{2n} = \sum_{(\alpha,\beta) \in I(1,n)} C_{\alpha,\beta}^1 x^\alpha y^\beta$$

avec

$$(26.5) \quad \sum_{(\alpha,\beta) \in I(1,n)} \mid C_{\alpha,\beta}^1 \mid \leq n(4p)^n.$$

On a donc

$$(26.6) \qquad \frac{1}{\mid x - y \mid^{2n}} = \frac{1}{\mid y \mid^{2n}} + \frac{1}{\mid x - y \mid^{2n}} \frac{1}{\mid y \mid^{2n}} \sum_{(\alpha,\beta) \in I(1,n)} C^1_{(\alpha,\beta)} x^\alpha y^\beta.$$

On a donc établi (26.1) à l'ordre 0 avec $C^0_{\alpha,\beta} = 1$.

On suppose maintenant (26.1) établie à l'ordre k.

On écrit

$$R(n,k) = \frac{1}{\mid y \mid^{2(k+1)n}} \sum_{(\alpha,\beta) \in J(k+1,n)} C^{k+1}_{(\alpha,\beta)} \frac{x^\alpha}{\mid x - y \mid^{2n}} y^\beta$$
$$+ \frac{1}{\mid x - y \mid^{2n}} \frac{1}{\mid y \mid^{2(k+1)n}} \sum_{(\alpha,\beta) \in I(k+1,n) \backslash J(k+1,n)} C^{k+1}_{(\alpha,\beta)} x^\alpha y^\beta.$$

On remplace dans la première somme $\frac{1}{\mid x - y \mid^{2n}}$ par son expression tirée de (26.6). On a

$$R(n,k) = \frac{1}{\mid y \mid^{2(k+2)n}} \sum_{(\alpha,\beta) \in J(k+1,n)} C^{k+1}_{(\alpha,\beta)} x^\alpha y^\beta$$
$$+ \frac{1}{\mid y \mid^{2(k+2)n}} \frac{1}{\mid x - y \mid^{2n}} \sum_{(\alpha,\beta) \in J(k+1,n)} C^{k+1}_{(\alpha,\beta)} x^\alpha y^\beta \sum_{(\alpha',\beta') \in I(1,n)} C^1_{(\alpha',\beta')} x^{\alpha'} y^{\beta'}$$
$$+ \frac{1}{\mid y \mid^{2(k+1)n}} \frac{1}{\mid x - y \mid^{2n}} \sum_{(\alpha,\beta) \in I(k+1,n) \backslash J(k+1,n)} C^{k+1}_{(\alpha,\beta)} x^\alpha y^\beta$$

Le premier terme du membre de droite de cette égalité s'ajoute à $P(n,k)$ pour former $P(n,k+1)$. On vérifie aisément, quitte à multiplier et à diviser le troisième terme par $\mid y \mid^{2n}$, que les deux derniers termes constituent le reste $R(n,k+1)$. Leur somme s'écrit en effet

$$\frac{1}{\mid y \mid^{2(k+2)n}} \frac{1}{\mid x - y \mid^{2n}} \sum_{(\alpha,\beta) \in I(k+2,n)} C^{k+2}_{(\alpha,\beta)} x^\alpha y^\beta.$$

On a

$$\sum_{(\alpha,\beta) \in I(k+2,n)} \mid C^{k+2}_{(\alpha,\beta)} \mid \ \leq \sum_{(\alpha,\beta) \in J(k+1,n)} \mid C^{k+1}_{(\alpha,\beta)} \mid \sum_{(\alpha',\beta') \in I(1,n)} \mid C^1_{(\alpha',\beta')} \mid$$
$$+ \sum_{(\alpha,\beta) \in I(k+1,n) \backslash J(k+1,n)} \mid C^{k+1}_{(\alpha,\beta)} \mid p^n.$$

On conclut, en utilisant l'hypothèse de récurrence,

$$\sum_{(\alpha,\beta) \in I(k+2,n)} \mid C^{k+2}_{(\alpha,\beta)} \mid \ \leq (n(4p)^n)^{k+2}.$$

Ceci achève la preuve du lemme.

27. PROPOSITION. *Soit K un compact arbitraire de $\mathbb{C}^n$. Soit f une fonction continue et à support compact dans $V(K)$ telle que $\bar{\partial}f$ soit continue dans $V(K)$.*

On suppose que,

(i) pour tout entier $p, p \geq 0$, il existe une constante $C(p)$ vérifiant

$$(27.1) \qquad \mid \bar{\partial}f(\zeta) \mid d(\zeta, K)^{-p} \leq C(p), \text{ pour tout } \zeta \text{ de } V(K)$$

et que

(ii) le jet F défini sur K, pour tout multi-indice L de $\mathbb{N}^{2n}$, par

$$(27.2) \qquad f^L(z) = -\frac{(n-1)!}{(2i\pi)^n} \int_{\zeta \in V(K)} \bar{\partial}_\zeta f \wedge D_z^L \omega'_\zeta \left(\frac{\bar{\zeta} - \bar{z}}{\mid \zeta - z \mid^2} \right) \wedge \omega(\zeta)$$

est identiquement nul.

Alors, pour tout entier $p, p \geq 0$, il existe une constante A_{13}, ne dépendant que de la dimension n et de $V(K)$, telle que l'on ait

$$(27.3) \qquad \mid f(\zeta) \mid d(\zeta, K)^{-(p+1)} \leq A_{13}^{p+1} C(p + 2n), \text{ pour tout } \zeta \text{ de } V(K).$$

Preuve. On sait, d'après la formule de Martinelli-Bochner [6], que, pour tout z de $\mathbb{C}^n$, on a

$$f(z) = -\frac{(n-1)!}{(2i\pi)^n} \int_{\zeta \in V(K)} \bar{\partial}_\zeta f \wedge \omega'_\zeta \left(\frac{\bar{\zeta} - \bar{z}}{\mid \zeta - z \mid^2} \right) \wedge \omega(\zeta).$$

Pour z dans $\mathbb{C}^n \backslash K$, on note $\hat{z}$ un point de K qui réalise $\mid z - \hat{z} \mid = d(z, K)$. Pour simplifier l'écriture, on supposera dans la suite que $\hat{z} = 0$. Ceci ne restreindra pas la généralité.

Soit k un entier. Puisque, d'après l'hypothèse (ii), le jet F est plat sur K on peut écrire, pour tout z de $V(K) \backslash K$, $z = (x_1, \cdots, x_{2n})$,

$$
\begin{aligned}
f(z) \qquad &= f(z) - \sum_{\{L; |L| \leq k\}} \frac{f^L(0)}{L!} x^L \\
(27.4) \qquad &= -\frac{(n-1)!}{(2i\pi)^n} \int_{\zeta \in V(K)} \bar{\partial}_\zeta f \wedge \left(\omega'_\zeta \left(\frac{\bar{\zeta} - \bar{z}}{\mid \zeta - z \mid^2} \right) \right. \\
&\qquad \left. \sum_{\{L; |L| \leq k\}} \frac{x^L}{L!} D_z^L \omega'_\zeta \left(\frac{\bar{\zeta} - \bar{z}}{\mid \zeta - z \mid^2} \right) \mid_{z=0} \right) \wedge \omega(\zeta)
\end{aligned}
$$

où le symbole $\mid_{z=0}$ signifie que les dérivées sont évaluées au point $z = 0$.

On reconnaît que chacune des composantes de la $(0,n-1)$ forme en $(\zeta, \bar{\zeta})$

$$\omega'_\zeta \left(\frac{\bar{\zeta} - \bar{z}}{\mid \zeta - z \mid^2} \right) - \sum_{\{L; |L| \leq k\}} \frac{x^L}{L!} D_z^L \omega'_\zeta \left(\frac{\bar{\zeta} - \bar{z}}{\mid \zeta - z \mid^2} \right) \mid_{z=0}$$

s'écrit en coordonnées réelles, si on pose $\zeta = (y_1, \cdots, y_{2n})$, comme une somme finie d'expresions $E_{n,k,i}(x,y)$ de la forme

$$\frac{x_i - y_i}{\mid x - y \mid^{2n}} - \sum_{\{L; \mid L \mid \leq k\}} \frac{x^L}{L!} D_x^L \left(\frac{x_i - y_i}{\mid x - y \mid^{2n}} \right) \mid_{z=0} .$$

La formule (26.1) que l'on a multipliée par $x_i - y_i$ conduit à

$$\begin{aligned} E_{n,k,i}(x,y) \quad &= \frac{1}{\mid y \mid^{2(k+1)n}} \sum_{(\alpha,\beta) \in J(k,n)} C^k_{(\alpha,\beta)} x_i x^\alpha y^\beta \\ &+ \frac{x_i - y_i}{\mid x - y \mid^{2n}} \frac{1}{\mid y \mid^{2(k+1)n}} \sum_{(\alpha,\beta) \in I(k+1,n)} C^{k+1}_{(\alpha,\beta)} x^\alpha y^\beta . \end{aligned}$$

En utilisant le lemme 26, il existe une constante A_{14}, ne dépendant que de n et de $V(K)$, telle que l'on ait, pour tout i,

$$\mid E_{n,k,i} \mid \leq A_{14}^{k+1} \left(1 + \frac{1}{\mid \zeta - z \mid^{2n-1}} \right) \frac{d(z,K)^{k+1}}{d(\zeta,K)^{2n+k}} .$$

En utilisant (27.1) et (27.4), on obtient (27.3).

28. PROPOSITION. *Soit K un compact de $\mathbb{C}^n$, s-H convexe et f une fonction continue et à support compact dans $V(K)$ telle que $\bar{\partial} f$ soit continue dans $V(K)$.*

On suppose que,

(i) pour tout entier $p, p \geq 0$, il existe une constante $C(p)$ telle que l'on ait

(28.1) $\qquad \mid \bar{\partial} f(\zeta) \mid d(\zeta, K)^{-p} \leq C(p), \text{pour tout } \zeta \text{ dans } \mathbb{C}^n$

et que

(ii) le jet défini sur K, pour tout multi-indice L de $\mathbb{N}^{2n}$, par

(28.2) $\qquad f^L(z) \quad = -\frac{(n-1)!}{(2i\pi)^n} \int_{\zeta \in V(K) \setminus K} \bar{\partial} f \wedge D_z^L \omega'_\zeta \left(\frac{w(\zeta,z)}{\Phi(\zeta,z)} \right) \wedge \omega(\zeta)$

est identiquement nul.

Alors, la fonction f vérifie

(28.3) $\qquad \mid f(\zeta) \mid d(\zeta, K)^{-(p+1)} \leq A_{16}^{p+1} C(p + 2n), \text{ pour tout } \zeta \text{ de } \mathbb{C}^n,$

où A_{16} ne dépend que de n et de $V(K)$.

Preuve. D'après la proposition 16, on a, pour tout z de K,

$$\begin{aligned} f^L(z) \quad &= -\frac{(n-1)!}{(2i\pi)^n} \int_{\zeta \in V(K)} \bar{\partial} f \wedge D_z^L \omega'_\zeta \left(\frac{w(\zeta,z)}{\Phi(\zeta,z)} \right) \wedge \omega(\zeta) \\ &= -\frac{(n-1)!}{(2i\pi)^n} \int_{\zeta \in V(K)} \bar{\partial} f \wedge D_z^L \omega'_\zeta \left(\frac{\bar{\zeta} - \bar{z}}{\mid \zeta - z \mid^2} \right) \wedge \omega(\zeta) \end{aligned}$$

et on utilise la Propositon 27 pour conclure.

29. Si le compact K est $1-H$ convexe et dans le cas des classes de Carleman, on a l'énoncé suivant.

PROPOSITION. *Soit* $\{p! M_p\}_{p \geq 0}$ *une suite régulière et* h, *la fonction associée à la suite* $\{M_p\}_{p \geq 0}$ *par la formule (4.3). Soit* K *un compact de* $\mathbb{C}^n$, 1-H *convexe et* f *une fonction continue et à support compact dans* $V(K)$ *telle que* $\bar{\partial} f$ *soit continue dans* $V(K)$

On suppose que,

(i) il existe des constantes D_1 *et* D_2 *positives telles que l'on ait*

$$(29.1) \qquad \mid \bar{\partial} f(\zeta) \mid \leq D_1 h[D_2 d(z,K)], \text{pour tout } \zeta \text{ dans } \mathbb{C}^n$$

et que

(ii) le jet défini sur K, *pour tout multi-indice* L *de* $\mathbb{N}^{2n}$, *par*

$$(29.2) \qquad f^L(z) \; = -\frac{(n-1)!}{(2i\pi)^n} \int_{\zeta \in V(K) \setminus K} \bar{\partial} f \wedge D_z^L \omega'_\zeta \left(\frac{w(\zeta,z)}{\Phi(\zeta,z)} \right) \wedge \omega(\zeta)$$

est identiquement nul.

Alors, il existe des constantes D'_1 *et* D'_2 *positives telles que la fonction* f *vérifie*

$$(29.3) \qquad \mid f(\zeta) \mid \leq D'_1 h[D'_2 d(z,K)], \text{ pour tout } \zeta \text{ de } \mathbb{C}^n.$$

Remarque. On a un résultat analogue pour les classes de Beurling.

Bibliographie.

1. Chaumat, J. & Chollet, A.-M., *Noyaux pour résoudre l'équation* $\bar{\partial} u = v$ *dans des classes indéfiniment différentiables*, C. R. Acad. Sci. Paris **306** (1988), 585-588.

2. _____ , *Noyaux pour résoudre l'équation* $\bar{\partial}$ *dans des classes ultradifférentiables sur des compacts irréguliers de* $\mathbb{C}^n$, Several complex variables. Proc. Mittag-Leffler Inst. 1987/1988, Math. Notes 38, Princeton Univ. Press, à paraître.

3. Droste, B., *Fortsetzung des holomorphen funktionalkalküls in mehren variablen auf algebren mit zerlegung der eins*, Dissertation, Mainz (1980).

4. _____ , *Holomorphic approximation of ultradifferentiable functions*, Math Ann. **257** (1981), 293-316.

5. Dynkin, E. M., *Pseudoanalytic extension of smooth functions. The uniform scale*, Amer. Math. Soc. Transl. **115** (1980), 33-58.

6. Henkin, G. M. & Leiterer, J., *Theory of functions on complex manifolds*, Monographs in mathematics **79**, Birkhaüser Verlag, 1984.

7. Komatsu, H., *An analogue of the Cauchy-Kowalesky theorem for ultradifferentiable functions and a division theorem of ultradistributions as its dual*, J. Fac. Sci. Univ. Tokyo **26** (1979), 239-254.

UNIVERSITÉ DE PARIS-SUD, MATHÉMATIQUES, BÂTIMENT 425, 91405 ORSAY, FRANCE

UNIVERSITÉ DE LILLE, U.F.R. DE MATHÉMATIQUES, 59655 VILLENEUVE D'ASCQ, FRANCE

Contemporary Mathematics
Volume **137**, 1992

A Class of Hypoelliptic PDE
Admitting Non-Analytic Solutions

MICHAEL CHRIST

ABSTRACT. Certain sums of squares of vector fields in $\mathbb{R}^3$, having real-analytic coefficients and known to be hypoelliptic in the C^∞ sense, are shown to have solutions which are not real-analytic. The proof is based on an asymptotic analysis of a two-parameter family of ordinary differential equations.

0. Introduction.

The phenomenon of analytic hypoellipticity for subelliptic partial differential operators with multiple characteristics is at present poorly understood, with the exception of the symplectic case [**Tr**], [**Ta1**], [**Ta2**], [**M**]. Results in both the positive and negative directions are fragmentary. In this paper we construct nonanalytic solutions to an infinite-dimensional family of operators, which generalize a discrete set of examples studied recently.

Let b be a real-valued, real-analytic function defined in an open subset of $\mathbb{R}$. Consider the vector fields

$$X = \partial_x, \qquad Y = \partial_y - b'(x)\partial_t$$

in $\mathbb{R}^3$, with coordinates (x, y, t), and consider further the partial differential operator

$$L = X^2 + Y^2.$$

Such operators have been extensively studied, and belong to a class known to be C^∞ hypoelliptic, provided that b'' does not vanish to infinite order at any point [**H**]. Moreover, L is analytic hypoelliptic, provided that b'' vanishes nowhere (the "symplectic" case) [**Tr**]. The smooth and analytic theories diverge when b'' vanishes to finite order, though; for $b(x) = x^m$, $m \in \{3, 4, 5, \dots\}$, L fails to be analytic hypoelliptic [**PR**], [**HH**], [**C5**]. Our purpose here is to extend the scope of the analysis of [**C5**] so as to obtain:

1991 *Mathematics Subject Classification*. Primary 35G05.
Research supported in part by National Science Foundation grants.
This paper is in final form and no version of it will be submitted for publication elsewhere.

155

THEOREM 0. *Suppose that $b''(x_0) = 0$ for some $x_0 \in \mathbb{R}$. Then for any $y_0, t_0 \in \mathbb{R}$, there exists a C^∞ function f, defined in a neighborhood of $(x_0, y_0, t_0) \in \mathbb{R}^3$, such that $Lf \equiv 0$ but f is not real-analytic.*

More precisely, if b'' vanishes to order exactly $m - 2$ at x_0, then f may be constructed so as to belong to the Gevrey class G^m, but not to G^s for any $s < m$.

For $b(x) \equiv x^m$, this was accomplished in several works by analyzing the one-parameter family of ordinary differential equations

$$\mathcal{L}_\zeta = -\frac{d^2}{dx^2} + (\zeta - mx^{m-1})^2, \qquad \zeta \in \mathbb{C}$$

and exploiting a dilation symmetry resulting from the homogeneity of b. For general b the dilation symmetry is broken, forcing one to study the two-parameter family

$$\mathcal{L}_{\zeta,\tau} = -\frac{d^2}{dx^2} + (\zeta - \tau b'(x))^2, \qquad \zeta \in \mathbb{C}, \tau \in \mathbb{R},$$

asymptotically as $\tau \to \infty$. The present article consists primarily of a comparison argument, in which certain solutions of $\mathcal{L}_{\zeta,\tau}$ are shown to behave asymptotically like dilated solutions of $\mathcal{L}_\zeta$, as $\tau \to \infty$.

In §1 we state the principal lemma concerning existence of solutions of the two-parameter family with good growth properties, and review the situation for the model case, $b \equiv x^m$. §2 contains the details of the comparison argument, and in §3 we review how the existence of nonanalytic solutions to the PDE follows from the existence of suitably tame solutions to the family of ODEs.

Thanks are owed to Daryl Geller for suggestions concerning the exposition.

1. Main lemma and rescaling.

The principal step will be to prove:

PROPOSITION 1. *Let $m \in \{3, 4, 5, \ldots\}$. Let $b : [-\eta, \eta] \mapsto \mathbb{R}$ be C^∞. Suppose that*

$$\frac{d^j}{dx^j}b(0) = 0 \qquad \forall\, 0 \le j < m$$

but

$$\frac{d^m}{dx^m}b(0) \ne 0.$$

Fix $\zeta_0 \in \mathbb{C}$ and suppose that there exists $f \in L^\infty(\mathbb{R})$, not identically vanishing, such that

$$-\frac{d^2}{dx^2}f + (\zeta_0 - [(m-1)!]^{-1}b^{(m)}(0)x^{m-1})^2 f \equiv 0.$$

Then there exist $\tilde{\eta}, C_0 \in \mathbb{R}^+$, such that for any $\tau > 0$, there exist $\zeta(\tau) \in \mathbb{C}$ and

$f_\tau \in C^\infty([-\tilde\eta, \tilde\eta])$ *satisfying*

$$\begin{cases} -\dfrac{d^2}{dx^2}f + (\zeta(\tau) - \tau b'(x))^2 f \equiv 0 & on \ [-\tilde\eta, \tilde\eta] \\[2mm] \tau^{-1/m}\zeta(\tau) \to \zeta_0 & as \ \tau \to +\infty \\[2mm] \max_{|x|\le\tilde\eta} |f_\tau(x)| \le C_0 \\[2mm] |f_\tau(0)| + \tau^{-1/m}|\dfrac{d}{dx}f_\tau(0)| \ge C_0^{-1}. \end{cases}$$

Furthermore, $\tilde\eta$ may be taken to be independent of ζ_0, but this is of no consequence for us.

By dilating coordinates, and replacing b by $-b$ and ζ_0 by $-\zeta_0$ as necessary, we may reduce to the situation where $b(x) = x^m + O(|x|^{m+1})$ near 0. If $b(x) \equiv x^m$, then it suffices to take $\zeta(\tau) = \tau^{1/m}\zeta_0$ and $f_\tau(x) \equiv f(\tau^{1/m}x)$; we aim to show that the term $O(|x|^{m+1})$ results only in a small perturbation for large τ. To this end we introduce

$$b_\tau(x) \equiv \tau b(\tau^{-1/m}x) \qquad \text{for } |x| \le \tau^{1/m}\eta.$$

Choose $\tilde\eta$ so that for all $j \le m$, for all $|x| \le \tilde\eta$,

$$|\frac{d^j}{dx^j}b(x)| \approx |x|^{m-j}.$$

Then

$$|\frac{d^j}{dx^j}(x^m - b_\tau(x))| \le C\tau^{-1/m}|x|^{m+1-j} \qquad \forall\, 0 \le j \le m, \ |x| \le \tau^{1/m}\tilde\eta.$$

Indeed,

$$\begin{aligned} \frac{d^j}{dx^j}(x^m - b_\tau(x)) &= \tau \cdot \tau^{-j/m}\frac{d^j}{dy^j}(y^m - b(y))\Big|_{y=\tau^{-1/m}x} \\ &= \tau^{(m-j)/m} \cdot O(\tau^{-1/m}|x|)^{m+1-j} \\ &= \tau^{-1/m} \cdot O(|x|^{m+1-j}). \end{aligned}$$

Thus b_τ is well approximated by x^m for $|x| \le \tau^\delta$, provided $\delta > 0$ is small enough.

Let us recall the situation for the case $b(x) \equiv x^m$. Set

$$\mathcal{L}_\zeta = -\frac{d^2}{dx^2} + (\zeta - mx^{m-1})^2.$$

Suppose for the moment that $m \in \{2, 4, 6, \dots\}$. Define

$$\psi_\zeta(x) = \zeta x - x^m + h(x)$$

where $h \in C^\infty(\mathbb{R})$, and

$$h(x) \equiv -\tfrac{1}{2}\ln(m|x|^{m-1}) \qquad \text{for } |x| \ge 1.$$

The following is proved in [C5], and is merely a more precise form of a standard result in the theory of ordinary differential equations with irregular singular points [CL].

LEMMA 2. *For each $\zeta \in \mathbb{C}$ there exist unique $f_\zeta^\pm \in C^\infty(\mathbb{R})$ satisfying*

(⋆) $\mathcal{L}_\zeta f_\zeta^\pm \equiv 0$

(1) $|f_\zeta^-(x) - e^{\psi_\zeta(x)}| \leq C(1 + |\zeta|)^{1/(m-1)}(1 + |x|)^{-1}|e^{\psi_\zeta(x)}|$

$$\text{for all } x \leq -C(1 + |\zeta|)^{1/(m-1)}$$

(⋆) $|f_\zeta^+(x) - e^{\psi_\zeta(x)}| \leq C(1 + |\zeta|)^{1/(m-1)}(1 + |x|)^{-1}|e^{\psi_\zeta(x)}|$

$$\text{for all } x \geq C(1 + |\zeta|)^{1/(m-1)}$$

Moreover $f_\zeta^\pm$ are holomorphic functions of ζ.

For $m \in \{3, 5, 7, \ldots\}$, the statement remains valid, provided ψ_ζ is replaced by $-\zeta x + x^m + h(x)$ in (1). Implicit in the proof are further estimates:

(2)
$$|f_\zeta^\pm(x)| \leq C|e^{\psi_\zeta(x)}| \qquad \text{for } x \in \mathbb{R}^\pm,$$
$$|\frac{d}{dx}f_\zeta^\pm(x)| \leq C(1 + |x|)^{m-1}|e^{\psi_\zeta(x)}| \qquad \text{for } x \in \mathbb{R}^\pm,$$

provided $|\zeta| \leq R$, for any fixed, finite R. For m odd, ψ_ζ should again be replaced by $-\zeta x + x^m + h(x)$ in the estimates for f_ζ^-.

It is readily seen [C5] that given $\zeta \in \mathbb{C}$, there exists $f \in L^\infty(\mathbb{R})$, not identically vanishing, such that $\mathcal{L}_\zeta f \equiv 0$, if and only if $W(\zeta) = 0$, where

$$W(\zeta) = [f_\zeta^+(f_\zeta^-)' - f_\zeta^-(f_\zeta^+)'](0).$$

Indeed, then $f = f_\zeta^-$ decays rapidly at $\pm\infty$. It was proved in [C5] that W is an entire function of finite and non-integral order, for $m \in \{3, 4, 5, \ldots\}$; hence W has infinitely many zeroes.

We require the analogous result for

$$\mathcal{L}_{\zeta,\tau} = -\frac{d^2}{dx^2} + (\zeta - b_\tau'(x))^2.$$

Define

$$\psi_{\zeta,\tau}(x) = \zeta x - b_\tau(x) + h(x)$$

where $h \in C^\infty(\mathbb{R})$ is independent of ζ, is C^∞ in $[-1, 1]$ uniformly in $\tau \in [1, \infty)$ and

$$h(x) = -\tfrac{1}{2}\ln(|b_\tau'(x)|) \qquad \text{for } |x| \geq 1.$$

Assume that b is normalized, and $\tilde{\eta}$ chosen, as above.

LEMMA 3. *Let $R < \infty$ and $m \in \{2, 4, 6, \ldots\}$ be given. Then there exists $C < \infty$ such that for all sufficiently large $\tau \in \mathbb{R}^+$ and $|\zeta| \leq R$, there exist functions $f_{\zeta,\tau}^{\pm}$, defined for all $|x| \leq \tau^{1/m}\tilde{\eta}$, such that*

$(\star)$ $\mathcal{L}_{\zeta,\tau} f_{\zeta,\tau}^{\pm} \equiv 0$

(3) $|f_{\zeta,\tau}^{+} - e^{\psi_{\zeta,\tau}(x)}| \leq C(1 + |x|)^{-1}|e^{\psi_{\zeta,\tau}(x)}|$ *for all $x \in [0, \tau^{1/m}\tilde{\eta}]$*

(4) $|f_{\zeta,\tau}^{-} - e^{\psi_{\zeta,\tau}(x)}| \leq C(1 + |x|)^{-1}|e^{\psi_{\zeta,\tau}(x)}|$ *for all $x \in [-\tau^{1/m}\tilde{\eta}, 0]$*

$(\star)$ $f_{\zeta,\tau}^{\pm}$ *depend holomorphically on ζ.*

Once again, the same holds for odd m, provided $\psi_{\zeta,\tau}$ is modified as indicated earlier in the asymptotic expression and estimates for $f_{\zeta,\tau}^{-}$.

To construct the $f_{\zeta,\tau}^{-}$ fix ζ, τ and let u be the solution of the initial-value problem

(5)
$$\begin{cases} \mathcal{L}_{\zeta,\tau} u = \mathcal{L}_{\zeta,\tau}(e^{\psi_{\zeta,\tau}}) \\ u(-\tau^{1/m}\tilde{\eta}) = u'(-\tau^{1/m}\tilde{\eta}) = 0. \end{cases}$$

Set $f_{\zeta,\tau}^{-} = e^{\psi_{\zeta,\tau}} - u$; $f_{\zeta,\tau}^{+}$ is constructed in the same way by imposing the boundary conditions instead at $+\tau^{1/m}\tilde{\eta}$. Certainly these are solutions of $\mathcal{L}_{\zeta,\tau}$, which depend holomorphically on ζ. The estimates (3) and (4) follow from the arguments of [**C5**]; the details are also very similar to computations below and are omitted.

2. Comparison.

Proposition 1 rests on the fact that $f_{\zeta,\tau}^{\pm}(x)$ tend to $f_{\zeta}^{\pm}(x)$ for x in a fixed bounded set, as $\tau \to \infty$. This is plausible since $f_{\zeta,\tau}^{\pm}$ satisfy an equation whose coefficients tend to the coefficients of the corresponding equation for $f_{\zeta}^{\pm}$, on compact sets. Unfortunately, these solutions are normalized by their behavior near infinity, and in general they will differ greatly for $|x| \geq \tau^{1/m}\tilde{\eta}/2$; both will be small there, yet their ratio could be quite large since $b_{\tau}(x)$ need not be close to x^m. We need to prove that nonetheless, their ratio is not large near the origin.

The following elementary estimate will be useful:

(6) $\left| \int_t^x e^{\psi_{\zeta,\tau}(s)}\, ds \right| \leq C(1 + |x|)^{1-m}|e^{\psi_{\zeta,\tau}(x)}|$ $\forall t \leq x \leq 0,$

provided ζ remains in a bounded subset of $\mathbb{C}$. Furthermore the same holds with $\psi_{\zeta,\tau}$ replaced by ψ_{ζ} in both occurrences, or by $2\psi_{\zeta,\tau}$ or $2\psi_{\zeta}$.

LEMMA 4. *If $\delta > 0$ is sufficiently small then for all sufficiently large $\tau \in \mathbb{R}^+$ and $|\zeta| \leq R$,*

(7) $|f_{\zeta,\tau}^{-}(y) - f_{\zeta}^{-}(y)| \leq C\tau^{-\delta}|e^{\psi_{\zeta}(y)}|$ $\forall y \in [-C_1\tau^{2\delta}, -\tau^{2\delta}]$

$(\star)$ $|(f_{\zeta,\tau}^{-})'(-\tau^{2\delta}) - (f_{\zeta}^{-})'(-\tau^{2\delta})| \leq C\tau^{-\delta}\tau^{2\delta(m-1)}|e^{\psi_{\zeta}(-\tau^{2\delta})}|.$

PROOF. Let C_1 be large and suppose that $y \in [-C_1\tau^{2\delta}, -\tau^{2\delta}]$. Then

$$|\psi_{\zeta,\tau}(y) - \psi_\zeta(y)| \le |b_\tau(y) - y^m| + \tfrac{1}{2}|\ln(-b'_\tau(y)) - \ln(-my^{m-1})|$$
$$\le C\tau^{-1/m}|y|^{m+1}.$$

Hence if δ is small,

$$|e^{\psi_{\zeta,\tau}(y)} - e^{\psi_\zeta(y)}| \le C|e^{\psi_\zeta(y)}|\tau^{-1/m}\tau^{2\delta(m+1)}$$
$$\le C\tau^{-\delta}|e^{\psi_\zeta(y)}|.$$

Invoking (1) and (4) yields (7).

To estimate the derivatives write

$$(f^-_{\zeta,\tau})'(-\tau^{2\delta}) - (f^-_\zeta)'(-\tau^{2\delta})$$
$$= (f^-_{\zeta,\tau})'(-\tau^{1/m}\tilde\eta) - (f^-_\zeta)'(-\tau^{1/m}\tilde\eta) + \int_{-\tau^{1/m}\tilde\eta}^{-\tau^{2\delta}} \frac{d^2}{dt^2}(f^-_{\zeta,\tau} - f^-_\zeta)(t)\, dt$$
$$= (f^-_{\zeta,\tau})'(-\tau^{1/m}\tilde\eta) - (f^-_\zeta)'(-\tau^{1/m}\tilde\eta)$$
$$+ \int_{-\tau^{1/m}\tilde\eta}^{-\tau^{2\delta}} \left[(\zeta - b'_\tau(t))^2 f^-_{\zeta,\tau}(t) - (\zeta - mt^{m-1})^2 f^-_\zeta(t) \right] dt.$$

The first two terms are $O(\exp(-\varepsilon\tau))$ for some $\varepsilon > 0$; for $f^-_{\zeta,\tau}$ this holds because of the boundary conditions in (5), while for f^-_ζ, it is implied by (1) and (2). And $\exp(-\varepsilon\tau)$ is much smaller than $|\exp(\psi_\zeta(-\tau^{2\delta}))|$ provided that $2\delta m < 1$.

As for the integral, we have

$$\left| \int_{-\tau^{1/m}\tilde\eta}^{-C_1\tau^{2\delta}} (\zeta - mt^{m-1})^2 f^-_\zeta(t)\, dt \right| \le C \int_{-\tau^{1/m}\tilde\eta}^{-C_1\tau^{2\delta}} |t|^{2m-2} |e^{\psi_\zeta(t)}|\, dt$$
$$\le C(\tau^{2\delta})^{m-1}|\exp(-\zeta C_1\tau^{2\delta} - C_1^m\tau^{2\delta m} - \tfrac{m-1}{2}\ln(m\tau^{2\delta}))|$$
$$\le Ce^{-\varepsilon\tau^{2\delta m}} |e^{\psi_\zeta(-\tau^{2\delta})}|$$

for some $\varepsilon > 0$, provided C_1 is chosen to be sufficiently large; we have invoked (6) in the second line. The same reasoning yields

$$\left| \int_{-\tau^{1/m}\tilde\eta}^{-C_1\tau^{2\delta}} (\zeta - b'_\tau(t))^2 f^-_{\zeta,\tau}(t)\, dt \right| \le Ce^{-\varepsilon\tau^{2\delta m}} |e^{\psi_{\zeta,\tau}(-\tau^{2\delta})}|$$
$$\le Ce^{-\varepsilon\tau^{2\delta m}} |e^{\psi_\zeta(-\tau^{2\delta})}|.$$

Lastly

$$\left| \int_{-C_1\tau^{2\delta}}^{-\tau^{2\delta}} \left[(\zeta - b_\tau'(t))^2 f_{\zeta,\tau}^-(t) - (\zeta - mt^{m-1})^2 f_\zeta^-(t) \right] dt \right|$$

$$\leq \int_{-C_1\tau^{2\delta}}^{-\tau^{2\delta}} |(\zeta - b_\tau'(t))^2 - (\zeta - mt^{m-1})^2| \cdot |f_{\zeta,\tau}^-(t)| \, dt$$

$$+ \int_{-C_1\tau^{2\delta}}^{-\tau^{2\delta}} |\zeta - mt^{m-1}|^2 \, |f_{\zeta,\tau}^-(t) - f_\zeta^-(t)| \, dt$$

$$\leq C \int_{-C_1\tau^{2\delta}}^{-\tau^{2\delta}} |t|^{m-1} \tau^{-1/m} |t|^m |e^{\psi_{\zeta,\tau}(t)}| \, dt$$

$$+ C \int_{-C_1\tau^{2\delta}}^{-\tau^{2\delta}} |t|^{2m-2} \tau^{-\delta} |e^{\psi_\zeta(t)}| \, dt$$

$$\leq C \int_{-C_1\tau^{2\delta}}^{-\tau^{2\delta}} \tau^{-\delta} |t|^{2m-2} |e^{\psi_\zeta(t)}| \, dt$$

$$\leq C\tau^{-\delta}(\tau^{2\delta})^{m-1} |e^{\psi_\zeta(-\tau^{2\delta})}|,$$

using (7) and then (6). The lemma is proved.

The next step is to push from $-\tau^{2\delta}$ to 0; success is to be anticipated since $\mathcal{L}_{\zeta,\tau}$ and $\mathcal{L}_\zeta$ have nearly identical coefficients on this interval, for large τ.

LEMMA 5. *For any finite R, there exist $\delta > 0$ and $C < \infty$ such that for all $|\zeta| \leq R$ and $\tau \geq 1$,*

$(\star)$ $$|(f_{\zeta,\tau}^- - f_\zeta^-)(0)| \leq C\tau^{-\delta}$$

(8) $$|((f_{\zeta,\tau}^-)' - (f_\zeta^-)')(0)| \leq C\tau^{-\delta}.$$

PROOF. Set $g = f_{\zeta,\tau}^- - f_\zeta^-$ and compute

$$\mathcal{L}_\zeta g(x) = [(\zeta - mx^{m-1})^2 - (\zeta - b_\tau'(x))^2] f_{\zeta,\tau}^-(x)$$

(9) $$= O\big(\tau^{-1/m} |x|^m ((1 + |x|)^{m-1} |e^{\psi_\zeta(x)}|\big)$$

for $x \in [-\tau^{2\delta}, 0]$. Set

$$D = \frac{d}{dx} + \psi_\zeta', \qquad \tilde{D} = -\frac{d}{dx} + \psi_\zeta'.$$

Then

$$D \circ \tilde{D} = \mathcal{L}_\zeta + E$$

where

$$Ef(x) = \left[(\psi_\zeta')^2 + \psi_\zeta'' - (\zeta - mx^{m-1})^2 \right] f(x)$$
$$= O((1 + |x|)^{-1} |f(x)|).$$

Set $y = -\tau^{2\delta}$. Two integrations yield:

$$
\begin{aligned}
g(x) &= e^{\psi_\varsigma(x)-\psi_\varsigma(y)}g(y) + e^{\psi_\varsigma(x)+\psi_\varsigma(y)}(g'(y) - \psi_\zeta{}'(y)g(y))\int_y^x e^{-2\psi_\varsigma(t)}\,dt \\
&\quad - e^{\psi_\varsigma(x)}\int_y^x e^{\psi_\varsigma(t)}\,D\tilde{D}g(t)\Big(\int_t^x e^{-2\psi_\varsigma(s)}\,ds\Big)\,dt.
\end{aligned}
$$

(10)

Indeed,

$$
\begin{aligned}
e^{\psi_\varsigma(s)}\tilde{D}g(s) &= e^{\psi_\varsigma(y)}\tilde{D}g(y) + \int_y^s \frac{d}{dt}e^{\psi_\varsigma(t)}\tilde{D}g(t)\,dt \\
&= -e^{\psi_\varsigma(y)}(g'(y) - \psi_\zeta{}'(y)g(y)) + \int_y^s e^{\psi_\varsigma(t)}\,D\tilde{D}g(t)\,dt
\end{aligned}
$$

and

$$
\begin{aligned}
e^{-\psi_\varsigma(x)}g(x) &= e^{-\psi_\varsigma(y)}g(y) + \int_y^x \frac{d}{ds}(e^{-\psi_\varsigma(s)}g(s))\,ds \\
&= e^{-\psi_\varsigma(y)}g(y) - \int_y^x e^{-\psi_\varsigma(s)}\tilde{D}g(s)\,ds \\
&= e^{-\psi_\varsigma(y)}g(y) - \int_y^x e^{-2\psi_\varsigma(s)}(-e^{\psi_\varsigma(y)})(g'(y) - \psi_\zeta{}'(y)g(y))\,ds \\
&\quad - \int_y^x e^{-2\psi_\varsigma(s)}\int_y^s e^{\psi_\varsigma(t)}\,D\tilde{D}g(t)\,dt\,ds \\
&= e^{-\psi_\varsigma(y)}g(y) + e^{\psi_\varsigma(y)}(g'(y) - \psi_\zeta{}'(y)g(y))\int_y^x e^{-2\psi_\varsigma(t)}\,dt \\
&\quad - \int_y^x e^{\psi_\varsigma(t)}\,D\tilde{D}g(t)\Big(\int_t^x e^{-2\psi_\varsigma(s)}\,ds\Big)\,dt.
\end{aligned}
$$

The first two terms on the right-hand side of (10) are harmless:

$$
|e^{\psi_\varsigma(x)-\psi_\varsigma(y)}g(y)| \le |e^{\psi_\varsigma(x)-\psi_\varsigma(y)}| \cdot C\tau^{-\delta}|e^{\psi_\varsigma(y)}|
$$

by (7). Likewise

$$
|g'(y)| + |\psi_\zeta{}'(y)g(y)| \le C\tau^{-\delta}|y|^{m-1}|e^{\psi_\varsigma(y)}|,
$$

while for $s \le x$

(11)
$$
|\int_s^x e^{-2\psi_\varsigma(t)}\,dt| \le C|s|^{1-m}|e^{-2\psi_\varsigma(s)}|,
$$

so that

$$
\Big|e^{\psi_\varsigma(x)+\psi_\varsigma(y)}(g'(y) - \psi_\zeta{}'(y)g(y))\int_y^x e^{-2\psi_\varsigma(t)}\,dt\Big| \le C\tau^{-\delta}|e^{\psi_\varsigma(x)}|.
$$

In the third term of (10) substitute

$$D\tilde{D}g(x) = (\mathcal{L}_\zeta + E)g(x)$$
$$= O\big(\tau^{-1/m}(1+|x|)^{2m-1}|e^{\psi_\zeta(x)}|\big) + O\big((1+|x|)^{-1}|g(x)|\big),$$

using (9). Let A be a large positive constant, and define

$$B = \max_{-\tau^{2\delta} \le x \le -A} |g(x)e^{-\psi_\zeta(x)}| < \infty.$$

Invoking (11), the third term in (10) is therefore majorized by

$$C|e^{\psi_\zeta(x)}| \int_y^x \big[\tau^{-1/m}(1+|t|)^{2m-1}|e^{2\psi_\zeta(t)}| + (1+|t|)^{-1}|e^{\psi_\zeta(t)}g(t)|\big]$$

$$(1+|t|)^{1-m}|e^{-2\psi_\zeta(t)}|\,dt$$

$$\le C\tau^{-1/m}|e^{\psi_\zeta(x)}| \int_y^x (1+|t|)^m\,dt + CB|e^{\psi_\zeta(x)}| \int_y^x (1+|t|)^{-m}\,dt$$

$$\le C(\tau^{-1/m}\tau^{2\delta(m+1)} + B(1+|x|)^{1-m})|e^{\psi_\zeta(x)}|$$

$$\le C(\tau^{-\delta} + B(1+|x|)^{1-m})|e^{\psi_\zeta(x)}|.$$

Taking the supremum over $x \in [y, -A]$ yields

$$B \le C\tau^{-\delta} + CB(1+A)^{1-m}$$
$$\le C\tau^{-\delta} + \tfrac{1}{2}B,$$

so that

(12) $$|g(x)| = |f_{\zeta,\tau}^-(x) - f_\zeta^-(x)| \le C\tau^{-\delta}|e^{\psi_\zeta(x)}|$$

for all $x \in [-\tau^{2\delta}, -A]$.

Changing the definition to

$$B = \max_{x \in [-\tau^{2\delta},\,-A+\varepsilon]} |g(x)e^{-\psi_\zeta(x)}|,$$

returning to (10) and using (12) to estimate $g(t)$ for $t \le -A$ yields

$$B \le C\tau^{-\delta} + C\varepsilon B,$$

so that choosing ε sufficiently small gives $B \le C\tau^{-\delta}$. Repeating this argument finitely many times yields finally

(13) $$|(f_{\zeta,\tau}^- - f_\zeta^-)(x)| \le C\tau^{-\delta}|e^{\psi_\zeta(x)}| \text{for all } x \in [-\tau^{2\delta}, +1],$$

and in particular for $x = 0$. The analogous estimate (8) for the difference of the derivatives then follows by differentiating (10), and using (13) to estimate the main terms resulting.

Consider the Wronskian

$$W_\tau(\zeta) = [f_{\zeta,\tau}^+(f_{\zeta,\tau}^-)' - f_{\zeta,\tau}^-(f_{\zeta,\tau}^+)'](0).$$

Combining Lemma 5 with (2) and with the analogue of (2) for $f_{\zeta,\tau}^\pm$, we obtain

LEMMA 6. *There exists $\delta > 0$ such that for any $R < \infty$, there exists $C < \infty$ such that*

$$|W_\tau(\zeta) - W(\zeta)| \leq C\tau^{-\delta} \qquad \forall \tau \geq 1, |\zeta| \leq R.$$

From Rouché's Theorem there follows (with perhaps a smaller value of δ)

COROLLARY 7. *Suppose that $W(\zeta_0) = 0$. Then for each sufficiently large $\tau \in \mathbb{R}^+$ there exists $\zeta(\tau) \in \mathbb{C}$ such that $W_\tau(\zeta(\tau)) = 0$ and $|\zeta(\tau) - \zeta_0| \leq C\tau^{-\delta}$.*

PROOF OF PROPOSITION 1. Assuming always that $\tau > 0$ is sufficiently large, fix a function $\tau \mapsto \zeta(\tau)$ satisfying the conclusions of Corollary 7. Set $g_\tau = f^-_{\zeta(\tau),\tau}$. Then $\mathcal{L}_{\zeta(\tau),\tau} \equiv 0$ on $[-\tau^{1/m}\tilde{\eta}, \tau^{1/m}\tilde{\eta}]$. Certainly

$$|g_\tau(x)| \leq C|e^{\psi_\zeta t(x)}| \leq C < \infty \qquad \forall x \in [-\tau^{1/m}\tilde{\eta}, +1],$$

uniformly in τ.

To treat positive x note that $|f^+_{\zeta(\tau)\tau}(x) - f^+_{\zeta(\tau)}(x)| \leq C\tau^{-\delta}$ for $-1 \leq x \leq 1$, and likewise for the difference of their derivatives, by the same arguments we employed for f^- and for $x \leq 1$. Moreover, since $\zeta(\tau) \to \zeta_0$, $f^+_{\zeta(\tau)}$ converges to $f^+_{\zeta_0}$ in the C^1 norm, on any bounded subset of $\mathbb{R}$; again this follows from a repetition of the arguments of this section. Thus $f^+_{\zeta(\tau),\tau} \to f^+_{\zeta_0}$ in the C^1 norm on any fixed, bounded neighborhood of the origin. Likewise $f^-_{\zeta(\tau),\tau} \to f^-_{\zeta_0}$.

Neither of the limiting functions $f^\pm_{\zeta_0}$ vanishes identically (since they have the asymptotics prescribed in Lemma 2), and they are linearly dependent by the choice of ζ_0, so $f^+_{\zeta_0} \equiv \lambda_0 f^-_{\zeta_0}$ where $\lambda_0 \neq 0$. Since $W_\tau(\zeta(\tau)) = 0$, we have also $f^+_{\zeta(\tau),\tau} \equiv \lambda(\tau)f^-_{\zeta(\tau),\tau}$; from the remarks of the preceding paragraph it is apparent that $\lambda(\tau) \to \lambda_0$ as $\tau \to \infty$. Thus $g_\tau \equiv \lambda(\tau)^{-1}f^+_{\zeta(\tau),\tau}$ and we know already that $|f^+_{\zeta(\tau),\tau}| \leq C|\exp(\psi_{\zeta(\tau),\tau}(x))| \leq C < \infty$ for $x \in [0, \tau^{1/m}\tilde{\eta}]$. Thus

$$|g_\tau(x)| \leq C < \infty \qquad \forall |x| \leq \tau^{1/m}\tilde{\eta},$$

for all sufficiently large $\tau > 0$.

Since $f^-_{\zeta_0}$ does not vanish identically, it cannot be that both $f^-_{\zeta_0}(0)$ and $(f^-_{\zeta_0})'(0)$ vanish. Therefore

$$|g_\tau(0)| + |g'_\tau(0)| \geq \varepsilon > 0$$

for large τ.

To conclude the proof of Proposition 1, still under the assumption that m is even, define

$$f_\tau(x) = g_\tau(\tau^{1/m}x) \qquad \forall |x| \leq \tilde{\eta}.$$

All assertions of the Proposition are now evident. The case of odd m is treated in exactly the same fashion, the sole difference being that $\psi_\zeta, \psi_{\zeta,\tau}$ must be replaced by $-\zeta x + x^m + h(x)$, $-\zeta x + b_\tau(x) + h(x)$ respectively in the definitions of $f^-_\zeta, f^-_{\zeta,\tau}$ and in the corresponding estimates.

3. Nonanalytic solutions.

Let b be given, satisfying the hypotheses of Theorem 0. By translating in the x coordinate we may assume that $x_0 = 0$. Only b' appears in the expression defining L, so it is no loss of generality to assume that $b(0) = 0$. A further coordinate transformation $(x, y, t) \mapsto (x, y, t - cy)$ reduces matters to the case where $b'(0) = 0$. If $b'' \equiv 0$, then by the theorem of Frobenius, the span of X, Y defines an integrable field of two-dimensional planes in an open subset of $\mathbb{R}^3$, in which case $X^2 + Y^2$ is certainly not even C^∞ hypoelliptic. Otherwise a dilation of the x coordinate, followed if necessary by a reflection about the origin in t, brings b into the form $b(x) = x^m + O(|x|^{m+1})$ near 0, with $m \in \{3, 4, 5, \dots\}$, without affecting the analytic hypoellipticity, or lack thereof, of L.

The function

$$e^{it\tau} e^{i\zeta y} \kappa(x)$$

is a solution of L, if and only if $\mathcal{L}_{\zeta,\tau}\kappa \equiv 0$. Fix ζ_0 such that $\mathcal{L}_{\zeta_0} = -\frac{d^2}{dx^2} + (\zeta_0 - mx^{m-1})^2$ admits a solution in $L^\infty(\mathbb{R})$, not vanishing identically; the existence of such a parameter was established in [PR] for $m = 3$, in [HH] for larger, odd m, and in [C5] for even $m \geq 4$. Necessarily $\zeta_0 \notin \mathbb{R}$, since in the real case, $\mathcal{L}_{\zeta_0}$ manifestly has a strictly positive lowest eigenvalue. Replacing ζ_0 by its conjugate if necessary, we may assume that $\zeta_0 = \sigma + i\lambda$ where $\lambda > 0$. For τ large and positive set

$$G_\tau(x, y, t) = e^{it\tau} e^{i\zeta(\tau)y} f_\tau(x)$$

where $\zeta(\tau)$, f_τ are as in Proposition 1. Then assuming always that $1 \leq y \leq 3$,

(14)
$$|G_\tau(x, y, t)| \leq C e^{-a(\tau)\tau^{1/m}\lambda y}$$

where $a(\tau) \to 1$ as $\tau \to \infty$,

$$|G_\tau(0, 2, 0)| + \tau^{-1/m}|\frac{\partial}{\partial x}G_\tau(0, 2, 0)| \geq 2\delta e^{-2\lambda a(\tau)\tau^{1/m}},$$

where $\delta > 0$ is some constant, and $LG_\tau \equiv 0$. Furthermore it is apparent that
(15)
$$|\frac{\partial^k}{\partial t^k}G_\tau(0, 2, 0)| + \tau^{-1/m}|\frac{\partial^k}{\partial t^k}\frac{\partial}{\partial x}G_\tau(0, 2, 0)| \geq 2\delta\tau^k e^{-2\lambda a(\tau)\tau^{1/m}} \qquad \forall k \geq 0.$$

Suppose for the sake of simplicity that there exists a sequence $\tau_j \to +\infty$ such that $|f_{\tau_j}(0)| \geq \delta > 0$, and moreover that each $\tau_j^{1/m} \in \mathbb{Z}^+$. Then set $F_j = G_{\tau_j}$ and

(16)
$$F = \sum_{j=1}^{\infty} F_j.$$

Replacing $\{\tau_j\}$ by a sufficiently rapidly increasing subsequence, the series converges uniformly on $[-\tilde{\eta}, \tilde{\eta}] \times [1, 3] \times \mathbb{R}$, by (14). Clearly $LF \equiv 0$, so $F \in C^\infty$, as

L is C^∞ hypoelliptic[1]. We have

$$\frac{\partial^k}{\partial t^k} F(0, 2, 0) = i^k \sum_j \tau_j^k e^{2i\sigma_j \tau_j^{1/m}} e^{-2\lambda_j \tau_j^{1/m}} f_{\tau_j}(0)$$

where $\sigma_j \to \sigma_0$, $\lambda_j \to \lambda_0 > 0$ as $j \to \infty$, and $\zeta(\tau_j) = \tau_j^{1/m}(\sigma_j + i\lambda_j)$.

Fix J and take $k = (\tau_J)^{1/m}$. Then by the triangle inequality,

$$|\frac{\partial^k}{\partial t^k} F(0, 2, 0)| \geq \delta k^{mk} e^{-2\lambda_J k} - C \sum_{j < J} \tau_j^k e^{-2\lambda_j \tau_j^{1/m}} - C \sum_{j > J} \tau_j^k e^{-2\lambda_j \tau_j^{1/m}}.$$

The first term on the right is bounded below by $k^{(m-\varepsilon)k}$ for all sufficiently large k, for any $\varepsilon > 0$. We may choose the subsequence $\{\tau_j\}$ recursively so that the first term on the right is more than four times the sum of the $J - 1$ summands in the second term, for each J. Since $\tau_j^k \exp(-2\lambda_j \tau_j^{1/m}) \to 0$ as $j \to \infty$, for each fixed k, we may further arrange, by repeatedly passing to a subsequence, that

$$C\tau_j^k e^{-2\lambda_j \tau_j^{1/m}} < 2^{-j-2} \delta k^{mk} e^{-2\lambda_J k}$$

for all $j > J$, for all J, where $k = (\tau_J)^{1/m}$. Then

$$\left| \frac{\partial^k}{\partial t^k} F(0, 2, 0) \right| \geq \frac{\delta}{4} k^{(m-\varepsilon)k},$$

for a sequence of values of k tending to ∞. Thus F fails to be real analytic.

If the simplifying assumption that $\tau_j^{1/m}$ be an integer is dropped, then the same argument applies, with k taken to be the integer closest to $(\tau_J)^{1/m}$. If there exists no sequence τ_j for which $|f_{\tau_j}(0)|$ is bounded away from zero, then (15) guarantees instead that $|f_\tau'(0)| \geq \delta \tau^{1/m}$, and the same reasoning applies to $\frac{\partial^k}{\partial t^k} \frac{\partial}{\partial x} F(0, 2, 0)$.

References.

[C1] M. Christ, *On the $\bar\partial$ equation in $\mathbb{C}^1$ with weights*, Jour. Geom. Analysis **1** (1991), 193-230.

[C2] _____, *Some non-analytic-hypoelliptic sums of squares of vector fields*, Bulletin AMS **16** (1992), 137-140.

[C3] _____, *Analytic hypoellipticity breaks down for weakly pseudoconvex Reinhardt domains*, International Mathematics Research Notices **1** (1991), 31-40.

[C4] _____, *Remarks on the breakdown of analyticity for $\bar\partial_b$ and Szegö kernels*, Harmonic Analysis, S. Igari ed., Springer-Verlag, Tokyo, 1991, pp. 61-78.

[C5] _____, *Certain sums of squares of vector fields fail to be analytic hypoelliptic*, Comm. PDE **16** (1991), 1695-1707.

[CG] M. Christ and D. Geller, *Counterexamples to analytic hypoellipticity for domains of finite type*, Annals of Math. (to appear).

---

[1] Alternatively, smoothness follows by differentiating the series (16) term-by-term and estimating the derivatives.

[CL] E. Coddington and N. Levinson, *Theory of Ordinary Differential Equations*, McGraw-Hill, New York, 1955.

[HH] N. Hanges and A. A. Himonas, *Singular solutions for sums of squares of vector fields*, Comm. PDE **16** (1991), 1503-1511.

[Ho] L. Hörmander, *Hypoelliptic second order differential equations*, Acta Math. **119** (1967), 147-171.

[M] G. Métivier, *Analytic hypoellipticity for operators with multiple characteristics*, Comm. PDE **6** (1981), 1-90.

[PR] Pham The Lai and D. Robert, *Sur un problème aux valeurs propres non linéaire*, Israel J. Math. **36** (1980), 169-186.

[Ta1] D. Tartakoff, *Local analytic hypoellipticity for $\Box_b$ on non-degenerate Cauchy-Riemann manifolds*, Proc. Nat. Acad. Sci. USA **75** (1978), 3027-3028.

[Ta2] _____ , *On the local real analyticity of solutions to $\Box_b$ and the $\bar{\partial}$-Neumann problem*, Acta Math. **145** (1980), 117-204.

[Tr] F. Trèves, *Analytic hypo-ellipticity of a class of pseudodifferential operators with double characteristics and applications to the $\bar{\partial}$-Neumann problem*, Comm. PDE **13** (1978), 475-642.

DEPARTMENT OF MATHEMATICS, UCLA, LOS ANGELES, CALIFORNIA 90024

Contemporary Mathematics
Volume **137**, 1992

Fourier Analysis Off Groups

W.C. CONNETT
A.L. SCHWARTZ

To Walter Rudin: Your gift for exposition formed our
mathematical tastes; your mathematics inspired our life's work.

1. Introduction.

When the second author was Walter's student in the late 60s, he was set
to investigating the harmonic analysis of radial functions and measures on R^n.
About that time Walter also put Charles Dunkl to work investigating zonal
functions and measures on the sphere S^n. Both systems admitted rich structures
very much like those associated with locally compact abelian groups. Indeed,
Walter's book *Fourier Analysis on Groups* [Ru] contains theorems and methods
which could often be adapted to these settings, hence the title of this paper. Of
course, this was not unexpected, since these objects inherited their properties
from the groups R^n and $SO(n + 1)$ and are members of the class of Gelfand
pairs.

The algebras of zonal and radial measures are examples of hypergroups as
were the structures associated with ultraspherical polynomials first observed by
I.I. Hirschman [Hi], though the term "hypergroup" did not gain currency until
much later. It is now a topic of some interest to workers in harmonic analysis,
probability, and special functions; a bibliography compiled by Kenneth Ross in
1988 contains 120 items. Indeed, the subject has reached the status of its own
MR classification number. The field has existed longer than the term "hypergroup"; that word does not appear in many of the works cited here or in Ross's
bibliography which are, after the fact, studies in various classes of hypergroups.

Our purpose here is to acquaint our readers with these objects, to give them
several examples, to direct them to some of the relevant sources (both recent
and ancient), and to make a preliminary announcement of a few results for a

---

1991 *Mathematics Subject Classification*. Primary 43A62.

The authors were partially supported by the National Science Foundation (grant no. DMS–
9005999).

The final (detailed) version of this paper will be submitted for publication elsewhere.

family of hypergroups on the unit disk. This is not intended to be a complete
survey of the subject; for a more complete review of the literature, the reader is
referred to the surveys of Heyer [H] and Litvinov [Li].

2. The algebra of radial measures on R^n.

As a concrete example we describe the hypergroup with which the authors first
became acquainted. We say $\lambda \in M(R^n)$ is *radial* if $\lambda(E) = \lambda(\tau E)$ for every Borel
set E and rotation τ; denote the set of radial measures by $M_r(R^n)$. Associated
with λ is a measure $P\lambda \in M([0,\infty))$ defined by $(P\lambda)(A) = \lambda(\{x : |x| \in A\})$.
Now if λ_1, and λ_2 are radial, it follows that $\lambda_1 \star \lambda_2$ is as well (where $\star$ denotes
the usual convolution associated with the group R^n). Consequently we define a
convolution $*_n$ on $M([0,\infty))$ by setting

$$\mu_1 *_n \mu_2 = P((P^{-1}\mu_1) \star (P^{-1}\mu_2)).$$

With this definition, the measure algebra $M([0,\infty), *_n)$ is isometrically isomorphic with $(M_r(R^n), \star)$.

In fact, the convolution can be given explicitly in terms of its action on the
unit point masses δ_x by requiring $\delta_0 *_n \delta_x = \delta_x$ and if $x, y > 0$

(1) $$(\delta_x *_n \delta_y)(E) = \int_E \Phi_\nu(x, y, z) dm_\nu(z)$$

where $\nu = (n-2)/2$, $dm_\nu(x) = c_\nu x^{2\nu+1} dx$, $c_\nu = [2^\nu \Gamma(\nu + 1)]^{-1}$, and

(2) $$\Phi_\nu(x, y, z) = \frac{2^{3\nu-1} \Gamma(\nu + 1)^2 \Delta(x, y, z)^{2\nu-1}}{\Gamma(\nu + \frac{1}{2}) \pi^{1/2} (xyz)^{2\nu}} \quad \text{or} \quad 0$$

the first value being assumed provided there is a non-degenerate triangle with
sides x, y and z and area $\Delta(x, y, z)$. (See [S1] and references cited there. A
similar construction is valid for the zonal measures on the sphere, see [D].)

The Fourier-Stieltjes transform is given by a Hankel-Stieltjes transform in the
sense that

$$\int_{R^n} e^{-ix \cdot y} d\lambda(x) = \int_0^\infty (c_\nu(r\rho)^\nu)^{-1} J_\nu(r\rho) \, d(P\lambda)(r)$$

where $\rho = |y|$. What is interesting here is that thanks to a formula of Gegen-
bauer [W, 11.41(16)] the formulas (1) and (2) actually define a convolution on
$M([0,\infty))$ for all values of ν exceeding $-1/2$. The resulting measure algebra denoted by M_ν is variously called a Hankel algebra or a Bessel-Kingman algebra.
With the inspiration of Rudin's book [Ru] it was possible to establish analogues
of many results in Fourier analysis for M_ν even though these measure algebras
are not associated with any group unless 2ν is integral.

3. Hypergroups and other measure algebras.

A hypergroup is a measure algebra which in its axioms captures many of the most useful properties associated with group measure algebras.

DEFINITION. $(H, *)$ *is a* hypergroup *if* H *is a locally compact Hausdorff space,* $(M(H), *)$ *is a Banach algebra and:*

i)   *If* μ *and* ν *are probability measures, so is* $\mu * \nu$.

ii)   *The mapping* $(\mu, \nu) \to \mu * \nu$ *is weakly continuous.*

iii)   *There is an element* $e \in H$ *such that* δ_e *is an identity, that is,* $\delta_e * \mu = \mu * \delta_e = \mu$ *for every* $\mu \in M(H)$.

iv) $(x, y) \to supp(\delta_x * \delta_y)$ *is a continuous mapping from* $H \times H$ *into the space of compact subsets of* H *with the topolgy described in [Mi].*

v)   *There is a continuous mapping (called involution)* $x \to x^\vee$ *on* H *such that* $x^{\vee\vee} = x$ *and* $e \in supp(\delta_x * \delta_y)$ *if and only if* $y = x^\vee$.

vi) $(\mu * \nu)^\vee = \nu^\vee * \mu^\vee$ *(where* $\int f(x) d\mu^\vee(x) = \int f(x^\vee) d\mu(x)$*).*

The *Hermitian characters* are those ϕ which are continuous on H and which satisfy $\int_H \phi\, d(\delta_x * \delta_y) = \phi(x)\phi(y)$ and $\phi(x^\vee) = \overline{\phi(x)}$. For example, the characters of M_ν are $\phi_0(r) = 1$ and $\phi_\rho(r) = (c_\nu(\rho r)^\nu)^{-1} J_\nu(\rho r)$, $(\rho > 0,\ r > 0)$.

The methods outlined in Section 2 can be used with many special functions to construct measure algebras and often hypergroups. Examples include those whose Hermitian characters are given by ultraspherical polynomials ([Hi], [D]), Jacobi polynomials ([G2], [CMS1], [CS4]), and the spheroidal wave functions [CMS3]. Many hypergroups arise in connection with Sturm-Liouville problems ([CMS2], [CS1], [CS2], [S3], [CS3]), in particular Fourier-Bessel series (of order 1/2, 3/2 and 5/2) [M].

In the case of the ultraspherical and Jacobi polynomials, there are dual hypergroup structures on $\mathbf{N}_0 = \{0, 1, 2, \dots\}$ analogous to the Pontryagin duality between $\mathbf{T}$ and $\mathbf{Z}$. These and other examples, related to orthogonal polynomials are collectively called *polynomial hypergroups* ([S2], [La], [G1]).

Another family of examples which we will discuss in some detail below consists of certain hypergroups with characters which are orthogonal polynomials on the unit disk $D = \{z : |z| \le 1\}$.

It is possible to do harmonic analysis in other classes of measure algebras. For instance Olivier Gebuhrer has defined the *Gelfand-Levitan spaces* in [Ge1] and [Ge2]. This category satisfies the hypergroup axioms with the following changes: The underlying space must be compact and $*$ is commutative, axiom (iv) is deleted, and (v) is replaced by $e \in supp(\delta_x * \delta_{x^\vee})$. The class includes all compact abelian hypergroups and will have:

a) A positive measure σ called *Haar measure* which satisfies $\delta_x * \sigma = \sigma$ for all $x \in H$.

b) $\widehat{H}$ (the set of Hermitian characters) is discrete in the weak$-*$ topology.

c) There is a Plancherel formula.

d) The linear span of $\widehat{H}$ (i.e., finite linear combinations of Hermitian characters) is dense in $C(H)$.

Examples of Gelfand-Levitan spaces which are not hypergroups include the generalized Chebyshev polynomials [L].

Finally, there are many interesting examples of measure algebras which are neither hypergroups nor Gelfand-Levitan spaces. These arise, for example, from Jacobi polynomials for certain values of the parameters [G2] and the continuous q-Jacobi polynomials [R]. See also [S2], [CMS2], [CMS3].

4. Hypergroups on the disc.

We will now discuss the hypergroups on D referred to above. For each $\alpha > 0$ there is a hypergroup which we denote D_α (Kanjin [K] uses the notation $M_\alpha(D)$) with characters

$$R_{m,n}^{(\alpha)}(x,y) = R_{m,n}^{(\alpha)}(\rho e^{i\theta}) = e^{i(m-n)\theta}\rho^{|m-n|}R_{m\wedge n}^{(\alpha,|m-n|)}(2\rho^2 - 1),$$

where $x + iy = \rho e^{i\theta}$, $m \wedge n = \min(m,n)$, and

$$R_n^{(\alpha,\beta)}(x) = P_n^{(\alpha,\beta)}(x)/P_n^{(\alpha,\beta)}(1).$$

Then $\{R_{m,n}^{(\alpha)} : m, n \in \mathbf{N}_0\}$ is a set of orthogonal polynomials with respect to the measure

$$dm_\alpha(x,y) = \frac{\alpha+1}{\pi}(1 - x^2 - y^2)^\alpha \, dx \, dy.$$

These polynomials satisfy a product formula

$$\int_D R_{m,n}^{(\alpha)}(\xi) \, K_\alpha(z,\zeta,\xi) \, dm_\alpha(\xi) = R_{m,n}^{(\alpha)}(z) \, R_{m,n}^{(\alpha)}(\zeta)$$

where

$$K_\alpha(z,\zeta,\xi) = \begin{cases} \dfrac{\alpha}{\alpha+1} \dfrac{(1 - |z|^2 - |\xi|^2 - |\zeta|^2 + 2\Re(z\bar{\xi}\zeta))^{\alpha-1}}{(1-|z|^2)^\alpha(1-|\xi|^2)^\alpha(1-|\zeta|^2)^\alpha} \\ 0 \end{cases}$$

with the first value assigned if and only if ξ is in the disk with center $z\zeta$ and radius $\sqrt{1 - |z|^2}\sqrt{1 - |\zeta|^2}$. This gives rise to a hypergroup D_α with identity element 1 and characters $R_{m,n}^{(\alpha)}$ by setting

$$d(\delta_z *_\alpha \delta_\zeta)(\xi) = K_\alpha(z,\zeta,\xi) \, dm_\alpha(\xi) \qquad |z| < 1 \text{ and } |\zeta| < 1$$

$$\delta_z *_\alpha \delta_\zeta = \delta_{z\zeta} \qquad |z| = 1 \text{ or } |\zeta| = 1.$$

If $\mu, \nu \in M(D)$ and $f \in C(D)$, then $\mu *_\alpha \nu$ is defined by

$$\int_D f \, d(\mu *_\alpha \nu) = \int_D \int_D \int_D f(\xi) \, d(\delta_z *_\alpha \delta_\zeta)(\xi) \, d\mu(z) \, d\nu(\zeta).$$

$D_\alpha = (D, *_\alpha)$ is a compact commutative hypergroup with identity δ_1, involution $z^\vee = \bar{z}$, and normalized Haar measure m_α.

For $\alpha > 0$ and $\mu \in M(D)$ define the Fourier-Stieltjes coefficients by

$$(3) \qquad \widehat{\mu}(m,n) = \widehat{\mu}^{(\alpha)}(m,n) = \int_D R^{(\alpha)}_{m,n}(\bar{z})\, d\mu(z)$$

so that if $\mu, \nu \in M(D)$,

$$(\mu *_\alpha \nu)\widehat{} = \widehat{\mu}\,\widehat{\nu}.$$

If $f \in L^1_\alpha = L^1(D, m_\alpha)$, define

$$\widehat{f}(m,n) = \widehat{f}^{(\alpha)}(m,n) = \int_D R^{(\alpha)}_{m,n}(\bar{z}) f(z)\, dm_\alpha(z).$$

Thus, if f is a polynomial in two variables, then $supp(\widehat{f})$ is compact and

$$f(z) = \sum_{m,n=0}^{\infty} \widehat{f}(m,n) h^{(\alpha)}_{m,n} R^{(\alpha)}_{m,n}(z)$$

where

$$h^{(\alpha)}_{m,n} = \left(\int_D \left| R^{(\alpha)}_{m,n}(z) \right|^2 dm_\alpha(z) \right)^{-1}.$$

We note for the sake of completeness that $\mathbf{N}_0 \times \mathbf{N}_0$ can also be given a hypergroup structure. There are coefficients such that

$$R^{(\alpha)}_{m,n} R^{(\alpha)}_{p,q} = \sum_{k,l} C^{k,l}_{m,n;p,q} R^{(\alpha)}_{k,l}$$

([Ko]) and it is possible to define a hypergroup on $\mathbf{N}_0 \times \mathbf{N}_0$ (see [BG]) by setting

$$\delta_{(m,n)} * \delta_{(p,q)} = \sum_{k,l} C^{k,l}_{m,n;p,q} \delta_{(k,l)}$$

The result is an example of a 2-dimension polynomial hypergroup, and since both D_α and $\widehat{D_\alpha}$ are hypergroups, this is an example of a *strong* hypergroup. We believe that such objects are rare. (See for instance [CS4]).

The study of the hypergroup D_α has met with some success (see [K]). One of the main reasons for this is that the measure algebra on the circle $\mathbf{T}$ is embedded as a subhypergroup in D_α. For instance, the maximal ideal space is only slightly more complicated than that of $M(\mathbf{T})$, the idempotents can be identified, and it is possible to establish an analog of the F. and M. Riesz Theorem.

5. Sidon sets and Riesz sets for D_α.

Recently the first author together with Olivier Gebuhrer obtained characterizations of Sidon sets for this example and a stronger theorem of F. and M. Riesz type than the one included in [K].

In the following we will be interested in Sidon sets for the hypergroup D_α and the group algebra $M(\mathbf{T})$ so it will be useful to state the definitions and some of the basic facts in a more general context: that of a compact commutative hypergroup. (Indeed this all makes sense in a Gelfand-Levitan space.)

Suppose $(H, *)$ is a compact commutative hypergroup. Let m be the Haar measure of $(H, *)$ normalized so that $m(H) = 1$ and for each Hermitian character χ let $h(\chi) = \left(\int_H |\chi|^2 dm\right)^{-1}$.

For $f \in L^1(H, dm)$ let

$$\widehat{f}(\chi) = \int_H f \, \overline{\chi} \, dm,$$

then if f is in the span of $\widehat{H}$ we have

$$f = \sum_{\chi \in \widehat{H}} \widehat{f}(\chi) \, h(\chi) \, \chi.$$

A set $E \in \widehat{H}$ is called an $(H, *)$-*Sidon set* if there is a constant B_E such that for every f in the span of $\widehat{H}$ with $supp(f) \subset E$ we have

$$\sum_{\chi \in E} |\widehat{f}(\chi)| h(\chi) \leq B_E \|f\|_\infty.$$

We will state a result for D_α-Sidon sets which relates to the corresponding results on a group, the proof of which depends heavily on methods in [Ru, Section 5.7]. Many of the relevant lemmas can actually be established in the more general setting of Gelfand-Levitan spaces.

Let $\#(E)$ denote the cardinality of the set E, let

$$d_k = \{(n + k, n) : n \in \mathbf{N}_0\},$$

and if $E \subset \mathbf{N}_0 \times \mathbf{N}_0$, let

$$E_\infty = \{k : E \cap d_k \neq \emptyset\}.$$

THEOREM 1. *Let* $\alpha > 0$. $E \subset \mathbf{N}_0 \times \mathbf{N}_0$ *is a* D_α-*Sidon set if and only if both of the following are true:*

(i) $\#(E \cap d_k) < \infty$ *for every* $k \in \mathbf{Z}$, *and there is* $k_0 \in \mathbf{N}_0$ *such that for every* $|k| \geq k_0$, $\#(E \cap d_k) \leq 1$.

(ii) E_∞ *is a* $M(\mathbf{T})$-*Sidon set.*

As a consequence of Theorem 1, knowledge about $M(\mathbf{T})$-Sidon sets yields knowledge about D_α-Sidon sets. See in particular [Ru, Sections 5.75 and 5.76] with $\Gamma = \mathbf{Z}$.

Riesz sets are sets which generalize the classical F. and M. Riesz Theorem [Ru, Theorem 8.2.1]. We say that $E \subset \mathbf{N}_0 \times \mathbf{N}_0$ is a *Riesz set* if $\mu \in M(D)$ and $supp(\widehat{\mu}) \subset E$ imply that μ is absolutely continuous (where $\widehat{\mu}$ is defined by (3)). The following two results list a necessary condition and a sufficient condition for Riesz sets.

THEOREM 2. *If* E *is a Riesz set then* $E^c \cap d_k$ *is an infinite set for every* k.

THEOREM 3. *Suppose $\{\phi(k)\}_{k=0}^{\infty}$ is a non-negative sequence such that*

$$\varlimsup_{k \to \infty} (log\ k/k)\phi(k) = 0,$$

then

$$\{(m, n) : m \le \phi(n) \quad or \quad n \le \phi(m)\}$$

is a Riesz set.

Proofs of these results will appear elsewhere.

References.

[BG] M. Bouhaik and L. Gallardo, *Une loi des grands nombres et un theoreme limite central pour les chaines de Markov sur associees aux polynomes discaux,* C.R. Acad. Sci. Paris **310**, Serie 1 (1990), 739–744.

[CMS1] W.C. Connett, C. Markett and A.L. Schwartz, *Jacobi polynomials and related convolution structures,* Probability Measures on Groups X, Proceedings of a Conference held in Oberwolfach, Nov. 4-10, 1990, (H. Heyer, ed.), Plenum, to appear.

[CMS2] W.C. Connett, C. Markett and A.L. Schwartz, *Convolution and hypergroup structures associated with a class of Sturm-Liouville systems,* Tran. Amer. Math. Soc. (to appear).

[CMS3] W.C. Connett, C. Markett and A.L. Schwartz, *Product formulas and convolutions for angular and radial spheroidal wave functions,* Trans. Amer. Math. Soc. (to appear).

[CS1] W.C. Connett and A.L. Schwartz, *A Hardy-Littlewood maximal inequality for Jacobi type hypergroups,* Proc. Amer. Math. Soc. **107** (1989), 137–143.

[CS2] _____, *Analysis of a class of probability preserving measure algebras on compact intervals,* Trans. Amer. Math. Soc. **320** (1990), 371–393.

[CS3] _____, *Positive product formulas and hypergroups associated with singular Sturm-Liouville problems on a compact interval,* Coll. Math. **LX/LXI** (1990), 525–535.

[CS4] _____, *Product formulas, hypergroups, and Jacobi polynomials,* Bull. Amer. Math. Soc. **22** (1990), 91–96.

[D] C.F. Dunkl, *Operators and harmonic analysis on the sphere,* Trans. Amer. Math. Soc. **125** (1966), 250–263.

[G1] G. Gasper, *Linearization of the product of Jacobi polynomials, II,* Can. J. Math. **32** (1970), 582–593.

[G2] _____, *Banach algebras for Jacobi series and positivity of a kernel,* Ann. of Math **95** (1972), 261–280.

[Ge1] O. Gebuhrer, *Analyse harmonique sur les espaces de Gelfand Levitan et applications a la theorie des semi-groupes de convolution,* Thesis, l'Institut de Recherche, Mathematique Avancee, Strasbourg, France, 1989.

[Ge2] _____, *Trigonometric polynomials on compact Gelfand Levitan Spaces,* in preparation.

[H] H. Heyer, *Probability theory on hypergroups: a survey*, Probability Measures on Groups VII, Proceedings Oberwolfach 1983 (H. Heyer, ed.), Springer Verlag, Lecture Notes in Mathematics, 1064, 481–550.

[Hi] I.I. Hirschman, Jr., *Harmonic analysis and ultraspherical polynomials*, Symposium on Harmonic Analysis and Related Integral Transforms,, Cornell University, 1956.

[K] Y. Kanjin, *A convolution measure algebra on the unit disc*, Tohoku Math. Journ. **28** (1976), 105–115.

[Ko] T. Koornwinder, *Positivity proofs for linearization and connection coefficients of orthogonal polynomials satisfying an addition formula*, J. London Math. Soc. **18** no. (2) (1978), 101–114.

[L] T.P. Laine, *The product formula and convolution structure for the genneralized Chebyshev polynomials*, SIAM J. Math. Anal. **11** (1980), 133–146.

[La] R.Lasser, *Orthogonal polynomials and hypergroups*, Rend. Mat. **3** no. (7) (1983), 185–209.

[Li] G.L. Litvinov, *Hypergroups and hypergroup algebras*, Journal of Soviet Mathematics **38** (1987), 1734–1761.

[M] C. Markett, *Product formulas and convolution structure for Fourier-Bessel series*, Constr. Approx. **5** (1989), 383–404.

[Mi] E. Michael, *Topologies on spaces of subsets*, Trans. Amer. Math. Soc. **71** (1951), 152–182.

[R] M. Rahman, *A product formula for the continuous q-Jacobi polynomials*, J. Math. Anal. and Appl. **118** (1986), 309–322.

[Ru] W. Rudin, *Fourier Analysis on Groups*, Interscience Publishers, 1962.

[S1] A.L. Schwartz, *The structure of the algebra of Hankel and Hankel-Stieltjes transforms*, Can. J. Math. **2** (1971), 236–246.

[S2] ———, *l^1-Convolution Algebras: Representation and Factorization*, Zeit. f. Wahr. und verw. Geb. **41** (1977), 161–176.

[S3] ———, *Classification of one-dimensional hypergroups*, Proc. Amer. Math. Soc. **103** (1988), 1073–1081.

[W] G. Watson, *A Treatise on the Theory of Bessel Functions*, Cambridge University Press, 1966.

DEPARTMENT OF MATHEMATICS AND COMPUTER SCIENCE, UNIVERSITY OF MISSOURI-ST. LOUIS, ST. LOUIS, MISSOURI 63121

Contemporary Mathematics
Volume **137**, 1992

Uniform Extendibility of Automorphisms

BERNARD COUPET

In honor of Walter Rudin.

This paper will be devoted to the uniform extendibility of automorphisms of a domain D. We shall suppose D be a bounded strictly pseudoconnex domain with real analytic boundary in $\mathbb{C}^n$.

Let us recall some facts.

In the 1970's, the Lewy [**Le**], S. Pinchuk [**Pi 1**] proved that every automorphism of D extends to an holomorphic mapping on a neighborhood of $\overline{D}$. A bit latter, in a short paper S. Webster [**We**] proved the same result by using the edge of the wedge theorem [**Ru**], associating a totally real manifold to D. Then a question was :

Does there exist a fixed neighborhood V of $\overline{D}$ such that any automorphism of D extends on V ?

Of course, as the automorphism group is explictly known for the ball and as it is easy to see that there are sequences of automorphisms with poles accumulating on the sphere, the answer is negative for the ball and any domain biholomorphic to the ball.

However, it is the only exception, and the answer is affirmative for the other domains. We have the following:

MAIN THEOREM. *Let D be a strictly pseudoconvex domain with real analytic boundary. If D is not biholomorphic to the ball, there exists a fixed neighborhood of $\overline{D}$ such that any automorphism of D extends on V.*

First of all, we would like to say that the result is not new and has been already proved by Vitushkin [**Vi**] by using Chern–Moser invariants in 1982, and S. Krantz [**Kr**] gave another proof in 1986. Their proofs rely on Fefferman's theorem. Their point of view is that C^∞-differentiability makes possible the study of automorphism group in terms of differential invariants. Another approach is due to by S. Bell [**Be**].

1991 *Mathematics Subject Classification*. Primary 32H02.
This paper is in final form and no version of it will be submitted for publication elsewhere.

All these methods rely on $\bar{\partial}$-Neuman theory and are not completely elementary. So, my purpose is to develop a proof accessible to a larger audience.

The strategy is very simple and naive :
1) Get a proof for a given automorphism.
2) Get uniformity.

What is the reason of the uniformity ?

The uniformity follows from a characterization of the ball by its automorphism group, according to Wong-Rosay's theorem : [**Ro**], [**Wo**] or [**Pi 3**].

THEOREM. *D is not biholomorphic to the ball if and only if the automorphism group $Aut(D)$ is compact.*

Indeed, all the estimates in our method depend on the following invariant :

$$C(f) = \sup_{z \in D} \frac{dist(f(z), bD)}{dist(z, D)}$$

and it is necessary to control it. In fact, by using Hopf lemma and Wong-Rosay's theorem, we have :

PROPOSITION. *The mapping $f \mapsto C(f)$ is bounded on $Aut(D)$.*

PROOF. By uniform version of Hopf lemma, there exists a positive real number C and a compact set K into D such that any negative plurisubharmonic function ρ on D, satisfies the estimate :

$$\forall z \in D \ \ |\rho(z)| \geq C dist(z, bD) \ \inf_K |\rho|$$

To prove this version, it is sufficient to take again the usual proof or to use the plurisubharmonic measure of compact sets with respect to D [**Sa**].

As the distance to the boundary is equivalent to the modulus of the defining function r, by the above estimate we get for any $f \in Aut(D)$

$$\forall z \in D \ \ dist(f^{-1}(z), bD) \geq C dist(z, bD) \ \inf_K |r \circ f^{-1}|$$

and so $C(f)^{-1} \geq C \inf_K |r \circ f^{-1}|$. Since $Aut(D)$ is compact, there exists a compact subset L of D containing the set $f^{-1}(K)$ for every f in $Aut(D)$. Therefore, we have : $C(f)^{-1} \geq C \inf_L |r|$. That proves the proposition.

1. Describing the method.

We have to make some auxilliary constructions. The model is S. Webster's paper. This method has been used by S. Pinchuk [**Pi 3**] or F. Forstneric [**Fo**] or B. Coupet [**Co 2**].

1 - Lifting of D.

The domain D is lifted in the domain $D \times P_{n-1}$ in $\mathbb{C}^n \times P_{n-1}$. Its boundary is lifted in the manifold M where :

$$M = \{(z,p) \in \mathbb{C}^n \times P_{n-1}/z \in bD \text{ and } p = T_z^{\mathbb{C}}\}$$

where $T_z^{\mathbb{C}}$ is the complex tangent space at z, that is to say the maximal complex space contained into the tangent space. It is also the hyperplane determined by $\frac{\partial r}{\partial z}(z)$ if r is a defining function of D.

These definitions are due to S. Webster [We] who has proved the following lemma :

LEMMA. *M is a real analytic and totally seul submanifold in $\mathbb{C}^n \times P_{n-1}$ of dimension $2n - 1$.*

2 - Wedge on M.

Fix a point a of the boundary of D and denoted by $\tilde{a}$ its lifting on M. We can assume that the tangent space to bD at a is $\mathbb{R}^{2n-1}$ and the complex tangent space is given by $z_n = 0$. If Φ is a parametrization for bD near a defined on a ball of $\mathbb{R}^{2n-1}$ and r a defining function for D, the mapping $T = T_a$ from $\mathbb{R}^{2n-1}$ to $\mathbb{C}^{2n-1}$ given by

$$T(X) = \left(\Phi(X), \frac{r_1(w)}{r_n(w)}, \ldots, \frac{r_{n-1}(w)}{r_n(w)}\right) \quad w = \Phi(X)$$

where $r_j = \frac{\partial r}{\partial z_j}$ is a parametrization of M. Since M is real analytic and totally real, T can be extended near the origin in a biholomorphic mapping, always denoted by T, on $\mathbb{C}^{2n-1}$.

The image under T of the real ray $z' = 0, x_n = 0$ and $0 < y_n$ is a curve transverse to $bD \times P_{n-1}$ and extending into the open set $D \times P_{n-1}$. So, there exists an open cone Λ in $\mathbb{R}^{2n-1}$ such that the image $T[(\mathbb{R}^{2n-1} + i\Lambda) \cap B(0,R)]$ is included into $D \times P_{n-1}$. We can compose with a real linear map of $\mathbb{C}^{2n-1}$ and assume that $\Lambda_0 = \{t \in \mathbb{R}^{2n-1}\backslash|t_j| < t_{2n-1}, 1 \le j \le 2n - 2\}$ is contained into Λ. The wedge W is this image and M is the edge. The following important estimates for the distances hold.

PROPOSITION. *In a neighborhood of 0 in $\mathbb{C}^{2n-1}$, for (z,p) in W, we have :*
1) $dist((z,p), M) \approx dist(z, bD)$
2) $dist((z,p), (z, T_z^{\mathbb{C}})) = 0(dist(z, bD))$

PROOF. First of all, T is a diffeomorphism of a neighborhood of 0 in $\mathbb{C}^{2n-1}$ on a neighborhood of $T(0)$, taking $\mathbb{R}^{2n-1}$ to M. Thus, if X and Y are small the distance of $T(X + iY)$ to M is equivalent to $|Y|$ and as $\frac{\partial}{\partial x_j} r_0 \Phi(0) = 0$ and $\frac{\partial}{\partial y_j} r_0 \Phi(0) = -\delta_{j,n}$ this distance is equivalent to Y_{2n-1} if Y is near 0 and in Λ_0. That proves the first assertion.

Now, writing $(z,p) = T(X + iY) = (\Phi(X + iY), P(X + iY))$, we have to compare $P(X + iY)$ to $T_z^{\mathbb{C}}$. Since these maps are smooth and equal for $Y = 0$, their difference is $0(|Y|)$.

3 - Lifting of an automorphism.

Every automorphism f of D is lifted in the mapping f^* defined by :

$$f^*(z,p) = (f(z),\ \text{Image of } p \text{ under } f'(z))$$
$$= (f(z), [^t g'(f(z))(p)])$$

if $g = f^{-1}$ and we identify complex hyperplanes with linear forms in $\mathbb{C}^n$ ([ ] being the class in P_{n-1}).

We extend the definition of f^* to M by the following :

$$f^*(z, T_z^{\mathbb{C}}) = (f(z), T_{f(z)}^{\mathbb{C}}).$$

This expression takes sense since any automorphism of D extends to a $\frac{1}{2}$-lipschitzian homeomorphism of $\overline{D}$ according to S. Pinchuk [**Pi 1**].

Of course, a priori, f^* is not continuous up to M. The method will be the continuity of this mapping. Indeed, if f^* is continuous, we can apply the wedge of the wedge theoem to the mapping $T_b^{-1} \circ f^* \circ T_a$ and get a full neighborhood where it extends. Returning to $\mathbb{C}^n$, we have extension near any point of bD. It is necessary to control the size of the neighborhood and for this, we must determine the dependance of this neighborhood under f (this one depends on the choice of $b = f(a)$ and so of f.).

2. Condition A – Continuity of the lifting.

To get the continuity, it is convenient to introduce a special basis $S(z)$ of $\mathbb{C}^n$ for z in D. $S(z)$ will be an orthonormal basis where the last vector is an orthogonal vector to the complex tangent space $T_z^{\mathbb{C}}$. We may assume that the mapping $z \longmapsto S(z)$ is smooth.

As a consequence of the usual estimates of Kobayashi metric in a strictly pseudoconvex domain [**Gr**] and [**Si**], we get the following.

PROPOSITION. *The matrix $A(z)$ of the automorphism $f'(z)$ into the bases $S(z)$ and $S(f(z))$ has the following form*

$$\begin{bmatrix} 0(C(f)^{1/2}) & 0(C(f)^{1/2}\delta^{-1/2}) \\ 0(C(f)\delta^{1/2}) & 0(C(f)) \end{bmatrix}$$

where $\delta = dist(z, bD)$.

As a consequence, we have :

PROPOSITION. *The automorphisms of D are uniformly 1/2-lipschitzian on D or equivalently, there exists a constant C such that for any $f \in Aut(D)$:*

$$|f'(z)| \le C dist(z, bD)^{-1/2}.$$

Roughly, $[^t(f^{-1})'(f(z))(p)]$ is as a quotient where the denominator is the term on the last row and the last column in the former matrix $A(z)$. This term is the quantity :

$$\nu(r \circ f)(z) = |\nabla r(z)|^{-1} \cdot \sum_{j=1}^{n} \frac{\partial r}{\partial \bar{z}_j}(z) \frac{\partial (r \circ f)}{\partial z_j}(z)$$

(ν is a normal complex vector field) and it is necessary to prove that it is bounded away 0. It is condition A introduced by Nirenberg, Webster and Yang [N. W. Y.]. Their initial proof is long and the dependance under f is not clear. A simpler proof can be done by using scaling method and moreover we get uniformity.

THEOREM. *Condition A is uniformly satisfied. There exists a constant A and a neighborhood V of bD such that for any z in V :*

$$A \leq \left| \sum_{j=1}^{n} \frac{\partial r}{\partial \bar{z}_j}(z) \frac{\partial (r \circ f)}{\partial z_j}(z) \right|$$

PROOF. The proof is by contradiction. If it is false, there exists a sequence (z^k) in D having a limit point on the boundary and a sequence (f^k) of automorphisms such that $\lim \nu(z^k)(r \circ f^k) = 0$.

Using the rescaling method for these sequences, we get a mapping $\tilde{f}$ on $\Sigma = \{('z, z_n) \in \mathbb{C}^{n-1} \times \mathbb{C} | 2Re(z_n) + |'z|^2 < 0\}$ of the form $\tilde{f}('z, z_n) = (U('z), z_n), U$ being a unitary transform of $\mathbb{C}^{n-1}$ [Pi 3] or [Co 2].

Thus, the condition $\lim \nu(z^k)(r \circ f^k) = 0$ gives $\frac{\partial \tilde{f}_n}{\partial z_n}('0, -1) = 0$. This is a contradiction and the theorem is proved.

As a consequence, we get

PROPOSITION. *f^* is continuous on W up to M.*

and following Webster's method, by the edge of the wedge theorem.

PROPOSITION. *Any automorphism of D extends on a neighborhood of $\overline{D}$.*

Remark : In his paper [Fo], F. Forstneric does not use scaling method (for a given automorphism) but only the theorem of Julia-Caratheodory. The author has not examined if it would be possible to get uniform condition A by this way.

3. Uniform extendibility.

Now, we can give the outlines of the proof of the main theorem in this note. Fix a point a of bD.

Let us recall that for every point b of bD, there exists a ball $B(0, r_b)$ into $\mathbb{C}^{2n-1}$ and a biholomorphic mapping T_b from $B(0, r_b)$ onto a relatively compact neighborhood of $\tilde{b}$ into $\mathbb{C}^n \times P_{n-1}$. Fix another neighborhood V_b of $\tilde{b}$ such that its closure is compact into $T_b[B(0, r_b)]$.

As the elements of $Aut(D)$ are uniformly $\frac{1}{2}$-lipschitzian on M, for every b, there exists a neighborhood V'_b of $\tilde{b}$ and a neighborhood U_b of a such that if $\tilde{f}(\tilde{a})$ belongs to V_b, then $\tilde{f}(\tilde{z})$ belongs to V_b for z in $U_b \cap bD$.

By compactness of M, we can get a finite covering of M by some

$V_b'(b \in \Lambda, \Lambda$ finite set). $U = \cap_{b \in \Lambda} U_b$ is a neighborhood of a such that if z belongs to $U \cap bD$ and $\tilde{f}(\tilde{a})$ is in V_b', then $\tilde{f}(\tilde{z})$ belongs to V_b.

Introduce the sets $G(b)$, for b in Λ, defined by :

$$G(b) = \{f \in Aut(D) \quad \tilde{f}(\tilde{a}) \in V_b'\}.$$

These sets are a finite covering of $Aut(D)$ and now we shall consider a fixed b in Λ.

An essential step in the proof is the choice of a wedge $W'(a) \subset W(a)$ constructed on M near $\tilde{a}$ such that if f belongs to $G(b)$, the image of $W'(a)$ under $\tilde{f}$ is contained into the domain of the same local chart near $\tilde{b}$.

We can assume that $\frac{\partial r}{\partial z}(b)$ is e_n. For w near b we can write $e_n = \alpha(w)\frac{\partial r}{\partial z}(w) + t(w)$ where $t(w)$ is a comptex tangent vector at w and α is a smooth map taking value 1 at b. Then, using the uniform $\frac{1}{2}$-lipschitzian extendability and proposition, for (z,p) in $W(a)$, we have :

$$< e_n, \; {}^tg'(f(z))(p) > = \alpha(f(z)) < \frac{\partial r}{\partial z}(f(z)), {}^tg'(f(z))(\frac{\partial r}{\partial z}) > +0(dist(z, bD)^{1/2}))$$

By the uniform condition A, the first term is bounded away 0 for z near a and so, we are in the local chart of domain $\mathbb{C}^n \times \{p \in (\mathbb{C}^n)^*/p_n \neq 0\}$.

In the same way, it is possible to verify that the mappings $\tilde{f}(f \in G(b))$ are uniformly bounded near $\tilde{a}$.

As the mappings $\tilde{f}$ are uniformly bounded and uniformly $\frac{1}{2}$-lipschitzian on M, it follows [Co 1] that they are also uniformly $\frac{1}{2}$-lipschitzian in every smaller wedge $W \subset W'(a)$ and so, shrinking if necessary $W'(a)$, we can consider the mappings $T_b^{-1} \circ \tilde{f} \circ T_a$ and apply the edge of the wedge theorem because they are continuous up to $\mathbb{R}^{2n-1}$ and real on $\mathbb{R}^{2n-1}$ [Ru]. The conclusion is easy.

References

[Be] S. Bell, *Uniform extendibility of sequences of biholomorphic mappings.*

[Co 1] B. Coupet, *Régularité de fonctions holomorphes sur des wedges*, Can. Jour. Math **XL** no. 3 (1988), 532-545.

[Co 2] _____ , *Precise regularity up to the boundary of proper holomorphic mappings*, (à paraitre).

[Fe] C. Fefferman, *The Bergman kernel and biholomorphic mappings of pseudoconvex domains*, Inv. Math. **26** (1974), 1-65.

[Fo] F. Forstneric, *An elementary proof of Fefferman's theorem* Expo. Math (1991), (to appear).

[Gr] I. Graham, *Boundary behavior of the Caratheodory and Kobayashi metrics on strongly pseudo-convex domains in $\mathbb{C}^n$ with smooth boundary*, T.A.M.S. **207** (1975), 219-240.

[Gr-Kr] R. Greene and S. Krantz, *A new invariant metric in complex analysis and some applications*, (à paraitre).

[Kr] S. Krantz, *Compactness Principle in Complex Analysis*, Seminarios I Volumen 3, Universidad autonóma de Madrid. División de Matematicas.

[Le] H. Lewy, *On the boundary behavior of holomorphic mappings*, Acad. Wa 2. Linei 35 (1977), 1-8.

[Ni. We. Ya] L. Nirenberg, S. Webster and P. Yang, *Local boundary regularity of holomorphic mappings*, Com. Pur. Appl. Math. **33** (1980), 305-328.

[Pi 1] S. Pinchuk, *On the analytic continuation of holomorphic mappings*, Math. USSR Sbornik **27** (1975), 375-392.

[Pi 2] ———, *Holomorphic inequivalence of some classes of domain in* $\mathbb{C}^n$, Math. USSR **39** (1981), 61-86.

[Pi 3] ———, *The scaling method and holomorphic mappings*, AMS Summer School Research Institute in Several Complex Variables and Complex Geometry. Santa-Cruz, California July 1989.

[Ro] J.P. Rosay, *Sur une caractérisation de la boule par son groupe d'automorphismes*, Ann. Inst. Fourier **XXIX** (1979), 91-97.

[Ru] W. Rudin, *Lectures on the edge of the wedge theorem*, AMS, Providence, 1971.

[Sa] A. Sadullaev, *A boundary uniqueness theorem in* $\mathbb{C}^n$, Math. USSR Sbornik **30** no. 4 (1976), 501-514.

[Si] N. Sibony, *A class of hyperbolic manifolds*, Recent Developments in several Complex Variables, J.E. Fornaess ed, Annals of Mathematics Studies, Princeton University Press, 1981.

[Vi] A.G. Vitushkin, *Holomorphic continuation of maps of compact hypersurfaces*, Math USSR t.2v. 20 (1983), 27-33.

[We] S. Webster, *On the reflection principle in several complex variables*, Proc. Ams. Math. Soc. **71** (1978), 26-28.

[Wo] B. Wong, *Characterization of the unit ball in* $\mathbb{C}^n$ *by its automorphism group*, Inv. Math. **41** (1977), 253-257.

UFR - MIM ET URA 225, 3, PLACE VICTOR HUGO, 13331 MARSEILLE CEDEX 3, FRANCE

Contemporary Mathematics
Volume **137**, 1992

The Class L log L With Weights

DAVID CRUZ-URIBE, SFO

1. Introduction.

This note establishes weighted versions of the following two theorems about the class L log L with respect to Lebesgue measure on the unit circle.

THEOREM A. [**Zygmund-Stein**] *If f is in L^1, then the Hardy-Littlewood maximal function of f, Mf, is in L^1 if and only if $|f|\log^+|f|$ is in L^1.*

THEOREM B. [**Zygmund**] *If f is in L^1 and $|f|\log^+|f|$ is in L^1 then $\tilde{f}$ is in L^1. Conversely, if $f \geq 0$ and f, $\tilde{f}$ are in L^1, then $|f|\log^+|f|$ is in L^1.*

(See [5] and [6] for proofs.)

Let $w \in L^1$, $w \geq 0$, and let $L^1(w) = L^1(wdx)$. The weight w satisfies the (A_1) condition (denoted by $w \in (A_1)$) if there exists a constant C such that $Mw \leq Cw$ almost everywhere on ∂D. In this case analogous results hold for $L^1(w)$:

THEOREM 1. *If w is in (A_1) and f is in $L^1(w)$, then Mf is in $L^1(w)$ if and only if $|f|log^+|f|$ is in $L^1(w)$.*

THEOREM 2. *Let w be in (A_1). If f is in $L^1(w)$ and $|f|\log^+|f|$ is in $L^1(w)$ then $\tilde{f}$ is in $L^1(w)$. Conversely, if $f \geq 0$ and f, $\tilde{f}$ are in $L^1(w)$, then $|f|\log^+|f|$ is in $L^1(w)$.*

The proofs of these theorems are in sections 2 and 3. The notation used throughout is standard and will be defined as needed. Throughout, C will denote a constant whose value may change at each appearance.

1991 *Mathematics Subject Classification*. Primary 42A50, 42B25.

This material is based upon work supported under a National Science Foundation Graduate Fellowship.

This paper is in final form and no version of it will be submitted for publication elsewhere.

2. Proof of Theorem 1

The proof that Mf is in $L^1(w)$ uses standard techniques involving distribution inequalities. It depends on a result proved by Muckenhoupt in [4]: If $w \in (A_1)$ then there exists a constant $C > 0$ such that for $x > 0$

$$w(\{Mf > x\}) \leq \frac{C}{x} \int_0^\infty w(\{|f| > t\}) \, dt$$

for all $f \in L^1(w)$. (Here, $w(E) = \int_E w \, d\theta$ for any measurable set $E \subset \partial D$.)

We also need the following equality, which is an immediate consequence of Fubini's theorem:

$$\int_{\partial D} |f| \, w \, d\theta = \int_0^\infty w(\{|f| > t\}) \, dt.$$

Now let $f \in L^1(w)$ be such that $|f| \log^+ |f| \in L^1(w)$. Fix $x > 0$ and define $f = g_x + h_x$, where

$$g_x = \begin{cases} f & \text{where } |f| > x/2 \\ 0 & \text{otherwise,} \end{cases}$$

and $h_x = f - g_x$. Then $\|h_x\|_\infty \leq x/2$, so $\|Mh_x\|_\infty \leq x/2$. By the subadditivity of the maximal function, $Mf \leq Mg_x + x/2$. Hence $\{Mf > x\} \subset \{Mg_x > x/2\}$. Applying Muckenhoupt's result to g_x gives

$$w(\{Mf > x\}) \leq \frac{C}{x} \int_0^\infty w(\{|g_x| > t\}) \, dt.$$

But

$$w(\{|g_x| > t\}) = \begin{cases} w(\{|f| > x/2\}) & t \leq x/2 \\ w(\{|f| > t\}) & t > x/2. \end{cases}$$

Therefore

$$w(\{Mf > x\}) \leq Cw(\{|f| > x/2\}) + \frac{C}{x} \int_{x/2}^\infty w(\{|f| > t\}) \, dt.$$

Using this, we can bound the $L^1(w)$ norm of Mf:

$$\int_0^\infty w(\{Mf > x\}) \, dx \leq 2\|w\|_1 + \int_2^\infty w(\{Mf > x\}) \, dx.$$

The second term on the right is bounded by

$$C \int_2^\infty w(\{|f| > x/2\}) \, dx + C \int_2^\infty \frac{1}{x} \int_{x/2}^\infty w(\{|f| > t\}) \, dt \, dx.$$

The first integral is bounded by $C\|f\|_{L^1(w)} < \infty$. By Fubini's theorem the second integral is bounded by

$$C \int_1^\infty \left(\frac{1}{x} \int_x^\infty \int_{\{|f|>t\}} w\,d\theta\,dt \right) dx = C \int_1^\infty \left(\frac{1}{x} \int_{\{|f|>x\}} \int_x^{|f(e^{i\theta})|} dt\,w\,d\theta \right) dx$$

$$\leq C \int_1^\infty \left(\frac{1}{x} \int_{\{|f|>x\}} \int_0^{|f(e^{i\theta})|} dt\,w\,d\theta \right) dx$$

$$= C \int_1^\infty \frac{1}{x} \int_{\{|f|>x\}} |f(e^{i\theta})|\,w\,d\theta\,dx$$

$$= C \int_{\{|f|>1\}} |f(e^{i\theta})| \int_1^{|f(e^{i\theta})|} \frac{dx}{x}\,w\,d\theta$$

$$= C \int_{\{|f|>1\}} |f| \log|f|\,w\,d\theta$$

$$= C \int_{\partial D} |f| \log^+ |f|\,w\,d\theta$$

$$< \infty.$$

Hence, $Mf \in L^1(w)$.

The proof of the converse follows the proof of the second half of Theorem A given by Stein in [5]. Suppose that f, $Mf \in L^1(w)$. Fix $t > 0$ and let $E = \{Mf > t\}$. Then E is open. Since $w \in (A_1)$, it is a "doubling measure" and the Calderon-Zygmund lemma extends to it. (See [3, pp. 143–4, 397] for details.) Hence there exists a collection of disjoint intervals $I_n \subset E$ and an absolute constant $C > 0$ such that for all n

$$\frac{1}{w(I_n)} \int_{I_n} |f|\,w\,d\theta \leq Ct$$

and $|f| \leq t$ almost everywhere in $E \backslash \bigcup I_n$. By Lebesgue's differentiation theorem, $|f| \leq t$ a.e. in $\partial D \setminus E$, so we have, by summing with respect to n,

$$\frac{1}{t} \int_{\{|f|>t\}} |f|\,w\,d\theta \leq Cw(\{Mf > t\}).$$

Integrating this with respect to t, we get

$$\int_1^\infty \frac{1}{t} \int_{\{|f|>t\}} |f|\,w\,d\theta\,dt \leq C\|Mf\|_{L^1(w)} < \infty.$$

By Fubini's theorem, the left-hand side equals

$$\int_{\{|f|>1\}} |f| \int_1^{|f(e^{i\theta})|} \frac{dt}{t}\,w\,d\theta = \int_{\partial D} |f| \log^+ |f|\,w\,d\theta. \qquad \square$$

3. Proof of Theorem 2

Suppose first that f and $|f|\log^+|f|$ are in $L^1(w)$. Then, by the first half of Theorem 1, $Mf \in L^1(w)$. In [2], Coifman and Fefferman proved the following inequality holds whenever $w \in (A_\infty)$:

$$\int_{\partial D} |\tilde{g}|\, w\, d\theta \le C \int_{\partial D} Mg\, w\, d\theta, \qquad g \in L^1(w).$$

Since $w \in (A_1)$ implies that $w \in (A_\infty)$ (see [3] for details) it follows immediately from this that $\tilde{f} \in L^1(w)$.

The proof of the converse again follows Stein's proof in [5]. Suppose that f is non-negative and that $f, \tilde{f} \in L^1(w)$. Since $w \in (A_1)$, $f \in L^1$ ([3, pp. 395-6]). Fix $e^{it} \in \partial D$ and let I be any interval containing it. Then there exists an absolute constant $C > 0$ such that

$$\frac{1}{|I|} \int_I f\, d\theta \le C \int_{\partial D} f P_z\, d\theta$$

where $z \in D$ is such that $z/|z|$ is the center of I, $1 - |z| = |I|/2\pi$, and P_z is the Poisson kernel at the point z. For $\alpha < \pi/2$ let $\Gamma_\alpha(e^{it})$ denote the convex hull of the disk $\{|z| \le \sin\alpha\}$ and the point e^{it}. Elementary geometric reasoning (using the law of sines and L'Hôpital's rule) shows that as $|I|$ tends to 0, the smallest α such that $z \in \Gamma_\alpha(e^{it})$ tends to $\operatorname{arccot}(1/\pi)$. Therefore we can find an $\alpha < \pi/2$ such that $z \in \Gamma_\alpha(e^{it})$ for all I.

Then, taking the supremum over all intervals I, the above inequality becomes $Mf(e^{it}) \le C N_\alpha f(e^{it})$, where $N_\alpha f$ is the non-tangential maximal function of f on the angle α. Therefore, by the second half of Theorem 1, it will suffice to show that $N_\alpha f \in L^1(w)$.

Since $w \in (A_1)$, w is log integrable. Hence there exists an outer function $h \in \mathrm{H}^1$ such that $|h| = w$. Since $g = f + i\tilde{f} \in \mathrm{H}^1(w)$, we must have $gh \in \mathrm{H}^1$. By the Riesz factorization theorem, there exists $v \in \mathrm{H}^2$ and a Blaschke product B such that $gh = Bv^2$, or $g = B(vh^{-1/2})^2$. Thus $vh^{-1/2} \in \mathrm{L}^2(w)$. Since $w \in (A_1)$, $w \in (A_2)$, so the Hardy-Littlewood maximal function is a bounded operator on $\mathrm{L}^2(w)$. Since the non-tangential maximal function is dominated pointwise by a constant multiple of the Hardy-Littlewood maximal function, it too must be bounded on $\mathrm{L}^2(w)$. Therefore, $N_\alpha(vh^{-1/2}) \in \mathrm{L}^2(w)$. But $N_\alpha f \le N_\alpha g \le N_\alpha(vh^{-1/2})^2 = (N_\alpha vh^{-1/2})^2 \in \mathrm{L}^1(w)$. $\square$

4. Remarks

The full strength of the hypothesis that $w \in (A_1)$ was used only in proving the first half of Theorem 1. If $w \in (A_\infty)$ it is in (A_p) for some $p \ge 1$ and it is a doubling measure, so the other arguments only need this weaker condition. Further, Carbery, Chang and Garnett ([1]) found a complicated necessary and sufficient condition for the first half of Theorem 1 to hold which is weaker than the (A_1) condtion. It is tempting to speculate that their condition is also necessary and sufficient for the first half of Theorem 2. I have no conjecture as to what are the weakest possible assumptions for the converse of either theorem to be true.

References.

1. A. Carbery, S.-Y. A. Chang, J. Garnett, *Weights and L logL*, Pacific J. Math. **120** (1985), 33–45.

2. R. Coifman and C. Fefferman, *Weighted Norm Inequalities for Maximal Functions and Singular Integrals*, Studia Math. **51** (1974), 241–250.

3. J. Garcia-Cuerva and J.L. Rubio de Francia, *Weighted Norm Inequalities and Related Topics*, Mathematical Studies **116**, North-Holland, New York, 1985.

4. B. Muckenhoupt, *Weighted Norm Inequalities for the Hardy Maximal Function*, Trans. Amer. Math. Soc. **165** (1972), 207–226.

5. E.M. Stein, *Note on the Class L logL*, Studia Math. **32** (1969), 305–310.

6. A. Zygmund, *Trigonometric Series*, Vols. I and II, 2nd ed., Cambridge University Press, London, 1959.

Contemporary Mathematics
Volume **137**, 1992

The Geometry of Proper
Holomorphic Maps between Balls

JOHN P. D'ANGELO

Introduction.

Walter Rudin asked in his book *Function Theory in the Unit Ball of* $\mathbb{C}^n$ what
are the proper holomorphic mappings between balls in (perhaps different dimensional) complex Euclidean spaces. The present paper, written for the symposium
held in his honor, contains two parts. In the first we summarize the progress on
this question to date. We omit most proofs, but give some new examples and
a complete description of proper rational maps with linear denominator. This
generalizes the author's result classifying polynomial proper maps, and gives a
glimpse of the general case. In the second part we pay particular attention to
invariant proper mappings. This part focuses mainly on questions about whether
there are rational proper maps that are invariant under finite unitary groups.
Here we observe a peculiar yet pleasing blend of analysis and algebra.

I. Proper Maps Between Balls

Let Ω_1, Ω_2 be domains in complex Euclidean spaces. Recall that a holomorphic map $f : \Omega_1 \to \Omega_2$ is proper if $f^{-1}(K)$ is a compact subset of Ω_1 whenever K
is a compact subset of Ω_2. For bounded domains this condition is equivalent to
the statement that, whenever $z_\nu \to b\Omega_1$, $f(z_\nu) \to b\Omega_2$. For $\Omega_j = B_{n_j}$, the unit
balls in C^{n_j}, this latter condition takes the simple form $||f(z_\nu)|| \to 1$ whenever
$||z_\nu|| \to 1$. For maps continuous up to the boundary sphere, we may replace the
limits with equality; we study therefore the consequences of the statement that
$||f(z)|| = 1$ whenever $||z|| = 1$.

Let us begin by recalling known results. In general there are no proper holomorphic maps between a given pair of domains. A simple necessary condition
is that the dimension of the image domain Ω_2 must be at least as large as the

1991 *Mathematics Subject Classification.* Primary 32H35.
Partially supported by the NSF.
This paper is in final form and no version of it will be submitted for publication elsewhere.

dimension of Ω_1. Henceforth we assume that $f : B_n \to B_N$ is a proper holomorphic map, and that $N \geq n$. The first case of interest is when $n = N$. Here the results are complete. When $n = 1$, f must be a finite "Blaschke product;" in particular it is rational and extends past the disk.

THEOREM 1. *Every proper holomorphic map f from the unit disc to itself is a finite Blaschke product. That is, there are finitely many points $\{a_j\}$ in the disc, positive integer multiplicities m_j, and a number $e^{i\theta}$ such that*

$$f(z) = e^{i\theta} \prod_{j=1}^{m} \left(\frac{a_j - z}{1 - \overline{a}_j z} \right)^{m_j} \tag{1}$$

A superficial first glance suggests that this result fails completely in higher dimensions. One has the theorem (Theorem 2 below), proved by H. Alexander [A], that proper holomorphic self maps of B_n, for $n \geq 2$, are necessarily automorphisms. The statement in two dimensions may have been known to Poincare; see [F2] for some historical remarks about this. The conclusion of Theorem 2 holds under considerably weaker hypotheses, such as for strongly pseudoconvex domains and even for certain weakly pseudoconvex domains. [B]. In order to find more examples of proper maps, one must consider target balls in higher dimensions. Investigation of this then enables us to discover what objects deserve the name "finite Blaschke products" in more variables.

THEOREM 2. *A proper holomorphic self map of B_n is necessarily an automorphism when $n \geq 2$.*

The automorphism group $\text{Aut}(B_n)$ of the ball has the following description. See Rudin's book [R1] for a beautiful discussion of this. Every f in $\text{Aut}(B_n)$ is of the form

$$f = U\xi_a \tag{2}$$

where U is unitary, and ξ_a is the linear fractional transformation

$$\xi_a(z) = \frac{a - L_a(z)}{1 - \langle z, a \rangle} \tag{3}$$

Here L_a is a linear transformation defined by

$$L_a(z) = sz + \frac{\langle z, a \rangle}{s + 1} a \tag{4}$$

and s is the positive number satisfying $s^2 = 1 - ||a||^2$. One could also describe $\text{Aut}(B_n)$ as the real Lie group $SU(n, 1)$.

We now turn to the positive codimension case; the codimension is the number $N - n$. There are many new points requiring consideration. One difficulty is that there are proper maps that are not rational. It is probably impossible to classify the non-rational maps; see [D,F2,H] to get an idea of how arbitrary such maps can be. It is an open problem to determine the minimal smoothness requirement at the sphere to guarantee that a proper map must be rational. We discuss this briefly at the conclusion of the paper. Forstneric [F3] has proved that a sufficiently differentiable proper map between balls is rational if $n \geq 2$, so we spend most of this paper discussing rational proper maps. He obtained the following result.

THEOREM 3. *(Forstneric). Suppose that $N > n > 1$, that*

$$\Omega_1 \subset\subset C^n$$
$$\Omega_2 \subset\subset C^N$$
(5)

are strongly pseudoconvex domains with real analytic boundaries, and that $f : \Omega_1 \to \Omega_2$ is a proper holomorphic mapping between them. If also $f \epsilon C^\infty (\overline{\Omega_1})$, then it extends holomorphically past a dense open subset of the boundary. If the second domain is a ball, the result holds when f is assumed only to be of class C^{N-n+1}. Furthermore, if both domains are balls, then f is a rational function.

REMARKS. In case f is a rational proper holomorphic map between balls, it extends holomorphically past the sphere. Thus there cannot be a point of indeterminacy on the sphere. This result was proved independently by Pinchuk (not published but see [P]) and by Cima-Suffridge [CS3]. A generalization (to arbitrary real analytic domains and some special ranges) of this result was proved also by Chiappari [C]. Forstneric also obtains a crude bound on the degree of the defining polynomials.

The proof of Theorem 3 involves certain varieties associated to the proper mapping. In case the map extends holomorphically past the boundary, these varieties have a pleasing definition. The idea of Forstneric's proof is to define these varieties by considering only a finite Taylor development at the boundary. Once one has these varieties, there remains considerable work to do to show that they are rational. The reader should consult [F3] and the survey article [F2].

The algebraic properties of rational proper maps are complicated by the possibility of composition with automorphisms. To account for this, it is natural to consider a pair of proper holomorphic maps $f, g : B_n \to B_N$ to be the same if there are automorphisms of the domain and target balls ϕ_n, ϕ_N such that $\phi_N f = g \phi_n$. One then says that f, g are spherically equivalent. The reader can verify for example that all Blaschke products (in one variable) of order two are spherically equivalent. As soon as Blaschke products have three factors, this no longer holds. The appropriate analog holds in higher dimensions. There are additional algebraic difficulties, some of which we confront in the rest of this paper.

We continue with our summary. The first results say that there are no interesting examples unless $N > 2n - 2$. Faran [Fa2], following up on work of Webster [W], proved the following theorem.

THEOREM 4. *(Faran) Suppose that $f : B_n \to B_N$ is proper, smooth on the closed ball, and that $N < 2n - 1$. Then f is spherically equivalent to the linear imbedding given by $z \to (z, 0)$.*

In case $N = 2n - 1$, one obtains a quadratic polynomial map.

EXAMPLE 1. *For each dimension n we have the map*

$$f(z_1, ..., z_n) = (z_1, z_2, ..., z_{n-1}, z_1 z_n, ... z_n^2) \tag{6}$$

It is a proper map between the unit balls; it is the simplest example of a tensor product.

Before proceeding to tensor products, we state another result of Faran [Fa1] in the special case $n = 2, N = 3$. This gives two other easy examples.

THEOREM 5. *(Faran) Suppose that $f : B_2 \to B_3$ is proper and twice continuously differentiable on the closed ball. Then f is spherically equivalent to one of the following four maps:*

$$\begin{aligned}
f(z, w) &= (z, w, 0) \\
f(z, w) &= (z, zw, w^2) \\
f(z, w) &= \left(z^2, \sqrt{2}zw, w^2\right) \\
f(z, w) &= \left(z^3, \sqrt{3}zw, w^3\right)
\end{aligned} \tag{7}$$

Faran proved Theorem 5 assuming three continuous derivatives, but Cima-Suffridge [CS2] reduced the differentiability hypothesis to two continuous derivatives. The fact that only two continuous derivatives are required follows also from Theorem 3.

In order to cast these specialized results into a general framework, we need to introduce tensor products. The author's motivation in doing this was quite naive. When he first heard the statement of Theorem 2, he wondered what happened to functions like z^m. The tensor product operation generalizes this map to higher dimensions, but increases the target dimension. It becomes necessary to study proper maps from a given ball to all possible larger dimensional balls in order to classify the rational maps.

Suppose g, h are holomorphic maps with the same domain Ω, but taking values in C^n and C^N respectively. We define the tensor product of the maps by

$$g \otimes h = (g_1 h_1, g_2 h_1, ..., g_n h_1, g_1 h_2, ..., g_n h_N) \tag{8}$$

Thus the tensor product of two maps is the map whose components are all possible products of the components of the two maps. Suppose that the range domain is a ball. Maps agreeing up to unitary transformations are then spherically equivalent, so it doesn't really matter what order we assign to the list of components. To include maps such as example 1 we need to allow the tensor product to be taken on a subspace. We will also require a version of its inverse.

Suppose that we have written a complex vector space as an orthogonal sum

$$C^N = A \oplus A^\perp \tag{9}$$

If f is a holomorphic mapping that takes values in C^N, it has then an induced decomposition

$$f = f_A \oplus f_{A^\perp} \tag{10}$$

Suppose also that g is any holomorphic mapping with the same domain as f, and with range contained in any finite dimensional complex vector space. Given an orthogonal summand A of the range, we can form the map

$$E(A, g)(f) = (f_A \otimes g) \oplus f_{A^\perp} \tag{11}$$

We also have the undoing map, defined on maps in the range of the above, by

$$E^{-1}(A, g)((f_A \otimes g) \oplus f_{A^\perp}) = f \tag{12}$$

The connection to proper maps is the following simple fact whose proof the reader can supply easily.

LEMMA 1. *Suppose that*

$$f : \Omega \subset\subset C^n \rightarrow B_N$$
$$g : \Omega \subset\subset C^n \rightarrow B_K \tag{13}$$

are proper holomorphic mappings. Suppose that the range of f has the orthogonal decomposition $C^N = A \oplus A^\perp$. Then the map

$$E(A, g)(f) = (f_A \otimes g) \oplus f_{A^\perp} \tag{14}$$

is a proper holomorphic map from Ω to the appropriate dimensional ball.

This operation generalizes ordinary multiplication in one dimension. It is interesting that one must also allow the "undoing" of a tensor product in higher dimensions; this analog of division is superfluous in one variable. Thus, if h is the map on the right side of (14), we put $E^{-1}(A, g) h = f$. In higher dimensions it arises because one may wish to "undo" on a different subspace from the one on which one "did".

Perhaps the most important proper polynomial map is the m-fold tensor product. Perform the tensor product of the identity mapping with itself, always on the full space, a total of m times. The resulting map is a homogeneous monomial map of degree m, with each component being listed the appropriate number of times. After performing a unitary change of coordinates that collects the same monomials together, and ignoring zero components, one obtains the map

$$H_m(z) = z \otimes z \otimes ... \otimes z \qquad \text{m-times} \tag{15}$$

In coordinates, the map has the formula

$$H_m(z) = (..., c_a z^a, ...)$$
$$|c_a|^2 = \binom{m}{a} = \frac{m!}{a_1! ... a_n!} \tag{16}$$

It turns out to be particularly important in the classification of proper mappings between balls. One can also form the tensor product of m distinct automorphisms, to obtain a "Blaschke product" in several variables. Such expressions do not exhaust the list of proper rational maps between balls. The reason involves the notion of "undoing". The generalization of Blaschke product will require "undoing". We discuss this briefly later in this section.

The importance of the map $z \to H_m(z)$ derives in part from the following simple result, noticed independently by Rudin [R2] and the author [D2]. This result implies that H_m is essentially the only homogeneous proper map between balls. This result, although easy, is important in the classification results, so we state and prove it.

THEOREM 6. *Suppose that*

$$f : B_n \to B_N \tag{17}$$

is a proper mapping between balls, the components of f are linearly independent, and each component is a homogeneous polynomial of degree m. Then, after composition in the range with a unitary map, f is the map H_m.

PROOF. As a proper map from ball to ball that extends past the boundary, f must map the sphere to itself. This yields the equations

$$
\begin{aligned}
||f(z)||^2 &= 1 \quad &\text{on} \quad ||z||^2 = 1 \Rightarrow \\
||f(z)||^2 &= ||z||^{2m} \quad &\text{on} \quad ||z||^2 = 1
\end{aligned}
\tag{18}
$$

The second equality is one of homogeneous polynomials, so it holds in all of C^n. Notice that

$$||z||^{2m} = \sum_{|a|=m} \binom{m}{a} |z|^{2a} = ||H_m(z)||^2 \tag{19}$$

by the multinomial expansion. Note the use of multi-index notation in the middle term. We have now two holomorphic functions with the same squared norm. This implies that they differ by a unitary matrix of constants. [D1] Thus, for some U,

$$f(z) = U H_m(z) \tag{20}$$

completing the proof. ∎

By a process of orthogonal homogenization, one can reduce the classification of proper polynomial mappings between balls to Theorem 6. The same technique applies, with considerable amplification, in the rational case. Let us suppose that

$f = \frac{p}{q}$ is a rational function that maps the unit sphere S^{2n-1} to S^{2N-1}. Write

$$p(z) = \sum_{|\alpha|=0}^{m} c_\alpha z^\alpha = \sum_{k=0}^{m} p_k$$
$$q(z) = \sum_{|\alpha|=0}^{m} d_\alpha z^\alpha = \sum_{k=0}^{m} q_k$$

(21)

The coefficients of p are elements of C^N, the coefficients of q are complex numbers, and the expressions on the right are expansions into homogeneous polynomials. Use the condition that $f : S^{2n-1} \to S^{2N-1}$ and clear denominators in the result. Next replace z by $re^{i\theta}$ (this is multi-index notation) and equate Fourier coefficients. We obtain the identities

$$0 = \sum_{|\beta|=0}^{m} \left(\langle c_{\beta+\gamma}, c_\beta \rangle - d_{\beta+\gamma} \overline{d_\beta} \right) r^{2\beta} \; \forall \gamma$$

(22)

on the set $\|r\|^2 = 1$. We can do the same for the homogeneous parts. Replace z by $ze^{i\theta}$ where now θ is just one variable. Using the homogeneity, and proceeding as above, we obtain

$$0 = \sum_{k=0}^{m} \left(\langle p_{k+j}, p_k \rangle - q_{k+j} \overline{q_k} \right) \; 0 \le j \le m$$

(23)

on the sphere. These identities can then be homogenized so as to hold everywhere.

In the polynomial case, we may assume without loss of generality that $q = 1$. Supposing that $f = \sum f_k$ and using (23), we obtain

$$\sum_{k=0}^{m-j} \langle f_{k+j}, f_k \rangle = 0 \quad j = 1, ..., m$$
$$\sum_{k=0}^{m} \|f_k\|^2 = 1$$

(24)

on the sphere. Homogenizing these give rise to identities on all of C^n.

$$\sum_{k=0}^{m-j} \langle f_{k+j}, f_k \rangle \, \|z\|^{2m-2k} = 0 \quad j = 1, ..., m$$
$$\sum_{k=0}^{m} \|f_k\|^2 \, \|z\|^{2m-2k} = \|z\|^{2m}$$

(25)

Using the definition of the tensor product, and the fact that $\|f \otimes g\|^2 = \|f\|^2 \|g\|^2$, we can rewrite the second equation in (25) as

$$\sum_{k=0}^{m} \|f_k \otimes H_{m-k}\|^2 = \|H_m\|^2$$

(26)

These identities have a remarkable consequence even in the polynomial case. Every polynomial proper map between balls can be constructed via the same algorithm; there are choices of subspaces involved that determine the map. This enables us to completely classify the polynomial proper maps between balls in all cases.

THEOREM 7. *Every proper holomorphic polynomial map f of degree m between balls is constructed in the following manner. Begin with the (essentially unique) homogeneous proper map H_m. Then there exists a finite list of subspaces such that*

$$f = \prod_{j=0}^{m} E^{-1}(A_j, z) H_m. \tag{27}$$

The formula in Theorem 7 does not preclude the possibility that some of the subspaces A_j may be trivial. We refer to [D3] for a proof, and give here only an example. Consider the map $f(z, w) = \left(z^5, \sqrt{5}z^4 w, \sqrt{10}z^2 w^2, \sqrt{5}z w^4, w^5\right)$. One can verify by computation that f defines a proper map from the two ball to the five ball. It is easier to note that it is obtained from the map H_5 by undoing a tensor product. More precisely, start with

$$H_5(z, w) = g(z, w) = \left(z^5, \sqrt{5}z^4 w, \sqrt{10}z^3 w^2, \sqrt{10}z^2 w^3, \sqrt{5}z w^4, w^5\right) \tag{28}$$

Consider the subspace A of C^6 that is generated by the pair of vectors $(0, 0, 1, 0, 0, 0)$, $(0, 0, 0, 1, 0, 0)$. Then we have $f = E^{-1}(A, z) g$. For examples of this sort it is typical that the undoing is required; in other words, one cannot find an alternate factorization involving tensor products alone.

COROLLARY 1. *Every proper polynomial map between balls admits the composition product factorization*

$$f(z) = \prod_{j=0}^{m} E^{-1}(A_j, z) \ L \ \prod_{j=1}^{m} E(V_j, z)(1) \tag{29}$$

Here 1 denotes the constant map to the point 1 on the unit circle, each V_j is the range of the map on which the tensor product is applied, L is linear, and each A_j is a subspace of the range of the map on which the tensor product is "undone".

PROOF. The corollary follows from the theorem and the fact that the homogeneous map H_m is itself an m-fold tensor product. ∎

For explicit polynomials it is possible to write down the factorization given by this proof. It is sometimes possible to write down a simpler one. For the convenience of the reader we give two "factorizations" for the simplest non-trivial case. Consider the map $f : B_2 \to B_3$ defined by

$$f(z_1, z_2) = \left(z_1^2, z_1 z_2, z_2\right) \tag{30}$$

To factorize it according to the proof, we write

$$C^3 = C^2 \oplus C = \{s, t, 0\} \oplus \{0, 0, u\} \tag{31}$$

Then the map is an orthogonal summand, and applying one tensor product we obtain that

$$f = \left(z_1^2, z_1 z_2, 0\right) \oplus (0, 0, z_2) \Rightarrow$$

$$
\begin{aligned}
E\left(A, z\right) f &= \left(z_1^2, z_1 z_2, 0\right) \oplus \left(\left(z_1, z_2\right) \otimes (0, 0, z_2)\right) \\
&= \left(z_1^2, z_1 z_2, 0, 0\right) \oplus \left(0, 0, z_1 z_2, z_2^2\right) \\
&= \left(z_1^2, z_1 z_2, z_1 z_2, z_2^2\right) \simeq H_2\left(z_1, z_2\right)
\end{aligned}
\tag{32}
$$

Note that $H_2\left(z\right)$ is a 2–fold tensor product. The proof of the theorem gives a factorization of the form

$$f = E\left(A_3, z\right)^{-1} E\left(A_2, z\right) E\left(A_1, z\right) \tag{33}$$

In fact, one can write

$$
\begin{aligned}
f &= E\left(A, z\right)\left(z\right) = \\
&\left(\left(z_1, 0\right) \otimes \left(z_1, z_2\right)\right) \oplus (0, 0, z_2)
\end{aligned}
\tag{34}
$$

for a simpler factorization.

The same ideas apply in the rational case. By allowing also tensor products of automorphisms that move the origin, undoing by these, and composition with such automorphisms, one can write a rational proper map between balls as a finite factorization. The proof is much more difficult to write down. The reason is that the homogeneous map H_m requires the following rational analog. It is a rational proper map of the form

$$\frac{p_{m-d} + p_{m-d+1} + \ldots + p_m}{1 + q_1 + \ldots + q_d} \tag{35}$$

where the individual terms are homogeneous polynomials. The reduction to this case is virtually identical to the reduction for a polynomial map to the homogeneous case. It is of course much harder to factorize and classify maps such as in formula (35) than to do the same for the homogeneous polynomial map of Theorem 6.

As an illustration, we offer a complete analysis of the case where the denominator is linear. This analysis requires some information on Hermitian forms. The idea is that real valued polynomials can be identified with Hermitian forms, and the positive semi-definiteness of such forms arises in the classification problem.

We begin with the necessary preliminary comments on Hermitian forms. If

$$Q\left(z, \bar{z}\right) = \sum_{|a|=|b|=m} c_{ab} z^a \bar{z}^b \tag{36}$$

is the indicated homogeneous real valued polynomial, linear algebra or the methods in [D1] show how to write it in the form

$$Q\left(z, \bar{z}\right) = \left\| Q^+\left(z\right)\right\|^2 - \left\| Q^-\left(z\right)\right\|^2 \tag{37}$$

where Q^+ and Q^- are holomorphic homogeneous vector-valued polynomials. The necessary and sufficient condition that one can write

$$Q(z, \bar{z}) = \left\| Q^+(z) \right\|^2 \tag{38}$$

is simply that the Hermitian matrix of coefficients (c_{ab}) be positive semi-definite. Simple examples show that this can fail even when $Q(z, \bar{z}) \geq 0 \quad \forall z$. Suppose that (38) holds, and we write

$$Q^+(z) = \sum_{|\alpha|=m} v_\alpha z^\alpha \tag{39}$$

It then follows that the matrix of coefficients for $\|Q^+\|^2$ satisfies

$$(c_{ab}) = (\langle v_a, v_b \rangle) \tag{40}$$

We identify a homogeneous polynomial of the form (36) with the corresponding Hermitian form. Then the form is positive semi-definite if and only if the polynomial is the squared norm of a holomorphic map, or equivalently, if and only if (40) holds for appropriate vectors.

Consider the vector space $V = V(n, m)$ of homogeneous polynomials of degree m in n variables. Let K denote its dimension. If we have a C^K—valued homogeneous polynomial map p_m, we can always write

$$p_m = L H_m \tag{41}$$

for a linear map L. Here $H_m = z \otimes z... \otimes z$ is the homogeneous proper map (15).

Discussion on maps with linear denominator.

We describe all proper rational maps with linear denominator. Suppose that $f = \frac{p}{q}$ is a proper rational map between balls, that the denominator is linear, and that the fraction is reduced to lowest terms. Thus q does not divide every component of p.

Step 1. We use the "undoing" operation to reduce to (44). We may suppose that the homogeneous expansions of the numerator and denominator are as follows, for some integer $k \geq 0$.

$$f = \frac{p_k + p_{k+1} + ... + p_{d+1}}{1 + q_1} \tag{42}$$

If $k < d$, it follows from (24) that $p_k \perp p_{d+1}$. If we let A denote the subspace spanned by the coefficient vectors of p_k, then we can write $f = f_A \oplus f_{A^\perp}$. Take now the tensor product on this subspace, to form the proper map $E(A, z) f = (f_A \otimes z) \oplus f_{A^\perp}$. An expansion into homogeneous parts of the result shows that

$$E(A, z) f = \frac{p_{k+1}^\# + ... + p_{d+1}^\#}{1 + q_1} \tag{43}$$

for new homogeneous polynomials. The point is that k is increased, but that $d+1$ is not. We proceed by induction and apply this process finitely many times, to obtain

$$\prod_{j=0}^{d-1} E\left(A_j, z\right) f = f^{\#} = \frac{p_d + p_{d+1}}{1 + q_1} \tag{44}$$

The denominator in (44) is the same as for the original map, but the homogeneous polynomials in the numerator are in general not the same.

Step 2. To classify all maps with linear denominators, it is therefore sufficient to classify all maps of the form

$$f^{\#} = \frac{p_d + p_{d+1}}{1 + q_1} \tag{45}$$

It is easy to verify that a denominator is possible if and only if $q_1 = \langle z, a \rangle$ where $\|a\| < 1$. We are assuming that $a \neq 0$. Our goal in step 2 is to determine the conditions on p_d and p_{d+1}, given the choice of q_1. The answer will be that p_d determines part of p_{d+1} exactly, and determines the rest up to a unitary map. By homogenizing (23) as before, we obtain the following pair of equations that hold everywhere:

$$\|p_{d+1}\|^2 + \|p_d \otimes H_1\|^2 = \|H_{d+1}\|^2 + \|q_1 H_d\|^2 \tag{46}$$

$$\langle p_{d+1}, p_d \rangle = q_1 \|H_d\|^2 \tag{47}$$

Let now A denote the span of the coefficients of p_d. We write $C^N = A \oplus A^\perp$. This enables us to write $p_{d+1} = g_{d+1} \oplus h_{d+1}$ where $g_{d+1} = (p_{d+1})_A$. This shows that our map has the following form:

$$f = \frac{(p_d + g_{d+1}) \oplus h_{d+1}}{1 + q_1} \tag{48}$$

We may assume that A is contained in a copy of C^{N_d}, where the dimension is the number of linearly independent monomials of degree d in n variables. As above we can write $p_d = L H_d$ for a linear map L and where $H_d = z \otimes z \dots \otimes z$ is the usual homogenous map.

Putting (48) into (47) shows that $g_{d+1} = q_1 g_d$ for some uniquely determined homogeneous map g_d. Using (47) and polarizing, we see that L must be invertible, and in fact that

$$\begin{aligned} p_d &= L H_d \\ g_d &= \left(L^{-1}\right)^* H_d \end{aligned} \tag{49}$$

Thus $p_d = L H_d$ determines $(p_{d+1})_A = q_1 g_d$ exactly. We also note also that (48) and (50) determine $\|h_{d+1}\|^2$, and hence determine h_{d+1} up to a unitary. From (49) we obtain that $p_{d+1} = q_1 \left(L L^*\right)^{-1} p_d \oplus h_{d+1}$ and we see precisely the extent to which p_{d+1} is determined by p_d and q_1.

Step 3. Next we obtain information on the size of L. We first claim that $\|L H_d\|^2 \leq \|H_d\|^2$ must hold. To see this start with

$$\|p_{d+1}\|^2 + \|p_d \otimes H_1\|^2 = \|H_{d+1}\|^2 + \|q_1 H_d\|^2 \tag{50}$$

Eliminate q_1 in (50) by using

$$q_1 = \frac{\langle p_{d+1}, p_d \rangle}{||z||^{2d}} \tag{51}$$

Using this and the Cauchy-Schwarz inequality yields

$$||p_{d+1}||^2 + ||p_d||^2 \, ||z||^2 \leq ||z||^{2d+2} + ||p_d||^2 \, ||p_{d+1}||^2 \, ||z||^{-2d} \tag{52}$$

and hence that

$$||p_{d+1}||^2 \left(||z||^{2d} - ||p_d||^2 \right) \leq \left(||z||^{2d} - ||p_d||^2 \right) ||z||^{2d+2} \tag{53}$$

If our desired inequality were false, we would obtain $||p_{d+1}||^2 \geq ||z||^{2d+2}$ somewhere. Putting this back into (50) yields a contradiction to the fact that $|q_1|^2 = |\langle z, a \rangle|^2 \leq ||z||^2$.

Thus we may assume that our map satisfies

$$f^{\#} = \frac{\left(LH_d + q_1 \left(L^* \right)^{-1} H_d \right) \oplus h_{d+1}}{1 + q_1} \tag{54}$$

where $L^*L \leq$ Id. Not every such L is possible because of (50). We need to use this information. The polar decomposition that $L = UP = U\sqrt{L^*L}$ simplifies the discussion. Here U is unitary, and P is symmetric and positive definite. Plugging the polar decomposition into (49), and replacing $f^{\#}$ by a unitary matrix times it, we see that it is sufficient to classify maps of the form

$$U^{\#} f^{\#} = \frac{\left(PH_d + q_1 P^{-1} H_d \right) \oplus h_{d+1}}{1 + q_1} \tag{55}$$

Here the positive operator P is not arbitrary. The key piece of information is that

$$- \left|\left| q_1 P^{-1} H_d \right|\right|^2 - ||z \otimes PH_d||^2 + ||H_{d+1}||^2 + ||q_1 H_d||^2 \tag{56}$$

must define a positive semi-definite form, as it equals $||h_{d+1}||^2$.

Step 4. We discuss the consequences of (56). Given q_1, the terms

$$||H_{d+1}||^2 + ||q_1 H_d||^2 = ||z||^{2d+2} + |q_1|^2 \, ||z||^{2d} \tag{57}$$

define an Hermitian form with matrix $(\langle w_a, w_b \rangle)$ (for appropriate vectors) on $V(n, d+1)$. Given a positive definite symmetric matrix P on $V(n, d)$, consider the matrix $(\langle v_a, v_b \rangle)$ defined on $V(n, d+1)$ by the coefficients of

$$\left|\left| q_1 P^{-1} H_d \right|\right|^2 + ||z \otimes PH_d||^2 \tag{58}$$

The matrix P is allowable if, as matrices on $V(n, d+1)$, we have

$$(\langle v_a, v_b \rangle) \leq (\langle w_a, w_b \rangle) \tag{59}$$

Conversely any such matrix defines a proper map of the form (55).

The condition that (56) is positive semi-definite is rather complicated when expressed in terms of coordinates. It gives however the necessary and sufficient condition on P for a proper map of the form (55) to exist.

The simplest case is when $n = 1$. There we may assume that P is a positive number. We obtain, assuming that $q_1 = z\bar{a}$, that $|a| \leq P \leq 1$. Suppose without loss of generality that $0 < a < 1$. Putting $P = a$ gives the map $z^d \otimes \frac{a+z}{1+az}$. Putting $P = 1$ gives the map $z^d = z^d \frac{1+az}{1+az}$; this is not in lowest terms. If we choose P so that $a < P < 1$, we obtain maps with higher dimensional range. For example, setting $P = ta$ and $b = \sqrt{\left(|t|^2 - 1\right)\left(|t|^{-2} - |a|^2\right)}$ we obtain a one-parameter family of maps:

$$\frac{\left(taz^d + t^{-1}z^{d+1}, bz^{d+1}\right)}{1 + az} \tag{60}$$

The maps are defined for $1 \leq t \leq \frac{1}{a}$ and map properly from B_1 to B_2.

In a similar fashion ones obtains families of maps with linear denominator, but depending on more parameters, when the domain dimension is larger than one. Given the denominator $q_1 = \langle z, a \rangle$ and the integer d, the positive semi-definiteness of (56) yields the conditions that must be met by the matrix P. It is worth noting that the maps resulting from different choices of P are not generally spherically equivalent.

One can obtain also a factorization for proper maps of the form (55), and hence for maps of the form (42), but we omit the detailed computations. This finishes our discussion of maps with linear denominator. ∎

We continue with the general summary. Observe that (even for polynomials) there are always infinitely many distinct spherical equivalence classes as soon as $N \geq 2n$. Thus $N = 2n - 1$ is the critical dimension when there is a non-trivial example, but not necessarily infinitely many non-trivial examples. The tensor product operation provides again an explanation.

Since the automorphism group of the ball is so large, superficially different maps can often be in fact spherically equivalent. On the other hand, in this section we will exhibit multi-dimensional families of polynomial maps that are mutually inequivalent. We restrict ourselves to the polynomial case for notational ease, but the rational case is not much more difficult. It is fairly easy to show [D2] that polynomial maps preserving the origin are spherically equivalent only if they are unitarily equivalent, so the verification that maps in our families are inequivalent becomes a matter of linear algebra.

Using the ideas on Hermitian forms we prove the following result that indicates how abundant proper polynomial maps are in the high codimension case.

PROPOSITION 1. *Let*

$$p_j : C^n \to C^{N_j} \qquad j = 1, ..., m - 1 \tag{61}$$

be homogeneous vector valued polynomial mappings of degree j. Suppose that the Hermitian form defined by

$$Q\left(z, \overline{z}\right) = ||z||^{2m} - \sum_{k=0}^{m-1} ||p_k\left(z\right)||^2 ||z||^{2m-2k}$$

$$= ||H_m||^2 - \sum_{k=0}^{m-1} ||p_k \otimes H_{m-k}||^2 \tag{62}$$

is positive semi-definite. Observe that this hypothesis is satisfied if the coefficients of the p_j are sufficiently small. Then there is a proper polynomial map p of degree m from B_n to some B_K such that

$$p = p_0 \oplus p_1 \oplus ... \oplus p_m. \tag{63}$$

Sketch of Proof. Write $p = p_0 + p_1 + ... + p_{m-1} + p_m$ for the desired map, where p_m is to be determined. We must satisfy the orthogonality relations (25). By decreeing that all the homogeneous components of different orders are orthogonal, we satisfy all the conditions except the condition that

$$\sum_{j=0}^{m} ||p_j||^2 = 1 \quad \text{on} \quad ||z||^2 = 1 \tag{64}$$

Homogenizing this equation as in (25) and solving the resulting equation for $||p_m||^2$ yields

$$||p_m\left(z\right)||^2 = ||z||^{2m} - \sum_{j=0}^{m-1} ||p_j\left(z\right)||^2 ||z||^{2m-2j} \tag{65}$$

According to (38), in order to find such a p_m, we need the form defined by the right hand side of (65) to be positive semi-definite. This is the hypothesis. If it holds, then (65) defines p_m up to a unitary matrix, and finishes the proof. ∎

This proposition has many corollaries. One of the simplest arises from choosing

$$\begin{aligned} p_0 &= 0 \\ p_1 &= L \end{aligned} \tag{66}$$

Then there is a quadratic polynomial proper map

$$Qz = Lz \oplus \left(\left(\sqrt{I - L^*L}\right) z \otimes z\right) \tag{67}$$

whenever the matrix $I - L^*L$ is positive semi-definite. There are in particular infinitely many inequivalent proper quadratic polynomial maps arising in this fashion. We state this as corollary 2 below. First we give a simple example.

EXAMPLE 2. *Consider the family of maps defined by*

$$f_\theta(z) = \left(\cos(\theta)z, \sin(\theta)z^2\right) \tag{68}$$

as θ varies in the closed interval $\left[0, \frac{\pi}{2}\right]$.

These maps are spherically inequivalent for each value of θ. One interpretation of this result is that the family of maps yields a homotopy between the maps $f_0(z) = z$ and $f_1(z) = z^2$. It is necessary to map into the two ball to have enough room to see this homotopy. It is a simple matter to replace the squared term with $z \otimes z$ (or by the tensor product on a subspace) if one wishes to extend this example to higher dimensions.

COROLLARY 2. *There are infinitely many spherically inequivalent proper quadratic polynomial mappings $f : B_n \to B_{2n}$.*

PROOF. Suppose that L is a linear map from C^n to itself, and that $I - L^*L$ is positive semi-definite. Then

$$Lz \oplus \left(\left(\sqrt{I - L^*L}\, z\right) \otimes z\right) \tag{69}$$

is always a proper map $f : B_n \to B_{n+n^2}$. The easiest way to create an infinite family of spherically inequivalent maps is to choose the matrix

$$L = \begin{pmatrix} I & 0 \\ 0 & t \end{pmatrix} \tag{70}$$

where the top left entry is the $(n-1)$ x $(n-1)$ identity matrix. Each value of $t\,\varepsilon\,[0,1]$ gives rise to an inequivalent map, because of the result mentioned above on unitary equivalence. After applying a linear transformation, the map resulting from (70) will map into $B_{2n} \subset B_{n+n^2}$. We leave the verification of this to the reader. Thus the range can be chosen to be $2n$ dimensional. ■

Note here that the parameter $t\,\varepsilon\,[0,1]$ gives rise to a homotopy between the maps

$$
\begin{aligned}
f(z) &= (z, 0) \\
g(z) &= \left(z_1, ..., z_{n-1}, z_1 z_n, ..., z_n^2, 0\right)
\end{aligned}
\tag{71}
$$

In fact every rational proper map between balls can be considered as being obtained by specialization of parameters of a proper map into a ball of perhaps much larger dimension. Thus the space of rational maps of any bounded degree to a sufficiently large dimensional ball is contractible. These considerations are not difficult, but we will not pursue them here.

We close this section by discussing a generalization of Proposition 1. The question is this. Given a polynomial

$$
\begin{aligned}
p &: C^n \to C^N \\
p &= p_0 + p_1 + ... + p_{m-1}
\end{aligned}
\tag{72}
$$

it is natural to ask whether it can be the Taylor polynomial of a proper holomorphic mapping between balls. Let us call such a Taylor polynomial an allowable jet of order $m-1$. Let us denote by $W(n, N, m-1)$ the vector space of polynomial mappings

$$p : C^n \to C^N \tag{73}$$

of degree at most $m-1$. It is not difficult to prove the following theorem on allowable jets.

THEOREM 8. *For given integers* $n, m-1$ *suppose that* N *is sufficiently large. Then there is an open subset* $A \subset W(n, N, m-1)$ *with the following properties:*

1. Each element of A *is an allowable jet of order* $m-1$ *of a proper polynomial map between balls of degree* m.

2. The origin (the trivial jet) is a boundary point of A *and is allowable.*

3. The set A *is defined by finitely many polynomial inequalities.*

Sketch of the proof. See [D3] for details. One must suppose first that the coefficients of the distinct monomials in a given jet are linearly independent, or else the jet may not be allowable. This is a generic condition, and requires that the target dimension be sufficiently large. Next one observes that the trivial jet is allowable, because one can choose the map H_m from Theorem 6. The orthogonality conditions (25) can be all met in case the coefficients are linearly independent, and the last condition there can be met in the same manner as in Proposition 1. This is a condition on positive semi-definiteness of an Hermitian form. The condition holds at the origin, and the matrix corresponding to $||z||^{2m}$ has minimum eigenvalue equal to unity. Therefore the condition remains true nearby. Furthermore it is determined by finitely many polynomial inequalities. ∎

These results demonstrate that there is an abundance of proper polynomial maps between balls if the codimension is large enough. One can get additional examples in the rational case. At each stage in the factorization of a rational map, allow also the possibility of tensoring or undoing by automorphisms that move the origin. It is not hard to see that there are rational examples not spherically equivalent to any polynomial. The easiest one is the tensor product of three automorphisms that send the origin to different points. Also, given a domain dimension n, and a degree m, all rational maps of degree m that map to a fixed target are obtained by specifying enough parameters in a proper rational map of this degree, but that maps to a sufficiently high dimensional ball. This kind of reasoning implies the statement above to the effect that, in large enough codimension, the set of all rational proper maps of a fixed degree is contractible. It is an open problem to compute the maximum degree of a rational map given the domain and target dimensions. Now that we know that there are many examples of rational proper maps, it is natural to seek some with additional properties.

II. Invariance under fixed point free finite unitary groups.

We now turn to the question whether there are proper maps that are invariant under fixed point free finite unitary groups. Consider first the simplest case, in one dimension.

EXAMPLE 2. *The proper holomorphic map $f(z) = z^m$ from the unit disc to itself is invariant under a cyclic group of order m generated by an m-th root of unity. In other words, let G denote the group of one by one matrices generated by ϵ where $\epsilon^m = 1$. If γ is an element of this group, then $f(\gamma z) = f(z)$. Notice that all finite subgroups of the space of one by one unitary matrices, that is, the circle, are of this form. Thus every such group occurs as the group of invariants of a proper map from the disc to itself.*

EXAMPLE 3. *Let G_m denote now the cyclic group of order m, but represented as n by n matrices as follows. Consider the powers of ϵI, where I is the identity and again ϵ is an m-th root of unity. Then the proper holomorphic map H_m defined by (15) is invariant under G_m.*

DEFINITION 1. *A fixed point free finite unitary group is a representation*

$$\pi : G \to U(n) \tag{74}$$

of a finite group G as a group of unitary matrices on C^n, such that 1 is never an eigenvalue of any $\pi(\gamma)$ for $\gamma \in G$ other than for $\gamma = $ identity.

THEOREM 9. *(Forstneric) [F1] Let*

$$\pi : G \to U(n) \tag{75}$$

be a fixed point free finite unitary group. Then there is a proper holomorphic map from B_n to some B_N that is invariant under G. One can assume that the map f is continuous on the closed ball.

Lichtblau [L] has proved that this result holds for any finite unitary group. The fixed point free groups are the most interesting, though, because of the following fact. If f is a proper holomorphic map between balls, smooth on the closed ball, and invariant under a finite unitary group, then that group must be fixed point free. In fact one can say much more.

For smooth proper maps, not all fixed point free groups occur. According to Theorem 3 such maps would have to be rational. It turns out to be impossible to find rational proper maps between balls that are invariant under most fixed point free groups. Forstneric discovered some restrictions in [F1]. In his thesis (see [L]), Lichtblau showed that the restrictions are very severe. The groups must be cyclic, and some representations are impossible. The author and Lichtblau [DL] then determined precisely which representations of cyclic groups actually arise. There are essentially three classes of examples. One is Example 3; we now describe a second class of examples, found by the author in [D2].

Let $\Gamma(p, q)$ denote the group of 2 by 2 matrices of the form

$$\begin{pmatrix} \epsilon & 0 \\ 0 & \epsilon^q \end{pmatrix}^k \tag{76}$$

for $k = 0, 1, ..., p-1$, where ϵ is a primitive root of unity with $\epsilon^p = 1$. We assume that the integers p and q are relatively prime. We also assume without loss of generality that the root of unity is chosen so that q is minimal.

It turns out (Theorem 13) that $\Gamma(p, q)$ occurs for a rational map only when $q = 1$ or $q = 2$. In the second case $\Gamma(2r + 1, 2)$ is a fixed point free representation of a cyclic group of odd order. This representation differs from the first case, namely $\Gamma(p, 1)$, that applies for homogeneous polynomials of degree p. We have seen already in Example 3 that the homogeneous maps are invariant under the representation $\Gamma(p, 1)$ of the cyclic group. The content of the next theorem is that there are maps invariant under the more complicated representation given by $\Gamma(2r + 1, 2)$.

THEOREM 10. *For each non-negative integer r, there is a proper polynomial map*

$$f : B_2 \to B_{2+r} \tag{77}$$

invariant under

$$\Gamma(2r + 1, 2) \tag{78}$$

SKETCH OF PROOF. We write the map down explicitly, referring to [D2] for the derivation. Call the variables (z, w). Consider a basis for the algebra of all polynomials invariant under the group. The basic polynomials are

$$z^{2r+1}, z^{2r-1}w, ..., z^{2(r-s)+1}w^s, ..., zw^r, w^{2r+1} \tag{79}$$

One considers the map

$$f(z, w) = \left(z^{2r+1}, ..., c_s z^{2(r-s)+1} w^s, ..., w^{2r+1} \right) \tag{80}$$

and seek to choose the coefficients to make the function map the sphere to the sphere. Computing $||f||^2$, making some surprising changes of variables, and doing a fair amount of manipulation yields

$$|c_s|^2 = \left(\frac{1}{4} \right)^{r-s} \sum_{k=s}^{r} \binom{2r+1}{2k} \binom{k}{s} \tag{81}$$

This formula determines the coefficients of the proper map up to a unitary transformation. ∎

Recall from topology that the quotient spaces of spheres by such groups are examples of Lens spaces.

DEFINITION. *The Lens space $L(p, q)$ is the smooth manifold defined by*

$$S^3 / \Gamma(p, q) \tag{82}$$

Since $L(p, q)$ is locally the same as the sphere, it is an example of a CR manifold. Thus it is of interest to know whether there are non-trivial CR mappings from Lens spaces to other CR manifolds.

COROLLARY 3. *There are non-trivial smooth CR immersions from the Lens spaces $L(p,1)$ and $L(2r+1,2)$ to sufficiently high dimensional spheres. Here p, r are arbitrary positive integers.*

PROOF. We have seen already that the homogeneous polynomial mapping H_p determines such a map in the first case. In the second case, the map in Theorem 10 produces an example. It is trivial that these maps are immersions, and since they are holomorphic and invariant, their restrictions to the Lens spaces are CR mappings. ∎

REMARK. If one analyzes the equations for the coefficients $|c_s|^2$ in a direct manner, it becomes difficult to verify that the solutions are non-negative. This difficulty is significant if one considers other groups of this sort. It is not hard to verify the following assertion. Suppose that one lists the basic monomials that generate the algebra of polynomials under the group $\Gamma(p,q)$. Then, it is impossible to find solutions to the equations for the absolute valued squared of the coefficients unless the parameter q equals one or two. This suggests Theorem 13, but the proof eliminating the other representations requires new ideas.

EXAMPLE 4. From the map in Theorem 10 it is possible to construct invariant maps from the 2–ball to much higher dimensional balls, by taking tensor products. It is much less obvious, and was noted in [C], that one can construct an invariant map from the 3–ball that is closely related to the special case $\Gamma(7,2)$. To do so, consider the cyclic group Q of order 7, now represented by the three by three matrices of the form

$$\begin{pmatrix} \epsilon & 0 & 0 \\ 0 & \epsilon^2 & 0 \\ 0 & 0 & \epsilon^4 \end{pmatrix} \tag{83}$$

and suppose that ϵ is a primitive seventh root of unity. Then there is a map

$$f : B_3 \to B_{17} \tag{84}$$

that is invariant under Q. To expedite the writing of this map, we recall a trick from [D4]. The map f is a monomial map, with 17 components. Let us write

$$\begin{aligned} x &= |z_1|^2 \\ y &= |z_2|^2 \\ u &= |z_3|^2 \end{aligned} \tag{85}$$

and $f^{\#}(x,y,u) = \|f(z)\|^2$. Since f is a monomial map, it is easy to read it off from the map $f^{\#}$, by taking square roots of each term and replacing the plus signs by commas. The condition that f is proper becomes the condition that

$$f^{\#}(x,y,u) = 1 \quad \text{on} \quad x + y + u = 1 \tag{86}$$

The reader can verify himself, or check on a computer, that the following map does the trick:

$$
\begin{aligned}
f^{\#}(x, y, u) = {} & x^7 + y^7 + u^7 + 14 \left(x^3 y^2 + x^2 u^3 + y^3 u^2 + xyu \right) \\
& + 7 \left(x^5 y + x u^5 + y^5 u + x y^3 + x^3 u + y u^3 \right) \\
& + 7 \left(x y^2 u^4 + x^2 y^4 u + x^4 y u^2 + x^2 y^2 u^2 \right)
\end{aligned} \tag{87}
$$

Thus we have exhibited a map that is invariant under the group Q and maps to the ball in 17 dimensions.

We continue to discuss the geometry of invariant maps. The three classes of examples given by Example 3, Theorem 10, and Example 4 (and trivial generalizations) are the only ones that can occur. We state this in Theorem 13. See [DL] for the proof. Here we give an indication of some of the ideas. We will indicate why proper rational maps can be invariant only under cyclic groups, and why there are restrictions on the possible representations. Let us study the homogeneous maps from the 2–ball in more detail. Recall that

$$
\Gamma(m, 1) \tag{88}
$$

is the group of 2 by 2 matrices generated by

$$
\begin{pmatrix} \epsilon & 0 \\ 0 & \epsilon \end{pmatrix} \tag{89}
$$

where $\epsilon^m = 1$. The Lens space $L(m, 1)$ is then the quotient of the sphere by this group. The homogeneous polynomial map H_m of degree m has the following interpretation. The image of the 2–ball under this map is a domain in the complex analytic variety determined by $C^{m+1} / \Gamma(m, 1)$. This variety has an isolated singularity at the origin; this follows because the group is fixed point free. The map H_m is then an m fold covering map on the complement of the origin. In particular, this shows that, for m at least 2, the corresponding variety is not locally Euclidean near the image of 0; if it were, removing a point would yield a simply connected space. The existence of this m-fold covering shows that this set is not simply connected. The map has full rank on the boundary, so its image is a strongly pseudoconvex domain with real analytic boundary. In this case the image imbeds explicitly as a ball in C^{m+1}. Since the map is a polynomial, it is smooth up to the boundary.

In the general case of a fixed point free group, one can try to imitate this. The difference is that the imbedding into some complex vector space cannot always be made smooth at the boundary. The proof of this is that a smooth proper map between balls must be rational, and there are groups that cannot occur for rational proper maps between balls.

We now present a simple proof that certain groups cannot occur. This result appears in [L] and in [DL]. A condition on the number of factors of two in a pair of given integers arises. If p is a positive integer, we write $v_2(p)$ for the number of times that two divides p.

THEOREM 11. *There is no proper holomorphic map* $f : B_2 \rightarrow B_N$ *that is smooth up to the boundary and is also invariant under the group* $\Gamma(p,q)$, *if*

$$v_2(p) > v_2(q-1) \tag{90}$$

PROOF. According to Theorem 3, a proper holomorphic map that is smooth up to the boundary must be a rational function. Let us suppose that

$$f = \frac{g}{h} \tag{91}$$

If the map is invariant under the group, we may assume that both the numerator and denominator are also invariant. After composing with an automorphism, we may assume that the origin in C^2 is mapped to the origin in C^N. As a consequence of the identities (25), we obtain that $\deg(g) > \deg(h)$. As usual we write

$$g(z, w) = \sum_{a,b} c_{ab} z^a w^b$$
$$h(z, w) = \sum_{a,b} d_{ab} z^a w^b \tag{92}$$

It follows from identity (22) that

$$\sum_{a,b} \left(||c_{ab}||^2 - |d_{ab}|^2 \right) |z|^{2a} |w|^{2b} = 0 \text{ on } |z|^2 + |w|^2 = 1 \tag{93}$$

Supposing that $\deg(g) = s$, we substitute $|w|^2 = 1 - |z|^2$ and consider the highest order term. Noting also that $d_{ab} = 0$ for $a + b = s$ reveals that

$$0 = \sum_{a+b=s} \left(||c_{ab}||^2 - |d_{ab}|^2 \right) (-1)^b = \sum_{a+b=s} ||c_{ab}||^2 (-1)^b \tag{94}$$

We must also consider the group invariance. Since the group is diagonal, the only monomials that can occur are invariant monomials. Therefore we have the condition that

$$\epsilon^{a+bq} = 1 \Rightarrow$$
$$a + bq = kp \tag{95}$$

for some positive integer k. Combining these two equations forces the integers a, b to satisfy

$$a + bq = kp$$
$$a + b = s \tag{96}$$
$$s = kp - b(q-1)$$

If the number of factors of two in p is greater than the number in $q-1$, it follows that p is even. Since $(p,q) = 1$, $q - 1$ is also even, and therefore so is s. Divide both sides of the last equation in (96) by 2^d, where $d = v_2(q-1)$. It follows from that equation and (90) that

$$b \equiv 2^{-d}s \mod 2 \tag{97}$$

Since s, the degree of the numerator, is a fixed integer, one sees that each b that occurs in the sum has the same parity. In either case, $(-1)^b$ is independent of b. This contradicts (94) because some coefficient is non zero. ∎

COROLLARY 4. *A smooth (necessarily rational) proper map between balls cannot be invariant under any fixed point free group that contains a subgroup generated by the matrix*

$$\begin{pmatrix} i & 0 \\ 0 & -i \end{pmatrix} \tag{98}$$

PROOF. Since $i^4 = 1$, we have $2 = v_2(4) > v_2(3-1) = 1$. The corollary thus follows from the theorem. ∎

Corollary 4 together with the classification of fixed point free finite unitary groups [Wo] can be used to eliminate all groups containing quaternionic subgroups. See [F1], who proved the corollary directly, without proving the general result of Lichtblau. Lichtblau has proved also the striking statement that the only groups that can occur for rational proper maps must be cyclic. We sketch his proof; [DL] contains a different proof.

THEOREM 12. *Suppose that $G \subset U(n)$ is a fixed point free finite unitary group. Suppose that $f : B_n \to B_N$ is a rational proper holomorphic map that is invariant under G. Then G must be cyclic.*

Sketch of proof. One begins with the classification of the fixed point free groups from [Wo]. It is shown there that all such groups, unless they are cyclic, that do not contain quaternionic subgroups must have a generator of the following form.

$$\begin{pmatrix} 0 & 1 & 0 & 0 & \ldots & 0 \\ 0 & 0 & 1 & 0 & \ldots & 0 \\ 0 & 0 & 0 & 1 & \ldots & 0 \\ 0 & 0 & 0 & 0 & \ldots & 0 \\ & & & \ldots & & \\ \delta & 0 & 0 & 0 & \ldots & 0 \end{pmatrix} \tag{99}$$

In (99) δ is a primitive root of unity. Invariance under this group forces certain (vector) coefficients of the numerator of a rational map to be equal. On the other hand, the orthogonality relations force these coefficients to be orthogonal. Thus this sort of generator cannot occur. Since Forstneric ruled out those groups that contain quaternionic subgroups, combining these results rules out all groups that are not cyclic. ∎

The final result determines precisely which representations of cyclic groups actually arise for rational maps. For simplicity we state the result from [DL] in the special case where the domain is two dimensional. See that paper for complete proofs of the assertions about invariant proper maps made in this section.

THEOREM 13. *Let $\Gamma(p,q) \subset U(2)$ be the fixed point free finite unitary group defined by (76), where without loss of generality q is minimal. Suppose that*

$f : B_2 \to B_N$ is a rational proper holomorphic map that is invariant under $\Gamma(p, q)$. Then $q = 1$ or $q = 2$.

It is also of interest to consider other finite unitary groups. It is not difficult to verify [F2] that only fixed point free groups occur for smooth mappings. More precisely, suppose that $f : B_n \to \Omega$ is a proper holomorphic map. Suppose that $b\Omega$ is smooth, that f is smooth up to the boundary, and also that f is invariant under a group G. Then G must be fixed point free. On the other hand, if one allows domains with non-smooth boundaries, many interesting examples arise. These include reflection groups. We refer to [R3] for general results, but give one nice example.

EXAMPLE 5. *Put*

$$f(z, w) = (z^2 - w^2, z^4 + w^4) \tag{100}$$

Then $f : B_2 \to \Omega$ is a proper holomorphic map with multiplicity eight from the ball to a domain Ω. It is not hard to check that it is invariant under the group D_4 of symmetries of the square. More precisely, consider the matrix group generated by

$$
\begin{aligned}
A &= \begin{pmatrix} 1 & 0 \\ 0 & -1 \end{pmatrix} \\
B &= \begin{pmatrix} 0 & i \\ i & 0 \end{pmatrix}
\end{aligned}
\tag{101}
$$

Then these generators satisfy the relations:

$$
\begin{aligned}
A^2 &= I \\
B^4 &= I \\
AB &= B^3 A
\end{aligned}
\tag{102}
$$

and hence define a group isomorphic to the group of the square. It is clear that the map (100) is invariant under these matrices.

We return briefly to proper maps between balls. Perhaps the most important open problem is the degree of differentiability required to ensure that a proper map must be rational. Even for rational maps there remain interesting problems. One area of interest involves extremal properties. If we assume without loss of generality that the denominator takes the value unity at the origin, then there are bounds for the coefficients of rational maps of a fixed degree. For a given degree it is tempting to ask whether maps that maximize appropriate coefficients have any special properties. Also, at the end of section I we noted that finding the maximum possible degree of a rational proper map given the domain and target dimensions is an open problem. Computer studies suggest that when the domain is two dimensional, the invariant maps from Theorem 10 are of maximal degree.

References.

[A] Alexander, H., *Proper holomorphic mappings in C^n*, Indiana Univ. Math. J. **26** (1977), 137-146.

[B] Bedford E., *Proper Holomorphic Maps*, Bulletin A.M.S. **10** no. No. 2 (1984), 157-175.

[C] Chiappari, S., *Proper holomorphic mappings of positive codimension in several complex variables*, Thesis, University of Illinois, 1990.

[CS1] Cima J. and Suffridge, T., *A reflection principle with applications to proper holomorphic mappings*, Math Annalen 265 (1983), 489-500.

[CS2] _____, *Proper mappings between balls in C^n*, Lecture Notes in Math. 1268, Springer-Verlag, Berlin, 1987, pp. 66-82.

[CS3] _____, *Boundary behavior of rational proper maps*, Duke Math J. **60** no. No 1 (1990), 135-138.

[D1] D'Angelo, J. *Real hypersurfaces, orders of contact, and applications*, Annals of Math **115** (1982), 615-637.

[D2] _____, *Proper polynomial mappings between balls*, Duke Math J. **57** (1988), 211-219.

[D3] _____, *Proper polynomial maps between balls, II*, Michigan Math J. **38** (1991), 53-65.

[D4] _____, *Proper holomorphic maps between domains in different dimensions*, Michigan Math J. **35** (1988), 83-90.

[D5] _____, *The structure of proper rational maps between balls*, to appear in Proceedings of year in Several Complex Variables at Mittag-Leffler Institute.

[DL] D'Angelo, J. and Lichtblau, D., *Spherical space forms, CR maps, and invariant proper holomorphic maps between balls*, Journal of Geometric Analysis (to appear).

[D] Dor A., *Proper holomorphic maps from strictly pseudoconvex domains in C^2 to the ball in C^3 and boundary interpolation by proper holomorphic maps*, thesis, Princeton University (1987).

[Fa1] Faran, J., *Maps from the two-ball to the three ball*, Inventiones Math. **68** (1982), 441-475.

[Fa2] _____, *On the linearity of proper maps between balls in the low codimension case*, J. Differential Geometry **24** (1986), 15-17.

[F1] Forstneric, F., *Proper holomorphic maps from balls*, Duke Math J. **53** no. No. 2 (1986), 427-441.

[F2] _____, *Proper holomorphic maps: a survey*, to appear in Proceedings of year in Several Complex Variables at Mittag-Leffler Institute.

[F3] _____, *Extending proper holomorphic mappings of positive codimension*, Inventiones Math. **95** (1989), 31-62.

[H] Hakim, M., *Applications holomorphes propres continues de domaines strictement pseudoconvexes de C^n dans les boule unite de C^{n+1}*, Duke Math. J. (1990).

[L] Lichtblau, D., *Invariant proper holomorphic maps between balls*, Indiana Univ. Math J. (to appear).

[P] Pincuk, S., *Holomorphic maps in C^n and the problem of holomorphic equivalence*, Pp. 173-200, Encyclopedia of Math. Sciences, Vol. 9, G.M. Khenkin, Editor, Several Complex Variables III, Springer-Verlag, Berlin, 1980.

[RS] Remmert, R. and Stein, K., *Eigentliche holomorphe abbildungen*, Math. Z. **73** (1960), 159-189.

[R1] Rudin, W., *Function theory in the unit ball of C^n*, Springer-Verlag, New York, 1980.

[R2] _____ , *Homogeneous proper maps*, Nederl. Akad. Wetensch. Indag. Math **46** (1984), 55-61.

[R3] _____ , *Proper holomorphic maps and finite reflection groups*, Indiana Univ. Math J **31** no. No 5 (1982), 701-719.

[W] Webster, S., *On mapping an n ball into an $n + 1$ ball in complex space*, Pacific J. Math **81** (1979), 267-272.

[Wo] Wolf, J., *Spaces of Constant Curvature*, McGraw-Hill, New York, 1967.

DEPARTMENT OF MATHEMATICS, UNIVERSITY OF ILLINOIS, URBANA, ILLINOIS 61801

Contemporary Mathematics
Volume **137**, 1992

Contractive Zero-Divisors in Bergman Spaces

PETER DUREN, DMITRY KHAVINSON,
HAROLD S. SHAPIRO, AND CARL SUNDBERG

This is a summary of work to appear elsewhere. For $0 < p < \infty$, the Bergman space A^p over the unit disk $\mathbf{D}$ is the class of analytic functions f for which

$$\| f \|_p^p = \int \int_{\mathbf{D}} | f(z) |^p \, d\sigma < \infty,$$

where $d\sigma = \frac{1}{\pi} dx \, dy$ is normalized area measure. An A^p zero-set is the set of zeros, repeated according to multiplicity, of some nonconstant function in A^p. Given an A^p zero-set $\{\zeta_j\}$, let N^p be the set of functions in A^p with zeros of at least the prescribed order at each point ζ_j. For $0 < p < \infty$, Horowitz [4] attached to each A^p zero-set a kind of generalized Blaschke product H with the property $\| f/H \|_p \leq C_p \| f \|_p$ for every function $f \epsilon N^p$, where C_p is a constant depending only on p. However, H need not belong to A^p.

The main result in [1] is the existence of a contractive divisor associated with every A^p zero-set $\{\zeta_j\} \subset \mathbf{D} \backslash \{0\}$. This is a function $G \epsilon A^p$ uniquely characterized by the properties $\| G \|_p = 1$, $G(0) > 0$, G vanishes (with the prescribed multiplicities) precisely on $\{\zeta_j\}$, and $\| f/G \|_p \leq \| f \|_p$ for all $f \epsilon N^p$. For $p = 2$ this was previously established by Hedenmalm [3], and the main outline of our proof is patterned after his. However, for $p \neq 2$ the Bergman kernel no longer plays a central role, and one must devise other methods. Our approach exploits the positivity of the biharmonic Green function of the disk and the regularity of the reproducing kernel in a certain weighted A^2 space.

The canonical divisor is constructed as follows. Given an A^p zero-set $\{\zeta_j\}$ not containing the origin, let N^p be the set of functions in A^p with zeros of at least the prescribed order at each point ζ_j. Pose the extremal problem of maximizing $f(0)$ among all $f \epsilon N^p$ with $\| f \|_p = 1$ and $f(0) > 0$. It turns out that this problem has a unique solution for $0 < p < \infty$, called the *canonical divisor* of $\{\zeta_j\}$ in A^p. (For $1 \leq p < \infty$ the uniqueness follows from strict convexity.)

Here are the main results:

1991 *Mathematics Subject Classification.* Primary 30D55.

The final (detailed) version of this paper will be submitted for publication elsewhere.

THEOREM 1. *The canonical divisor has no extraneous zeros; it vanishes precisely at the prescribed points with the prescribed multiplicities.*

THEOREM 2. *For $0 < p < \infty$, let G be the canonical divisor in A^p of a zero-set $\{\zeta_j\}$ in $\mathbf{D}\backslash\{0\}$. Then $\| f/G \|_p \leq \| f \|_p$ for each $f \epsilon N^p$. Furthermore, G is the only function in A^p with the given normalization which has this contractive property.*

THEOREM 3. *Let G be the canonical divisor of an arbitrary A^p zero-set, $0 < p < \infty$. Then $\| Gf \|_p \geq \| f \|_p$ for all $f \epsilon A^p$.*

THEOREM 4. *For $0 < p < \infty$, let G be the canonical divisor in A^p of a finite zero-set $\{\zeta_j\}$ in $\mathbf{D}\backslash\{0\}$, and let B be the associated finite Blaschke product. Let $J(z,\zeta)$ be the reproducing kernel of the weighted Bergman space A_w^2 with weight $w = | B |^p$. Then*

$$G(z) = J(0,0)^{-1/p} B(z) J(z,0)^{2/p}.$$

Furthermore, G has an analytic continuation to a disk $| z | < R$ for some $R > 1$, and it has no zeros there except for those of B. In particular, $J(z,0)$ has an analytic continuation to $| z | < R$ and $J(z,0) \neq 0$ there.

The expansive multiplier property of G (Theorem 3) follows immediately from its contractive divisor property (Theorem 2) . However, the proof of Theorem 2 depends on the special case of Theorem 3 where the prescribed zero-set is finite. In that case, Theorem 4 shows that G has a smooth extension to the boundary, allowing an application of Green's theorem to establish the formula

$$\int \int_{\mathbf{D}} (| G |^p - 1) | f |^p \, d\sigma$$
$$= \pi \int \int_{\mathbf{D}} \int \int_{\mathbf{D}} \Gamma(z,\zeta) \Delta(| G(z) |^p) \Delta(| f(\zeta) |^p) d\sigma(z) d\sigma(\zeta),$$

where

$$\Gamma(z,\zeta) = \frac{1}{16\pi} \left\{ | z - \zeta |^2 \log \left| \frac{z - \zeta}{1 - \bar{\zeta}z} \right|^2 + (1 - | z |^2)(1 - | \zeta |^2) \right\}$$

is the biharmonic Green function of $\mathbf{D}$. But

$$\Delta(| f |^p) = p^2 | f |^{p-2} | f' |^2 \geq 0,$$

and similarly $\Delta(| G |^p) \geq 0$. Because $\Gamma(z,\zeta) > 0$ in $\mathbf{D}$, this proves Theorem 3 for finite zero-sets. Theorem 2 then follows with the help of Theorem 1 and the fact that $G(z) \neq 0$ on the boundary. The result is extended to infinite zero-sets by a normal family argument.

The analytic continuation of the kernel function $J(z,\zeta)$ is valid for general domains Ω with analytic boundary and for general positive real-analytic weights bounded away from zero near the boundary. It holds for each fixed $\zeta \epsilon \Omega$. This result was found by Garabedian [2] with the aid of integral representations. A much shorter proof can be based on the phenomenon of elliptic regularity.

For a single point $\alpha \epsilon \mathbf{D}$, the canonical divisor can be computed explicitly. It is

$$G_\alpha(z) = C \frac{z - \alpha}{1 - \bar{\alpha} z} \left\{ 1 + \frac{p}{2}(1 + \bar{\alpha}\frac{z - \alpha}{1 - \bar{\alpha} z}) \right\}^{2/p},$$

where

$$C = -\frac{\bar{\alpha}}{|\alpha|}[1 + \frac{p}{2}(1 - |\alpha|^2)]^{-1/p}.$$

For $1 \leq p < \infty$ this formula was already found by Osipenko and Stessin [5].

The question arises whether the contractive divisor G is in fact isometric. For the Hardy spaces H^p the zero-sets are independent of p and are described by the Blaschke condition $\sum(1 - |\zeta_j|) < \infty$. A simple but very useful theorem of F. Riesz says that if $f \epsilon H^p$ and B is the Blaschke product of its zeros, then f/B is a nonvanishing H^p function with the same norm. In contrast, however, the Bergman spaces A^p have no isometric zero-divisors. In fact, the canonical divisor (the only possible candidate) is properly contractive when applied to every admissible function except a constant multiple of itself. This follows at once from the above integral formula involving $\Gamma(z, \zeta)$. The formula remains valid for infinite zero-sets and can be extended (with appropriate modifications) to more general domains in the plane.

There is an apparent defect in the theory caused by excluding the origin from the prescribed zero-set. However, as Hedenmalm [3] has remarked (for $p = 2$), the defect is easily remedied. If a zero of multiplicity m is prescribed at the origin, one poses the extremal problem of maximizing $|f^{(m)}(0)|$ among all $f \epsilon N^p$ with $\|f\|_p = 1$.

The canonical divisor may be expected to play a significant role in the theory of Bergman spaces, now under active development. One interesting open problem is to describe the functions $f \epsilon A^p$ for which the polynomial multiples of f/G are dense in A^p. More broadly, the problem is to describe the invariant subspace generated by f/G, as in Beurling's theory for H^p spaces.

References.

1. P. Duren, D. Khavinson, H. S. Shapiro and C. Sundberg, *Contractive zero-divisors in Bergman spaces*, Pacific J. Math. (to appear).

2. P. R. Garabedian, *A partial differential equation arising in conformal mapping*, Pacific J. Math. **1** (1951), 485-524.

3. H. Hedenmalm, *A factorization theorem for square area-integrable analytic functions*, J. Reine Angew. Math. **422** (1991), 45-68.

4. C. Horowitz, *Zeros of functions in the Bergman spaces*, Duke Math. J. **41** (1974), 693-710.

5. K. Yu. Osipenko and M. I. Stessin, *On recovery problems in Hardy and Bergman spaces*, Mat. Zametki **49** (1991), 95-104. (Russian).

DEPARTMENT OF MATHEMATICS, UNIVERSITY OF MICHIGAN, ANN ARBOR, MICHIGAN 48109

DEPARTMENT OF MATHEMATICS, UNIVERSITY OF ARKANSAS, FAYETTEVILLE, ARKANSAS 72701

ROYAL INSTITUTE OF TECHNOLOGY, S–100 44 STOCKHOLM, SWEDEN

DEPARTMENT OF MATHEMATICS, UNIVERSITY OF TENNESSEE, KNOXVILLE, TENNESSEE 37916

Contemporary Mathematics
Volume **137**, 1992

The Theorem of F. and M. Riesz
for Unbounded Measures

FRANK FORELLI

To my friend Walter Ru~~ ~n the occasion of his retirement.

ABSTRACT. This is the text of a talk on whether the Riesz theorem is meaningful for unbounded measures (it is), and if so, whether it holds for such measures (it does if the measure is not too large).

[The referee, in a gracious and subtle way, has told me of Beurling's lectures on quasi-analyticity [2]. In one of these (*Applications to Harmonic Analysis*) there is the simple idea which is the endgame of the proof of the theorem here. The referee goes on to say that the infinite product with a priori majorant, which is in the proof of my theorem, is in Mandelbrojt's book [4], and that there are similiar things in papers of Agmon [1] and Marchenko [5]. Although I was unaware of these works, I was aware of the use of such products (or similiar things) for purposes other than mine. E.g., in the proof of the Denjoy-Carleman theorem [8]. And although I was not aware of the lectures on quasi-analyticity, the knowledgeable reader will recognize my debt to Beurling.]

I

The Riesz theorem, which has occupied a central place in harmonic and complex analysis since its discovery in 1916, says that certain measures and linear measure have the same null sets. Before giving the precise statement, let me briefly explain several terms, each involving the word **measure**.

In harmonic analysis, measures live on a variety of groups, often the circle or the line. In this talk, measures live on the line. A complex measure (or simply a measure) is any complex linear combination of Lebesgue-Stieltjes measures. Alternatively, μ is a complex measure if $d\mu = \varphi\, dF$ where $F(x)$ is non-decreasing and $\varphi(x)$ is both unitary and measurable with respect to the Lebesgue-Stieltjes

---

1991 *Mathematics Subject Classification*. 42A99.

This paper is in final form and no version of it will be submitted for publication elsewhere.

measure dF. Since $|\varphi| = 1$, the variation of μ, in symbols, $|\mu|$ or $d|\mu|$, is the Lebesgue-Stieltjes measure dF. The complex measure μ is **bounded** if

$$\int_{-\infty}^{\infty} d|\mu| < \infty,$$

otherwise it is **unbounded**. Since the measure that vanishes everywhere is without interest, it is understood that always

$$\int_{-\infty}^{\infty} d|\mu| > 0.$$

Suppose μ is bounded. Then there is a Fourier transform. This is the function

$$\widehat{\mu}(t) = \int_{-\infty}^{\infty} e^{-itx} \, d\mu(x),$$

defined for $-\infty < t < \infty$. The theorem of F. and M. Riesz is this:

If the Fourier transform of μ vanishes on the negative half-line, then μ and linear measure have the same null sets. In symbols, if

(1) $$\widehat{\mu}(t) = 0 \quad \text{for} \quad t < 0,$$

then

(2) $$\int_E d|\mu| = 0 \quad \text{iff} \quad \int_E dx = 0.$$

The two Riesz did not express their theorem in quite this way. In particular, their measures lived on the circle $|z| = 1$, where one speaks of Fourier coefficients instead of transforms.

There is a hypothesis alternative to (1). This involves the Poisson integral of μ. The Poisson kernel of the half-plane $y > 0$ is, essentially, the imaginary part of $1/z$:

(3) $$P(z) = -\frac{1}{\pi} \Im \frac{1}{z}.$$

The Poisson kernel is positive (provided $y > 0$) and harmonic (in the punctured plane $z \neq 0$). The Poisson integral of the bounded complex measure μ, in symbols, $\mu^\star$, is the convolution of P with μ:

$$\mu^\star(z) = \int_{-\infty}^{\infty} P(z - \xi) \, d\mu(\xi).$$

Since $P(z)$ is harmonic in the half-plane $y > 0$, so is $\mu^\star(z)$. The hypothesis alternative to (1) is that $\mu^\star(z)$ is holomorphic in the half-plane. This is to say,

(4) $$\widehat{\mu}(t) \text{ vanishes for } t < 0 \text{ iff } \mu^\star(z) \in \mathcal{O}(y > 0).$$

To see that the two imply one another, we use this test:

$$\mu^\star \text{ is holomorphic in the half-plane } y > 0 \text{ iff } \frac{\partial \mu^\star}{\partial \bar{z}} \text{ vanishes there.}$$

By (3), the partial derivative in the test (times $-2\pi i$) is the convolution of $1/\bar{z}^2$ with μ. But the first term of this latter convolution is the Laplace transform of $-t$:

$$\frac{1}{\bar{z}^2} = -\int_0^\infty te^{-i\bar{z}t}\,dt;$$

so

$$\frac{\partial\mu^\star}{\partial\bar{z}} = \frac{1}{2\pi i}\int_0^\infty te^{-i\bar{z}t}\widehat{\mu}(-t)\,dt.$$

This gives (4). (Here we use: If the Fourier transform of a bounded measure vanishes everywhere, so does the measure.)

Over the years the Riesz theorem has been extended in a variety of ways. Among other workers, by Bochner, Helson & Lowdenslager, deLeeuw & Glicksberg, Stein & Weiss, Gårding & Hörmander, myself, and ultimately by Glicksberg, König, & Seever. The earliest of these extensions, and the easiest to describe, is Bochner's (1944). He works with measures in two or more variables, say two. So here (but only here) the measure μ is a bounded complex measure in the plane. Bochner's hypothesis is that the Fourier transform of μ vanishes outside a sector of opening less than π, and his conclusion is that μ and spatial measure have the same null sets. Bochner, like the two Riesz, did not express his theorem in this way. His measures lived on the product of two or more circles, and so like they, he spoke of Fourier coefficients, not transforms.

Let's return to the Riesz theorem. The conclusion (2) is in two parts: if E is of linear measure 0, then $|\mu|(E) = 0$; and vice versa. In other words, the first half is that μ is absolutely continuous with respect to linear measure, so

$$d\mu = f(x)\,dx.$$

In terms of $f(x)$, the second half is that $f(x)$ cannot vanish on any set of positive linear measure. There is a stronger property, which loosely speaking is that $f(x)$ cannot be too small. The precise statement is this:

$$(5) \qquad \int_{-\infty}^\infty \log\left(1 + \frac{1}{|f(x)|}\right)\frac{dx}{1+x^2} \quad < \quad \infty.$$

The strict inequality (5) follows from the hypothesis alternative to (1), plus a certain version of the Fatou-Lebesgue lemma [7].

If we take (1) for our hypothesis, then the Riesz theorem may be proved by the discrete Hilbert space methods of Helson & Lowdenslager, while if we take the hypothesis alternative to (1), the proof is function theoretic.

There is a second hypothesis alternative to (1). This third hypothesis involves the Laplace (or Fourier-Carleman) transform of the measure μ; in symbols, μ^*. The Laplace transform of μ lives in the $s = \sigma + it$ plane, and comes in two pieces, one holomorphic in the half-plane $\sigma > 0$, the other holomorphic in the half-plane $\sigma < 0$. For $\sigma > 0$, let

$$F_+(s) = \frac{1}{2}\mu(0) + \int_0^\infty e^{-sx}\,d\mu(x),$$

and for $\sigma < 0$, let

$$F_-(s) = -\frac{1}{2}\mu(0) + \int_0^{-\infty} e^{-sx}\,d\mu(x).$$

($\int_0^\infty$ means $\int_{x>0}$ and $\int_0^{-\infty}$ means $-\int_{x<0}$.) Then $F_+ \in \mathcal{O}(\sigma > 0)$ and $F_- \in \mathcal{O}(\sigma < 0)$. By $\mu^*(s)$ we mean $F_+(s)$ if $\sigma > 0$, and $F_-(s)$ if $\sigma < 0$, so $\mu^* \in \mathcal{O}(\sigma \neq 0)$. Since μ is bounded, $F_+(s)$ exists not only in the open half-plane $\sigma > 0$, but in the closed half-plane $\sigma \geq 0$, where it is continuous. Likewise, $F_-(s)$ is continuous in the closed half-plane $\sigma \leq 0$. On the line $\sigma = 0$, the difference $F_+ - F_-$ is the Fourier transform of μ:

(6) $$F_+(it) - F_-(it) = \widehat{\mu}(t).$$

By (6), we see that (1) holds iff $\mu^*(s)$ continues holomorphicly to the s-plane with the half-line $s = it$, $t \geq 0$ removed, or in symbols, iff

(7) $$\mu^* \in \mathcal{O}(s \neq it,\ t \geq 0).$$

The difference $F_+(\varepsilon + it) - F_-(-\varepsilon + it)$ is the harmonic transform of μ (Beurling's term). So writing $\widehat{\mu}(\varepsilon, t)$ for the harmonic transform, we have

$$\widehat{\mu}(\varepsilon, t) = \int_{-\infty}^\infty e^{-\varepsilon|x| - itx}\,d\mu(x).$$

If (1) holds, then

(8) $$\lim_{\varepsilon \to 0} \widehat{\mu}(\varepsilon, t) = 0 \text{ uniformly in the half-line } t < 0.$$

(And vice versa.) Although the harmonic transform and its limit are easier to grasp (and perhaps easier to use) than the Laplace transform and its continuation, I will speak of the latter more than the former, at least for a while. After all, the latter seems to cut more deeply.

The property (7) has meaning not just for bounded measures, but for certain unbounded ones as well. Of course, the original hypothesis alternative to (1) is also meaningful for unbounded measures, but then the condition is more stringent:

$$\int_{-\infty}^\infty \frac{1}{1+x^2}\,d|\mu|(x) < \infty.$$

The Laplace transform of μ exists provided

(9) $$\int_{-\infty}^\infty e^{-\varepsilon|x|}\,d|\mu|(x) < \infty \quad \text{for} \quad \varepsilon > 0.$$

If (9) holds, we will say that μ is of zero exponential type. The set of measures of zero exponential type is a locally convex linear topological space with semi-norms the integrals in (9). It is also the dual of a linear topological space. Namely, the inductive limit of normed spaces $X_\varepsilon =$ the set of continuous functions $\varphi(x)$ such that

$$\|\varphi\|_\varepsilon = \sup_{-\infty < x < \infty} |\varphi(x)| e^{\varepsilon|x|} < \infty.$$

Neither the X_ε and their inductive limit, nor the locally convex dual of the inductive limit play a rôle in this talk. Only the vector called μ counts.

Let $V(y)$ be the variation of the open interval with endpoints $-y$ and y:

$$V(y) = \int_{-y}^{y} d|\mu|.$$

Then (9) holds iff

$$\lim_{y \to \infty} \frac{\log(1 + V(y))}{y} = 0.$$

Alternatively, iff to each $\varepsilon > 0$ there is a positive number $A(\varepsilon)$ such that

$$V(y) \le A(\varepsilon)e^{\varepsilon y},$$

which explains why we use the term *zero exponential type* for measures which satisfy (9).

All of this makes evident (I hope) this question:

PROBLEM. *Let the measure μ be of zero exponential type and suppose the Laplace transform of μ continues holomorphicly to the s-plane with the half-line $s = it$, $t \ge 0$ removed, i.e., suppose (7) holds. With this hypothesis, must μ and linear measure have the same null sets, i.e., does (2) hold?*

If (7) holds, then (8) holds provided we replace "uniformly" there with "locally uniformly", and vice versa. This is to say, suppose μ is of zero exponential type. Then the Laplace transform continues in the way we like iff

(10) $\displaystyle\lim_{\varepsilon \to 0} \int_{-\infty}^{\infty} e^{-\varepsilon|x|-itx}\, d\mu(x) = 0$

locally uniformly in the half-line $t < 0$.

Here is a lemma which supplies measures that satisfy the hypothesis.

LEMMA. *Let $f(z)$ be holomorphic in the half-plane $y > 0$, and suppose $f(z)$ is of zero exponential type there, i.e., suppose that to each $\varepsilon > 0$ there is a positive number $A(\varepsilon)$ such that*

(11) $|f(z)| \le A(\varepsilon)e^{\varepsilon|z|}$ for $y > 0$.

Then (7) holds for $d\mu = f(x)dx$.

Alternatively, (10) holds, but I cannot prove this directly. Let me make explicit $f(x)$ and its size, although this may be hardly necessary. By $f(x)$ we mean

$$\lim_{y \to 0} f(x + iy).$$

By (11) and $f(z) \in \mathcal{O}(y > 0)$, the limit exists for almost all x, and by (11) once more, $|f(x)| \le A(\varepsilon)\exp(\varepsilon|x|)$, so μ is of zero exponential type. More or less, I owe the lemma to Macintyre [3].

PROOF OF THE LEMMA. The Laplace transforms $F_+(s)$ and $F_-(s)$ were defined earlier. The first is holomorphic in the half-plane $\sigma > 0$, the second in the half-plane $\sigma < 0$. Besides $F_+(s)$ and $F_-(s)$, or what is the same, $f(x)\,dx$ on the half-lines $x > 0$ and $x < 0$, we have $f(iy)\,d(iy)$ on the half-line $y > 0$. Its Laplace transform

$$F_i(s) = \int_0^{i\infty} e^{-sz} f(z)\,dz$$

exists in the half-plane $t < 0$, but more to the point, $F_i(s)$ is holomorphic there. We will show that $\mu^*(s)$ continues holomorphicly to the s-plane less the half-line $s = it$, $t \geq 0$ by showing that

(12) $F_+(s) = F_i(s)$ in the 4th quadrant $\sigma > 0$, $t < 0$,

and

(13) $F_-(s) = F_i(s)$ in the 3rd quadrant $\sigma < 0$, $t < 0$.

Let s be in the 4th quadrant, so $\sigma > 0$ and $t < 0$, and let $b > 0$. By Cauchy (plus dominated convergence),

(14) $$\int_\square e^{-sz} f(z)\,dz = 0$$

where $\square$ is the square $(0, b, b + ib, ib, 0)$. The left side of (14) is

$$\int_0^b \;+\; \int_b^{b+ib} \;+\; \int_{b+ib}^{ib} \;+\; \int_{ib}^0 .$$

The second integral is bounded by $A(\varepsilon)b$ times $\exp[(2\varepsilon - \sigma)b]$, and the third by $A(\varepsilon)b$ times $\exp[(2\varepsilon + t)b]$, so

$$\int_0^\infty \;+\; \int_{i\infty}^0 \;=\; 0$$

which is (12). Likewise, (13) holds. $\square$

An afterword. The proof yields more than the property (7). If $t < 0$, then

$$\widehat{\mu}(\varepsilon, t) = F_i(\varepsilon + it) - F_i(-\varepsilon + it) = \int_{-\varepsilon+it}^{\varepsilon+it} F_i'(v)\,dv,$$

and so $\widehat{\mu}(\varepsilon, t) = O(\varepsilon)$ uniformly in each half-line to the left of 0. In particular,

$$\lim_{\varepsilon \to 0} \widehat{\mu}(\varepsilon, t) = 0 \text{ uniformly in such half-lines.}$$

Let $B(z) \in \mathcal{O}(y > 0)$. If $B(z)$ is bounded in the half-plane $y > 0$, then it is of zero exponential type there, so by the lemma,

(15) $$\lim_{\varepsilon \to 0} \int_{-\infty}^\infty e^{-\varepsilon|x| - itx} B(x)\,dx = 0$$

locally uniformly in the half-line $t < 0$.

In brief, if $B(x) \in H^\infty(dx)$, then (15) holds. Vice versa, let $B(x) \in L^\infty(dx)$. If (15) holds, then by a certain statement involving the Poisson integral of $B(x)$ and a certain part of the proof of the theorem to follow, $B(x) \in H^\infty(dx)$. I have not seen this elsewhere. [The *certain statement* is the test following (4), but with $d\mu = B(x)\,dx$.]

There is a similar lemma for functions of positive type: Let $f(z)$ (once more) be holomorphic in the half-plane $y > 0$, and suppose that

$$|f(z)| \le Ae^{b|z|}.$$

Let $f(x) = \lim_{y \to 0} f(x + iy)$, and suppose the measure $f(x)\,dx$ is of zero exponential type. If p is such that $|f(iy)| \le A\exp(py)$, then (7) holds for $d\mu = \exp(ipx)f(x)\,dx$. (For example, $p = b$ works, but so might certain $p < b$.) Alternatively, if $d\mu = f(x)\,dx$, then μ^* continues holomorphicly to the s-plane with the half-line $s = it,\ t \ge -p$ removed.

Let's return to the problem and prove that the answer to the question is yes provided the measure is not too large (or not too unbounded).

THEOREM. *Let the measure μ be of zero exponential type and suppose its Laplace transform μ^* continues holomorphicly to the s-plane less the half-line $s = it,\ t \ge 0$. If $\log(1 + V)$ is Poisson summable (Beurling's term), i.e., if*

$$(16) \qquad \int_0^\infty \log(1 + V(y))\,\frac{dy}{1 + y^2} \quad < \quad \infty,$$

then μ and linear measure have the same null sets:

$$(2) \qquad \int_E d|\mu| = 0 \quad \text{iff} \quad \int_E dx = 0.$$

The idea of the proof is to use the hypothesis on the variation $V(y)$ to construct functions f which vanish outside an interval and whose Fourier transforms satisfy

$$\int |\widehat{f}|\,d|\mu| < \infty.$$

If $d\nu$ is the measure $\widehat{f}\,d\mu$, then ν^* (the Laplace transform of ν) is the convolution of μ^* with f, and so ν^*, like μ^*, continues holomorphicly to the s-plane with a certain half-line removed. But then by the Riesz theorem, ν and dx have the same null sets, and so μ does too (provided we ignore the zeros of $\widehat{f}$).

II

PROOF OF THE THEOREM. The hypothesis on the variation $V(y)$ is just a way of saying there is a positive even function $\varphi(x)$, increasing for $x > 0$, such that

$$(i) \qquad \int_{-\infty}^\infty e^{-\varphi}\,d|\mu| < \infty,$$

(ii)
$$\int_0^\infty \varphi(x)\,\frac{dx}{1+x^2} < \infty, \quad \text{and}$$

(iii)
$$\lim_{x\to\infty} \varphi(x) = \infty.$$

To see this, let
$$e^{-\varphi(x)} = \frac{1}{(1+x^2)\big(1+V(|x|+1)\big)}.$$

Then (ii) and (iii) hold. Also,

$$\int_{-\infty}^\infty e^{-\varphi}\,d|\mu| = \frac{|\mu(0)|}{1+V(1)} + \sum_{n=0}^\infty \int_{n<|x|\le n+1} e^{-\varphi(x)}\,d|\mu|(x),$$

while

$$\int_{n<|x|\le n+1} e^{-\varphi(x)}\,d|\mu|(x) \le \frac{1}{1+n^2}\cdot\frac{V((n+1)+)}{1+V((n+1)+)} \le \frac{1}{1+n^2},$$

so (i) holds as well.

For $x > 0$, $\varphi(x)$ is continuous from the left, i.e., $\varphi(x-) = \varphi(x)$, but this is unimportant, and we will ignore it. We will use the points where φ is an integer to construct the functions mentioned at the outset. (This is to say, at the end of part I.) To each $n > \varphi(0+)$ there is $\mu_n > 0$ such that

$$\varphi(\mu_n-) \le n \le \varphi(\mu_n+). \tag{17}$$

Here we use (iii). (Since φ increases strictly, μ_n is unique, but this is unimportant.) By (17), the μ_n increase (to ∞), but not necessarily strictly. It is less obvious that

$$\sum_{n>\varphi(0+)} \frac{1}{\mu_n} < \infty. \tag{18}$$

To see this, begin with

$$\int_{\mu_\ell}^\infty \varphi(x)\,\frac{dx}{x^2} = \sum_{n=\ell}^m \int_{\mu_n}^{\mu_{n+1}} \varphi(x)\,\frac{dx}{x^2} + \int_{\mu_{m+1}}^\infty \varphi(x)\,\frac{dx}{x^2}.$$

By the second of the inequalities (17), the integral to the right is bounded below by

$$n\int_{\mu_n}^{\mu_{n+1}} \frac{dx}{x^2} = n\left(\frac{1}{\mu_n} - \frac{1}{\mu_{n+1}}\right),$$

and the integral to the far right by $(m+1)/\mu_{m+1}$, so summing by parts, we see that the integral on the left is bounded below by

$$\frac{\ell}{\mu_\ell} + \sum_{n=\ell+1}^{m+1} \frac{1}{\mu_n}.$$

This proves (18).

Here is an alternative proof of (18). To each $y > \varphi(0+)$ there is $\xi > 0$ such that
$$\varphi(\xi-) \leq y \leq \varphi(\xi+)$$
(cf. (17)). The point ξ is unique, and we let $G(y) = 1/\xi$. Then $G(y)$ decreases, but not necessarily strictly, $G(y)$ is continuous, but this is unimportant, and if $b > 0$, we have
$$\int_b^\infty \varphi(x) \frac{dx}{x^2} = \int_0^{1/b} \varphi\left(\frac{1}{u}\right) du = \frac{\varphi(b\mp)}{b} + \int_{\varphi(b\mp)}^\infty G(y)\, dy.$$
But by (17), $G(n) = 1/\mu_n$, so
$$\sum_{n=m}^\infty \frac{1}{\mu_{n+1}} \leq \int_m^\infty G(y)\, dy \leq \sum_{n=m}^\infty \frac{1}{\mu_n}.$$
This is more than we need, but it proves (18), and it shows that the hypothesis is necessary as well. (I.e., if (18) holds, then so does (ii), or what is the same, (16).)

Let
$$\varepsilon_n = \frac{e}{\mu_n}, \quad \vartheta(x) = \frac{\sin x}{x}, \quad \text{and} \quad F_\ell(x) = \prod_{n=\ell+1}^\infty \vartheta(\varepsilon_n x).$$
The difference $1 - \vartheta(x)$ is bounded by $|x|/2$, so by (18), the product converges locally uniformly and $F_\ell(x)$ is continuous. (The F_ℓ are the desired $\widehat{f}$ s.) We have
$$|F_\ell(x)| \leq e^{\ell+1} e^{-\varphi(x)}, \tag{19}$$
provided $x \geq$ the first of the μ_n, i.e., $\geq \mu_{[1+\varphi(0+)]}$.

To see this, let m be such that
$$\mu_m \leq x < \mu_{m+1}. \tag{20}$$
Since $\vartheta(x)$ is bounded by 1 and by $1/x$,
$$|F_\ell(x)| \leq \prod_{n=\ell+1}^m |\vartheta(\varepsilon_n x)| \leq \prod_{n=\ell+1}^m \frac{1}{\varepsilon_n x}. \tag{21}$$
By the first of the inequalities (20), the denominators in (21) are bounded below by e, so
$$|F_\ell(x)| \leq \left(\frac{1}{e}\right)^{m-\ell} = e^\ell e^{-m}.$$
But by the second of the inequalities (20), and the first of the inequalities (17),
$$\varphi(x) \leq \varphi(\mu_{m+1}-) \leq m+1,$$
so (19) holds.

We have yet to truly use (18). Let b_ℓ be e times the tail of the series there:
$$b_\ell = \sum_{n=\ell+1}^\infty \varepsilon_n.$$

We need:

$\widehat{F}_\ell(y)$, *the Fourier transform of* $F_\ell(x)$,

$$\text{vanishes off the interval } -b_\ell < y < b_\ell. \quad (22)$$

Granted (22), we have $\widehat{F}_\ell \in L^1$ (here we use (18)), so F_ℓ is $1/2\pi$ times the transform of $\widehat{F}_\ell$:

$$F_\ell(x) = \frac{1}{2\pi} \int_{-b_\ell}^{b_\ell} e^{ixy} \widehat{F}_\ell(y) \, dy. \quad (23)$$

(Here we use the continuity of $F_\ell(x)$.) There are various ways, all well-known, to see that (22) holds. Here is one.

The quotient ϑ is the Fourier transform of a rectangle $\square$ whose base is the interval $-1 < y < 1$, so the second of the three products in the inequalities (21), but with varying sign, namely,

$$F_{\ell m}(x) = \prod_{n=\ell+1}^{m} \vartheta(\varepsilon_n x),$$

is the Fourier transform of a nonnegative (even) function $f_{\ell m}(y)$ which vanishes off the interval

$$-\sum_{n=\ell+1}^{m} \varepsilon_n < y < \sum_{n=\ell+1}^{m} \varepsilon_n,$$

and so off the interval in (22). But $F_{\ell m} \in L^1$ (provided $m > \ell + 1$), so $f_{\ell m}$ is $1/2\pi$ times the transform of $F_{\ell m}$:

$$f_{\ell m}(y) = \frac{1}{2\pi} \int_{-\infty}^{\infty} e^{-iyx} F_{\ell m}(x) \, dx. \quad (24)$$

(Here we use continuity once more: $f_{\ell m}(y)$ is continuous provided $m > \ell + 1$.) By (24), plus dominated convergence,

$$\frac{1}{2\pi} \widehat{F}_\ell(y) = \lim_{m \to \infty} f_{\ell m}(y),$$

so (22) holds. We also see that $\widehat{F}_\ell$ is nonnegative.

Since $F_\ell(x)$ is the tail of a convergent product,

$$\lim_{\ell \to \infty} F_\ell(x) = 1. \quad (25)$$

We will use this in a moment.

The proof is now easily completed (although this may be hardly necessary). By (i) and (19),

$$\int_{-\infty}^{\infty} |F_\ell| \, d|\mu| < \infty,$$

so the measure $d\nu_\ell = F_\ell \, d\mu$ is bounded, while by (23), plus Fubini,

$$\nu_\ell^*(s) = \frac{1}{2\pi} \int_{-b_\ell}^{b_\ell} \mu^*(s - iy) \widehat{F}_\ell(y) \, dy \quad \text{for} \quad \sigma \neq 0, \quad (26)$$

so $\nu_\ell^*(s)$ continues holomorphicly to the s-plane less the half-line $s = it$, $t \geq -b_\ell$. (Here it is crucial that the interval of integration is bounded since we know nothing of the size of $\mu^*(s)$ on the half-line $s = it$, $t < 0$.) Alternatively, the Fourier transform of ν_ℓ vanishes on a half-line:

$$\widehat{\nu}_\ell(t) = 0 \quad \text{for} \quad t < -b_\ell. \tag{27}$$

Suppose E is a linear null set:

$$\int_E dx = 0. \tag{28}$$

Then by (27), it is a null set for the measure ν_ℓ:

$$\int_E |F_\ell|\, d|\mu| = 0. \tag{29}$$

(Here we use the Riesz theorem.) But then by (25), plus the Fatou-Lebesgue lemma, E is a μ null set:

$$\int_E d|\mu| = 0. \tag{30}$$

Vice versa, suppose (28) does not hold, but that (30) does. Then so does (29) (i.e., E is a null set for the measure ν_ℓ), so by the Riesz theorem once more, the measure ν_ℓ vanishes everywhere:

$$\int_{-\infty}^{\infty} |F_\ell|\, d|\mu| = 0.$$

But then by (25) again, μ vanishes everywhere. $\square$

III

Where do we go from here? Once more, suppose (7) holds. If μ is the measure in the lemma, i.e., if $d\mu = f(x)\, dx$ with $f(z)$ holomorphic and of zero exponential type in the half-plane $y > 0$, then

$$f(z) = \frac{1}{2\pi i} \int_{\sqcup} e^{zs} \mu^*(s)\, ds. \tag{31}$$

This is an identity of Macintyre [3], and could be said to hold because the Cauchy kernel $1/z$ is the Laplace transform of 1. The cup $\sqcup$ is the infinite triangle

$$(-\varepsilon + i\infty, -\varepsilon - i\varepsilon, \varepsilon - i\varepsilon, \varepsilon + i\infty).$$

So if $\varphi(x)$ is of compact support, then

$$\int_{-\infty}^{\infty} \varphi\, d\mu = \lim_{y \to 0} \int_{-\infty}^{\infty} \varphi(x)\mu^{*\circ}(x + iy)\, dx$$

(provided μ is the measure in the lemma). The piece $\mu^{*\circ}(z)$ is the right side of (31). What if μ is not the measure in the lemma? Since (7) holds, $\mu^{*\circ}(z)$ is meaningful, and holomorphic in the half-plane $y > 0$. And it is of zero exponential type there except that it may be unbounded near the line $y = 0$. The second

item, namely, $\mu^{*\circ} \in \mathcal{O}(y > 0)$, brings to mind the Poisson integral, and in fact
if $d\mu = f(x)\,dx$ with f bounded, then $\mu^{*\circ} = \mu^*$, the Poisson integral of μ (or f).
For general μ there is no Poisson integral, and one wonders if $\mu^{*\circ}$ is a worthy
substitute. Can we relate μ to $\mu^{*\circ}$? If yes, then with luck the upshot could be
that $d\mu \ll dx$, which is half of (2). But this is pie in the sky. Nothing precludes

$$\mu^{*\circ}(z) = 0 \quad \text{for} \quad y > 0. \tag{32}$$

And in fact this occurs for certain measures in [6]. There Nyman finds functions
$f(x)$ of zero exponential type $(|f(x)| \le A(\varepsilon)\exp(\varepsilon|x|))$, or more to the point,
measures $d\mu = f(x)\,dx$ such that $\mu^*(s)$ is entire and bounded in the half-plane
$t < 0$. By Cauchy, (32) holds for these μ provided we replace cup $\sqcup$ with cap
$\sqcap$ and $y > 0$ with $y < 0$. Perhaps we should forget the Laplace transform of μ,
after all its rôle in the proof is minor, and concentrate instead on the harmonic
transform of μ. [The Laplace transform does serve however, via the lemma, to
provide measures that satisfy (8) (locally uniformly in the half-line).]

With this in mind, let $K(x)$ be such that

  (i) $K(0) = 1$,
  (ii) 0 is a point of continuity of $K(x)$, and
  (iii) $K(x)$ is bounded.

Suppose

$$\int_{-\infty}^{\infty} |K(\varepsilon x)|\,d|\mu|(x) < \infty \quad \text{for} \quad \varepsilon > 0, \tag{33}$$

and let

$$\widehat{\mu}(\varepsilon, t) = \int_{-\infty}^{\infty} K(\varepsilon x) e^{-itx}\,d\mu(x).$$

The notation does not reflect the dependence on the kernel K, but this is unim-
portant. If $K(x)$ is the Laplace kernel $\exp(-|x|)$, then μ is of zero exponential
type, and $\widehat{\mu}(\varepsilon, t)$ is Beurling's harmonic transform, which is harmonic in the half-
plane $\varepsilon + it$, $\varepsilon > 0$. Otherwise $\widehat{\mu}(\varepsilon, t)$ is not harmonic, nor is it anything else.
If we paraphrase the endgame of the proof of the theorem, we find the theorem
holds for general kernels, not just the Laplace kernel. Specificly:

THEOREM. *Suppose*

$$\lim_{\varepsilon \to 0} \widehat{\mu}(\varepsilon, t) = 0 \text{ locally uniformly in the half-line } t < 0. \tag{34}$$

If $\log(1 + V)$ *is Poisson summable, then* μ *and linear measure have the same
null sets.*

The paraphrase of the endgame is simply this: In place of the identity (26),
write

$$\widehat{\nu}_\ell(\varepsilon, t) = \frac{1}{2\pi} \int_{-b_\ell}^{b_\ell} \widehat{\mu}(\varepsilon, t - y)\widehat{F}_\ell(y)\,dy.$$

Then by the hypotheses on $K(x)$, we see once more that the Fourier transform
of the bounded measure ν_ℓ vanishes on a half-line.

As before, there is an evident gap in the theorem:

PROBLEM. *Does* (34) *imply* (2) *without hypotheses on the variation* $V(y)$ (*aside from those implicit in* (33))?

IV

Several more words. These involve (5), which loosely speaking says that if μ is bounded, then μ cannot be too small. We can do better than this. By the proof of the theorem, plus (5), we see that if μ is not too large, then μ is not too small. Speaking precisely:

COROLLARY. *Let* $d\mu = f(x)dx$, *and suppose*

$$\lim_{\varepsilon \to 0} \widehat{\mu}(K, \varepsilon, t) = 0 \text{ locally uniformly in a half-line.} \tag{35}$$

(*Here the notation reflects the dependence on the kernel.*) *Then* $\log(1 + 1/|f|)$ *is Poisson summable* (*i.e.*, (5) *holds*) *provided* $\log(1 + V)$ *is Poisson summable* (*i.e., provided* f *is not too large*).

However, if f is large, then all bets are off. To see this, let $f(x) = \exp(x)$ and let $K(x)$ be the Gauss kernel $\exp(-x^2)$. Then (35) holds, but nevertheless, (5) does not. To see that (35) holds in this case, we simply calculate $\widehat{\mu}(\varepsilon, t)$, or we may use a second lemma which like the first supplies measures that satisfy the hypothesis (34), but this time for the Gauss kernel. Here is the statement of the lemma, but we omit the proof.

LEMMA. *Let* $f(z)$ *be holomorphic in the half-plane* $y > 0$, *and suppose* $f(z)$ *is of exponential type there, i.e., suppose that*

$$|f(z)| \leq A e^{b|z|} \quad \text{for} \quad y > 0.$$

Let c *be such that* $|f(x)| \leq A \exp(c|x|)$, *and* p *such that* $|f(iy)| \leq A \exp(py)$. *For example,* $c = p = b$ *works, but so might smaller choices. If* $d\mu = f(x)\, dx$, *then*

$$\lim_{\varepsilon \to 0} \widehat{\mu}(\varepsilon, t) = 0 \text{ locally uniformly in the half-line } t < -c - p.$$

Actually, $\widehat{\mu}(\varepsilon, t) = O(\varepsilon)$ *uniformly in each half-line to the left of* $-c - p$, *and so the convergence is uniform in such half-lines.* (*This last is not surprising. We find it in the afterword to the first lemma, but here it holds for different reasons.*)

The number p, or for that matter c, may be negative, but if c is, then $f(z)$ vanishes everywhere. If we assume that $f(z)$ is entire (of exponential type), then the proof requires less gymnastics, and the limit vanishes in the half-line $t > c+q$ as well (provided q is such that $|f(iy)| \leq A\exp(-qy)$ for $y < 0$).

Postscript. January 17, 1992. A month or so ago, Alexander Volberg and Wayne Smith constructed functions $f(x)$ such that $\mu^*(s)$ is entire, where $d\mu = f(x)\, dx$, while $f(x)$ vanishes in an interval, so $dx \not\ll d\mu$. Volberg's $f(x)$ grows like

$$\exp\left(\frac{c|x|}{\log\log|x|}\right),$$

while Smith's simply satisfies

$$\int_{-\infty}^{\infty} |f(x)|^2 e^{-\varepsilon|x|}\, dx < \infty \quad \text{for} \quad \varepsilon > 0.$$

References.

1. S. Agmon, *A composition theorem for Dirichlet series*, J. Analyse Math. **1** (1951), 232-243.
2. A. Beurling, *Collected Works*, vol. 1, Birkhäuser, Boston, 1989.
3. A. J. Macintyre, *Laplace's transformation and integral functions*, Proc. London Math. Soc. **45** no. 2 (1939), 1-20.
4. S. Mandelbrojt, *Dirichlet Series*, Reidel, Dordrecht, 1972.
5. V. Marchenko, *On certain questions of approximation of continuous functions on the line*, Proc. Kharkov Math. Soc. **22** (1950). (Russian)
6. B. Nyman, *On the one-dimensional translation group and semi-group in certain function spaces*, Uppsala, 1950.
7. H. L. Royden, *Real Analysis*, Macmillan, New York, 1988.
8. W. Rudin, *Real and Complex Analysis, Third Edition*, McGraw-Hill, New York, 1987.

DEPARTMENT OF MATHEMATICS, UNIVERSITY OF WISCONSIN-MADISON, MADISON, WISCONSIN 53706

Contemporary Mathematics
Volume **137**, 1992

Intersections of Analytic and Smooth Discs

FRANC FORSTNERIČ

Dedicated to Walter Rudin.

ABSTRACT. If A is an embedded analytic disc in $\mathbf{C}^2$ and M is an embedded smooth disc in $\mathbf{C}^2$ with isolated complex tangents that has the same boundary as A and is totally real near the boundary, then the intersection number $A \cdot M$ is related to the number $I_+(M)$ of positively oriented complex tangents (counted with algebraic multiplicities) by the formula $A \cdot M = I_+(M) - 1$. In particular, if M is totally real or if it only has hyperbolic complex tangents, then the discs A and M must also intersect at an interior point.

Introduction.

Several years ago Vitushkin [14] raised the following question:

Does there exist a bounded strongly pseudoconvex domain $D \subset \mathbf{C}^2$ homeomorphic to the ball and an embedded analytic disc $A \subset \mathbf{C}^2 \backslash \overline{D}$, with the boundary of A contained in the boundary of D ?

An embedded analytic disc in $\mathbf{C}^2$ is the image of a smooth embedding $F: \overline{\Delta} = \{z \in \mathbf{C}: |z| \leq 1\} \to \mathbf{C}^2$ that is holomorphic in the open disc Δ.

It is essential to require that D be a topological cell (or at least to be simply connected), for one could otherwise take D to be a suitable tubular neighborhood of any smooth curve in $\mathbf{C}^2$ that bounds an analytic disc.

While it is easy to find such pairs (D, A) in $\mathbf{C}^3$, Vitushkin conjectured that there are no such configurations in $\mathbf{C}^2$. He believed that knowing this may perhaps be useful in solving the well-known Jacobian conjecture.

Later M. Gromov claimed that such a configuration existed. This was communicated to me by N. Sibony and Kruzhlin in May 1991. According to Gromov one should simply find an embedded analytic disc $A \subset \mathbf{C}^2$ and an embedded

---

1991 *Mathematics Subject Classification*. Primary 32F99.

Supported by the Research Council of Republic of Slovenia.

This paper is in final form and no version of it will be submitted for publication elsewhere.

totally real disc $M \subset \mathbf{C}^2$ such that $\overline{A} \cap \overline{M} = bA = bM$, i.e., the two discs have common boundary but they do not intersect at any interior point. Then, by taking D to be a suitable strongly pseudoconvex tubular neighborhood of M and by replacing A by a slightly smaller disc, one would obtain the required configuration.

Recall that a smooth embedded real surface $M \subset \mathbf{C}^2$ (with or without boundary) is said to be *totally real* at a point $p \in M$ if the tangent space $\mathrm{T}_p M$ spans $\mathrm{T}_p \mathbf{C}^2$ over the field $\mathbf{C}$. If this fails than $\mathrm{T}_p M$ is a complex line in $\mathrm{T}_p \mathbf{C}^2$, and p is said to be a *complex tangent* of M. Recall also [9] that a totally real submanifold $M \subset \mathbf{C}^n$ is the zero set of a strongly plurisubharmonic function $\rho \geq 0$ defined in a neighborhood U of M, with $d\rho \neq 0$ on $U \backslash M$, hence the sublevel sets $\{\rho < \epsilon\}$ for sufficiently small $\epsilon > 0$ are strongly pseudoconvex domains homotopic to M.

It seems that the first example of an embedded analytic disc $A \subset \mathbf{C}^2$ and an embedded totally real disc $M \subset \mathbf{C}^2$ with common boundaries was found by Wermer [17]. It suffices to take $A = \{(z, 0) : |z| \leq 1\}$ and $M = \{(z, f(z)) : |z| \leq 1\}$, where f is a smooth function that vanishes on the circle $|z| = 1$ and such that $\partial f / \partial \overline{z}$ is zero-free; for instance, one may take

$$f(z) = (z\overline{z} - 1)\overline{z} \exp(iz\overline{z}).$$

The problem with this particular example is that the two discs A and M also intersect at the interior point $(0,0)$, so when we fatten M to obtain a strongly pseudoconvex domain D, the complement $D \backslash A$ is an annulus with boundary in D rather than a disc.

At first this just seemed an unfortunate choice of the example. In [5] we showed that this is not so by proving

THEOREM. *An embedded analytic disc A and an embedded totally real disc M in $\mathbf{C}^2$ with common boundary $bA = bM$ must intersect at an interior point.*

We explicitly constructed a deformation of the analytic disc A into a totally real disc $\tilde{A}$, without introducing new intersections with M, such that the two discs glue smoothly along bM into an immersed totally real sphere $S = \tilde{A} \cup M$ in $\mathbf{C}^2$. As it is well known that every totally real embedded n-manifold in $\mathbf{C}^n$ must have Euler number zero [16] while $\chi(S^2) = 2$, we conclude that our sphere must have self-intersections and therefore A and M intersect at an interior point.

Totally real discs are not the only ones that have a Stein neighborhood basis: according to [8] the same is true for discs with isolated *hyperbolic* complex tangents (in the sense of Bishop [3]). Call such discs hyperbolic. It is well known that *elliptic* complex tangents are not allowed since near such points the surface has a nontrivial local envelope of holomorphy, see [3] and [12].

Thus one might hope that the construction suggested by Gromov is possible using a pair of an embedded analytic disc and an embedded hyperbolic disc that intersect transversely along their boundaries but have disjoint interiors.

In this article we show that this is not the case, by proving a formula (1) that relates the intersection number $A \cdot M$ and the number $I_+(M)$ of positive complex tangents of M, counted with appropriate algebraic multiplicities. If M is totally real, then $A \cdot M = -1$ (with an appropriate choice of orientations). If M has m isolated positive hyperbolic complex tangents and no elliptic complex tangents, then $A \cdot M = -m - 1 \leq -1$, hence the two discs intersect at an interior point as well.

Our proof here is substantially simpler than the one in [5].

While we are not able to settle Vitushkin's question at this moment, we show that the existence of such a configuration (with just slightly stronger hypotheses) would have the following interesting consequence (Theorem 2): One would be able to find an embedded analytic disc $A \subset \mathbf{C}^2$ and an embedded disc $M \subset \mathbf{C}^2$ such that

(i) $\overline{M} \cap \overline{A} = bA = bM$,

(ii) M is totally real except at an elliptic complex tangent $p \in M$, and

(iii) the envelope of holomorphy of $\overline{M}$ does not contain A.

Whether or not this is possible in $\mathbf{C}^2$ seems to be another interesting question.

Results.

Let M be an embedded real surface in $\mathbf{C}^2$ and let $p \in M$ be an isolated complex tangent. We recall the definition of *index* of p in M. (See [15] and [7] for details.) Locally near p we orient M coherently with the standard orientation of $\mathrm{T}_p M$ as a complex line. Choose an oriented neighborhood $U \subset M$ of p, U homeomorphic to the two-disc, and let $\tau\colon U \to \mathcal{G}$ be the Gauss map $\tau(q) = \mathrm{T}_q M$ into the Grassman manifold of oriented real 2-planes in $\mathbf{C}^2 = \mathbf{R}^4$. It is well-known (see Chern and Spanier [4]) that $\mathcal{G}$ is the product of two spheres $\mathcal{G} = S_1 \times S_2$ such that the set of complex lines in $\mathbf{C}^2$ equals $\mathrm{H}_+ \cup \mathrm{H}_-$, where $\mathrm{H}_+ = \{\pi_1\} \times S_2$ is the set of positively oriented lines and $\mathrm{H}_- = \{\pi_2\} \times S_2$ is the set of negatively oriented complex lines in $\mathbf{C}^2$.

DEFINITION. *(Notation as above.) The index $I(p)$ of an isolated complex tangent $p \in M$ is defined to be the local intersection number of $\tau\colon U \to \mathcal{G}$ with H_+ at the point $\tau(p) \in \mathrm{H}_+$.*

There are several equivalent definitions of the index, see [7]. The simplest way to compute $I(p)$ is to write M in suitable local holomorphic coordinates (z, w) near p as a graph $w = f(z)$, with p corresponding to $(0, 0)$. The condition that p is an isolated complex tangent of M is equivalent to $z = 0$ being an isolated zero of the function $\overline{\partial} f = \partial f / \partial \overline{z}$; the index $I(p)$ then equals the winding numbers of $\overline{\partial} f$ around the origin. See [15] or [7] for the details. Recall that every elliptic complex tangent $p \in M$ (in the sense of Bishop [3]) has index $+1$, and every hyperbolic complex tangent has index -1. A totally real point has index zero.

Suppose now that the surface M is orientable and choose an orientation on M. Denote by $\tau\colon M \to \mathcal{G}$ the induced Gauss map. If M has boundary bM, we shall assume that M is totally real along bM. Then we have well-defined global intersection numbers $I_+(M)$ resp. $I_-(M)$ of the map τ with submanifolds H_+

resp. H$_-$ of $\mathcal{G}$. Clearly $I_+(M)$ is just the sum of the indices of all positively oriented complex tangents $p \in M$, and similarly $I_-(M)$ is the sum of indices of negatively oriented complex tangents. Their sum

$$I(M) = I_+(M) + I_-(M)$$

is called the *index* of M, and their difference

$$I(bM; M) = I_+(M) - I_-(M)$$

is the *Maslov index* of the boundary bM in M, see [7] and the definition below. For a closed surface $M \subset \mathbf{C}^2$ we have $I_+(M) = I_-(M) = I(M)/2$ [6], [7].

It is obvious how to extend these definitions to immersions of a given surface M into $\mathbf{C}^2$. If $F_t: M \to \mathbf{C}^2$, $0 \le t \le 1$, is a regular homotopy of immersions (or embeddings) such that each F_t is totally real near the boundary of M (this condition is vacuous if $bM = \emptyset$), then the numbers I_+, I_-, I are the same for each immersion in the family.

Suppose now that A and M are embedded closed discs in $\mathbf{C}^2$ of class $\mathcal{C}^1$ with the common boundary $bA = bM$, and such that the interior of A is complex-analytic while M is totally real near bM. (This implies that A and M intersect transversely along their joint boundary.) We choose on A the canonical orientation induced by its complex structure; this uniquely determines an orientation on M such that the pairs (A, bA) and $(M, bM) = (M, bA)$ are oriented manifolds with boundary. Denote by $A \cdot M$ the oriented intersection number of A and M; this number is invariant under smooth deformations of A and M for which the two discs intersect transversely along $bA = bM$ at each step of the deformation, but the interior of each disc does not cross the boundary.

THEOREM 1. *With the notation as above we have*

$$A \cdot M = I_+(M) - 1. \tag{1}$$

COROLLARY. *If the disc M is hyperbolic in the sense that it only has isolated complex tangents p of index $I(p) \le 0$, then $A \cdot M \le -1$, hence A and M intersect at an interior point. In particular, we have $A \cdot M = -1$ whenever the disc M is totally real.*

REMARK. There is an apparent lack of symmetry in the formula (1) above since $I_-(M)$ does not appear. We shall see in the proof of Theorem 1 that

$$A \cdot M = I_-(M) + I_+(\tilde{A}) - 1,$$

where $\tilde{A}$ is a disc obtained by gluing A smoothly with a thin collar of M near bA.

The Corollary implies that one can not construct a pair (D, A) of Vitushkin's type using Gromov's suggestion. Although we do not know at present what is the answer to Vitushkin's question, the following result shows that it is rather unlikely for a configuration (D, A) of the required type to exist in $\mathbf{C}^2$.

THEOREM 2. *Let $D \subset \mathbf{C}^2$ be a domain whose closure $\overline{D}$ is diffeomorphic to the closed four-ball, and let $A \subset \mathbf{C}^2 \backslash D$ be an embedded analytic disc of class $\mathcal{C}^k$ up to the boundary ($k \geq 1$) intersecting $\overline{D}$ transversely along the curve $bA \subset bD$.*

Then there exists an embedded $\mathcal{C}^\infty$ disc $M \subset D$ of class $\mathcal{C}^k$ up to the boundary, with $bM = bA$, such that M is totally real except at one elliptic complex tangent $p \in M$. Moreover, we can choose M such that locally near p it is given in suitable affine complex coordinates by $w = z\bar{z}$.

QUESTION. *If A and M are as in the conclusion of Theorem 2, does the envelope of holomorphy of $\overline{M}$ necessarily contain A ?*

If the answer is positive, then the domain D in Theorem 2 cannot be strongly pseudoconvex for the obvious reason that the envelope of holomorphy of $\overline{M}$ would then be contained in $\overline{D}$. Thus a positive answer to this question proves Vitushkin's conjecture that there are no such configurations in $\mathbf{C}^2$.

Recall that near the elliptic complex tangent $p \in M$ the local envelope of holomorphy consists of a one-parameter family of analytic discs with boundaries in M [12], [3]. One would expect that this family of discs continues all the way to the disc A. Results of this type have been proved by Bedford and Gaveau [1], Bedford and Klingenberg [2], and Gromov [11] in the case when M is a part of a 'generic' smooth 2-sphere contained in the boundary of a strongly pseudoconvex domain.

Our last result is about the existence of totally real discs in $\mathbf{C}^2$ with prescribed one-jet at the boundary. Given a smooth oriented simple closed curve $\gamma \subset \mathbf{C}^2$ and a pair of smooth vector fields X, Y along γ such that $X(p)$ and $Y(p)$ are $\mathbf{C}$-linearly independent at each $p \in \gamma$, we define

DEFINITION. *The Maslov index $I(X,Y)$ equals the winding number of the function $p \to \det\big(X(p), Y(p)\big)$ as p traces γ once in the positive direction.*

Although the sign of $I(X,Y)$ depends on the orientation of γ, the condition $I(X,Y) = 0$ is independent of the orientation. Notice that the two-by-two determinant above is nonzero at p precisely when the two vectors are complex-linearly independent at p.

If $M \subset \mathbf{C}^2$ is an embedded disc with boundary $bM = \gamma$, we let X be the unit tangent to γ and let Y be the unit inward radial tangent vector field to M along γ. If M is a totally real disc, a simple argument with winding numbers shows that $I(X,Y) = 0$. This also follows from the formula $I(X,Y) = I_+(M) - I_-(M)$ proved in [7]. The converse of this is also true, and it may be useful in constructions of totally real discs:

PROPOSITION 3. *If $\gamma \subset \mathbf{C}^2$ is a smooth simple closed curve with tangent vector field X, and if Y is a smooth vector field along γ such that $I(X,Y) = 0$ (this requires that X and Y are $\mathbf{C}$-independent), then there exists a smooth embedded totally real disc $M \subset \mathbf{C}^2$ such that $bM = \gamma$ and Y is the inner radial tangent vector field to M along γ.*

PROOF OF THEOREM 1.

Step 1: Reduction to the case when A is analytic past bA.

Let X be a vector field along bA that is tangent to A and points towards the interior of A. We extend X to a vector field on $\mathbf{C}^2$ that is smooth on $\mathbf{C}^2 \backslash bA$ and is supported in a small neighborhood U of bA. We now flow the disc M for a short time $t > 0$ in the direction of X to obtain a new disc $\tilde{M}$ with boundary $b\tilde{M}$ contained in the interior of A. If $t > 0$ is sufficiently small, we do not introduce any new intersections of A and M and we do not affect the indices $I_\pm(M)$. Thus we may replace M by $\tilde{M}$ and assume that A is contained in a larger analytic disc A_0. After an additional small perturbation of M we may also assume that $bM = bA$ is a smooth real-analytic curve.

Step 2: Reduction to the case when M is a graph.

Denote by $\Delta(r) = \{z \in \mathbf{C} : |z| < r\}$ the open unit disc of radius r. There is a biholomorphic mapping $\Phi : U_\epsilon \subset \mathbf{C}^2 \to V \subset \mathbf{C}^2$ from a polydisc $U_\epsilon = \Delta(1+\epsilon) \times \Delta(1)$ onto an open neighborhood V of A such that $\Phi(\Delta(1) \times \{0\}) = A$ and $\Phi(T \times \{0\}) = bA = bM$, where $T = b\Delta(1)$ is the unit circle.

Let $M_0 = \Phi^{-1}(M \cap V)$. If the neighborhood V is chosen sufficiently small, then M_0 is a totally real annular region in U_ϵ, with one of its boundary components equal to $T \times \{0\}$. Moreover, its intersection with every real 3-plane $\Pi_\theta = \mathbf{R}e^{i\theta} \times \mathbf{C}$ is a disjoint union of two arcs that project one-to-one onto the plane $z = 0$.

The tangent space to M_0 at $(e^{i\theta}, 0)$ is spanned by vectors $X(\theta) = (ie^{i\theta}, 0)$ and $Y(\theta) = (a(\theta)e^{i\theta}, b(\theta))$, where $a(\theta)$ is real-valued and $b(\theta) \neq 0$. We can assume that $|b(\theta)| = 1$ for all θ.

Choose a smooth function $k \geq 0$ on $[0, \infty) \subset \mathbf{R}$ such that $k'(0) + a(\theta) > 0$ for all θ, $\operatorname{supp} k \subset [0, 1/2]$, and $\max k < \epsilon$. Denote by $\Psi_t : \mathbf{C}^2 \to \mathbf{C}^2$ the transformation

$$\Psi_t(z, w) = \big(z(1 + tk(|w|)), w\big), \qquad 0 \leq t \leq 1.$$

Then $M_t = \Psi_t(M_0)$ is a smooth deformation of M_0 within the polydisc U_ϵ such that the tangent space to the surface M_1 at $(e^{i\theta}, 0) \in bM_1$ is spanned by the vectors $X(\theta)$ and $Y_1(\theta) = \big(a_1(\theta)e^{i\theta}, b_1(\theta)\big)$, where $a_1(\theta) > 0$ for all θ. Moreover, M_t is totally real near the circle $T \times \{0\} \subset bM_t$ for each t and it coincides with M_0 near the boundary of U_ϵ.

The condition $a_1(\theta) > 0$ implies that near $T \times \{0\}$, M_1 is a totally real graph over the annulus $A_\delta = \Delta(1 + \delta) \backslash \Delta(1)$ for a suitable $\delta > 0$. Denote this part of M_1 by N. Let $\gamma \subset M_1$ be the curve that projects onto the circle $|z| = 1 + \delta$, so $bN = \gamma \cup (T \times \{0\})$.

Step 3: Gluing M with A.

Now we choose a smooth function $g : \overline{\Delta}(1 + \delta) \to \mathbf{C}$ such that its graph

$$K = \{(z, g(z)) : |z| \leq 1 + \delta\}$$

glues smoothly with M_1 along the curve γ, i.e., the set $M' = (M_1 \backslash N) \cup K$ is a smooth disc in U_ϵ. We leave out the obvious details of the construction of g.

Now we go back to the neighborhood V of A. Set

$$\tilde{A} = \Phi(K), \qquad \tilde{M} = (M \backslash V) \cup \Phi(M_1 \backslash N), \qquad \lambda = \Phi(\gamma).$$

By construction, $\tilde{M}$ and $\tilde{A}$ are smoothly embedded discs in $\mathbf{C}^2$ that are smoothly glued along their joint boundary λ into an immersed 2-sphere $S = \tilde{M} \cup \tilde{A}$ in $\mathbf{C}^2$. By construction we have $A \cdot M = \tilde{A} \cdot \tilde{M}$ and $I_{\pm}(\tilde{M}) = I_{\pm}(M)$ since the modification of M into $\tilde{M}$ was totally real near the boundary. After a generic small perturbation we may assume that S only has isolated complex tangents.

Step 4: An application of the index formula.

If S is a smoothly immersed closed oriented real 2-surface in $\mathbf{C}^2$ with isolated complex tangents and with transverse self-intersections, then we have the *index formula*

$$I(S) = \chi(S) - 2d(S),$$

where $I(S) = I_+(S) + I_-(S)$ is the total index of S (the sum of indices of all complex tangents), $\chi(S)$ is the Euler number, and $d(S)$ is the (Whitney's) oriented self-intersection number of S. (See [15], [7], and [6].) If S is non-orientable, the formula above holds modulo 2.

We apply this to our immersed sphere $S = \tilde{M} \cup \tilde{A}$. First we recall that

$$I_+(\tilde{M}) - I_-(\tilde{M}) = I_+(\tilde{A}) - I_-(\tilde{A})$$

since both numbers equal the Maslov index of the curve λ in S. Now comes the important point: Since K is a graph over the z-axis, all complex tangents of K and therefore of $\tilde{A} = \Phi(K)$ are positively oriented, so $I_-(\tilde{A}) = 0$. Using the last formula we thus get

$$I(S) = I_+(\tilde{M}) + I_-(\tilde{M}) + I_+(\tilde{A}) = 2I_+(\tilde{M}) = 2I_+(M).$$

Inserting this into the index formula and noting that $\chi(S) = 2$ we conclude

$$I_+(M) = 1 - d(S) = 1 + \tilde{A} \cdot \tilde{M} = 1 + A \cdot M.$$

The change of sign is due to the fact that the chosen orientations on $\tilde{A}$ and $\tilde{M}$ do not add up to an orientation on S.

This completes the proof of Theorem 1.

PROOF OF THEOREM 2.

First we prove that there exist a smooth embedded disc $M_0 \subset D \cup bA$ intersecting bD transversely along the curve γ.

Let S^4 be the standard one point compactification of $\mathbf{C}^2 = \mathbf{R}^4$ into a four-sphere. We represent S^4 as the union $B_+ \cup B_-$ of two closed four-balls that intersect in a 3-sphere (the equator). Since $\overline{D} \subset S^4$ is diffeomorphic to $B_+ \subset S^4$, Theorem 3.1 in [10, p.185] implies that there is a diffeomorphism $\Psi : S^4 \to S^4$ carying $\overline{D}$ onto B_+, hence Ψ carries the complement $S^4 \backslash D$ onto B_-.

Let $\tau : S^4 \to S^4$ be the reflection about the equator S^3 that interchanges B_+ with B_-. Then $M_0 = \Psi^{-1} \circ \tau \circ \Psi(A) \subset \overline{D}$ is the embedded disc in $\overline{D}$ with the required properties. This justifies our claim.

Recall that the disc A is the image of a $\mathcal{C}^k$ embedding $F:\overline{\Delta} \to A \subset \mathbf{C}^2$ that is holomorphic on the open disc Δ. We can extend F to a diffeomorphic mapping $\Phi:U_\epsilon \to V$ of class $\mathcal{C}^k$ from a polydisc $U_\epsilon = \Delta(1 + \epsilon) \times \Delta(1) \subset \mathbf{C}^2$ onto a neighborhood V of A such that Φ is biholomorphic in a neighborhood of the open disc $\Delta(1) \times \{0\}$.

The map $\gamma(\theta) = F(e^{i\theta})$ parametrizes the boundary curve $bA = bM_0$. Let $X(\theta)$ be the unit vector at $\gamma(\theta)$ that is tangent to M_0 and real orthogonal to $\gamma'(\theta)$; among the two possible choices we take the *inner* radial vector to M_0 pointing into D. Also let

$$Y(\theta) = D\Phi(e^{i\theta},0)\cdot (e^{i\theta},\delta) \in \mathrm{T}_{\gamma(\theta)}\mathbf{C}^2.$$

Choose $\delta > 0$ sufficiently small that $Y(\theta)$ points into D for each θ.

We now deform the vector field X continuously into Y by setting $X_t(\theta) = (1 - t)X(\theta) + tY(\theta)$ for $0 \leq t \leq 1$. By construction $X_t(\theta)$ points into D for each $t \in [0,1]$. We can follow the deformation $\{X_t : t \in [0,1]\}$ by an isotopy $\{M_t : t \in [0,1]\}$ consisting of embedded $\mathcal{C}^k$ discs in $D \cup bM_0$ such that M_t is tangent to X_t along the boundary $bM_t = bM_0$. We leave out the simple details.

Consider now the disc M_1. Its tangent bundle along the boundary is spanned by the vector fields Y and γ'. We claim that the Maslov index $I(bM_1, M_1)$ of the boundary curve bM_1 in M_1 equals one. Recall from [7] that the Maslov index is just the winding number m of the function $\theta \to \det\bigl(Y(\theta),\gamma'(\theta)\bigr)$, that is, $I(bM_1, M_1) = I(Y,\gamma')$. Since these vectors are the images of the vectors $\mu(\theta) = (e^{i\theta},\delta)$ resp. $\nu(\theta) = (ie^{i\theta},0)$ by the derivative $D\Phi(e^{i\theta},0)$, and since the derivative $D\Phi$ is $\mathbf{C}$-linear and nonsingular at every point of the disc $\overline{\Delta}\times\{0\} \subset \mathbf{C}^2$, it follows that m equals the winding number of the function

$$\theta \to \det\bigl(\mu(\theta),\nu(\theta)\bigr) = \det \begin{pmatrix} e^{i\theta} & ie^{i\theta} \\ \delta & 0 \end{pmatrix} = -i\delta e^{i\theta}$$

which equals one. This verifies the claim.

The calculation above also shows that the disc M_1 is totally real near the boundary $bM_1 = bA$. After a small generic perturbation of M_1 we can assume that M_1 only has isolated complex tangents.

We choose orientations on A and M_1 as in Theorem 1. Since $A\cdot M_1 = 0$ by construction, Theorem 1 implies $I_+(M_1) = 1$. Also, $I_+(M_1) - I_-(M_1) = I(bM_1, M_1) = 1$, hence $I_-(M_1) = 0$. .

The main result of [7] implies that we can find a $\mathcal{C}^0$-small perturbation M of M_1 in D which coincides with M_1 near the boundary of M_1 and such that M has precisely one positive complex tangent of index one (elliptic complex tangent) and no negative complex tangents. This is achieved by cancelling complex tangents of the same sign in pairs as explained in [7]. Moreover, we can specify the local form of M near the elliptic complex tangent.

This completes the proof of Theorem 2.

The proof shows that for every integer m we can find an embedded disc $M \subset D$ as above with one positive elliptic complex tangent and with one negative complex tangent of index m. Alternatively, M can have exactly m negative complex tangents that are elliptic if $m > 0$ and hyperbolic if $m < 0$.

PROOF OF PROPOSITION 3.

Choose an embedded totally real disc $M_0 \subset \mathbf{C}^2$. Let $\gamma_0 = bM_0$ and let Y_0 be the vector field tangent to M_0 along γ_0 and inner radial to M_0.

There is a smooth isotopy $\{\gamma_t : 0 \le t \le 1\}$ of $\gamma_0 = bM_0$ into $\gamma_1 = \gamma$ in $\mathbf{C}^2$. Let X_t be the tangent vector field to γ_t, depending smoothly on t. We can also find a vector field Y_t along γ_t, depending smoothly on $t \in [0,1]$, such that $Y_1 = Y$ and $I(X_t, Y_t) = 0$ for all t. (This requires in particular that X_t and Y_t are $\mathbf{C}$-independent for each t.)

The isotopy (γ_t, Y_t) can be extended to an isotopy A_t of a small annular collar $A_0 \subset M_0$ of γ_0 in M_0 into a similar annular collar $A = A_1$ along γ.

According to the isotopy extension theorem [10, p.180], the isotopy $t \to A_t \subset \mathbf{C}^2$ can be realized by a smooth diffeotopy $\Phi_t : \mathbf{C}^2 \to \mathbf{C}^2$, with Φ_0 the identity. Let $M_t = \Phi_t(M_0)$. By construction every M_t is totally real near the boundary $bM_t = \gamma_t$, hence the indices $I_\pm(M_t)$ are independent of t. Since M_0 is totally real, we conclude that $I_\pm(M_1) = 0$.

Using the results of [7] we can find a $\mathcal{C}^0$-small smooth deformation M of M_1 that agrees with M_1 near γ such that M is totally real. Clearly M is the required disc.

References.

1. E. Bedford and B. Gaveau, *Envelopes of holomorphy of certain 2-spheres in* $\mathbf{C}^2$, Amer. J. Math. **105** (1983), 957–1009.

2. E. Bedford and W. Klingenberg, *On the envelope of holomorphy of a 2-sphere in* $\mathbf{C}^2$, J. Amer. Math. Soc. **4** (1991), 623-646.

3. E. Bishop, *Differentiable manifolds in complex Euclidean spaces*, Duke Math. J. **32** (1965), 1–21.

4. S.S. Chern and E. Spanier, *A theorem on orientable surfaces in four-dimensional space*, Comm. Math. Helv. **25** (1951), 205–209.

5. T. Duchamp and F. Forstnerič, *Intersections of analytic and totally real discs*, Preprint, 1991.

6. Y. Eliashberg, *Filling by holomorphic discs and applications*, Preprint, 1989.

7. F. Forstnerič, *Complex tangents of real surfaces in complex surfaces*, to appear, Duke Math. J.

8. F. Forstnerič and E.L. Stout, *A new class of polynomially convex sets*, Arkiv för Mat. **29** (1991), 51–62.

9. F.R. Harvey and R.O. Wells, *Zero sets of non-negative strongly plurisubharmonic functions*, Math. Ann. **201** (1973), 165–170.

10. M. Hirsch, *Differential Topology*, Graduate Texts in Math. 33, Springer, New York-Heidelberg-Berlin.

11. M. Gromov, *Pseudoholomorphic curves in symplectic manifolds*, Invent. Math. **82** (1985), 307–347.

244

FRANC FORSTNERIČ

12. C.E. König and S.M. Webster, *The local hull of holomorphy of a surface in the space of two complex variables*, Invent. Math. **67** (1982), 1–21.
13. S. Smale, *Classification of immersions of two-sphere*, Trans. Amer. Math. Soc. **90** (1958), 281–290.
14. A. Vitushkin, oral communication, Institute Mittag-Leffler, 1987.
15. S.M. Webster, *Minimal surfaces in a Kähler surface*, J. Diff. Geom. **20** (1984), 463–470.
16. R.O. Wells, *Compact real submanifolds of a complex manifold with nondegenerate holomorphic tangent bundles*, Math. Ann. **179** (1969), 123–129.
17. J. Wermer, *Polynomially convex discs*, Math. Ann. **158** (1965), 6–10.

DEPARTMENT OF MATHEMATICS, UNIVERSITY OF WISCONSIN, MADISON, WISCONSIN 53706

Contemporary Mathematics
Volume **137**, 1992

Critically Finite Rational Maps on $\mathbb{P}^2$

JOHN ERIK FORNÆSS
NESSIM SIBONY

1. Introduction.

In one complex variable, the study of rational maps $R : \mathbb{P}^1 \to \mathbb{P}^1$ such that every critical point is preperiodic is quite central in the theory, see ([Th]). In particular it follows from a result of Sullivan, ([Su]), that the Julia set for such a map is $\mathbb{P}^1$. M. Rees, ([Re]), has studied the ergodic theory of rational maps close to critically finite rational maps.

Classical examples of such maps are

$$R_\lambda = \lambda \left(\frac{z-2}{z} \right)^d$$

with λ such that $(\frac{\lambda-2}{\lambda})^d = 1$. The critical points are 0 and 2 and the critical orbits are $2 \to 0 \to \infty \to \lambda \to \lambda$.

In ([B]) Mc Mullen raised the question of constructing similar examples in $\mathbb{P}^k$. More precisely, the problem is to construct

$$f : \mathbb{P}^k \to \mathbb{P}^k$$

such that the complement of the post critical set is Kobayashi hyperbolic and to study the dynamics. The purpose of this paper is to study the dynamics of two such examples in $\mathbb{P}^2$. In particular we construct a holomorphic map $g : \mathbb{P}^2 \mapsto \mathbb{P}^2$ such that the critical set is preperiodic and whose Julia set is $\mathbb{P}^2$.

In a forthcoming paper we will study general questions concerning iterations of holomorphic maps from $\mathbb{P}^2$ to $\mathbb{P}^2$ and in particular we will give a structure theorem for critically finite holomorphic maps in $\mathbb{P}^2$. For background concerning iteration of rational maps on $\mathbb{P}^1$ we refer to Milnor ([Mi]) or Carleson ([Ca]).

1991 *Mathematics Subject Classification*. Primary 32H50.
The first author supported by an NSF grant.
This paper is in final form and no version of it will be submitted for publication elsewhere.

2. Calculation of the Critical Orbits

We introduce here at first a class of rational maps on $\mathbb{P}^k$. And we describe their critical orbits. After that we describe more precisely two examples in $\mathbb{P}^2$ which we will investigate in more detail.

Define the rational map h on $\mathbb{P}^k$ by

$$h : \mathbb{P}^k \quad\quad \longrightarrow \mathbb{P}^k$$
$$[z_0 : z_1 : ... : z_k] \mapsto [z_0^2 : (z_0 - 2z_1)^2 : ... : (z_0 - 2z_k)^2].$$

Our maps will be compositions of h with a permutation of the variables. So let $\sigma : \{0, 1, ..., k\} \mapsto \{0, 1, ..., k\}$ be an arbitrary permutation. We then set $h_\sigma := \sigma \circ h$.

Let M be a compact complex manifold and $g : M \mapsto M$ a holomorpic map with discrete fibers.

DEFINITION 2.1. *We say that g is critically finite if* $V := \bigcup_{n=0}^{n=\infty} g^n(C)$ *is a closed complex hypersurface of M, where C denotes the critical set consisting of those points where the Jacobian determinant vanishes. In that case we define* $W = g^{-1}(V)$, *which is a closed complex hypersurface containing V.*

DEFINITION 2.2. *A critically finite map is preperiodic if for every irreducible component C_i of the critical set $g^n(C_i) \not\subset C_i$ $\forall n \geq 1$.*

We recall the definition of the infinitesimal Kobayashi metric.

DEFINITION 2.3. *If M is a complex manifold and ξ is a non zero tangent vector at $p \in M$, the infinitesimal Kobayashi metric $F_M(p, \xi) = \inf\{\lambda > 0; \exists f : \{|z| < 1\} \mapsto M, holomorphic, f(0) = p, f'(0)(1) = \xi/\lambda\}$. We say that M is hyperbolic if F_M is locally bounded below by a strictly positive number.*

For background concerning hyperbolicity in the above sense see ([L]).

PROPOSITION 2.4. *For all σ, h_σ is critically finite. The map $h_\sigma : \mathbb{P}^k \setminus W \mapsto \mathbb{P}^k \setminus V$ is a covering map. The sets $\mathbb{P}^k \setminus W$ and $\mathbb{P}^k \setminus V$ are Kobayashi hyperbolic and $h_\sigma : \mathbb{P}^k \setminus W \mapsto \mathbb{P}^k \setminus V$ is noncontracting in the infinitesimal Kobayashi metric of $\mathbb{P}^k \setminus V$.*

PROOF. Observe at first that the critical set of h (and hence h_σ) is the union of the complex hyperplanes $(z_0 = 0)$, $(z_0 - 2z_j = 0)$, j=1, ...,k. In the case of the identity permutation , we get the critical orbits $(z_0 = 0) \hookleftarrow$, $(z_0 - 2z_j = 0) \mapsto (z_j = 0) \mapsto (z_0 = z_j) \hookleftarrow$. Observe also that in general $(z_j = z_i) \hookleftarrow$. For a general h_σ, $(z_0 = 0) \mapsto (z_{\sigma(0)} = 0) \mapsto (z_{\sigma^2(0)} = z_{\sigma(0)}) \mapsto ... \mapsto (z_{\sigma^{l+1}(0)} = z_{\sigma^l(0)}) \mapsto ...$ which clearly becomes periodic. Similarly for $j \geq 1$, $(z_0 - 2z_j = 0) \mapsto (z_{\sigma(j)} = 0) \mapsto (z_{\sigma^2(j)} = z_{\sigma(0)}) \mapsto ...$ if $\sigma(j) \neq 0$ and otherwise continues as the preceeding one. In any case h_σ is critically finite.

Notice that from the description of the critical orbit it follows that V always contains the $2k + 1$ hyperplanes $(z_j = 0), j = 0, ..., k$ and $(z_0 - 2z_j = 0), j = 1, ..., k$. We can use coordinates so that $(z_0 = 0)$ is the hyperplane at ∞. That is, we set the homogeneous coordinate $z_0 = 1$ and consider the remaining $\mathbb{C}^k$ with coordinates $(z_1, ..., z_k)$. Then V contains the affine complex hyperplanes $z_j = 0, z_j = 1/2, j = 1, ..., k$. Hence $\mathbb{P}^k \setminus V$ and $\mathbb{P}^k \setminus W$ both are contained in a product of k copies of $\bar{\mathbb{C}} \setminus \{0, 1/2, \infty\}$. The universal covering of this set is the unit polydisc in $\mathbb{C}^k$ and is therefore Kobayashi hyperbolic. Since holomorphic maps are distance decreasing (also inclusion maps), any open subset is also hyperbolic. Recall also that covering maps are isometries.

As $h_\sigma : \mathbb{P}^k \setminus W \mapsto \mathbb{P}^k \setminus V$ is a covering map, we obtain that the infinitesimal Kobayashi metric F at points x and tangent vectors ξ satisfy

$$F_{\mathbb{P}^k \setminus V}(h_\sigma(x), h'_\sigma(x)(\xi)) = F_{\mathbb{P}^k \setminus W}(x, \xi) \geq F_{\mathbb{P}^k \setminus V}(x, \xi).$$

REMARK 2.5. *It is also possible to define*

$$h_\lambda : \mathbb{P}^k \longrightarrow \mathbb{P}^k$$
$$[z_0 : z_1 : ... : z_k] \mapsto [z_0^d : \lambda(z_0 - 2z_1)^d : ... : \lambda(z_0 - 2z_k)^d],$$

$d \geq 2$, *with* $(1 - 2\lambda)^d = 1$ *and let* $h_\sigma := \sigma \circ h_\lambda$. *The same result holds.*

We will make a detailed study of two special cases. For the rest of this paper, let f, g be the maps

$$f([z : w : t]) = [(z - 2w)^2 : z^2 : (z - 2t)^2]$$
$$g([z : w : t]) = [(z - 2w)^2 : (z - 2t)^2 : z^2]$$

For ease of reference we list the critical orbits of f and g.

PROPOSITION 2.6. *The critical orbits for f are*

$$(z = 0) \mapsto (w = 0) \mapsto (z = w) \hookleftarrow$$

$$(z = 2w) \mapsto (z = 0) \mapsto ...$$

$$(z = 2t) \mapsto (t = 0) \mapsto (t = w) \hookleftarrow (z = t)$$

The critical orbits for g are

$$(z = 0) \mapsto (t = 0) \mapsto (w = t) \mapsto (z = w) \mapsto (z = t) \mapsto (w = t) \mapsto ...$$

$$(z = 2w) \mapsto (z = 0) \mapsto ...$$

$$(z = 2t) \mapsto (w = 0) \mapsto (z = t) \mapsto ...$$

We also list for convenience the irreducible components of W that are not already in V.

PROPOSITION 2.7. *For the map f, $\overline{W \setminus V}$ consists of* $(z = -2(\sqrt{2}+1)w), (z = 2(\sqrt{2}-1)w), (z = w+t), (z = 2(1+\sqrt{2})(\sqrt{2}t-w))$ *and* $(z = 2(\sqrt{2}-1)(w+\sqrt{2}t))$.

PROPOSITION 2.8. *For the map g, $\overline{W \setminus V}$ consists of $(z = 2(\sqrt{2} + 1)(-w +$ $\sqrt{2}t))$, $(z = 2(\sqrt{2} - 1)(w + \sqrt{2}t))$, $(z = -2(\sqrt{2} + 1)w)$, $(z = 2(\sqrt{2} - 1)w)$ and $(z = w + t)$.*

Let $\{C_i\}$ be the irreducible components of the critical set. For some l_i, $n_i \geq 1$ $f^{l_i + n_i}(C_i) = f^{l_i}(C_i)$. Define $V'' := \cup_{i,k} f^{l_i + k}(C_i)$. Next we discuss the orbit of the points on the critical set C of f which are also in V''. These are of particular importance for the dynamics.

LEMMA 2.9. *The point $[0 : 0 : 1]$ is a superattractive fixed point. The other 5 points in $C \cap V''$ all are preperiodic to the repelling fixed point $[1 : 1 : 1]$.*

PROOF. The first part is immediate. We list the orbits of the other 5 points:

$$[0 : 1 : 1] \mapsto [1 : 0 : 1] \mapsto [1 : 1 : 1] \hookleftarrow$$

$$[1 : 1/2 : 1] \mapsto [0 : 1 : 1] \mapsto \ldots$$

$$[0 : 1 : 0] \mapsto [1 : 0 : 0] \mapsto [1 : 1 : 1] \hookleftarrow$$

$$[2 : 1 : 1] \mapsto [0 : 1 : 0] \mapsto \ldots$$

$$[2 : 2 : 1] \mapsto [1 : 1 : 0] \mapsto [1 : 1 : 1] \hookleftarrow .$$

Direct computations of the eigenvalues at the fixed point give the values $4, -4$ so this point is repelling.

In general we can define strictly preperiodic maps inductively in the following way:

DEFINITION 2.10. *Let $F : M^n \mapsto M^n$ be a finite holomorphic self map of a complex manifold M of dimension n. Denote by C the critical set. For simplicity we will assume that C consists of a finite number of nonsingular complex hypersurfaces C_i. We assume that each C_i is preperiodic, $C_i \mapsto F(C_i) \mapsto \ldots F^{l_i}(C_i) \mapsto \ldots F^{l_i + n_i}(C_i) = F^{l_i}(C_i)$, $n_i, l_i \geq 1$ and that all the $F^j(C_i)$ are nonsingular complex hypersurfaces. We say that F is strictly preperiodic if all the maps $F^{n_i} : F^{l_i + j}(C_i) \mapsto F^{l_i + j}(C_i)$ are strictly preperiodic. In dimension $n = 1$, F is strictly preperiodic if F is preperiodic.*

We will only consider the nontrivial cases when $C \neq \emptyset$. Since $[0 : 0 : 1]$ is a superattractive fixed point for f, (Lemma 2. 9), f is not strictly preperiodic. We will show that g is strictly preperiodic. This makes the dynamics of the two similar looking maps rather different as we shall see.

3. The Domain of Attraction of $[0 : 0 : 1]$

In the next few sections we will study the dynamics of the map f. After that we will discuss g. Recall that f has a superattractive fixed point at $p = [0 : 0 : 1]$.

We will next describe the basin of attraction of p.

The first remark is that the line $t = 0$ does not intersect the basin of attraction because the map f sends this to the forward invariant set $(t = w) \cup (t = z)$ which does not contain p. Hence we can use (z, w) as coordinates on the basin U of attraction of p. In these coordinates, we set $t = 1$ and then the map looks like

$$f(z, w) = \left(\frac{(z - 2w)^2}{(z - 2)^2}, \frac{z^2}{(z - 2)^2} \right).$$

From our description of V it is clear that U contains no point on the complex lines $(z = 1), (z = 2), (w = 1)$. Define $U_0 = \{|z| < 1/10, |w| < 1/10\}$. It is clear that $U_0 \subset U$. We define $U_n = f^{-n}(U_0)$. Then it is clear that $U_n \subset U_{n+1} \forall n$ and that $U = \bigcup_{n=0}^{n=\infty} U_n$.

THEOREM 3.1. *The domain U is a domain of holomorphy which is Kobayashi hyperbolic. Moreover for every complex line L through $(z, w) = (0, 0)$ in the $t = 1$ coordinates the intersection $U \cap L$ is biholomorphic to a disc.*

PROOF. Notice that the map $f : U_{n+1} \mapsto U_n$ is a 4 to 1 branched covering. It follows inductively that each U_n is a connected domain of holomorphy. Hence U is also a connected domain of holomorphy. We show next that each $U_n \cap L$ is simply connected. For $n = 0$ this is obvious.

Assume this is known for n and consider a complex line L and the intersection $U_{n+1} \cap L$. Assume at first that $L = \{z = \alpha w\}$, $\alpha \neq 0$. Then $(z, w) \in U_{n+1} \cap L$ if and only if $f(z, w) \in L' \cap U_n$ where $L' = \{(Z, W); Z = \frac{(\alpha - 2)^2}{\alpha^2} W\}$. Now the map $f' : \mathbb{C} \mapsto \mathbb{C}; w \mapsto \alpha^2 w^2 / (\alpha w - 2)^2$ is a branched covering of order 2, mapping the origin to itself and being branched there. Furthermore the map is branched at $w = 2/\alpha$. However this point is mapped after two iterations to the line $(z, 1)$ which is periodic and does not contain the superattractive fixed point. Hence $w = 2/\alpha$ is not in $U_n \cap L'$. Hence the inverse image of the simply connected region $U_n \cap L'$ is a simply connected region also. The argument for L being the $z-$ or $w-$ axis is the same. Since $U \cap L$ is a proper subset of L it follows that the sets $U \cap L$ are biholomorphic to the unit disc.

It remains to show that U is Kobayashi hyperbolic. Observe at first that the inverse image of the line $z = 1$ cannot intersect U. Hence, the complement of U contains the complex lines $(w = 1), (z = 1), (z = 2)$ and also $w = z - 1$. We can easily find a biholomorphic map between $\mathbb{C}^2 \setminus (z = 1) \cup (z = 2) \cup (w = 1) \cup (w = z - 1)$ and $(\mathbb{C} \setminus \{1, 2\})^2$. Hence U is hyperbolic.

4. The Complement J of the Basin of Attraction

We denote by J the complement of the basin of attraction U. In this section we will discuss the structure of J. It is convenient to introduce the sets $J_L := J \cap L$ and $J_{\bar{L}} := J \cap \bar{L}$ where L denotes any complex line through $(0, 0)$ in the (z, w) coordinates and $\bar{L}$ denotes it's closure in $\mathbb{P}^2$.

Notice that the point at infinity in $\bar{L}$ is always in $J_{\bar{L}}$. Let $X := (t = 0)$ be the line at ∞. So $X \subset J$. We have the following description of J:

THEOREM 4.1. *For every L, the set $J_{\bar{L}}$ is connected. The set J has no interior.*

PROOF. The connectedness of $J_{\bar{L}}$ follows immediately from the simply connectedness of $U \cap L$. Note at first that f maps the complex lines L to complex lines. And if we parametrize the lines as $L = \{z = \alpha w\}$ then this map on lines $\bar{f}$ maps the line with slope α to the line with slope $\frac{(\alpha-2)^2}{\alpha^2}$. This is a rational map on $\mathbb{P}^1$ whose critical points are preperiodic to repelling fixed points, hence from a theorem of Sullivan, ([Su]), the Julia set of this map is $\mathbb{P}^1$. In particular it has a fixed point 1 and the preimages of 1 are dense in $\mathbb{P}^1$.

Now we can consider the dynamics of f restricted to the line $\alpha = 1$, i.e. $z = w$. On that line the map f has the form $z \mapsto \frac{z^2}{(z-2)^2}$. Under the coordinate change $z \mapsto 4/z - 2$ this map becomes the well known map $z \mapsto z^2 - 2$. The Julia set of this map is the interval $[-2, 2]$ and the complement is mapped under iteration to ∞. In our coordinates that translates into the interval $J_{(z=w)} = [1, \infty)$. Also for every line L' such that $\bar{f}^n(L') = L$, we have that $J_{\bar{L}'} = f^{-n}(J_{\bar{L}}) \cap \bar{L}'$. Hence for all such lines the set $J_{\bar{L}'}$ has no relative interior in $\bar{L}'$. Since these lines as we observed are dense in all lines, it follows that J has no interior.

Observe that from this Theorem it immediately follows that J is the Julia set of F in the classical sense that J consists of those points where the iterates of f do not constitute a normal family on any neighborhood.

PROPOSITION 4.2. *Let $X = \{t = 0\}$. Define $Y := \overline{\bigcup_{n=-\infty}^{n=2} f^n(X)}$. Then $Y = J$.*

PROOF. Note that Y contains $(t = 0), (w = t), (z = t), (z = 2t)$ and $(z = w + 1)$. Hence the complement of Y is hyperbolic. This was already used above to prove that U is hyperbolic. We have of course that $Y \subset J$ so we only need to consider the opposite inclusion. Let $\Omega := \mathbb{P}^2 \setminus Y$. Then $f : \Omega \mapsto \Omega$ and since Ω is hyperbolic $\{f^n\}$ is a normal family. Hence the iterates converge to the only possible candidate, the superattractive fixed point p. But then Ω is contained in the basin of attraction of p, which provides the other inclusion. Notice that we used implicitly that J has no interior.

REMARK 4.3. *It is interesting to observe that every critical point of*

$$f^2([z : 1 : 1]) = [((z-2)^2 - 2z^2)^2 : (z-2)^2 : (z-2)^2]$$

is preperiodic. So the Julia set of $f^2|(z = t)$ is $(z = t)$.

5. The Green's Function on Slices

For any complex line L through $[0 : 0 : 1]$, let G_L denote the Green's function of $J_{\bar{L}}$ in L with pole at $[0 : 0 : 1]$ and asymptotic behavior $-log\|(z, w)\|$. Since $J_{\bar{L}}$ is connected and (not a point) , the Green's function is a continuous nonnegative extended real valued function with value 0 precisely on $J_{\bar{L}}$.

We will denote by G also the function which agrees on $\mathbb{P}^2$ with $G_{\bar{L}}$ on each line L.

THEOREM 5.1. *The function G is continuous. The basin of attraction U is a topological cell.*

We prove at first two lemmas.

Let $U_L := U \cap L$ and $Y_{\bar{L}} := \overline{\bigcup_{m,n \geq 0} \{f^{-m} f^n (t = 0) \cap \bar{L}\}}$.

LEMMA 5.2. *We have the inclusions $\partial U_L \subset Y_{\bar{L}} \subset J_{\bar{L}}$.*

PROOF. Notice that f maps $Y_{\bar{L}}$ to $Y_{\bar{L}'}$ and $\bar{L} \setminus Y_{\bar{L}}$ to $\bar{L}' \setminus Y_{\bar{L}'}$ whenever $\bar{f} : L \mapsto L'$. Since $\bigcup_{n \in \mathbb{Z}} \{f^n (t = 0)\}$ contains the line at ∞, $(t = 0)$, $z = 1, z = 2, w = 1, z = w + 1$, it follows that the family $f^n | L \setminus Y_{\bar{L}}, n = 1, 2, \ldots$ is a normal family. Hence it follows that the component of $L \setminus Y_{\bar{L}}$ containing $[0 : 0 : 1]$ is contained in U_L. This implies the left inequality. The right inclusion is immediate.

For $(z, w) \in L, \|(z, w)\| = 1$, let $G_{[z:w]}(\tau) := G(z\tau, w\tau)$. Then we can write the asymptotic expansion of $G_{[z:w]}(\tau) = -\log|\tau| + \gamma([z : w]) + o(1)$ where $\gamma([z : w])$ is the logarithmic capacity of $J_{\bar{L}}$ in $\bar{L} \setminus (0,0)$, if we let L be the complex line through (z, w). Obviously γ depends only on L and not on the particular choice of $(z, w) \in L$.

LEMMA 5.3. *The function $L \mapsto \gamma(L)$ is continuous.*

PROOF. Since J is closed, given any neighborhood Ω of $J_{\bar{L}}$, then Ω contains $J_{L'}$ for all L' close to L. So if $\epsilon > 0$,

$$G_{[z:w]}(\tau) - \epsilon < G_{[z':w']}(\tau)$$

if $(z', w'), \|(z', w')\| = 1$ is close enough to (z, w), since $G_{[z:w]} - \epsilon$ is a competitor for $G_{[z':w']}$. Hence the map $[z : w : t] \mapsto G([z : w : t])$ is lower semicontinuous. And also the map $L \mapsto \gamma(L)$ is lower semicontinuous.

For the reverse direction we will use the concept of logarithmic capacity. By definition, ([Ts]), the capacity of $J_{\bar{L}}$ (in the $1/\tau$ coordinates) is given by $cap(J_{\bar{L}}) := e^{-\gamma([z:w])}$. Since $J_{\bar{L}}$ is connected and since $J_{\bar{L}}$ contains the intersections of L with $(z = 1), (z = 2), (w = 1), (z = w + 1)$, it follows that ([Ts])

$$1/C \leq diam J_{\bar{L}} \leq 4 cap(J_{\bar{L}}) \leq C (*)$$

for a fixed constant C.

Assume that $L_n \to L$. Taking a subsequence, we may assume that $J_{\bar{L}_n} \to X$, for a compact set X. Clearly $X \subset J_{\bar{L}}$ and $\bigcup_{m,n \geq 0} \{f^{-m} f^n (t = 0) \cap \bar{L} \subset X$. Hence $Y_{\bar{L}} \subset X$. From Lemma 5.2 it follows that $\partial U_{\bar{L}} \subset X$.

Let h_n be the conformal representation of U_{L_n} as a disc of radius $e^{-\gamma(L_n)}$ around zero sending the origin to the origin and with derivative one. By $(*)$ it follows that the sequence h_n is uniformly bounded, so if h is some limit, then h is defined on U_L and is injective there. But we can also take limits of the inverses. This shows that there is a conformal map of U_L to the disc of radius $limsup \ e^{-\gamma(L_n)}$ with derivative one. Hence γ is upper semicontinuous as well. Therefore G and γ are both continuous. This finishes the proof of the Lemma.

To finish the proof of the theorem we need to prove that U is a topological cell. We will construct a homeomorphism between U and the unit ball $\mathbb{B}$ in $\mathbb{C}^2$. Let h_L denote the conformal representation of U_L to the disc of radius $e^{-\gamma(L)}$ in the line L sending 0 to 0 with derivative 1 at 0. Define

$$\phi : U \mapsto \mathbb{B}, \phi|L = h_L e^{\gamma(L)}.$$

Then ϕ is clearly continuous and bijective and proper. Hence ϕ is a homeomorphism.

REMARK 5.4. *The open set U is hyperbolic but is not biholomorphic to a bounded symmetric domain.*

Indeed, $f : U \mapsto U$ is a proper holomorphic map, $f(p) = p, f'(p) = 0$. If $U \simeq \Delta^2$, f would be conjugate to the map $(z, w) \mapsto (z^2, w^2)$ which is easy to rule out from the orbit of the critical set. It is not biholomorphic to the ball since $\mathbb{B}^2$ has no proper holomorphic selfmaps which are not biholomorphic. Similarly one proves that U is not biholomorphic to a product of two domains.

6. Topological transitivity of f

In this section we start investigating the dynamics of f on J. In the next section we construct an ergodic measure. Here we prove the map is toplogically transitive.

At first, let $Y := \bigcup_{m,n\geq 0}\{f^{-m}f^n(t = 0)\}$. Then Y is a completely invariant subset of J and the complement of $\bar{Y}$ in $\mathbb{P}^2$ is hyperbolic. Hence:

LEMMA 6.1. *The Julia set J is the closure of Y.*

Recall that if $R : \mathbb{P}^1 \mapsto \mathbb{P}^1$ is critically finite then the only Fatou components are preperiodic to the superattractives ones.

Note that the map $\bar{f}$ has a dense collection of repelling orbits. For each line L on a repelling orbit, except possibly finitely many lines contained in the critical orbit, the critical points in the line except the fixed point $[0 : 0 : 1]$ is preperiodic. Hence the Julia sets of the map restricted to the line L has connected complement and hence the repelling points are dense in $J_{\bar{L}}$. Therefore these sets $J_{\bar{L}}$ have no interior, so by Lemma 5.2, $Y \cap \bar{L}$ is dense in $J_{\bar{L}}$.

Let $\{L_n\}$ denote these lines. Then, since each irreducible branch of Y intersects these lines in a dense set, it follows that:

LEMMA 6.2. *The Julia set is the closure of the set $\cup J_{\bar{L}_n}$.*

Now we can prove the topological transitivity of f.

THEOREM 6.3. *The map f is topologically transitive on J. Repelling periodic points are dense in J.*

PROOF. Recall that f is topologically transitive on S if for every pair of nonempty relatively open sets ω, ω' in S there exists some positive integer k

such that $f^k(\omega) \cap \omega' \neq \emptyset$. So pick ω, ω'. Let $\Omega = \cup_{n \geq 0} f^n(\omega)$. By the above lemma, ω must intersect some $J_{\bar{L}_n}$. Hence the union of the forward orbits of ω must contain all of $J_{\bar{L}_n}$ and hence $J_{\bar{L}}$ for all L in the space of lines in a neighborhood of L_n. Since the critical points of $\bar{f}$ are all preperiodic, and $\bar{f}$ is not exceptional, it follows that the forward orbit of any open set of lines covers all lines. This implies that Ω contains J, so must intersect ω'.

The fact that the repelling points are dense was already proved in the previous lemma.

REMARK 6.4. *The proof shows that the forward orbit of any open set that intersects J covers $\mathbb{P}^2 \setminus p$, and covers p if and only if p already belongs to the open set.*

COROLLARY 6.5. *Let $D := \{x \in J; \{f^n(x)\}$ is dense in $J\}$. Then D is a dense G_δ set.*

PROOF. For any relatively open set W in J, let $\omega := \bigcup_{n \geq 1} f^{-n}(W)$. By the topological transitivity of f, ω is dense. Let W_k be a basis for the topology, then $D = \bigcap_{k=1}^{\infty} (\bigcup_{n=1}^{\infty} f^{-n}(W_k))$, so is a dense G_δ.

7. An Invariant Ergodic Measure

Let ν be an ergodic invariant measure on the space of lines L, $\mathbb{P}^1$, for $\bar{f}$. Such a measure can have as support all of $\mathbb{P}^1$, ([Ly]), ([FLM]). For every L, let $\mu_L := \Delta_\tau G_L(1/\tau)$, $\tau \in \mathbb{C}$. The Laplacian is with respect to τ. Then μ_L is the harmonic measure on ∂U_L. Define the measure $\tilde{\nu}$: If $\phi \in \mathcal{C}_0^\infty(\mathbb{P}^2 \setminus [0:0:1])$, then

$$\int \phi d\tilde{\nu} := \int_{[z:w] \in \mathbb{P}^1, \, \|(z,w)\|=1} \int_{\tau \in \mathbb{C}} \phi([z:w:\tau]) d\mu_{[z:w]} \, d\nu([z:w]) =$$

$$\int_{[z:w] \in \mathbb{P}^1, \, \|(z,w)\|=1} \int_{\tau \in \mathbb{C}} G([z:w:\tau]) \Delta_\tau \phi([z:w:\tau]) d\lambda(\tau) \, d\nu([z:w]).$$

The definition makes sense since G is a continuous function and $\int_\tau G \Delta \phi$ is continuous in L.

THEOREM 7.1. *The measure $\tilde{\nu}$ is invariant, ergodic and supported precisely on J.*

PROOF. The support question is clear because of the continuity of $\int_\tau G \Delta \phi$. Next we prove the invariance. This means that we will show that the measure of a set is the same as the measure of the inverse of the set.

The Green function satisfies the functional equality $G[\frac{(z-2w)^2}{(z-2)^2} : \frac{z^2}{(z-2)^2} : 1] = 2G[z:w:1]$. Let $\phi \in \mathcal{C}_0^\infty(\mathbb{P}^2 \setminus \{p\})$. Fix a line $L = [z:w]$, $\|(z,w)\| = 1$. We set $L' := \bar{f}(L)$. For $d := \|((z-2)^2, z^2)\|$ we define $[z':w':\tau'] := [(z-2w)^2/d : z^2/d : (z-2\tau)^2/d]$ and obtain $L' = [z':w']$. Integrating we obtain:

$$\int \phi \circ f d\mu_L = \int \phi \circ f([z:w:\tau])d\mu_{[z:w]}(\tau)$$

$$= \int G([z:w:\tau])\Delta_\tau(\phi \circ f)[z:w:\tau]d\lambda(\tau)$$

$$= 1/2 \int (G \circ f)([z:w:\tau])\Delta_\tau(\phi \circ f)[z:w:\tau]d\lambda(\tau)$$

$$= \int G([z':w':\tau'])\Delta_{\tau'}\phi[z':w':\tau']d\lambda(\tau')$$

$$= \int \phi d\mu_{L'}$$

In the second to the last equality we used the fact that f is a $2-1$ cover from L to L'. Now we integrate with respect to ν to obtain:

$$\int \phi \circ f d\widetilde{\nu} = \int_{[z:w]\in\mathbb{P}^1,\ \|(z,w)\|=1} \left(\int_{\tau\in\mathbb{C}} \phi \circ f d\mu_{[z:w]}(\tau) \right) d\nu([z:w])$$

$$= \int_{[z:w]\in\mathbb{P}^1,\ \|(z,w)\|=1} \left(\int_{\tau'\in\mathbb{C}} \phi d\mu_{[z':w']}(\tau') \right) d\nu([z:w])$$

$$= \int_{[z':w']\in\mathbb{P}^1,\ \|(z',w')\|=1} \left(\int_{\tau'\in\mathbb{C}} \phi d\mu_{[z':w']}(\tau') \right) d\nu([z':w'])$$

$$= \int \phi d\widetilde{\nu}$$

This proves the invariance of $\widetilde{\nu}$.

It remains to prove that $\widetilde{\nu}$ is ergodic. Suppose that E is a measurable forwards and backwards invariant set which has strictly positive $\widetilde{\nu}$ measure, but strictly less than full $\widetilde{\nu}$ measure. That is, we assume that the measure is not ergodic. We will show that this leads to a contradiction. Let χ denote the characteristic function of E. Since χ can be approximated by $\mathcal{C}_0^\infty$ functions, we obtain from the above that $\int \chi \circ f d\mu_L = \int \chi d\mu_{L'}$ for every line $L' = \bar{f}(L)$.

We define for each line $L = [z:w], \|(z,w)\| = 1$,

$$\psi(L) = \int_{\tau\in\mathbb{C}} \chi([z:w:\tau])d\mu_L(\tau)$$

$$= \int_{\tau\in\mathbb{C}} \chi \circ f([z:w:\tau])d\mu_L(\tau)$$

$$= \int_{\tau'\in\mathbb{C}} \chi([z':w':\tau'])d\mu_{L'}(\tau')$$

$$= \psi(L')$$

Hence ψ is an invariant function on $\mathbb{P}^1$. Since ν is ergodic, it follows that there exists a strictly positive constant c such that $\psi(L) = c$ $a.e.d\nu$. Since $d\mu(J_{\bar{L}}) = 1 \forall L$, our assumption of nonergodicity implies that $c < 1$. We will show that this leads to a contradiction.

For a line $L' = [z' : w']$, $\|(z', w')\| = 1$, let $L_j = [z_j : w_j]$, $\|(z_j, w_j)\| = 1$, $j= 1, 2$ be the preimages under $\bar{f}$. Also, for $\tau' \in \mathbb{C}$, let $\tau_{j,i}$ give the preimages of $[z' : w' : \tau']$, $[z_j : w_j : \tau_{j,i}]$ under f. If ϕ is in $\mathcal{C}_0^\infty$, we define the push forwards by $\phi_j^+([z' : w' : \tau']) = 1/2[\sum_i \phi([z_j : w_j : \tau_{j,i}])]$.

LEMMA 7.2. $\int \phi([z_j : w_j : \tau_j]) d\mu_{L_j}(\tau_j) = \int \phi_j^+([z' : w' : \tau']) d\mu_{L'}(\tau')$

PROOF. Computing, we get

$$\int \phi([z_j : w_j : \tau_j]) d\mu_{L_j}(\tau_j) = \int \Delta_{\tau_j} \phi([z_j : w_j : \tau_j]) G([z_j : w_j : \tau_j]) d\lambda(\tau_j)$$

$$= 1/2 \int \Delta_{\tau_j} \phi([z_j : w_j : \tau_j]) G([z' : w' : \tau']) d\lambda(\tau_j)$$

$$= 1/2 \int \phi \Delta_{\tau'} G([z' : w' : \tau']) |\frac{\partial \tau'}{\partial \tau_j}|^2 d\lambda(\tau_j)$$

$$= 1/2 \sum_i \int \phi([z_j : w_j : \tau_{j,i}]) \Delta_{\tau'} G d\lambda(\tau')$$

$$= \int \phi_j^+([z' : w' : \tau']) d\mu_{L'}(\tau')$$

We define the measure σ_L by $d\sigma_L(\tau) := \chi d\mu_L(\tau)$. We then have a similar lemma as above

LEMMA 7.3. $\int \phi([z_j : w_j : \tau_j]) d\sigma_{L_j}(\tau_j) = \int \phi_j^+([z' : w' : \tau']) d\sigma_{L'}(\tau')$

PROOF. Assume that $\{\psi_n\} \to \chi$. Applying the above lemma we obtain that

$$\int \phi([z_j : w_j : \tau_j]) d\sigma_{L_j}(\tau_j) =_{a.e.} \int \phi([z_j : w_j : \tau_j]) \chi([z_j : w_j : \tau_j]) d\sigma_{L_j}(\tau_j)$$

$$=_{a.e.} \lim_{n \to \infty} \int \phi \psi_n([z_j : w_j : \tau_j]) d\sigma_{L_j}(\tau_j)$$

$$=_{a.e.} \lim_{n \to \infty} \int (\phi \psi_n)_j^+([z' : w' : \tau']) d\mu_{L'}(\tau')$$

$$=_{a.e.} \int (\phi \chi)_j^+([z' : w' : \tau']) d\mu_{L'}(\tau')$$

$$=_{a.e.} \int \phi_j^+([z' : w' : \tau']) \chi([z' : w' : \tau']) d\mu_{L'}(\tau')$$

$$=_{a.e.} \int \phi_j^+([z' : w' : \tau']) d\sigma_{L'}(\tau')$$

With this invariance proved we proceed to prove that $\sigma_L = \mu_L$. First we discuss the harmonic potential $H[z : w : \tau]$ of σ_L/c. Then we will show that this is almost everywhere equal to G.

LEMMA 7.4. *The function H satisfies the functional equation $H \circ f = 2H + h(L)$ for some function h for almost every L.*

PROOF. We show at first that $\Delta(H \circ f) = 2\Delta H$.

$$\int \Delta H([z_j : w_j : \tau_j])\phi([z_j : w_j : \tau_j])d\lambda(\tau_j)$$

$$=_{a.e.} \int \phi([z_j : w_j : \tau_j])d\sigma(\tau_j)/c$$

$$=_{a.e.} \int \phi_j^+([z' : w' : \tau'])d\sigma(\tau')/c$$

$$=_{a.e.} \int \Delta H\phi_j^+ d\lambda(\tau')$$

$$=_{a.e.} 1/2 \int \Delta(H \circ f)\phi d\lambda(\tau_j)$$

Since ϕ is arbitrary it follows that $\Delta(H \circ f) = 2\Delta H$. Since asymptotically, $H \circ f = -2log|\tau| + ...$ and $H = -log|\tau| + ...$, the lemma follows.

LEMMA 7.5. *The function $H^\sim := H + \gamma(L)$ satisfies the functional equation $H^\sim \circ f = 2H^\sim$*

PROOF. Recall that $G([z : w : \tau]) = -log|\tau| + \gamma(L) + o(1)$. Hence it follows from the functional equation for G that $\gamma(L') = 2\gamma(L) - log(4/\|(z - 2w)^2, z^2\|)$. Using the formula $H = \int log|\tau^{-1} - \zeta|d\sigma(\zeta)$ we obtain from the previous Lemma that $H(L) = -log(4/\|(z - 2w)^2, z^2\|)$. Hence the lemma follows.

$H^\sim$ is bounded above on J since $H^\sim \leq G$, it follows that $H^\sim \leq 0$ on J. We next prove that $H^\sim = 0$ on J. From the functional equation it follows that it suffices to prove that $H^\sim$ is bounded below uniformly. We choose a fixed positive constant M such that $log(|\tau^{-1} - \zeta|/M) \leq 0$ on each line L whenever τ, ζ are in $J_{\bar{L}}$. Observe that $\int log|\tau^{-1} - \zeta|d\mu_L(\zeta) \geq -\gamma(L)$. We obtain from the negativity of the integrand that

$$H^\sim(\tau) = \int log|\tau^{-1} - \zeta|d\sigma(\zeta)/c$$

$$= \int (logM + log|\tau^{-1} - \zeta|/M)\chi(\zeta)/cd\mu(\zeta)$$

$$\geq logM + \int log|\tau^{-1} - \zeta|/M d\mu(\zeta)/c$$

$$= -\gamma(L)/c + logM(1 - 1/c)$$

$$\geq M'$$

$$> -\infty.$$

It follows that $H^\sim = G$ so $\Delta_\tau H^\sim = \Delta_\tau G$, hence $c = 1$.

8. The Dynamics of g

We now consider the map g.

THEOREM 8.1. *The map g has no periodic attractive cycle.*

PROOF. Recall that $\mathbb{P}^2 \backslash V$ is hyperbolic and that the map $g : \mathbb{P}^2 \backslash W \mapsto \mathbb{P}^2 \backslash V$ is an expanding map in the Kobayashi metric on $\mathbb{P}^2 \setminus V$. Hence any periodic attractive cycle must be contained in V. From the description of V it follows that any attractive cycle is in the cycle $(w = t) \mapsto (z = w) \mapsto (z = t) \mapsto (w = t) \mapsto \dots$. Hence $g^3|(w = t)$ must have an attractive cycle. Computing we obtain on that line that

$$g^3(z) = \left(\frac{R(z)}{S(z)} \right)^2, \quad R(z) = (z-2)^4 - 2(z^2 + 4z - 4)^2, \quad S(z) = (z-2)^4.$$

Observe at first that $2 \mapsto \infty \mapsto 1 \hookleftarrow$ and that 1 is a repelling fixed point. We will show that all critical points are (strictly) preperiodic. This implies that all periodic orbits are repelling. First consider the zeroes of R. Since $g^3(0) = 1$, these critical points are preperiodic. It remains to consider the critical points of $R/S = 1 - 2(P/Q)^2$, $P = z^2 + 4z - 4$, $Q = (z-2)^2$. First we look at the zeroes of P. Since g^3 maps these points to 1, these points are again preperiodic. Finally consider the points where $(P/Q)' = -8z/(z-2)^3 = 0$. So $z = 0$ which also is preperiodic.

The expansive property of the Kobayashi metric also gives:

THEOREM 8.2. *For periodic cycles not in $(z = w) \cup (z = t) \cup (w = t)$, both eigenvalues of the orbit are ≥ 1.*

The map g is critically finite in the sense that the forward image of the critical set is contained in a periodic variety. However, because of dimension reasons, the critical set still intersects these periodic varieties in finitely many points. We say that the map g is strictly preperiodic if these finitely many points are themselves strictly preperiodic.

PROPOSITION 8.3. *The map g is strictly preperiodic.*

PROOF. This is done by straightforward computations of the intersection points and their orbits.

$$[0 : 0 : 1] \mapsto [0 : 1 : 0] \mapsto [1 : 0 : 0] \mapsto [1 : 1 : 1] \hookleftarrow$$
$$[0 : 1 : 1] \mapsto [1 : 1 : 0] \mapsto [1 : 1 : 1] \hookleftarrow$$
$$[1 : 1/2 : 1] \mapsto [0 : 1 : 1] \mapsto \dots$$
$$[2 : 1 : 1] \mapsto [0 : 0 : 1] \mapsto \dots$$
$$[2 : 2 : 1] \mapsto [1 : 0 : 1] \mapsto [1 : 1 : 1] \hookleftarrow.$$

So all the intersections points are mapped finally to the point $[1 : 1 : 1]$. This is a fixed point for g and a direct computation shows that both eigenvalues for the derivative of g are strictly larger than 1 at that point. So this point is a repelling fixed point.

We show next that the map g is chaotic. Recall that the Julia set of a map is the complement of the largest open set where the iterates $\{g^n\}$ is a normal family.

THEOREM 8.4. *The Julia set J for g is all of $\mathbb{P}^2$.*

PROOF. We compute at first the map g^3 restricted to the line $(w = t)$ which is mapped to itself. We know that this is a rational map of $\mathbb{P}^1$ to itself which is strictly preperiodic. Hence the Julia set of this map is all of $\mathbb{P}^1$ ([Su]).

Assume next that there is a nonempty open set U on which g^n is a normal family. Clearly this implies that $g^n(U) \cap V = \emptyset$ for all n. Here V denotes the critical set together with it's orbit. So in fact $g^n(U) \cap W = \emptyset$, $W := g^{-1}(V)$ also.

We show at first that $g^n \to V$ uniformly on compact subsets of U. Suppose not, then there exists a subsequence n_k, a nonempty open subset $U' \subset\subset U$ and an open set N containing W such that $g^{n_k}(U') \cap N = \emptyset$ for all k.

Taking a thinner subsequence if necessary, we might even assume that the sequence converges to a limit map h. Since g^{n_k} must be noncontracting in the Kobayashi volume form of $\mathbb{P}^2 \setminus W$, it follows that h is nondegenerate. Hence we can assume that $g^{n_{k+1} - n_k}$ converges to the identity map on the image of h. This then follows on the whole component Ω of the Fatou set containing the image of h. Also it follows that the domain Ω is periodic for g.

Let l be the smallest period of g on Ω. Then it follows that the map g^l is a biholomorphic map of Ω to itself. Hence g^l is an isometry in the Kobayashi volume form in Ω.

Notice that g^{kl} is a covering map of $\mathbb{P}^2 \setminus g^{-kl}(V)$ to $\mathbb{P}^2 \setminus V$. Hence it is an isometry in the Kobayashi volume form of the respective domains. Since the first set is contained in the second it follows that the map is noncontracting using the Kobayashi volume form of the first for both image and domain for all points in Ω. But since again the iterates converge to the identity, it must be an isometry. This has the remarkable consequence that for points in Ω, the Kobayashi volume form of $\mathbb{P}^2 \setminus g^{-kl}(V)$ is identical to the Kobayashi volume form of $\mathbb{P}^2 \setminus V$. Namely, for any point q in Ω, if the Kobayashi volume form at q with respect to $\mathbb{P}^2 \setminus g^{-kl}(V)$ is strictly larger than the Kobayashi volume form of $\mathbb{P}^2 \setminus V$, then the iterates cannot converge to the identity at q. Using a normal families argument it follows that on Ω, the Kobayashi volume form of $\mathbb{P}^2 \setminus V$ agrees with the Kobayashi volume form on Ω. This is impossible because we will prove that the latter must blow up at some boundary point of Ω which is an interior point of $\mathbb{P}^2 \setminus V$: Pick a boundary point of Ω in $\mathbb{P}^2 \setminus V$.

We will consider two cases. There must be a regular point p of some $W_n := g^{-n}(V)$ arbitrarily close. Suppose at first that there is a ball centered at p of positive radius contained in $\mathbb{P}^2 \setminus \Omega$. We may assume that some boundary point of this ball also is a boundary point of Ω. Then it suffices by Schwarz' lemma and monotonicity of the Volume form to show that if U is the open set $\mathbb{B}(0; 2) \setminus \overline{\mathbb{B}}(0; 1)$

in $\mathbb{C}^2$, then it's volume form blows up when you approach the boundary of the inner ball. To prove this consider a holomorphic map (f, g) from the unit ball in $\mathbb{C}^2$ to U which maps the origin to the point $(1 + \delta, 0)$. We will show that the Jacobian of (f, g) at the origin must go to zero when δ goes to zero. To see this, we can assume at first that $(\partial g/\partial z) = 0$ at the origin. Then the one variable disc $z \mapsto (f(z, 0), g(z, 0))$ is a competitor for the Kobayashi infinitesimal metric on U in the z direction.

The second case is when there are points in Ω arbitrarily close to p. In that case we can reduce the situation to asking whether the Kobayashi volume form blows up on the domain $\Delta^2 \setminus \{(z = 0)\}$ at points of the form $(\delta, 0)$ when $\delta \mapsto 0$. This follows by the same argument as for the above case, using that the infinitesimal Kobayashi metric blows up at the origin for the punctured disc ([L]).

Hence we know that g^n converges uniformly on compact subsets of U to V. This implies that in fact g^n converges to $V' := (z = w) \cup (z = t) \cup (w = t)$. There are finitely many points in $\overline{W \setminus V'} \cap (w = t)$. It is easy to see that one can remove a tiny disc about each of these points in $(w = t)$ and ensure that there is a subsequence g^{n_k} converging uniformly on a small ball B in U to the complement of these balls in $(w = t)$. Since these maps are nondecreasing for the Kobayashi metric on $\mathbb{P}^2 \setminus W$, we can show by considering vectors that are mapped by the derivative to vectors parallell to the $z-$ axis that an even thinner subsequence has a nonconstant limit. Hence we can find a disc in B and on that a limit map of the form $\tau \mapsto [f(\tau) : 1 : 1]$ where f is a function with a nonzero derivative. But then the image of this map must contain a repelling orbit of g^3 on $(w = t)$. Hence considering iterates of the form $g^{n_k + 3r}$ for arbitrarily large r, we produce mappings with arbitrarily large derivatives, contradicting the normality of the family g^n.

References.

[B] Bielefeld, B, *Conformal Dynamics Problem List*, Preprint 1, SUNY Stony Brook.

[Ca] Carleson, L, *Complex dynamics*, preprint UCLA.

[FLM] Freire, Lopes, Mane, *An invariant Measure for Rational Maps*, Bol. Soc. Bras. Mat. **6** (1983), 45–62.

[HP] Hubbard, J. H, Papadopol, P, *Superattractive fixed points in $\mathbb{C}^n$*, preprint 1991.

[K] Kobayashi, S, *Hyperbolic manifolds and Holomorphic Mappings*, Marcel Dekker, New York, 1970.

[L] Lang, S, *Introduction to complex hyperbolic spaces*, Springer Verlag, New York, 1987.

[Ly] Lyubich, M, *Entropy properties of rational endomorphisms of the Riemann sphere*, Ergodic Theory and Dynamical Systems **3** (1983), 351–385.

[Mi] Milnor, J, *Notes on complex dynamics*, Preprint MSI, Suny Stony Brook.

[Re] Rees, M, *Positive measure sets of ergodic rational maps*, Ann. Scient. Ec. Norm. Sup. **19** (1986), 383–407.

[Su] Sullivan, D, *Quasiconformal homeomorphisms and dynamics I*, Ann. Math. **122** (1985), 401–418.

[Th] Thurston, W, *On the combinatorics and dynamics of iterated rational maps*, Preprint.

[Ts] Tsuji, *Potential theory in Modern function theory*, Mazuren, Tokyo, 1959.

MATHEMATICS DEPARTMENT, THE UNIVERSITY OF MICHIGAN, ANN ARBOR, MICHIGAN 48109

UNIVERSITE PARIS SUD, BAT. 425. MATHEMATIQUES, 91405 ORSAY, FRANCE

Contemporary Mathematics
Volume **137**, 1992

A Disc in the Ball Approaching the
Boundary Non-Nontangentially

JOSIP GLOBEVNIK

Dedicated to Walter Rudin on the occasion of his retirement.

Introduction and the main result.

Let Δ be the open unit disc in C and let B_N be the open unit ball in C^N. Let $f: \Delta \to B_N$ be a proper holomorphic map. Then $f^*(\zeta) = \lim_{t \nearrow 1} f(t\zeta)$ exists for almost all $\zeta \in b\Delta$. E.Poletsky has asked whether there are points $\zeta \in b\Delta$ such that $f(t\zeta)$ approaches $f^*(\zeta)$ nontangentially as t increases to 1. It is known that this is the case if f is sufficiently regular near $b\Delta$, for instance, if $f^*(\zeta)$ exists and if $t \mapsto f'(t\zeta)$ is bounded on $(0,1)$ [GS1]. In the present paper we show that this is not the case in general and thus enlarge the list of examples of bad boundary behavior of proper holomorphic maps of discs [G1, G2, G3, G4, GS1, GS2, J].

If $D \subset \Delta$ is a convex domain whose closure contains 1 and which satisfies $\overline{D} \setminus \{1\} \subset \Delta$ we call D an approach region for the point 1. We then define $D(\zeta) = \zeta D$ ($\zeta \in b\Delta$). We also need conical approach regions in B_N: given $x \in bB_N$ and R, $0 < R < 1$, we denote by $K(x, R)$ the interior of the convex hull of $RB_N \cup \{x\}$. Clearly $K(x, R) \subset B_N$ is a conical approach region for the point x.

THEOREM. *Let $N \geq 2$ and let $D_n \subset \Delta$ be an increasing sequence of approach regions for the point 1. There is a nonconstant continuous map $f: \overline{\Delta} \to \overline{B_N}$, holomorphic on Δ which satisfies $f(b\Delta) \subset bB_N$ and which has the following property: If $p: [0,1] \to \overline{\Delta}$ is a path, $p(1) \in b\Delta$, such that $p([0,1)) \subset D_n(p(1))$ for some $n \in N$ then there is no $R < 1$ such that $f(p([0,1))) \subset K(f(p(1)), R)$.*

1991 *Mathematics Subject Classification*. Primary 32A40. Secondary 30B30.

This work was supported in part by a grant from the Ministry of research and technology of the Republic of Slovenia.

This paper is in final form and no version of it will be submitted for publication elsewhere.

COROLLARY. *Let $N \geq 2$. There is a nonconstant continuous map $f: \overline{\Delta} \to \overline{B_N}$, holomorphic on Δ which satisfies $f(b\Delta) \subset bB_N$ and is such that for no $\zeta \in b\Delta$ approaches $f(t\zeta)$ the boundary bB_N nontangentially as t increases to 1.*

Our result is in the same general spirit as a result of Rudin [R] who proved that there is an inner function F on Δ such that for almost all $\zeta \in b\Delta$, $F(t\zeta)$ approaches $F^*(\zeta)$ non-nontangentially as t increases to 1.

Note that it is enough to prove the theorem in the case $N = 2$.

2. Two lemmas

Denote the Hermitian inner product in C^2 by $\langle \; | \; \rangle$ and for each $y \in C^2 \setminus \{0\}$ write $H(y) = \{z \in C^2 : \Re\langle z \mid y \rangle = |y|^2\}$. Thus, $H(y)$ is the affine real hyperplane tangent to the sphere $b(|y|B_2)$ at y.

LEMMA 1. *Let $0 < R < r < 1$. There are $\delta > 0$, $\lambda > 0$ and r_1, $r < r_1 < 1$, such that if $x \in bB_2$, if $z \in K(x, R)$, $|z| > r_1$, and if $w \in H(y)$, $|w| > r_1$ for some $y \in [K(x, R) + \delta B_2] \cap b(rB_2) + \delta B_2$, then $|z - w| > \lambda$.*

PROOF. Let $y \in b(rB_2) \cap \overline{K(x, R)}$. There is a segment joining a point in $R\overline{B_2}$ with x that intersects $b(rB_2)$ at y transversely and thus intersects H(y) transversely at y. Since $x \neq y$ it follows that $x \notin H(y)$. One completes the proof by using the compactness of the set of hyperplanes $\{H(y): y \in b(rB_2) \cap \overline{K(x, R)}\}$.

LEMMA 2. *[G1, p.145] Let $p: C \to C^2$ be a polynomial whose first component has no zero on $b\Delta$. There is a polynomial $q: C \to C^2$ such that q has no zero on $b\Delta$ and such that $\langle p(\zeta) \mid q(\zeta) \rangle = 0$ $(\zeta \in b\Delta)$.*

3. Proof of the theorem, Part 1.

Let R_n be an increasing sequence of positive numbers converging to 1. We construct inductively
- increasing sequences r_n, r'_n of positive numbers converging to 1 and satisfying $r_n < r'_n < r_{n+1}$ for each n
- increasing sequences ρ_n, ρ'_n of positive numbers converging to 1 and satisfying $\rho'_n < \rho_n < \rho'_{n+1}$ for each n
- decreasing sequences λ_n, δ_n, ε_n of positive numbers converging to 0
- a sequence p_n of polynomials from C to C^2
- a sequence f_n of continuous maps from $\overline{\Delta}$ to C^2, holomorphic on Δ, $f_1(0) = 0$, such that for each $n \in N$
(i) $|f_n(\zeta)| = r_n$ $(\zeta \in b\Delta)$
(ii) $3\varepsilon_n < \delta_n$
(iii) for each $\zeta \in b\Delta$ the oscillation of f_n on $\{\xi \in \overline{D_n(\zeta)}: |\xi| \geq \rho'_n\}$ is smaller that $\lambda_{n-1}/3$
(iv) for each $\zeta \in b\Delta$ the oscillation of f_n on $\{\xi \in \overline{D_n(\zeta)}: |\xi| \geq \rho_n\}$ is smaller than ε_n
(v) $|f_{n+1} - f_n| < \min\{\varepsilon_n/2^n, 1/2^n\}$ on $\rho_n\overline{\Delta}$

(vi) $|f_{n+1} - f_n| < \lambda_{n-1}/(3.2^{n-1})$ on $\overline{\Delta}$

(vii) $|p_n - f_n| < 2^{-1} \min\{\varepsilon_n/2^n, 1/2^n\}$ on $\overline{\Delta}$

(viii) $f_{n+1}(\zeta) \in H(p_n(\zeta))$ $(\zeta \in b\Delta)$

(ix) $2^{1/2}(1 - r_n)^{1/2} < \lambda_{n-1}/(3.2^{n-1})$

(x) $\varepsilon_n + (2 + \varepsilon_n)^{1/2}(1 + \varepsilon_n - r_n)^{1/2} < \lambda_{n-1}/(3.2^{n-1})$

(xi) if $x \in bB_2$, $z \in K(x, R_n)$, $|z| > r'_n$, and if $w \in H(y)$, $|w| > r'_n$ where $y \in [K(x, R_n) + \delta_n B_2] \cap b(r_n B_2) + \delta_n B_2$ then $|z - w| > \lambda_n$.

Suppose for a moment that we have constructed all the quantities and functions with the properties above. By (vi), the sequence f_n converges uniformly on $\overline{\Delta}$. Its limit f is continuous on $\overline{\Delta}$ and holomorphic on Δ. Since $f_1(0) = 0$ it follows by (v) that $|f(0)| < 1$. Since r_n increases to 1 it follows by (i) that $|f(\zeta)| = 1$ $(\zeta \in b\Delta)$. In particular, f is not a constant.

We now show the following:

If $n \in N$ and if $\xi \in D_n(\zeta), |\xi| = \rho_n$, is such that $f(\xi) \in K(x, R_n)$ for some $x \in bB_2$ then $\eta \in D_n(\zeta), |\eta| \geq \rho_{n+1}$ implies that $f(\eta) \notin \{z \in K(x, R_n): |z| > r'_n\}$. (1)

To see this, assume that for some $n \in N$ there is $\xi \in D_n(\zeta), |\xi| = \rho_n$, such that $f(\xi) \in K(x, R_n)$ for some $x \in bB_2$. By (iv), $|f_n(\xi) - f_n(\zeta)| < \varepsilon_n$. By (v) we have $|f(\xi) - f_n(\xi)| \leq |f_{n+1}(\xi) - f_n(\xi)| + |f_{n+2}(\xi) - f_{n+1}(\xi)| + \cdots < \varepsilon_n/2^n + \varepsilon_{n+1}/2^{n+1} + \cdots < \varepsilon_n$. It follows that $|f(\xi) - f_n(\zeta)| < 2\varepsilon_n$ which, by (i) implies that $f_n(\zeta) \in [K(x, R_n) + 2\varepsilon_n B_2] \cap b(r_n B_2)$. By (vii) and by (ii) it follows that $p_n(\zeta) \in [K(x, R_n) + 2\varepsilon_n B_2] \cap b(r_n B_2) + \varepsilon_n B_2 \subset [K(x, R_n) + \delta_n B_2] \cap b(r_n B_2) + \delta_n B_2$. By (viii), $f_{n+1}(\zeta) \in H(p_n(\zeta))$; by (i)), $|f_{n+1}(\zeta)| = r_{n+1} > r'_n$. It follows by (xi) that

$$|f_{n+1}(\zeta) - z| > \lambda_n \quad (z \in K(x, R_n), |z| > r'_n). \qquad (2)$$

Suppose now that $\eta \in D_n(\zeta)$, $|\eta| \geq \rho_{n+1}$. Since $\rho'_{n+1} < \rho_{n+1}$ it follows from (iii) that $|f_{n+1}(\eta) - f_{n+1}(\zeta)| < \lambda_n/3$. Further, by (vi), $|f_{n+1}(\eta) - f(\eta)| \leq 3^{-1}(\lambda_n/2^n + \lambda_{n+1}/2^{n+1} + \cdots) < \lambda_n/3$. Now, if $z \in K(x, R_n)$, $|z| > r'_n$, (2) implies that $|f(\eta) - z| \geq |f_{n+1}(\zeta) - z| - |f_{n+1}(\eta) - f_{n+1}(\zeta)| - |f_{n+1}(\eta) - f(\eta)| > \lambda_n - \lambda_n/3 - \lambda_n/3 = \lambda_n/3 > 0$. This proves (1).

With (1) in hand one completes the proof of the theorem as follows. Suppose that, contrarily to what we want to prove, $p: [0, 1] \to \overline{\Delta}$ is a path, $p(1) \in b\Delta$, such that $p([0, 1)) \subset D_n(p(1))$ for some $n \in N$, and that there is an $R < 1$ such that $f(p([0, 1))) \subset K(f(p(1)), R)$. Since R_n increases to 1 it follows that there is some n_0 such that

$$f(p([0, 1))) \subset K(f(p(1)), R_m) \quad (m \geq n_0). \qquad (3)$$

Since the sequence D_n is increasing one may, passing to a larger n_0 if necessary, assume that $p([0, 1)) \subset D_m(p(1))$ $(m \geq n_0)$. Since p terminates at $p(1) \in b\Delta$ there are $m > n_0$ and τ, $0 < \tau < 1$, such that $|p(\tau)| = \rho_m$. Since $f \circ p$ terminates at $f(p(1)) \in bB_2$ there is a $t < 1$ such that $|f(p(t))| > r'_m$. Since $p(\tau) \in D_m(p(1))$

and $f(p(\tau)) \in K(f(p(1)), R_m)$ it follows by (1) that $f(p(t)) \notin K(f(p(1)), R_m)$ which contradicts (3).

4. Proof of Theorem, Part 2

It remains to prove the existence of $r_n, r'_n, \rho_n, \rho'_n, \lambda_n, \delta_n, \varepsilon_n$ and f_n, p_n with the properties listed in Part 1. We use an induction process. To start the induction, put $\lambda_0 = \delta_0 = \varepsilon_0 = 20$, $\rho'_1 = 1/2$. Choose $r_1, R_1 < r_1 < 1$, and put $f_1(\zeta) = \zeta(r_1, 0)$ $(\zeta \in \overline{\Delta})$. Clearly f_1 satisfies (iii) for $n = 1$, and (ix) is also satisfied for $n = 1$. We also have $f_1(0) = 0$.

Suppose that $m \in N$ and that we have already constructed positive numbers λ_{m-1}, ε_{m-1}, δ_{m-1}, numbers $\rho'_m < 1$, r_m, $R_m < r_m < 1$, and a continuous map $f_m: \overline{\Delta} \to C^2$, holomorphic on Δ, such that (i), (iii) and (ix) are satisfied for $n = m$. Lemma 1 gives r'_m, λ_m, δ_m such that $r_m < r'_m < 1, 0 < \lambda_m < \lambda_{m-1}, 0 < \delta_m < \delta_{m-1}$, such that (xi) holds for $n = m$. Since (ix) holds for $n = m$ there is an $\varepsilon_m, 0 < \varepsilon_m < \varepsilon_{m-1}, \varepsilon_m + r_m < r'_m$, such that (x) and (ii) hold for $n = m$. Choose ρ_m, $\rho'_m < \rho_m < 1$, so close to 1 that (iv) holds for $n = m$. Approximate f_m by a polynomial p_m so that (vii) holds for $n = m$, and that $|p_m(\zeta)| < r'_m$ $(\zeta \in \overline{\Delta})$. With no loss of generality we may assume that the first component of p_m has no zero on $b\Delta$. By Lemma 2 there is a polynomial $q_m: C \to C^2$ such that q_m has no zero on $b\Delta$ and such that $\langle p_m(\zeta) \mid q_m(\zeta) \rangle = 0$ $(\zeta \in b\Delta)$. Choose r_{m+1}, $r'_{m+1} < r_{m+1} < 1$, $R_{m+1} < r_{m+1}$, so that (ix) holds for $n = m + 1$. Let ω be a function in the disc algebra such that for $\zeta \in b\Delta$, $|\omega(\zeta)| = (r_{m+1}^2 - |p_m(\zeta)|^2)^{1/2}/|q_m(\zeta)|$. Replacing $\omega(\zeta)$ by $\zeta^M \omega(\zeta)$ where $M \in N$ is large enough we may assume that

$$|\omega(\zeta)| \cdot |q_m(\zeta)| < 2^{-1} min\{\varepsilon_m/2^m, 1/2^m\} \; (|\zeta| \le \rho_m) \qquad (4)$$

Put $f_{m+1} = p_m + \omega q_m$. Since $\langle \omega(\zeta) q_m(\zeta) \mid p_m(\zeta) \rangle = 0$ $(\zeta \in b\Delta)$ it follows that $f_{m+1}(\zeta) = p_m(\zeta) + \omega(\zeta) q_m(\zeta) \in H(p_m(\zeta))$ $(\zeta \in b\Delta)$ so that (viii) is satisfied for $n = m$. If $\zeta \in b\Delta$ then $|f_{m+1}(\zeta)|^2 = |p_m(\zeta)|^2 + |\omega(\zeta)|^2 |q_m(\zeta)|^2 = r_{m+1}^2$ so (i) is satisfied for $n = m + 1$. Further, (vii) and (i) for $n = m$ imply that for $\zeta \in b\Delta$ we have $|\omega(\zeta) q_m(\zeta)|^2 = r_{m+1}^2 - |p_m(\zeta)|^2 = (r_{m+1} - |p_m(\zeta)|) \cdot (r_{m+1} + |p_m(\zeta)|) \le (r_{m+1} - r_m + \varepsilon_m) \cdot (r_{m+1} + r_m + \varepsilon_m)$ and by (vii) it follows that $|f_{m+1}(\zeta) - f_m(\zeta)| = |p_m(\zeta) + \omega(\zeta) q_m(\zeta) - f_m(\zeta)| \le |p_m(\zeta) - f_m(\zeta)| + |\omega(\zeta) q_m(\zeta)| \le \varepsilon_m + (2 + \varepsilon_m)^{1/2}(1 - r_m + \varepsilon_m)^{1/2}$, which, by (x) implies that (vi) holds for $n = m$. Moreover, (vii) and (4) imply (v) for $m = n$. Now choose ρ'_{m+1}, $\rho_m < \rho'_{m+1} < 1$ so that (iii) holds for $n = m + 1$. This completes the proof of the induction step.

Notice that at each induction step we were able to choose r_{m+1} and ρ_m arbitrarily close to 1. Thus we may arrange that the sequences r_m and ρ_m both converge to 1. The theorem is proved.

References.

[G1] J. Globevnik, *Boundary interpolation and proper holomorphic maps from the disc to the ball*, Math. Z. **198** (1988), 143–150.

[G2] _____, *Discs in the ball containing given discrete sets*, Math. Ann. **281** (1988), 87–96.

[G3] _____, *Relative embeddings of discs into convex domains*, Invent. Math. **98** (1989), 331–350.

[G4] _____, *A disc in the ball whose end is an arc*, Indiana Univ. Math. J. **40** (1991), 967–973.

[GS1] J. Globevnik and E. L. Stout, *The ends of discs*, Bull. Soc. Math. France **114** (1986), 175–195.

[GS2] _____, *The ends of varieties*, Amer. J. Math. **108** (1986), 1355–1410.

[J] P.W.Jones, *A complete bounded complex submanifold of C^3*, Proc. Amer. Math. Soc. **76** (1979), 305–306.

[R] W.Rudin, *Inner function images of radii*, Math. Proc. Cambr. Phil. Soc. **85** (1979), 357–360.

INSTITUTE OF MATHEMATICS, PHYSICS AND MECHANICS, UNIVERSITY OF LJUBLJANA, LJUBLJANA, SLOVENIA (YUGOSLAVIA)

Contemporary Mathematics
Volume **137**, 1992

On hypoellipticity for sums
of squares of vector fields

A. ALEXANDROU HIMONAS

Introduction

Let Ω be an open subset of $\mathbb{R}^n$ and $X_0, X_1, ..., X_r$ be r+1 real $\mathcal{C}^\infty$ vector fields in Ω, i.e.

$$(1) \qquad X_j = \sum_{k=1}^{n} a_{jk}(x) \frac{\partial}{\partial x_k}, \; j = 0, 1, ..., r,$$

with $a_{jk} \in \mathcal{C}^\infty(\Omega)$ and real-valued. We consider the following second order partial differential operator

$$(2) \qquad P = \sum_{j=1}^{r} X_j^2 + X_0 + c,$$

where $c \in \mathcal{C}^\infty(\Omega)$. Here we shall describe some $\mathcal{C}^\infty$ and analytic regularity results for the solutions u of the equation

$$(3) \qquad Pu = f,$$

where f is a given function. The $\mathcal{C}^\infty$ case has been well understood, at least when P has analytic coefficients. The analytic case is more complicated. If P is elliptic then it is well known that all solutions u to equation (3) are analytic. If P is not elliptic and its characteristic set is symplectic then it was proved by Tartakoff [**26**] and Treves [**27**] that again all solutions to (3) are analytic. When the characteristic set of P is not symplectic then no general condition is known for the analytic hypoellipticity of P. However there exist both positive and negative results in the form of families of examples. Some of the most recent of these examples will be described below. We start with the $\mathcal{C}^\infty$ case.

1991 *Mathematics Subject Classification*. Primary 35H05, 35H35. Secondary 58G15, 58G58.

The author was partially supported by NSF.

This paper is in final form and no version of it will be submitted for publication elsewhere.

$\mathcal{C}^\infty$ Hypoellipticity

The operator P is said to be $\mathcal{C}^\infty$ *hypoelliptic* in Ω if, given any open subset V of Ω and any function $f \in \mathcal{C}^\infty(V)$, then all solutions u to the equation (3) are in $\mathcal{C}^\infty(V)$.

The most well known examples of $\mathcal{C}^\infty$ hypoelliptic operators of the form (2) is the Laplacian

$$\Delta = \left(\frac{\partial}{\partial x_1}\right)^2 + \cdots + \left(\frac{\partial}{\partial x_n}\right)^2$$

and the heat operator

$$H = \left(\frac{\partial}{\partial x_1}\right)^2 + \cdots + \left(\frac{\partial}{\partial x_{n-1}}\right)^2 - \frac{\partial}{\partial t}.$$

In these examples the operators have contstant coefficients. A very simple example of an operator with variable coefficients of the form (2) which is $\mathcal{C}^\infty$ hypoeliptic is the Kolmogorov operator

$$K = \left(\frac{\partial}{\partial x}\right)^2 + x\frac{\partial}{\partial y} - \frac{\partial}{\partial t}.$$

In the last example we have

$$X_0 = x\frac{\partial}{\partial y} - \frac{\partial}{\partial t}, X_1 = \frac{\partial}{\partial x} \text{ and } c = 0.$$

Notice that the commutator of X_1 and X_0

$$[X_1, X_0] = X_1 X_0 - X_0 X_1 = \frac{\partial}{\partial y}$$

and that X_0, X_1 and $[X_1, X_0]$ generate the tangent space at every point $(x, y, t) \in \mathbb{R}^3$. This condition is responsible for the hypoellipticity of K.

More generally a point $x \in \Omega$ is of *finite type* (for the vector field $X_0, X_1, \ldots, X_r$) if there exists a positive integer k such that the vectors

$$X_0, ..., X_r, [X_{j_1}, X_{j_2}], ..., [X_{j_1}, [X_{j_2}, ..., [X_{j_{k-1}}, X_{j_k}]...]],$$

where $X_{j_i} \in \{X_0, \ldots, X_r\}$, generate the tangent space $T_x\mathbb{R}^n$. The smallest such k is called the *type* of the point x.

In 1967 Hörmander [16] generalized Kolmogorov's example into the following

THEOREM 1. *If all points* $x \in \Omega$ *are of finite type for the vector fields* $X_0, X_1, \ldots, X_r$, *then the operator P in (2) is $\mathcal{C}^\infty$ hypoelliptic in* Ω.

Another proof of this theorem was given by Kohn [17], Radkevic [22], and Rothschild-Stein [24].

In the case that the Lie algebra generated by $X_0, \ldots, X_r$ has constant rank $d < n$ near a point $x_0 \in \Omega$ then P is not C^∞ hypoelliptic. Then by the Frobenius Theorem there exist new coordinates near x_0 in which

$$X_j = \sum_{k=1}^{d} a_{jk}(x_1, \ldots, x_d) \frac{\partial}{\partial x_k}$$

If $c = 0$ and $u = u(x_{d+1}, \ldots, x_n)$ is any function independent of $x_1, \ldots, x_d$ then $Pu = 0$ and u may not be C^∞. In [6] Derridj proved the following

THEOREM 2. *If the vector fields $X_0, \ldots, X_r$ and the function c are analytic in Ω then P is C^∞ hypoelliptic in Ω iff every point $x \in \Omega$ is of finite type.*

In the C^∞ case the operator P may still be hypoelliptic although the finite type condition is violated in a certain subset lying on a finite union of $(n - 1)$-dimensional smooth manifolds in Ω. Such situations have been studied by Oleinik-Radkevic [21], and by Fedii [11]. For example if every point in $\Omega - S$ is of finite type and S is a set of isolated points $x \in \Omega$ at which at least one of the vector fields $X_0, \ldots, X_r$ is non-zero, then P is hypoelliptic in Ω. Another example in $\mathbb{R}^2$ is the following. Let $a(x) \in C^\infty(\mathbb{R})$ with $\frac{d^j a}{dx^j}(0) = 0, j = 0, 1, 2, \ldots$, and $a(x) \neq 0$ for $x \neq 0$. Then the operator

$$P = \left(\frac{\partial}{\partial x} \right)^2 + \left(a(x) \frac{\partial}{\partial y} \right)^2$$

is C^∞ hypoelliptic in $\mathbb{R}^2$ although the points $(0, y)$ are not of finite type.

Analytic Hypoellipticity

From now on we will assume that the vector fields $X_0, X_1, \ldots, X_r$ and the function c are analytic(real analytic). Then the operator P in (2) is analytic hypoelliptic in Ω if, given any open subset V of Ω and any function f which is analytic in V, then all solutions to equation (3) are analytic in V.

In 1972 Baouendi and Goulaouic [1] gave the following now well known example that shows the finite type assumption in Ω is not sufficient for P to be analytic hypoelliptic in Ω. In $\mathbb{R}^3$ let

$$X_1 = \frac{\partial}{\partial x}, \, X_2 = \frac{\partial}{\partial t}, \, X_3 = x \frac{\partial}{\partial y} \text{ and } P = X_1^2 + X_2^2 + X_3^2.$$

Then $[X_1, X_3] = \frac{\partial}{\partial y}$ and the vector fields $X_1, X_2,$ and $[X_1, X_2]$ generate $\mathbb{R}^3$ at any point of $\mathbb{R}^3$. If

$$u(x, t, y) = \int_0^\infty e^{i\rho^2 y + \rho t} e^{-\frac{1}{2}\rho^2 x^2} e^{-\rho} d\rho$$

then u is a C^∞ solution to $Pu = 0$. We also have that

$$\frac{\partial^j u}{\partial y^j}(0, 0, 0) = i^j \int_0^\infty \rho^{2j} e^{-\rho} d\rho = i^j (2j)!$$

Therefore u is not analytic at $0 \in \mathbb{R}^3$ since it violates the Cauchy estimates there. Therefore P is not analytic hypoelliptic. In this example all the points of the form $(0, t, y)$ are of finite type $k = 2$. All other points are of type $k = 1$, i.e. elliptic points.

It was shown by Tartakoff [26] and Treves [27], independently, that if $k = 2$ and the Levi matrix is nondegenerate (i.e. the characteristic set of P is symplectic) then P is analytic hypoelliptic in Ω. More precisely they proved

THEOREM 3. *Let Ω be an open set in $\mathbb{R}^{2n+1}$ and $X_1, \ldots, X_{2n}$ be real analytic vector fields in Ω. Let T be any real analytic vector field in Ω such that $X_1, \ldots, X_{2n}$ and T span the tangent space $T_x\Omega$ at every point $x \in \Omega$. If the Levi matrix (c_{jk}) defined by*

$$[X_j, X_k] = c_{jk}T \; modulo \; \{X_1, \ldots, X_{2n}\}$$

is nondegenerate, i.e.

(4) $$det(c_{jk}(x)) \neq 0, x \in \Omega,$$

then the operator

$$P = \sum_{j=1}^{r} X_j^2$$

is analytic hypoelliptic in Ω.

Notice the relation between the dimension of Ω and the number of the vector fields in Theorem 3. Last example does not satisfy this relation. The simplest example in $\mathbb{R}^3$ to which Theorem 3 is applicable is the operator

$$P = \left(\frac{\partial}{\partial x}\right)^2 + \left(\frac{\partial}{\partial t} - x\frac{\partial}{\partial y}\right)^2.$$

In this example

$$X_1 = \frac{\partial}{\partial x}, \; X_2 = \frac{\partial}{\partial t} - x\frac{\partial}{\partial y} \; \text{and} \; [X_1, X_2] = -\frac{\partial}{\partial y}.$$

The Levi matrix is the constant -1 and thus condition (4) is satisfied. The simplest example in $\mathbb{R}^3$ where condition (4) is not satisfied is the operator

$$P = \left(\frac{\partial}{\partial x}\right)^2 + \left(\frac{\partial}{\partial t} - x^2\frac{\partial}{\partial y}\right)^2.$$

Here the Levi matrix is equal to $-2x$ and it vanishes when $x = 0$. This operator is not analytic hypoelliptic. In fact we have the following

THEOREM 4. *If $k \in \{3, 4, 5, \ldots\}$ then the operator*

(5) $$P = \left(\frac{\partial}{\partial x}\right)^2 + \left(\frac{\partial}{\partial t} - x^{k-1}\frac{\partial}{\partial y}\right)^2$$

is not analytic hypoelliptic in $\mathbb{R}^3$.

The case $k = 3$ is due to Helffer [15] and Pham The Lai-Robert [23]. The case $k = 5, 7, 9, \ldots$ is due to Hanges-Himonas [14], and the case $k = 4, 6, 8, \ldots$ is due to Christ [3,4].

Next we shall describe the proof of Theorem 4. With P given by (5) let $u(x, t, y)$ be a solution to the equation

$$(6) \qquad Pu = \frac{\partial^2 u}{\partial x^2} + \frac{\partial^2 u}{\partial t^2} - 2x^{k-1}\frac{\partial^2 u}{\partial t \partial y} + x^{2(k-1)}\frac{\partial^2 y}{\partial y^2} = 0.$$

If we take Fourier Transform with respect to (t, y) in equation (6) we obtain the equation

$$(7) \qquad \hat{u}_{xx} - (x^{k-1}\eta - \tau)^2 \hat{u} = 0, \ \hat{u} = \hat{u}(x, \tau, \eta).$$

Let the change of variables

$$\eta = \rho^k, \ x = s\rho^{-1}, \ \tilde{\tau} = \rho^{-1}\tau.$$

Then equation (7) is transformed into

$$(8) \qquad v_{ss} - \left(s^{k-1} - \tilde{\tau}\right)^2 v = 0, \ v(s, \tilde{\tau}, \rho) = \hat{u}(s\rho^{-1}, \rho\tilde{\tau}, \rho^k).$$

Assuming we have a solution v to the above ordinary differential equation then by taking inverse Fourier Transform (formally) we obtain

$$u(x, t, y) = \int e^{iy\eta + it\tau}\hat{u}(x, \tau, \eta)d\tau d\eta$$

or

$$(9) \qquad u(x, t, y) = \int e^{i\rho^k y + i\rho\tilde{\tau}t}v(\rho x, \tilde{\tau}, \rho)k\rho^k d\tilde{\tau}d\rho$$

which formally satisfies equation (6).

In fact if we can solve equation (8) for a fixed $\tilde{\tau}$ then we can construct a solution to $Pu = 0$ of the form (9). More precisely if there exist a $\lambda \in \mathbb{C}$ for which the differential equation

$$(10) \qquad \frac{d^2 v}{dx^2} - (x^{k-1} - \lambda)^2 v = 0$$

has a nonzero solution in the Schwartz space $\mathcal{S}(\mathbb{R})$ then the function

$$(11) \qquad u(x, t, y) = \int_0^\infty e^{i\rho^k y + i\lambda\rho t}v(\rho x)e^{-\rho}d\rho$$

is well defined for $|t|$ small enough, and it is a solution to $Pu = 0$. A short computation gives

$$(12) \qquad \frac{\partial^j u}{\partial y^j}(0, 0, 0) = i^j v(0)\int_0^\infty \rho^{kj}e^{-\rho}d\rho = i^j v(0)(kj)!$$

Since we can assume $v(0) \neq 0$, relation (12) shows that u cannot be analytic at 0 since it violates the Cauchy estimates there. In fact u is in the Gevrey class of order k.

It is easy to see that equation (10) cannot have a solution in $\mathcal{S}(\mathbb{R})$ if λ is real. It was first shown in [23] that for $k = 3$ equation (10) has a nonzero solution $v \in \mathcal{S}(\mathbb{R})$ for some $\lambda \in \mathbb{C} - \mathbb{R}$. This proof was generalized in [14] for $k = 5, 7, 9, \ldots$. The case $k = 4, 6, 8, \ldots$ was proved in [3], where also another proof of $k = 5, 7, 9, \ldots$ is provided. For $k = 2$ equation (8) cannot have a solution in $\mathcal{S}(\mathbb{R})$ (see [3]) which of course must be the case since P is analytic hypoelliptic for $k = 2$.

We remark that in the integral that expresses our singular solution (11) we can use other weights instead of $e^{-\rho}$. If for example use the weight $e^{-\rho^2}$ then we obtain a global $\mathcal{C}^\infty$ solution u to $Pu = 0$ which is not analytic at 0.

The operator in (5) is an example of the more general class of operators in $\mathbb{R}^3$ of the form

$$(13) \qquad P = \left(\frac{\partial}{\partial x} \right)^2 + \left(\frac{\partial}{\partial t} - a(x,t) \frac{\partial}{\partial y} \right)^2,$$

where $a(x,t)$ is a real-valued function which is analytic near 0. By Theorem 4 if $a = x^{k-1}, k \geq 3$, then P is not analytic hypoelliptic near 0. M. Christ has announced [4] that if $a = a(x) = x^{k-1} + cx^k + \ldots, k \geq 3$ then again P is not analytic hypoelliptic at 0. When a depends on both variables x and t then P can be in some cases analytic hypoelliptic and in others not analytic hypoelliptic. For example the operator

$$(14) \qquad P = \left(\frac{\partial}{\partial x} \right)^2 + \left(\frac{\partial}{\partial t} - xt^{k-2} \frac{\partial}{\partial y} \right)^2,$$

is not analytic hypoelliptic for $k \geq 3$. In fact if we perform the following change of variables

$$\tilde{x} = x, \tilde{t} = t \text{ and } \tilde{y} = y + \frac{1}{k-1} t^{k-1} x$$

then the operator (P) in (13) takes the form (5).

On the other hand the operator

$$(15) \qquad P = \left(\frac{\partial}{\partial x} \right)^2 + \left(\frac{\partial}{\partial t} + [\frac{4}{3} x^3 + 4xt^2] \frac{\partial}{\partial y} \right)^2,$$

has been shown by Grigis and Sjöstrand [13] to be analytic hypoelliptic at 0. The basic difference between examples (5) and (15) is the geometry of the characteristic set $\mathrm{Char} P$. In (5) we have

$$\mathrm{Char} P = \{\xi = \tau - x^{k-1}\eta = 0\} \text{ and } \{\xi, \tau - x^{k-1}\eta\} = (k-1)x^{k-2}\eta,$$

where $\{.,.\}$ denotes Poisson brackets. If $k \geq 3$ then $\text{Char}P$ is non-symplectic on $x = 0$. The tangent space of $\text{Char}P$ is generated by the vector fields

$$\frac{\partial}{\partial t}, \frac{\partial}{\partial y}, \frac{\partial}{\partial \eta} + x^{k-1}\frac{\partial}{\partial \tau} \text{ and } \frac{\partial}{\partial x} + (k-1)x^{k-2}\eta\frac{\partial}{\partial \tau}.$$

If σ denotes the fundamental symplectic form and T any of the above vector fields then we have

$$\sigma(\frac{\partial}{\partial t}, T) = 0 \text{ for } x = 0.$$

Therefore the t-axis is a smooth curve inside $\text{Char } P \cap \{x = 0\}$ which is orthogonal, for the fundamental symplectic form σ, to the tangent space of $\text{Char}P$. In such a situation Treves has conjectured in [27] that P cannot be analytic hypoelliptic. On the other hand in (15) we have

$$\text{Char } P = \{\xi = \tau - (\frac{4}{3}x^3 + 4xt^2)\eta = 0\} \text{ and}$$

$$\{\xi, \tau - (\frac{4}{3}x^3 + 4xt^2)\eta\} = 4(x^2 + t^2)\eta.$$

$\text{Char}P$ is non-symplectic on $\{x = t = 0\}$. In this case there is no curve inside $\text{Char } P \cap \{x = t = 0\}$ which is orthogonal to the tangent space of $\text{Char}P$ at every point of the curve. It may be useful to mention here that P in (5) is the principal part of Kohn's Laplacian for the following hypersurface M in $\mathbb{C}^2$

$$M = \{(z, w) \in \mathbb{C}^2 : \text{Im}w = \frac{1}{k}(Rez)^k\},$$

while the corresponding hypersurface for P in (15) is given by

$$M = \{(z, w) \in \mathbb{C}^2 : \text{Im}w = |z|^4\}.$$

The last example was generalized by Tartakoff and Derridj in [7]. More results on analytic hypoellipticity and related topics can be found in [2], [5], [8], [9], [10], [12], [18], [19], [20] and [25].

References.

1. M.S. Baouendi and C. Goulaouic, *Nonanalytic-hypoellipticity for some degenerate elliptic operators*, Bull. AMS 78 **78** (1972), 483-486.

2. M. Christ, *Analytic hypoellipticity breaks down for weakly pseudoconvex Reinhardt domains*, International Mathematics Research Notices (1991, No. 3), 31-40.

3. M. Christ, *Existence of decaying solutions of ordinary differential equations*, Preprint (1991).

4. M. Christ, *A class of hypoelliptic PDE admitting non-analytic solutions*, Preliminary Draft (1991).

5. M. Christ and D. Geller, *Counterexamples to analytic hypoellipticity for domain of finite type*, Preprint (1990).

6. M. Derridj, *un probleme aux limites pour une classe d'operateurs du second ordre hypoelliptiques*, Ann. Inst. Fourier, Grenoble **21** (1971), 99-148.

7. M. Derridj and D.S. Tartakoff, *Local analyticity for $\square_b$ and the $\bar{\partial}$-Neumann problem at certain weakly pseudoconvex domains*, Comm. in P.D.E **12** (1988), 1521-1600.

8. M. Derridj and D.S. Tartakoff, *Local analyticity for $\square_b$ for a class of model domains not satisfying maximal estimates*, Preprint (1990).

9. M. Derridj and D.S. Tartakoff, *Local analyticity in the $\bar{\partial}$-Neumann problem for some model domains without maximal estimates*, Preprint (1990).

10. M. Derridj and C. Zuily, *Regularite analytique et Gevrey d'operateurs ellip-tiques degeneres*, J. Math. pures et appl. **52** (1973), 65-80.

11. E.V. Fedii, *Estimates in $H_{(s)}$ norms and hypoellipticity,*, Soviet Math. Dokl. **11, No. 4** (1970), 940-942.

12. A. Grigis and L.P. Rothschild, *A criterion for analytic hypoellipticity of a class of differential operators with polynomial coefficients*, Annals of Mathematics **118** (1983), 443-460.

13. A. Grigis and J. Sjöstrand, *Front d'onde analytique et sommes de carres de champs de vecteurs*, Duke Math J. **52** (1985), 35-51.

14. N. Hanges and A.A. Himonas, *Singular solutions for sums of squares of vector fields*, Comm. in PDE **16** (1991), 1503-1511.

15. B. Helffer, *Conditions necessaires d'hypoanalyticite pour des operateurs in-variants a gauche homogenes sur un groupe nilpotent gradue*, Journal of Differential Equations **44** (1982), 460-481.

16. L. Hörmander, *Hypoelliptic second order differential equations*, Acta Math. **119** (1967), 147-171.

17. J.J. Kohn, *Pseudo-differential operators and hypoellipticity*, Proceedings of symposia in pure mathematics **XXIII** (1973), 61-70.

18. T. Matsuzawa, *Sur les equations $u_{tt} + t^\alpha u_{xx} = f$ ($\alpha \geq 0$)*, Nagoya Math. J. **42** (1971), 43-55.

19. G. Metivier, *Une class d'operateurs non hypoelliptiques analytiques*, Indiana Univ. Math J. **29** (1980), 823-860.

20. G. Metivier, *Analytic hypoellipticity for operators with multiple characteris-tics*, Comm. in PDE **1** (1981), 1-90.

21. O.A. Oleinik and E.V. Radkevic, *Second order equations with nonnegative characteristic form*, AMS and Plenum Press (1973).

22. E.V. Radkevic, *Hypoelliptic operators with multiple characteristics*, Math. USSR Sbornik **8, No. 2** (1969), 181-205.

23. Pham The Lai and D. Robert, *Sur un problem aux valeurs propres non lin-eaire*, Israel J. of Math **36** (1980), 169-186.

24. L.P. Rothschild and E.M. Stein, *Hypoelliptic differential operators and nilpo-tent groups*, Acta Math **137** (1977), 247-320.

25. J. Sjöstrand, *Analytic wavefront sets and operators with multiple character-istics*, Hokkaido Mathematical Journal **12** (1983), 392-433.

26. D.S. Tartakoff, *On the local real analyticity of solutions to $\Box_b$ and the $\bar{\partial}$-Neumann problem*, Act. Math. **145** (1980), 117-204.

27. F. Treves, *Analytic hypo-ellipticity of a class of pseudodifferential operators with double characteristics and applications to $\bar{\partial}$-Neuman problem*, Comm. in P.D.E. **3** (1978), 475-642.

DEPARTMENT OF MATHEMATICS, UNIVERSITY OF NOTRE DAME, NOTRE DAME, INDIANA 46556

Contemporary Mathematics
Volume **137**, 1992

Composition Operators in Bergman Spaces on bounded symmetric domains

F. JAFARI

ABSTRACT. The study of composition operators in various function spaces on classical domains has provided many results specific to the geometry of that domain. In this paper a necessary and sufficient (hyperbolic) geometric condition for a composition operator to be bounded or compact on bounded symmetric domains in $\mathbb{C}^n$ is established. This condition is applied to obtain a computational criterion for a composition operator to be bounded or compact. A result on the boundary behavior of holomorphic self-maps of bounded symmetric domains necessary and sufficient for continuity of composition operators is also deduced.

Introduction.

The interplay between geometry of various domains and estimates of norms of operators defined on function spaces on these domains are of considerable interest. This interplay plays a particularly elegant role on the weighted Bergman spaces. In the study of composition operators in the unit disk [13], [15], [16], in the unit ball of $\mathbb{C}^n$ [3], [12], [20], [21] and in the unit polydisks [7], [17] the use of specific geometries have generated results specific to these domain geometries. Carleson measures have played an essential role in the study of composition operators. MacCluer and Shapiro [13] use Carleson measures successfully to characterize the compact composition operators in the Hardy and weighted Bergman spaces of the disc, and MacCluer [12] gives a Carleson measure characterization of the bounded and compact composition operators on the Hardy spaces of the unit ball of $\mathbb{C}^n$. In addition, they translate their Carleson measure characterizations in these settings to tangible geometric properties of the mapping that induces the composition operator. For the weighted Bergman spaces on the disk, these results are nicely described in terms of the nonexistence of

1991 *Mathematics Subject Classification*. Primary 47A30, 32A35.
Research supported in part by a Basic Research Grant from University of Wyoming.
This paper is in final form and no version of it will be submitted for publication elsewhere.

the angular derivative at the boundary of the disk for the maps inducing the composition operators.

In this paper we will study composition operators on bounded symmetric domains in $\mathbb{C}^n$, and attempt to establish criteria for boundedness and compactness of these operators in terms of the hyperbolic geometry of these regions. It is well known that the (weighted) Bergman reproducing kernel on a bounded symmetric domain induces a hyperbolic metric which gives rise to the usual topology on these domains. In classical settings the hyperbolic family of balls in this metric are equivalent to the Carleson sets, and therefore these hyperbolic family of balls, or more specifically the geometry induced by the Bergman reproducing kernel provides the natural candidate for generalizing the results from the above specific settings to bounded symmetric domains.

In section 2 of this paper we present an extensive introduction to the hyperbolic geometry of bounded symmetric domains, and establish various norm estimates about the Bergman kernel which will be used throughout the rest of this paper. Although the results of this section are not completely new, these results appear in extensive references, and presenting them here directs attention to these more extensive references (see [1], [2], [10], [18], [22], for example). In section 3 we give necessary and sufficient conditions for a composition operator to be bounded or compact in terms of various criteria on the hyperbolic family of balls on the bounded symmetric domain. In Theorem 3.1 we provide a Carleson measure type criterion for the boundedness and compactness of composition operators, and in Theorem 3.2 we translate this characterization into an algebraic criterion to be satisfied by the normalized reproducing kernel of these domains. The result of this theorem combined with the explicit expressions for the reproducing kernel of bounded symmetric domains (see [6] or [11], for example) provide computational criteria for determining the bounded and compact composition operators. Immediately following the statement of Theorems 3.1 and 3.2, we state and prove several remarks that attempt to interpret, restate, and apply these results to various specific situations. Finally, in Proposition 3.5 and Corollary 3.6 we give a result on the boundary behavior of the holomorphic self-maps of the domain which will be necessary and sufficient to generate bounded composition operators. These results are expressed in terms of the normalized reproducing kernel of the bounded symmetric domain. Further translation of these results into specific tangible geometric properties of the inducing map of the composition operators on these domains remain under investigation.

Notation and Preliminaries

Let Ω be a bounded symmetric domain in $\mathbb{C}^n$, and let ϕ denote a holomorphic map of Ω into itself. Let $K(z, w)$ be the Bergman kernel on Ω in its standard Cartan representation (see [6] or [11], for example) and let V be the probability (volume) measure on Ω. By [19], there exists an $\epsilon > 0$, depending on Ω, so that for all $\alpha < \epsilon$

$$c_\alpha = \int_\Omega K^\alpha(z,z)dV(z) < \infty.$$

For each $\alpha < \epsilon$ define $dV_\alpha = c_\alpha^{-1}K^\alpha(z,z)dV(z)$. Then $\{dV_\alpha\}$ defines a weighted family of probability measures on Ω. Define the weighted Bergman space on Ω, $A_\alpha^p(\Omega)$, as the set of all holomorphic functions f on Ω so that $\|f\|_{\alpha,p} = (\int_\Omega |f|^p dV_\alpha)^{\frac{1}{p}} < \infty$, and note that $A_\alpha^p(\Omega)$ is a closed subspace of $L^p(\Omega, dV_\alpha)$. For $p = 2$, there is a cannonical projection from $L^2(\Omega, dV_\alpha)$ onto $A_\alpha^2(\Omega)$ given by

$$P_\alpha f(z) = \int_\Omega f(w)K_\alpha(z,w)dV_\alpha(w)$$

where $K_\alpha(z,w) = K^{1-\alpha}(z,w)$. If f is holomorphic on Ω, then P_α is the identity map on $A_\alpha^2(\Omega)$. Continuity of this projection operator in the unit ball of $\mathbb{C}^n$ is proved by [5], and similar ideas are used by [19] to prove continuity of this projection on bounded symmetric domains. The following proposition is well-known (see [2] or [18] for example):

PROPOSITION 2.1. *For each $\alpha < \epsilon$ the weighted Bergman kernel $K_\alpha(z,.)$ is holomorphic in Ω, $K_\alpha(.,w)$ is conjugate holomorphic in Ω, $K_\alpha(z,.) \in A_\alpha^2$, and $K_\alpha(z,w)$ is the reproducing kernel for A_α^2.*

$K_\alpha(z,z)$ induces a pseudometric on Ω called the hyperbolic (Kobayashi) metric, β, which gives rise to the usual topology of Ω. The Hermitian metric form, ds^2, on Ω generated by the Bergman kernel and

$$g_{ij} = \frac{1}{2}\frac{\partial^2 \log K(z,z)}{\partial z_i \partial \overline{z_j}},$$

is $ds^2 = \sum_{i,j=1}^n g_{ij}dz_i d\overline{z_j}$. If $\rho(z,w)$ is the pseudohyperbolic distance on Ω given by the norm of the automorphism of Ω onto itself which interchanges z and w, $z,w \in \Omega$, then the hyperbolic metric induced by the Bergman metric is

$$\beta(z,w) = \frac{1}{2}\log\frac{1+\rho(z,w)}{1-\rho(z,w)}.$$

Let

$$E(z,r) = \{w \in \Omega : \beta(z,w) < r\}$$

be the hyperbolic family of balls induced by this metric on Ω. Note that β is only defined with respect to the unweighted Bergman kernel. For the unit disk, β is exactly the Poincaré metric on the disk. The properties of β and the family of hyperbolic balls $\{E(z,r) : z \in \Omega, r > 0\}$ may be summarized as follows:

PROPOSITION 2.2. *Let Ω be a bounded symmetric domain in $\mathbb{C}^n$ and ψ be an automorphism of Ω into itself. Then*
(i) $\beta(\psi(a),\psi(b)) = \beta(a,b)$ $\forall a,b \in \Omega$
(ii) Given $r > 0$ there exist constants C_1 and C_2 so that for every $z \in \Omega$

$$C_1 V^{1-\alpha}(E(z,r)) \leq V_\alpha(E(z,r)) \leq C_2 V^{1-\alpha}(E(z,r)).$$

PROOF. (i) Since holomorphic maps are distance decreasing with respect to the Kobayashi metric, we have

$$\beta(\psi(a), \psi(b)) \leq \beta(a, b).$$

Since ψ is an automorphism of Ω, we also have

$$\beta(a, b) = \beta(\psi^{-1}(\psi(a)), \psi^{-1}(\psi(b))) \leq \beta(\psi(a), \psi(b)).$$

So every automorphism of Ω is an isometry in terms of the Bergman metric.

(ii) We demonstrate that given $r > 0$ there exist constants C_1 and C_2 so that for every $z \in \Omega$

$$C_1 \leq V(E(z, r))K(z, z) \leq C_2. \tag{1}$$

Given (1), since

$$c_\alpha^{-1} C_1 (V(E(z, r)))^{1-\alpha} \leq \int_{E(z,r)} dV_\alpha = V_\alpha(E(z, r))$$

$$= c_\alpha^{-1} \int_{E(z,r)} K^\alpha(z, z) dV \leq c_\alpha^{-1} C_2 (V(E(z, r)))^{1-\alpha},$$

the desired result follows by renaming the constants. By (i), since the automorphisms of Ω are isometries in the Bergman metric, and if ψ_z is the automorphism of Ω that interchanges z and 0, $\psi_z(E(z, r)) = E(0, r)$. Therefore, by a change of variable

$$V(E(z, r)) = \int_{E(z,r)} dV(w) = \int_{E(0,r)} \frac{|K(w, z)|^2}{K(z, z)} dV(w).$$

Therefore

$$K(z, z)V(E(z, r)) = K(z, z) \int_{E(0,r)} \frac{|K(w, z)|^2}{K(z, z)} dV(w)$$

$$= \int_{E(0,r)} |K(w, z)|^2 dV(w).$$

Since the integrand is plurisubharmonic, $K(w, z)$ is a nonvanishing smooth function in both variables, and $\overline{E(0, r)}$ is compact, there exist C_1 and C_2 (depending on r but not on z) so that

$$C_1 \leq K(z, z)V(E(z, r)) \leq C_2. \ \blacksquare$$

For each $z \in \Omega$, define

$$k_z(w) = \frac{K(w, z)}{\sqrt{K(z, z)}}.$$

The following proposition is an immediate corollary of Proposition 1, and the above definition.

PROPOSITION 2.3. *(i) For each $z \in \Omega$, $k_z^{\cdot}(w)$ is a unit vector in $A_0^2(\Omega)$.*
(ii) If ψ_z is the automorphism of Ω which interchanges z and 0, then

$$|k_z(w)|^2 = |J_c\psi_z(w)|^2.$$

(iii) For each $z \in \Omega$, $k_z^{1-\alpha}(w)$ is a unit vector in $A_\alpha^2(\Omega)$.
(iv) $k_z^{1-\alpha}$ are the normalized reproducing kernels in $A_\alpha^2(\Omega)$.
(v) Given $r > 0$ there exist constants C_1 and C_2 so that for every $z \in \Omega$

$$C_1 \le |k_z(w)|^2|E(z,r)| \le C_2 \qquad \forall w \in E(z,r).$$

PROOF. (i) This is an immediate consequence of the definition of $k_z(w)$.
(ii) If ψ_z is an automorphism of Ω that interchanges z and 0, since (see [18] for example)

$$(J_c\psi_z(w))(\overline{J_c\psi_z(w)})K(\psi_z(w),0) = K(w,z), \tag{1}$$

setting $w = z$ gives

$$K(z,z) = K(0,0)|J_c\psi_z(z)|^2.$$

Taking the square of the modulus of both sides of equation (1), and noting that

$$|K(\psi_z(w),0)|^2 = K(\psi_z(w),0)\overline{K(\psi_z(w),0)} =$$
$$= K(\psi_z(w),0)K(0,\psi_z(w)) = K(0,0),$$

we obtain

$$|K(w,z)|^2 = K(z,z)|J_c\psi_z(w)|^2.$$

(iii) Since $K_\alpha(w,z) = K^{1-\alpha}(w,z)$, and since $k_z(w)$ is a unit vector in $A_0^2(\Omega)$, we have

$$\|k_z^{1-\alpha}(w)\|_\alpha^2 = \frac{1}{K^{1-\alpha}(z,z)} \int_\Omega |K^{1-\alpha}(w,z)|^2 dV_\alpha =$$
$$= \frac{1}{K_\alpha(z,z)} \int_\Omega |K_\alpha(w,z)|^2 dV_\alpha.$$

By Proposition 2.1, since K_α is the reproducing kernel for A_α^2,

$$\int_\Omega |K_\alpha(w,z)|^2 dV_\alpha = K_\alpha(z,z).$$

Hence for each $z \in \Omega$, $k_z^{1-\alpha}$ is a unit vector in A_α^2.
(iv) This result follows from an argument similar to part (iii).
(v) Since by composition with an automorphism of Ω sending z to 0 one obtains

$$|E(z,r)| = \int_{E(0,r)} \frac{|K(w,z)|^2}{K(z,z)} dV,$$

the expression $|k_z(w)|^2|E(z,r)|$ is equal to $\int_{E(0,r)} |K(w,z)|^2 dV$. Since $K(w,z)$ is a smooth nonvanishing function on Ω, and the closure of $E(0,r)$ is compact the desired result follows. ∎

Remarks 2.4. (i) The reproducing kernels for the weighted Bergman spaces $A_\alpha^2(\Omega)$ are related to the reproducing kernel of the unweighted Bergman spaces by a very simple relationship. Since the unweighted kernels generate the Bergman metric, it follows that the Bergman metric of weighted Bergman spaces are simply scalar multiples of the metric of the unweighted spaces.

(ii) The Bergman kernel and the Jacobian of the transitive automorphism group of bounded symmetric domains are intimately related. This relationship may be used to give the following nice geometric description of the normalized reproducing kernel of $A_\alpha^2(\Omega)$:

$$\int_{E(z,r)} |k_z^{1-\alpha}(w)|^2 dV_\alpha(w) \sim V_\alpha(E(z,r)).$$

Note that given $r > 0$ we say $A(z,r) \sim B(z,r)$ if there exist constants C_1 and C_2 (constants may depend on r) so that $C_1 B(z,r) \le A(z,r) \le C_2 B(z,r)$ for every $z \in \Omega$. The proof of this equivalence is essentially contained in the proof of Propositions 2.3 (v), but for clarity we retrace the argument.

$$V_\alpha(E(z,r)) = \int_{E(z,r)} dV_\alpha = c_\alpha^{-1} \int_{E(z,r)} K^\alpha(z,z) dV(z)$$

$$= c_\alpha^{-1} \int_{E(0,r)} \frac{|K(w,z)|^2}{K_\alpha(z,z)} dV(z).$$

Therefore $K_\alpha(z,z)V_\alpha(E(z,r))$ is proportional to $\int_{E(0,r)} |K(w,z)|^2 dV$. Therefore the desired equivalence follows.

(iii) The Bergman metric is positive definite, and for the four different Cartan domains the Bergman kernels are ingeniously and explicitly computed by L.K. Hua [6]. These explicit forms can be used to obtain specific cases of the results given in the next section of this paper.

Bounded and Compact Composition Operators

Let ϕ be a holomorphic mapping of Ω into itself and define C_ϕ to be the linear operator that sends $f \mapsto f \circ \phi$. We would like to know for which symbols ϕ the operator C_ϕ leaves $A_\alpha^p(\Omega)$ invariant. In Theorem 3.1 we shall formulate the answer to this question in a geometric result expressed in terms of the family $\{E(z,r) : z \in \Omega, r > 0\}$ of hyperbolic balls, and in Theorem 3.2 we furnish an equivalent algebraic result with respect to the normalized reproducing kernel of these domains. These results will be used to arrive at constraints on the approach of ϕ to the boundary of Ω in Proposition 3.5 and Corollary 3.6. The corresponding results for the map ϕ to induce compact composition operators are also discussed.

For each $\alpha < \epsilon$, define the nonnegative Borel measure μ_α on Ω as a measure satisfying $\int_\Omega h(z) d\mu_\alpha = \int_\Omega (h \circ \phi) dV_\alpha$ for every $h \in C(\Omega)$. Then for a Borel set $E \subseteq \Omega$, $\mu_\alpha(E) = \int_{\phi^{-1}(E)} dV_\alpha$.

THEOREM 3.1. *Let* $1 \le p < \infty$. *Then*

(i) C_ϕ is a bounded composition operator on $A_\alpha^p(\Omega)$ if and only if for some $r > 0$

$$\sup_{z \in \Omega} \frac{\mu_\alpha(E(z,r))}{|E(z,r)|^{1-\alpha}} \leq M < \infty.$$

(ii) C_ϕ is compact on $A_\alpha^p(\Omega)$ if and only if for some $r > 0$

$$\lim_{z \to \partial\Omega} \frac{\mu_\alpha(E(z,r))}{|E(z,r)|^{1-\alpha}} = 0.$$

THEOREM 3.2.. *Let $1 \leq p < \infty$. Then*
(i) C_ϕ is bounded on $A_\alpha^p(\Omega)$ if and only if

$$\sup_{z \in \Omega} \int_\Omega |k_z(w)|^{2(1-\alpha)} d\mu_\alpha(w) < \infty. \tag{1}$$

(ii) C_ϕ is compact on $A_\alpha^p(\Omega)$ if and only if

$$\lim_{z \to \partial\Omega} \int_\Omega |k_z(w)|^{2(1-\alpha)} d\mu_\alpha(w) = 0. \tag{2}$$

Before proving these two theorems, we provide several remarks which attempt to interpret, restate, and apply the above two theorems.

Remarks 3.3. (i) The condition of Theorem 3.2 can be rephrased by saying that the operator C_ϕ is bounded on $A_\alpha^p(\Omega)$, $1 \leq p < \infty$, if and only if

$$\sup_{z \in \Omega} \int_\Omega |k_{\phi(z)}(w)|^{2(1-\alpha)} dV_\alpha(w) < \infty,$$

and C_ϕ is compact on $A_\alpha^p(\Omega)$, $1 \leq p < \infty$, if and only if

$$\lim_{z \to \partial\Omega} \int_\Omega |k_{\phi(z)}(w)|^{2(1-\alpha)} dV_\alpha(w) = 0.$$

(ii) Theorem 3.1 is a statement about the regularity of the measures μ_α induced by the map ϕ. If $d\mu_\alpha(z) = g(z)dV_\alpha(z)$, $g \in L^1(\Omega)$, then Theorem 3.1(i) asserts that C_ϕ is bounded if and only if for some $r > 0$

$$\sup_{z \in \Omega} \frac{\int_{E(z,r)} g(z)dV_\alpha}{\int_{E(z,r)} dV_\alpha} \leq C,$$

and C_ϕ is compact if and only if for some $r > 0$

$$\lim_{z \to \partial\Omega} \frac{\int_{E(z,r)} g(z)dV_\alpha}{\int_{E(z,r)} dV_\alpha} = 0.$$

We shall also show that if C_ϕ is bounded on $A_\alpha^p(\Omega)$, $1 \leq p < \infty$, then for each $r > 0$ the average of the absolutely continuous part of μ_α (with respect to V_α) on the hyperbolic family of balls $\{E(z,r) : z \in \Omega\}$ is bounded.

We provide the proofs of these remarks and further discussion following the proofs of Theorems 3.1 and 3.2.

PROOF OF THEOREM 3.1. (i) Fix $\alpha < \epsilon$, $p > 0$, and suppose that C_ϕ is a bounded composition operator on $A_\alpha^p(\Omega)$. By Proposition 2.3, since for each $z \in \Omega$, $k_z^{1-\alpha}(w)$ is a unit vector in $A_\alpha^2(\Omega)$, and since $k_z(w)$ is non-vanishing, if we let $f(w) = k_z^{\frac{2(1-\alpha)}{p}}(w)$, then $f \in A_\alpha^p$. Since C_ϕ is bounded, for every $r > 0$ and for every $z \in \Omega$

$$\int_{E(z,r)} |f(z)|^p d\mu_\alpha = \int_{E(z,r)} |k_z(w)|^{2(1-\alpha)} d\mu_\alpha \leq C.$$

By Proposition 2.3 (v), since for each $r > 0$ there exists a constant $C_1 < \infty$ so that $|E(z,r)||k_z(w)|^2 \leq C_1$ for every $z \in \Omega$ and $w \in E(z,r)$

$$C_1^{1-\alpha} \frac{\mu_\alpha(E(z,r))}{|E(z,r)|^{1-\alpha}} \leq \int_{E(z,r)} |k_z(w)|^{2(1-\alpha)} d\mu_\alpha \leq C.$$

Therefore, for each $r > 0$ there exists a $C < \infty$ so that

$$\sup_{z \in \Omega} \frac{\mu_\alpha(E(z,r))}{|E(z,r)|^{1-\alpha}} \leq C. \tag{1}$$

Note that the proof in this direction is valid for $p > 0$ and for every $r > 0$.

Conversely, suppose (1) holds for some $r > 0$. We will show that the composition operator C_ϕ is bounded. Since the topology induced by the hyperbolic metric β on Ω is equivalent to the usual topology on Ω, given $E(0,r)$ let $B(0,d)$ be a Euclidean ball centered at 0 and which is contained in $E(0,r)$. If $f \in A_\alpha^p(\Omega)$, $1 \leq p < \infty$ by the n-subharmonicity of $|f|^p$

$$|f(0)|^p \leq \frac{1}{|B(0,d)|} \int_{B(0,d)} |f(w)|^p dV(w) \leq \frac{1}{|B(0,d)|} \int_{E(0,r)} |f(w)|^p dV(w).$$

If ψ_z is an automorphism of Ω that interchanges z and 0 replacing f by $f \circ \psi_z$ gives

$$|f(z)|^p \leq \frac{1}{|B(0,d)|} \int_{E(z,r)} |f(w)|^p |J_c\psi_z(w)|^2 dV(w).$$

By Proposition 2.3, since $|J_c\psi_z(w)|^2 = |k_z(w)|^2$ and $|k_z(w)|^2 \leq \frac{C}{|E(z,r)|}$,

$$\begin{aligned}
|f(z)|^p &\leq \frac{C}{|B(0,d)|} \frac{1}{|E(z,r)|} \int_{E(z,r)} |f(w)|^p dV(w) \\
&= \frac{C}{|E(z,r)|} \int_{E(z,r)} |f(w)|^p dV(w).
\end{aligned} \tag{2}$$

Note that the constant C will in general depend on r and n and is independent of z and p. This is the important (pseudo) n-subharmoninicity property of the functions in A_0^p, $1 \leq p < \infty$ with respect to the hyperbolic metric. By a standard covering lemma [see [2] or [8] for example], for each fixed $r > 0$ there exists a sequence of points $\{z_j\} \subset \Omega$ and a positive integer N such that $\cup_{j=1}^\infty E(z_j, r)$ covers Ω, and any $z \in \Omega$ belongs to at most N of the sets $E(z_j, r)$. Integrating

(2) we use this covering lemma to decompose the integral over Ω into the integrals over $\cup_{j=1}^{\infty} E(z_j, r)$.

$$\int_{\Omega} |f(z)|^p d\mu_\alpha \leq \sum_{j=1}^{\infty} \int_{E(z_j, r)} |f(z)|^p d\mu_\alpha$$

$$\leq \sum_{j=1}^{\infty} \sup\{|f(z)|^p : z \in E(z_j, r)\} \mu_\alpha(E(z_j, r))$$

$$\leq C \sum_{j=1}^{\infty} \frac{\mu_\alpha(E(z_j, r))}{|E(z_j, r)|} \int_{E(z_j, r)} |f(z)|^p dV$$

$$\leq C \sum_{j=1}^{\infty} \frac{\mu_\alpha(E(z_j, r))}{|E(z_j, r)|^{1-\alpha}} \int_{E(z_j, r)} |f(z)|^p dV_\alpha$$

$$\leq C \sup_{z_j \in \Omega} \frac{\mu_\alpha(E(z_j, r))}{|E(z_j, r)|^{1-\alpha}} \sum_{j=1}^{\infty} \int_{E(z_j, r)} |f(z)|^p dV_\alpha.$$

By hypothesis, given $r > 0$ since $\frac{\mu_\alpha(E(z, r))}{|E(z, r)|}$ is bounded for every $z \in \Omega$, and by the covering lemma since $\cup_{j=1}^{\infty} E(z_j, r)$ covers Ω at most N times, we conclude

$$\int_{\Omega} |f(z)|^p d\mu_\alpha \leq C N \int_{\Omega} |f(z)|^p dV_\alpha.$$

(ii) As in part (i), fix $p > 0$, $\alpha < \epsilon$, and suppose C_ϕ is a compact composition operator on $A_\alpha^p(\Omega)$. Let $f_z(w) = k_z^{\frac{2(1-\alpha)}{p}}(w)$. Then $\{f_z\}$ is a family of unit vectors in $A_\alpha^p(\Omega)$. Since C_ϕ is compact, for every $r > 0$

$$\lim_{z \to \partial\Omega} \int_{E(z, r)} |f_z(w)|^p d\mu_\alpha(w) = 0.$$

Now appealing to Proposition 2.3 exactly as in the proof of part (i) of this theorem, we obtain

$$\lim_{z \in \Omega} \frac{\mu_\alpha(E(z, r))}{|E(z, r)|^{1-\alpha}} = 0. \tag{3}$$

Conversely, suppose that (3) holds for some $r > 0$. We show that the composition operator C_ϕ is compact by demonstrating that if $\{f_n\}$ is a norm bounded sequence in $A_\alpha^p(\Omega)$ converging to zero uniformly on compact subsets of Ω, then $C_\phi f_n$ converges to zero in norm. Given $r > 0$ and $\epsilon > 0$, choose $R > 0$ so that

$$\frac{\mu_\alpha(E(z, r))}{|E(z, r)|^{1-\alpha}} < \epsilon$$

whenever $\sup\{\beta(z, \zeta) < R : \zeta \in \partial\Omega\}$, i.e. whenever z is sufficiently close to $\partial\Omega$ [2, Lemma 4]. Denote the set of all such points in Ω by Ω_1 and let $\Omega_2 = \Omega \setminus \Omega_1$. Since the sequence $\{f_n\}$ converge to zero uniformly on compact subsets of Ω, and since $\overline{\Omega_2}$ is compact,

$$\lim_{n \to \infty} \int_{\Omega_2} |f_n|^p d\mu_\alpha = 0.$$

On the other hand, by an argument exactly as in the proof of part (i), for $1 \leq p < \infty$

$$\int_{\Omega_1} |f_j(z)|^p d\mu_\alpha \leq C \sup_{z \in \Omega_2} \frac{\mu_\alpha(E(z,r))}{|E(z,r)|^{1-\alpha}} \int_\Omega |f_j(z)|^p dV_\alpha$$

$$< C\epsilon \int_\Omega |f_j|^p dV_\alpha.$$

Since $\{f_j\}$ is a norm bounded family in A^p_α, the desired result follows. ∎

PROOF OF THEOREM 3.2. (i) To prove the sufficiency of condition (1) in Theorem 3.2, suppose that $\int_\Omega |k_z(w)|^{2(1-\alpha)} d\mu_\alpha(w) \leq C$ for every $z \in \Omega$. By Proposition 2.3 (v), given $r > 0$ since there exist constants C_1 and C_2 so that $C_1 \leq |k_z(w)|^2 |E(z,r)| \leq C_2$ for every $z \in \Omega$ and $w \in E(z,r)$,

$$C_1^{1-\alpha} \leq |k_z(w)|^{2(1-\alpha)} |E(z,r)|^{1-\alpha} \leq C_2^{1-\alpha} \ \forall z \in \Omega.$$

Hence, given $r > 0$, for every $z \in \Omega$

$$C_1^{1-\alpha} \frac{\mu_\alpha(E(z,r))}{|E(z,r)|^{1-\alpha}} \leq \int_{E(z,r)} |k_z(w)|^{2(1-\alpha)} d\mu_\alpha \leq \int_\Omega |k_z(w)|^{2(1-\alpha)} d\mu_\alpha \leq C.$$

By Theorem 3.1 the boundedness of C_ϕ on $A^p_\alpha(\Omega)$ for $1 \leq p < \infty$ follows.

Conversely, suppose that C_ϕ is bounded on $A^p_\alpha(\Omega)$, for $p \in [1, \infty)$. Then by Theorem 3.1, C_ϕ is bounded on $A^2_\alpha(\Omega)$. By Proposition 2.3, since for each z, $k_z^{1-\alpha}$ is a unit vector in $A^2_\alpha(\Omega)$, by the boundedness of C_ϕ

$$\int_\Omega |k_z(w)|^{2(1-\alpha)} d\mu_\alpha \leq C \ \forall z \in \Omega.$$

(ii) Proof of compactness is similar. ∎

Remarks 3.4. (i) Note that the conditions of Theorem 3.1 and 3.2 are independent of p for $1 \leq p < \infty$. This important fact can be restated as saying that if C_ϕ is a bounded (compact) composition operator on $A^p_\alpha(\Omega)$ for some p, then C_ϕ is a bounded (compact) composition operator on $A^p_\alpha(\Omega)$ for every $1 \leq p < \infty$.

(ii) The role played by $\alpha < \epsilon$ is displayed by Theorem 3.2. By Remark 3.3 (i), since the condition of Theorem 3.2 may be restated in terms of the uniform boundedness of the A^2_α-norm of an associated family of kernels (by a change of variable), α plays a much more critical role than p in general. Furthermore, using specific estimates on the normalized Bergman kernels as in [5] or [19] more detailed information may be obtained.

We deduce a criterion for a map ϕ to induce a bounded composition operator:

PROPOSITION 3.5. *Fix $\alpha < \epsilon$, and let*

$$m_z(w) = \frac{k_{\phi(z)}^{1-\alpha}(w)}{k_z^{1-\alpha}(w)}.$$

Then C_ϕ is a bounded composition operator on A_α^p, $1 \le p < \infty$, if and only if, for each $z \in \Omega$, m_z maps the unit vector $k_z^{1-\alpha}$ into A_α^2 under pointwise multiplication.

PROOF. Let m_z map the unit vector $k_z^{1-\alpha}$ into $A_\alpha^2(\Omega)$ under pointwise multiplication. Then $k_z^{1-\alpha}(w)m_z(w) \in A_\alpha^2(\Omega)$ for every $z \in \Omega$. Therefore $k_{\phi(z)}^{1-\alpha} \in A_\alpha^2$ for every $z \in \Omega$. By the uniform boundedness principle, and Remark 3.3 (i) the boundedness of C_ϕ follows.

Conversely, suppose that C_ϕ is bounded. Then $k_{\phi(z)}^{1-\alpha} \in A_\alpha^2$ for every $z \in \Omega$. Hence $k_z^{1-\alpha}(w)m_z(w) \in A_\alpha^2(\Omega)$ for every $z \in \Omega$. ∎

COROLLARY 3.6. *Let m_z be defined as in 3.5. If m_z is essentially bounded on Ω, i.e. if there exists a $C < \infty$ (C may depend on z, but is independent of w) so that $|m_z(w)| \le C$ for almost every $w \in \Omega$, then C_ϕ is a bounded composition operator on $A_\alpha^p(\Omega)$ for every $p \in [1,\infty)$ and for every $\alpha < \epsilon$.*

PROOF. For each $z \in \Omega$, if $m_z \in H^\infty(\Omega)$ then m_z is a multiplier of $A_\alpha^p(\Omega)$ for every $\alpha < \epsilon$. Now by Proposition 3.5, since m_z maps $k_z^{1-\alpha}$ into A_α^p, C_ϕ is bounded on $A_\alpha^p(\Omega)$ for every $\alpha < \epsilon$ and every $1 \le p < \infty$. ∎

Note that if we show that the family of unit vectors $k_z^{1-\alpha}$ span $A_\alpha^2(\Omega)$, then the statement of Proposition 3.5 can be replaced by the statement that C_ϕ is bounded on $A_\alpha^p(\Omega)$ if and only if the family of functions m_z are multipliers of $A_\alpha^2(\Omega)$. In this direction various results have been proved by [4].

Proposition 3.5 and Corollary 3.6 imply various geometric constraints on ϕ in order for ϕ to induce a bounded composition operator. In particular, the boundedness of m_z in Corollary 3.6 suggests that $k_{\phi}(z)$ and k_z must have similar orders of growth as z approaches $\partial\Omega$. A study of multipliers of $A_\alpha^2(\Omega)$ would provide further detailed information.

PROOF OF REMARKS 3.3. (i) Fix $z \in \Omega$ and $\alpha < \epsilon$. Suppose C_ϕ is bounded on A_α^2 and let C_ϕ^* denote the adjoint of the operator C_ϕ. Since $||C_\phi||_{\alpha,2} = ||C_\phi^*||_{\alpha,2}$, applying C_ϕ^* to the unit vector $k_z(w)$ we get (see [7] for example)

$$\int_\Omega |k_{\phi(z)}(w)|^{2(1-\alpha)}dV_\alpha(w) \le C < \infty.$$

On the other hand, suppose

$$\sup_{z \in \Omega} \int_\Omega |k_{\phi(z)}|^{2(1-\alpha)}(w)dV_\alpha(w) \le C < \infty$$

For $f \in A_\alpha^2(\Omega)$, since

$$C_\phi f(z) = f(\phi(z)) = (f, k_{\phi(z)}^{1-\alpha}).$$

by the Cauchy-Schwarz inequality we obtain

$$\int_\Omega |C_\phi f(z)|^2 dV_\alpha(z) \le$$

$$\leq \sup_{z \in \Omega} \int_\Omega k_{\phi(z)}(w)|^{2(1-\alpha)} dV_\alpha(w) \int_\Omega |f(w)|^2 dV_\alpha(w) \leq C \int_\Omega |f(w)|^2 dV_\alpha(w).$$

That is, C_ϕ is a bounded operator on A_α^2. Hence we may replace condition (1) of Theorem 3.2 by

$$\sup_{z \in \Omega} \int_\Omega |k_{\phi(z)}(w)|^{2(1-\alpha)} dV_\alpha(w) < \infty.$$

Likewise, we may replace condition (2) in Theorem 3.2 by

$$\lim_{z \to \partial\Omega} \int_\Omega |k_{\phi(z)}(w)|^{2(1-\alpha)} dV_\alpha(w) = 0.$$

In fact, since the reproducing kernels $k_z^{1-\alpha}$ are unit vectors in A_α^2, the condition for equality in the Cauchy-Schwarz inequality shows that the A_α^2-norm of the composition operator C_ϕ is exactly $\sup_{z \in \Omega} \|k_{\phi(z)}^{1-\alpha}\|_{\alpha,2}$. This concludes the proof of Remark 3.3(i).

(ii) The proof of the first part of this statement follows directly from Theorem 3.1 for measures μ_α satisfying

$$d\mu_\alpha(z) = g(z) dV_\alpha(z)$$

for some $g \in L^1(\Omega)$. In particular, for these cases C_ϕ is a bounded composition operator on $A_\alpha^p(\Omega)$ if and only if given $r > 0$ the averages of the function g over the hyperbolic balls $E(z,r)$ is uniformly bounded for every $z \in \Omega$. Similarly, applying Theorem 3.1 (ii) to these class of measures implies that C_ϕ is a compact composition operator on $A_\alpha^p(\Omega)$ if and only if given $r > 0$ the averages of the function g over the hyperbolic balls $E(z,r)$ tends to zero as z approaches $\partial\Omega$.

For each α, since μ_α is a bounded nonnegative measure on Ω, by the Radon-Nikodym theorem we may decompose $d\mu_\alpha(z) = g(z) dV_\alpha + d\nu_\alpha(z)$ where ν_α is a singular measure with respect to the volume mesaure V_α. Since for each $E(z,r) \subset \Omega$

$$\int_{E(z,r)} g(w) dV_\alpha(w) \leq \mu_\alpha(E(z,r)),$$

unboundedness of the averages of g on the hyperbolic balls on Ω implies the unboundedness of the composition operators on $A_\alpha^p(\Omega)$. This completes the proof of statement (ii) in Remark 3.3.

Theorems 3.1 and 3.2, together with restatement of these Theorems in Remarks 3.3, provide alternative geometric and computational criteria to determine for which holomorphic maps ϕ of a bounded symmetric domains into itself the change of variable $z \mapsto \phi(z)$ is a continuous or compact linear operator on weighted Bergman spaces. Theorem 3.1 states that this operator is bounded if the measures induced by the map ϕ have uniformly bounded averages over the hyperbolic balls in Ω, and that this operator is compact if these averages tend to zero as z approaches the boundary of Ω. On the other hand, Theorem 3.2 combined with the explicit descriptions of the Bergman kernel given by [6], [11] provide integration formulas for computation. In Proposition 3.5 a criterion is obtained to express the behavior of ϕ near the boundary of Ω which would be necessary and sufficient to provide continuity of the composition operator. Corollary 3.6 provides an important sufficiency case deduced from Proposition 3.5. Writing the specific form of the Bergman kernel in the disk, polydisks or the unit ball of $\mathbb{C}^n$ would also show that the conditions of Theorem 3.2 and Proposition 3.5 may be expressed in terms of Julia-Carathéodory type criteria [9], [14]. Examples of results suggested by these theorems in the specific cases of the disk, polydisks and the unit ball agree with the corresponding results obtained with Carleson measures and Carleson sets [7],[12],[13]. Further geometric results would require a characterization of the multipliers of $A_\alpha^p(\Omega)$, and connection of these results to theorems on atomic decompositions of the weighted Bergman spaces [4]. This and related work for the Hardy spaces remain under investigation.

References.

1. D. Békollé, C.A. Berger, L.A. Coburn, K.H. Zhu, *BMO in the Bergman metric on bounded symmetric domains*, J. Functional Anal. **93** (1990), 310-350.

2. C.A. Berger, L.A. Coburn, K.H. Zhu, *Function theory on Cartan Domains and the Berezin-Toeplitz symbol Calculus*, Amer. J. Math. **110** (1988), 921-953.

3. J.A. Cima, W.R. Wogen, *Unbounded composition operators on $H^2(B_n)$*, Proceedings of Amer. Math. Soc. **99** (1987), 477-483.

4. R.R. Coifman, R. Rochberg, *Representation theorems for holomorphic and harmonic functions in L^p*, Astérisque **77** (1980), 1-66.

5. F. Forelli, W. Rudin, *Projections on spaces of holomorphic functions in balls*, Indiana Univ. Math. J. **24** (1974), 593-602.

6. L.K. Hua, *Harmonic analysis of functions of several complex variables in the classical domains*, Transl. of Math. Monographs **6** (1963), American Mathematical Society.

7. F. Jafari, *On bounded and compact composition operators in polydiscs*, Canad. Math. J. **XLII** (1990), 869-889.

8. _____, *Carleson measures in Hardy and weighted Bergman spaces of polydiscs*, Proceedings of Amer. Math. Soc. **112** (1991), 771-781.

9. _____ Angular derivatives in polydiscs, preprint.

10. S. Kobayashi, *Hyperbolic manifolds and holomorphic mappings*, Marcel Dekker, 1970.

11. A. Korányi, *The Poisson integral for generalized half-planes and bounded symmetric domains*, Ann. of Math. **82** (1965), 332-350.

12. B.D. MacCluer, *Compact composition operators on $H^p(B_N)$*, Michigan Math. J. **32** (1985), 237-248.

13. B.D. MacCluer, J.H. Shapiro, *Angular derivatives and compact composition operators on the Hardy and Bergman spaces*, Canad. J. Math. **38** (1986), 878-906.

14. W. Rudin, *Function theory in the unit ball of* $\mathbb{C}^n$, Grundlehren der Math. Wiss., vol. 241, Springer, 1980.

15. J.H. Shapiro, *The essential norm of a composition operator*, Ann. of Math. **125** (1987), 375-404.

16. J.H. Shapiro, P.D. Taylor, *Compact, Nuclear, and Hilbert-Schmidt composition operators*, Indiana Univ. Math. J. **23** (1973), 471-496.

17. S.D. Sharma, R.K. Singh, *Composition operators and several complex variables*, Bull. Austral. Math. Soc. **23** (1981), 237-247.

18. E.M. Stein, *Boundary behavior of holomorphic functions of several complex variables*, Math. Notes, Princeton Univ. Press, 1972.

19. M. Stoll, *Mean value theorems for harmonic and holomorphic functions on bounded symmetric domains*, J. reine angew Math. **283** (1977), 191-198.

20. W.R. Wogen, *Composition operators acting on spaces of holomorphic functions on domains in* $\mathbb{C}^n$, Proceedings of Symposia in Pure Mathematics **51** (1990), 361-366.

21. _____, *The smooth mappings which preserve the Hardy space* $H^2(B_2)$, Operator theory: Adv. and Appl. **35** (1988), 249-263.

22. K.H. Zhu, *VMO, ESV, and Toeplitz operators on the Bergman space*, Trans. of Amer. Math. Soc. **302** (1987), 617-646.

DEPARTMENT OF MATHEMATICS, UNIVERSITY OF WYOMING, LARAMIE, WYOMING 82071-3036

Contemporary Mathematics
Volume **137**, 1992

Isotopic Embeddings of Affine Algebraic Varieties into $\mathbf{C}^n$

SHULIM KALIMAN

1. Let X be a closed affine subvariety of $\mathbf{C}^n$. We shall study proper algebraic embeddings of X into Euclidean spaces. For every point $x \in X$ we shall denote the tangent space to X at x by $T_x X$. Recall that the Zariski's tangent bundle TX is the set $\{(x, v) | x \in X, \quad v \in T_x X\}$. The Zariski's tangent bundle can be viewed as an affine variety and, if X is smooth, $\dim TX = 2 \dim X$. The following theorem holds [K].

THEOREM 1. *Let f, $g : X \to \mathbf{C}^n$ be two proper algebraic embeddings of an affine algebraic variety X into $\mathbf{C}^n$ with $n > \max(1 + 2 \dim X, \dim TX)$. Then there exists a polynomial automorphism α of $\mathbf{C}^n$ such that $f = \alpha \circ g$.*

It is worth mentioning that from the explicit construction of α one can choose this automorphism to be tame. Theorem 1 implies that every polynomial embedding of $\mathbf{C}$ into $\mathbf{C}^n$ with $n \geq 4$ is equivalent to a linear embedding up to a polynomial automorphism of $\mathbf{C}^n$ (see [J] as well). It is interesting to study similar problems in the case when $n \leq \max(1 + 2 \dim X, \dim TX)$ (for example, polynomial embeddings of $\mathbf{C}$ into $\mathbf{C}^3$). In [BN] S. Bell and R. Narasimhan present the following unpublished theorem of M.V. Nori.

THEOREM 3.11. *Let X be a smooth affine variety of dimension k and let $f, g : X \to \mathbf{C}^n$ be two (proper) algebraic embeddings of X into $\mathbf{C}^n$ with $n \geq 2k+2$. Then f, g are isotopic via proper algebraic embeddings.*

This fact also follows from theorem 1 (see lemma 2 in this paper). By theorem 3.11, every two algebraic embeddings of $\mathbf{C}$ into $\mathbf{C}^n$ with $n \geq 4$ are isotopic via algebraic embeddings. S. Bell and R. Narasimhan supposed that for $n = 3$ the similar fact did not hold.

1991 *Mathematics Subject Classification*. Primary 32C10.

This paper is in final form and no version of it will be submitted for publication elsewhere.

CONJECTURE 3.13. *([BN]). There exists an algebraic embedding of* $\mathbf{C}$ *into* $\mathbf{C}^3$ *which is not isotopic via proper holomorphic embeddings to the linear embedding.*

We shall show that theorem 3.11 holds for $n \geq 2k + 1$ as well (this estimate cannot be improved) and, therefore, conjecture 3.13 is not true. Moreover we shall prove that every algebraic embedding of $\mathbf{C}$ into $\mathbf{C}^3$ is equivalent to a linear embedding up to a holomorphic automorphism of $\mathbf{C}^3$.

The referee has informed me that M.V. Nori knew theorem 1 and theorem 4 in the case when X is smooth. He also knew that α in theorem 1 was tame , but his argument was different.

2. We shall need two simple statements.

PROPOSITION 2. *(Makar–Limanov). The group* $Aut\,\mathbf{C}^n$ *of polynomial automorphisms is linearly connected.*

PROOF. One can easily see that the subgroup of linear automorphisms is linearly connected. Thus it suffices to show that every $\alpha \in Aut\,\mathbf{C}^n$ is linearly connected to some linear automorphism. Let $x = (x_1, \ldots, x_n)$ be a coordinate system in $\mathbf{C}^n$ and $\alpha(x) = (p_1(x), \ldots, p_n(x))$. Put $a_k = p_k(0)$ and $q_k(x) = p_k(x) - a_k$. Consideration of the polynomial automorphisms $(q_1(x) + a_1 t, \ldots, q_n(x) + a_n t)$ (where $t \in [0,1]$) shows that α is linearly connected to $\alpha'(x) = (q_1(x), \ldots, q_n(x))$. Put $\beta_\lambda(x) = (\lambda^{-1} q_1(\lambda x), \ldots, \lambda^{-1} q_n(\lambda x))$, where $\lambda \in [0,1]$. Then $\beta_1 = \alpha'$ and β_0 is a linear automorphism, which is the desired conclusion.

For every point x of a closed affine subvariety X of $\mathbf{C}^m$ there exists the natural embedding of the tangent space $T_x X$ into the space $W \cong \mathbf{C}^m$ of constant vector fields on $\mathbf{C}^m$. Consider the mapping of the Zariski's tangent bundle $TX = \{(x,v) | x \in X, \, v \in T_x X\}$ into $\mathbf{C}^m$ given by the formula $(x,v) \to x + v$, where we treat $v \in W$ as a vector in $\mathbf{C}^m$. We shall denote the image of TX under this mapping by $T^\circ X$. Recall that the chord variety CX of X is the closure in $\mathbf{C}^m$ of the set of lines crossing X in at least at two points (we have to fix a coordinate system in $\mathbf{C}^m$ first in order to determine lines uniquely). Let $LX = CX \cup T^\circ X$. Note that LX is again a closed affine subvariety of $\mathbf{C}^m$ and that $\dim LX \leq \max(1 + 2 \dim X, \dim TX)$. Since we have fixed the coordinate system in $\mathbf{C}^m$, there is the natural embedding $\mathbf{C}^m \hookrightarrow \mathbf{CP}^m$. Put $E = \mathbf{CP}^m - \mathbf{C}^m$, then $E \cong \mathbf{CP}^{m-1}$ and to each point $e \in E$ we can assign the orthogonal projection $\rho_e : \mathbf{C}^n \to \mathbf{C}^{n-1}$ with center at e. Let $\overline{LX}$ be the closure of LX in $\mathbf{CP}^m$. The following fact is well-known (e.g., see [K], proposition 8).

LEMMA 3. *Let* $e \in E - \overline{LX}$. *Then* $Y = \rho_e(X)$ *is a closed affine subvariety of* $\mathbf{C}^{m-1}$ *and the restriction of* ρ_e *to* X *is an isomorphism between* X *and* Y.

3. The main result of this paper is the following theorem.

THEOREM 4. *Let* $f, g : X \to \mathbf{C}^n$ *be two proper algebraic embeddings of an affine algebraic variety* X *into* $\mathbf{C}^n$ *with* $n \geq \max(1 + 2 \dim X, \dim TX)$. *Then* f, g *are isotopic via proper algebraic embeddings.*

PROOF. Fix a coordinate system $(x_1, \ldots, x_{n+1})$ in $\mathbf{C}^{n+1}$ and, therefore, the natural embedding $\mathbf{C}^{n+1} \hookrightarrow \mathbf{CP}^{n+1}$. Let H be the hyperplane $x_{n+1} = 0$ and let $i : \mathbf{C}^n \to H$ be the linear isomorphism given by the formula $(x_1, \ldots, x_n) \to (x_1, \ldots, x_n, 0)$. Put $F = i \circ f$, $G = i \circ g$, $X_0 = F(X)$, and $X_1 = G(X)$. By theorem 1, there exists a automorphism $\alpha \in \operatorname{Aut} \mathbf{C}^{n+1}$ such that $G = \alpha \circ F$. By proposition 2, one can choose a continuous mapping $r : [0, 1] \to \operatorname{Aut} \mathbf{C}^{n+1}$ so that $r(0) = id$ and $r(1) = \alpha$. Put $\alpha_t = r(t)$, $X_t = \alpha_t(X)$, and $F_t = \alpha_t \circ F$ (clearly, $F_0 = F$ and $F_1 = G$). Put $E = \mathbf{CP}^{n+1} - \mathbf{C}^{n+1}$. Let the set LX_t be the same as in lemma 3. Then $\overline{LX}_t$ is a closed algebraic subvariety of $\mathbf{CP}^{n+1}$, the dimension of $\overline{LX}_t \leq \max(1 + 2 \dim X_t, \dim TX_t) \leq n$, and by construction there is no irreducible component of $\overline{LX}_t$ that belongs to E. Hence $\dim \overline{LX}_t \cap E \leq n - 1 < \dim E$, and the set $E - \overline{LX}_t$ is a connected manifold. Obviously, the variety X_t depends on t continuously. Consider $M = E \times [0, 1]$ and the two natural projections $\tau : M \to [0, 1]$ and $\mu : M \to E$. Let L be the subset of M so that $\tau^{-1}(t) \cap L = \overline{LX}_t \cap E$ for every $t \in [0, 1]$. Then $M - L$ is a connected real manifold. Introduce the coordinate system $(y_0 : y_1 : \ldots : y_{n+1})$ in $\mathbf{CP}^{n+1}$ for which $x_k = y_k / y_0$. Then the point $e = (0 : 0 : \ldots 0 : 1) \in E$ corresponds to the projection $\rho_e : \mathbf{C}^{n+1} \to \mathbf{C}^n$ given by the formula $(x_1, \ldots, x_{n+1}) \to (x_1, \ldots, x_n)$. Clearly, $f = \rho_e \circ F$ and $g = \rho_e \circ G$. Since $X_0, X_1 \subset H$, the sets LX_0 and $LX_1 \subset H$. Hence $e \notin \overline{LX}_0 \cup \overline{LX}_1$. Put $u_0 = (e, 0)$ and $u_1 = (e, 1)$. By construction $u_0, u_1 \in M - L$. Since $M - L$ is connected, there exists a continuous curve u_s in $M - L$ with $s \in [0, 1]$. Let ρ_s be the projection from $\mathbf{C}^{n+1}$ to $\mathbf{C}^n$ that corresponds to the point $\mu(u_s)$ and let $Y_s = \alpha_{\tau(u_s)} \circ F(X)$. By lemma 3, the restriction of ρ_s to Y_s is an isomorphism and $\rho_s(Y_s)$ is a closed subvariety of $\mathbf{C}^n$. Thus $\rho_s \circ \alpha_{\tau(u_s)} \circ F : X \to \mathbf{C}^n$ is a proper embedding for every $s \in [0, 1]$. This is the desired isotopy.

We would like to note that the estimate $n \geq 2k + 1$ in the smooth case cannot be improved. Consider two different coprime pairs of natural numbers (p_1, q_1) and (p_2, q_2). Then the $\mathbf{C}^\star$-curves $x^{p_1} y^{q_1} = 1$ and $x^{p_2} y^{q_2} = 1$ are not isotopically equivalent in $\mathbf{C}^2$ [N], since they have different links at infinity.

COROLLARY 5. *Every algebraic embedding of* $\mathbf{C}$ *into* $\mathbf{C}^n$ *is isotopic to the linear embedding via algebraic embeddings.*

PROOF. For $n = 1$ it is obvious, for $n = 2$ it is follows from the Abhyankar-Moh-Suzuki theorem ([AM], [S]), for $n \geq 3$ it is a special case of theorem 3.

4. In this section we shall show that algebraic embeddings of $\mathbf{C}$ have an additional property.

THEOREM 6. *Let* $F : \mathbf{C} \to \mathbf{C}^n$ *be an algebraic embedding. Then there exists a holomorphic automorphism* α *of* $\mathbf{C}^n$ *such that* $\alpha \circ F : \mathbf{C} \to \mathbf{C}^n$ *is a linear embedding.*

PROOF. If $n \neq 3$, one can even choose an algebraic automorphism α so that theorem 6 holds ([AM], [S], [J], [K]). Thus it is enough to consider $n = 3$. Let (x, y, z) be a coordinate system in $\mathbf{C}^3$, and let $\rho : \mathbf{C}^3 \to \mathbf{C}^2$ be the projection given by the formula $(x, y, z) \rightarrow (x, y)$. Without loss of generality, one may suppose that the restriction ρ to Γ is one-to-one everywhere except for a finite number of points and that the curve $\Gamma^1 = \rho(\Gamma) \subset \mathbf{C}^2$ has normal singularities only. Let t be a coordinate in $\mathbf{C}$ and $F(t) = (f(t), g(t), h(t))$. Note that Γ^1 is the image of $\mathbf{C}$ in $\mathbf{C}^2$ under the mapping $F_1 = (f, g)$. Let $F_1(t_{i1}) = F_1(t_{i2})$ for $i = 1, \ldots, N$. Since Γ^1 has normal singularities only, the ring R of holomorphic functions that have a representation $\varphi \circ F_1$ consists of all holomorphic functions $q(t)$ such that $q(t_{i1}) = q(t_{i2})$. Put $\lambda_{im} = h(t_{im})$ $(m = 1, 2)$. Note that $\lambda_{i1} \neq \lambda_{i2}$, since the curve Γ has no double points. Choose the set of numbers $\{a_i \mid i = 1, \ldots, N\}$ so that $[t_{i1} - e^{a_i}\lambda_{i1} = t_{i2} - e^{a_i}\lambda_{i2}.]$ Consider a holomorphic function $\varphi_1 = \psi_1 \circ F_1 \in R$ for which $\varphi_1(t_{im}) = a_i$ $(m = 1, 2)$. Let α_1 be the holomorphic automorphism of $\mathbf{C}^3$ given by the formula $[(x, y, z) \rightarrow (x, y, e^{\psi_1(x,y)}z).]$ Then the mapping $G = \alpha_1 \circ F$ can be represented as $G = (f, g, h_1)$, where the holomorphic function h_1 satisfies the condition: $t_{i1} - h_1(t_{i1}) = t_{i2} - h_1(t_{i2})$ for every $i = 1, \ldots, N$. Thus the function $t - h_1(t)$ belongs to R, i.e., $t - h_1(t) = \psi_2 \circ F_1(t)$. Put $[\alpha_2(x, y, z) = (x, y, \psi_2(x, y) + z).]$ Then the proper holomorphic embedding $H = \alpha_2 \circ G : \mathbf{C} \rightarrow \mathbf{C}^3$ is given by the formula $t \rightarrow (f(t), g(t), t)$. Consider the automorphism $[\alpha_3(x, y, z) = (x - f(z), y - g(z), z).]$ Clearly the embedding $\alpha_3 \circ H = \alpha_3 \circ \alpha_2 \circ \alpha_1 \circ F$ is linear. Thus $\alpha_3 \circ \alpha_2 \circ \alpha_1$ is the desired holomorphic automorphism.

We do not know if it is possible to use a polynomial automorphism instead of a holomorphic one in the formulation of this theorem. It is worth mentioning that according to recent results of W. Rudin and J.-P. Rosay, theorem 6 does not hold for proper holomorphic embeddings of $\mathbf{C}$ into $\mathbf{C}^n$.

THEOREM 7. *For every $n \geq 3$ there exists a proper holomorphic embedding of $\mathbf{C}$ into $\mathbf{C}^n$ which is not equivalent to a linear embedding up to holomorphic automorphisms of $\mathbf{C}^n$.*

PROOF. Recall that a discrete subset B of $\mathbf{C}^n$ is called tame if it can be mapped by holomorphic automorphism onto an arithmetic progression. W. Rudin and J.-P. Rosay proved that there exists a discrete subset of $\mathbf{C}^n$ that is not tame [RR1]. Another theorem of the same authors says that for $n \geq 3$ and every discrete subset B in $\mathbf{C}^n$ there is a proper holomorphic embedding $F : \mathbf{C} \rightarrow \mathbf{C}^n$ with $F(\mathbf{C}) \supset B$ [RR 2]. Suppose that B is not tame. Assume that α is a holomorphic automorphism for which $\alpha \circ F(\mathbf{C})$ coincides with a coordinate axis. Then $\alpha(B)$ is a discrete subset of the coordinate axis. Since every discrete subset of a proper linear subspace of $\mathbf{C}^n$ is tame [RR1], the sets $\alpha(B)$ and , therefore, B must be tame. This contradiction shows that α does not exist.

It is a pleasure to thank L. Makar–Limanov for useful discussion which helped to simplify the proof of Theorem 4.

References.

[AM] S.S. Abhyankar, T.T. Moh, *Embeddings of the line in the plane*, J. Reine Angew. Math. **276** (1975), 148–166.

[BN] S.R. Bell, R. Narasimhan, *Proper holomorphic mappings of complex spaces*, Several Complex Variables VI (Encyclopaedia of Mathematical Sciences, v. 69), Springer-Verlag, Berlin, Heidelberg, 1990.

[J] Z. Jelonek, *The extension of regular and rational embeddings*, Math. Ann. **277** (1987), 113–120.

[K] Sh. Kaliman, *Extension of isomorphisms between algebraic subvarieties of k^n to automorphisms of k^n*, Proc. AMS **113** (1991), 325–334.

[N] W.D. Neumann, *Complex algebraic plane curves via their links at infinity*, Invent. Math. **98** (1989), 445–489.

[RR1] J.-P. Rosay, W. Rudin, *Holomorphic maps from $\mathbf{C}^n$ to $\mathbf{C}^n$*, Trans. AMS **310** (1988), 47–86.

[RR2] J.-P. Rosay, W. Rudin, *Holomorphic embeddings of $\mathbf{C}$ in $\mathbf{C}^n$*, preprint.

[S] M. Suzuki, *Propiétes topologiques des polynômes de deux variables complexes, et automorphismes algébrigue de l'espace $\mathbf{C}^2$*, J. Math. Soc. Jpn. **26** (1974), 241–252.

UNIVERSITY OF MIAMI, DEPARTMENT OF MATHEMATICS AND COMPUTER SCIENCE, CORAL GABLES, FLORIDA 33124

Contemporary Mathematics
Volume **137**, 1992

Polynomial solutions of the Fueter-Hurwitz equation

WIESLAW KRÓLIKOWSKI AND ENRIQUE RAMÍREZ DE ARELLANO

ABSTRACT. One important result derived from the theory of Hurwitz pairs is an analogue to the classical Cauchy–Riemann equations, the so called generalised Fueter equation, or Fueter–Hurwitz equation. In this paper the Hurwitz multiplication structure is used to find a special class of solutions of this equation.

Introduction

The theory of Euclidean and non-Euclidean Hurwitz pairs developed by Lawrynowicz and Rembieliński (e.g., [5–8]) is based on the papers by A. Hurwitz [1–2]. Hurwitz proved in [2] that any normed division algebra over $\mathbb{R}$, with unity, is isomorphic to $\mathbb{R}$, $\mathbb{C}$, $\mathbb{H}$ and $\mathbb{Q}$, the real, complex, quaternion or octonion number. In particular, Hurwitz showed that all positive integers n and all systems $c_{j\alpha}^k \in \mathbb{R}$, $j, k, \alpha = 1, \ldots, n$ such that the collection of bilinear forms $\eta_j = x_\alpha c_{j\alpha}^k y_k$ satisfies the condition

$$\sum_j \eta_j^2 = \left(\sum_\alpha x_\alpha^2\right)\left(\sum_k y_k^2\right),$$

are restricted to the cases $n = 1, 2, 4$ or 8.

In the theory of Hurwitz-pairs where $j, k = 1, \ldots, n$ and $\alpha = 1, \ldots, p$, an analogue to the classical Cauchy–Riemann equations can be defined, the so-called generalised Fueter equation. This equation first appeared in the quaternionic analysis in order to obtain a proper notion of quaternionic holomorphicity. The theory of mappings which are regular in the sense of Fueter is still being developed. In this paper we find a special class of solutions of the Fueter–Hurwitz equation.

1991 *Mathematics Subject Classification*. Primary 32A30, 30G35; Secondary 32K15.

Key words and phrases. Generalized Cauchy–Riemann equations, Fueter–Hurwitz equation, hypercomplex regular functions.

Research partially supported by CONACyT A128 CCOE-910622(MT-7) and COSNET 053.K.91 grants.

This paper is in final form and no version of it will be submitted for publication elsewhere.

The Fueter–Hurwitz equation

Consider two real unitary vector spaces S and V of dimensions p and n, respectively. In both cases we denote the usual norms by $\| \ \|$. Let (ϵ_α) be an orthonormal basis in S and (e_j) in V. Define a mapping $\circ : S \times V \to V$ (*multiplication* of vectors in S by vectors in V with values in V) with the properties:

(1) $(a + b) \circ f = a \circ f + b \circ f$ and $a \circ (f + g) = a \circ f + a \circ g$ for $f, g \in V$ and $a, b \in S$;
(2) $\|a\|_S \|f\|_V = \|a \circ f\|_V$ (the Hurwitz condition);
(3) there exists the unit element ϵ_p in S with respect to the multiplication $\circ : \epsilon_p \circ f = f$.

If "$\circ$" does not leave invariant nondegenerate proper subspaces of V, the corresponding pair (S, V) is said to be *irreducible**. In such a case we call (S, V) a *Hurwitz pair*.

In a more general setting, non degenerate pseudo-Euclidean real scalar products in S and V may be considered and pseudo-Euclidean Hurwitz pairs (S, V) defined (see e.g [4–8]). Let us regard a continuously differentiable V-valued mapping f from an open domain Ω in S with values in V ($f : \Omega \to V$). It can be expresed as

$$f(\mathbf{x}) = f(x^\alpha \epsilon_\alpha) = f^k(\mathbf{x})e_k.$$

Lawrynowicz and Rembieliński [5–8] defined the generalised Fueter operator D^+ as

$$D^+ := \epsilon_\alpha \partial^\alpha, \qquad \partial^\alpha = \frac{\partial}{\partial x_\alpha}, \qquad \alpha = 1, \dots, p.$$

The mapping f introduced above is said to be *regular* in its domain if it satisfies the *Fueter–Hurwitz equation*

(1) $$D^+ \circ f = 0,$$

where "$\circ$" denotes the *Hurwitz multiplication*, which is characterised by the multiplication scheme for the base vectors:

(2) $$\epsilon_\alpha \circ e_j = C_{\alpha j}^k e_k.$$

Since we defined only a left-side multiplication of the elements of S by those of V, we define only the left-regular mappings, called *regular* for short. We have [8, Thm. 1]:

Any generalised Fueter operator D^+ has a conjugate $D := \epsilon_\alpha^+ \partial^\alpha$, where the ϵ_α^+ are defined by

$$\epsilon_\alpha^+ \circ e_j = C_{\alpha j}^{+k} e_k, \qquad C_\alpha^+ := \kappa C_\alpha^T \kappa^{-1},$$

$$\kappa \equiv [\kappa_{jk}] := [\langle e_j, e_k \rangle_V];$$

---

*This property is called *unsplittable* in [9] in the noneuclidean case.

$\langle \, , \, \rangle_V$ denotes the (non-degenerate) pseudo-Euclidean real scalar product in V. Moreover, if f is regular, then $D \circ f$ is regular, $DD^+ \circ f = D^+ D \circ f$ and $DD^+ = D^+ D = \eta_{\alpha\beta} \partial^\alpha \partial^\beta$, where

$$\eta_{\alpha\beta} := \langle \epsilon_\alpha, \epsilon_\beta \rangle_S;$$

$\langle \, , \, \rangle_S$ denotes the (non-degenerate) pseudo-Euclidean real scalar product in S.

Special polynomial solutions of the Fueter–Hurwitz equation

Let us express the Fueter equation (1) in a more convenient form. By (2) we obtain

$$(D^+ \circ f = 0) \iff (\sum_{\alpha=1}^{p} \sum_{k=1}^{n} (\epsilon_\alpha \partial^\alpha) \circ (f^k e_k) = 0) \iff$$

$$(\sum_{\alpha,k} (\partial^\alpha f^k)(\epsilon_\alpha \circ e_k) = 0) \iff (\sum_{\alpha,k,m} (\partial^\alpha f^k) C^m_{\alpha k} e_m = 0) \iff$$

$$\sum_{k=1}^{n} \sum_{m=1}^{n} \sum_{\alpha=1}^{p-1} [(\partial^\alpha f^k C^m_{\alpha k} e_m + (\partial^p f^k) C^m_{pk} e_m] = 0.$$

By a convenient change of coordinates we can take $C_p = I_n$, where I_n is the identity $(n \times n)$-matrix (see, e.g., [8]); then the last equation is equivalent to

$$\sum_{k=1}^{n} \sum_{\alpha=1}^{p-1} (\partial^\alpha f^k) \epsilon_\alpha \circ e_k + \sum_{k=1}^{n} (\partial^p f^k) e_p \circ e_k = 0.$$

Hence

$$(D^+ \circ f = 0) \iff ([\sum_{\alpha=1}^{p-1} \epsilon_\alpha \partial^\alpha + \epsilon_p \partial^p] \circ f = 0).$$

Henceforth, the following equation will be also called the *Fueter–Hurwitz equation (F–H equation)*

$$(3) \qquad (\sum_{\alpha=1}^{p-1} \epsilon_\alpha \partial^\alpha + \epsilon_p \partial^p) \circ f = 0.$$

Define a mapping $F_1 : S \to V$ given by

$$(4) \qquad F_1(\mathbf{x}, \mathbf{t}) = F_1(\mathbf{x}, \vec{t}_1, \ldots, \vec{t}_n) := \sum_{k=1}^{n} (\sum_{\beta=1}^{p-1} t^\beta_k \epsilon_\beta) \circ x^p e_k - \sum_{k=1}^{n} (\sum_{\gamma=1}^{p-1} t^\gamma_k x^\gamma) e_k,$$

where $\vec{t}_k = (t^\beta_k)_{\beta=1,\ldots,p-1} \in \mathbb{R}^{p-1}$; $k = 1, \ldots, n$, are arbitrary given parameters. Since $\epsilon_\beta \circ e_k = C^m_{\beta k} e_m$, the mapping F_1 is well-defined for any system of parameters $(\vec{t}_1, \ldots, \vec{t}_n) \in \underbrace{\mathbb{R}^{p-1} \times \cdots \times \mathbb{R}^{p-1}}_{n \text{ times}}$.

Let us notice that

$$\partial^p F_1 = \sum_{k=1}^{n} \left(\sum_{\beta=1}^{p-1} t_k^\beta \epsilon_\beta \right) \circ e_k,$$

$$\partial^\alpha F_1 = -\sum_{k=1}^{n} t_k^\alpha e_k, \qquad \alpha = 1, \ldots, p-1.$$

Hence

$$D^+ \circ F_1 = \left(\sum_{\alpha=1}^{p-1} \epsilon_\alpha \partial^\alpha + \epsilon_p \partial^p \right) \circ F_1$$

$$= \sum_{\alpha=1}^{p-1} \epsilon_\alpha \circ \left(-\sum_{k=1}^{n} t_k^\alpha e_k \right) + \sum_{k=1}^{n} \left(\sum_{\beta=1}^{p-1} t_k^\beta \epsilon_\beta \right) \circ e_k$$

$$= -\sum_{\alpha=1}^{p-1} \sum_{k=1}^{n} \epsilon_\alpha t_k^\alpha \circ e_k + \sum_{\beta=1}^{p-1} \sum_{k=1}^{n} \epsilon_\beta t_k^\beta \circ e_k = 0,$$

so F_1 satisfies the F–H equation. We say that the mapping F_1 is a *special solution of the Fueter–Hurwitz equation*. The solution F_1 is a homogenous polynomial of degree 1 in the variables x^γ, $\gamma = 1, \ldots, p$. We now look for polynomials F_m of degree $m > 1$ in the variables x^γ, which are solutions of the F–H equations.

First of all notice that if we define

$$\vec{t}_k := \sum_{\alpha=1}^{p-1} t_k^\alpha \epsilon_\alpha, \qquad k = 1, \ldots, n; \qquad \vec{x} := \sum_{\beta=1}^{p-1} x^\beta \epsilon_\beta,$$

then $F_1(\mathbf{x}, \vec{t}_1, \ldots, \vec{t}_n)$ can be rewritten in the form

$$(5) \qquad F_1(\mathbf{x}, \vec{t}_1, \ldots, \vec{t}_n) = \sum_{k=1}^{n} (x^p \vec{t}_k - \langle \vec{t}_k, \vec{x} \rangle \epsilon_p) \circ e_k,$$

where $\langle\ ,\ \rangle$ stands for the usual scalar product in $\mathbb{R}^{p-1}$. Using the above notation define the following polynomial of degree 2:

$$(6) \qquad F_2(\mathbf{x}, \vec{t}_1, \ldots, \vec{t}_n) := \sum_{k=1}^{n} \left[(x^p)^2 \vec{t}_k \vec{t}_k - 2x^p \vec{t}_k \langle \vec{t}_k, \vec{x} \rangle + \langle \vec{t}_k, \vec{x} \rangle^2 \epsilon_p \right] \circ e_k,$$

where

$$\vec{t}_k \vec{t}_k \circ e_k := \vec{t}_k \circ (\vec{t}_k \circ e_k), \qquad \vec{t}_k \in S.$$

PROPOSITION. *F_2 is a well-defined mapping from S into V for any system $(\vec{t}_1, \ldots, \vec{t}_n) \in S \times \cdots \times S$ and satisfies the F–H equation.*

PROOF. Indeed, we have

$$\partial^\alpha F_2 = \sum_{k=1}^{n} \left[2\langle \vec{t}_k, \vec{x} \rangle t_k^\alpha \epsilon_p - 2x^p \vec{t}_k t_k^\alpha \right] \circ e_k,$$

$$\partial^p F_2 = \sum_{k=1}^{n} \left[2x^p \vec{t}_k \vec{t}_k - 2\vec{t}_k \langle \vec{t}_k, \vec{x} \rangle \right] \circ e_k.$$

Hence

$$D^+ \circ F_2 = \sum_{\alpha=1}^{p-1} \epsilon_\alpha \partial^\alpha F_2 + \epsilon_p \partial^p F_2$$

$$= \sum_{\alpha=1}^{p-1} \sum_{k=1}^{n} \left[2\langle \vec{t}_k, \vec{x} \rangle \epsilon_\alpha t_k^\alpha - 2x^p \epsilon_\alpha t_k^\alpha \vec{t}_k \right] \circ e_k + \sum_{k=1}^{n} \left[2x^p \vec{t}_k \vec{t}_k - 2\vec{t}_k \langle \vec{t}_k, \vec{x} \rangle \right] \circ e_k$$

$$= \sum_{k=1}^{n} \left[2\langle \vec{t}_k, \vec{x} \rangle \vec{t}_k - 2x^p \vec{t}_k \vec{t}_k \right] \circ e_k + \sum_{k=1}^{n} \left[2x^p \vec{t}_k \vec{t}_k - 2\vec{t}_k \langle \vec{t}_k, \vec{x} \rangle \right] \circ e_k = 0. \quad \square$$

For each $m = 1, 2, ...$, introduce

$$(\vec{t}_k)^{\circ m} := \overbrace{\vec{t}_k \cdots \vec{t}_k}^{m \text{ times}},$$

(7)
$$(\vec{t}_k)^{\circ m} \circ e_k := \overbrace{\vec{t}_k \cdots \vec{t}_k}^{m-1 \text{ times}} \circ (\vec{t}_k \circ e_k),$$

$$(\epsilon_p)^m = \epsilon_p, \ \epsilon_p \circ e_k = e_k, \ \vec{t}_k \epsilon_p = \vec{t}_k, \ k = 1, \ldots, n.$$

Define

(8)
$$F_m(\mathbf{x}, \vec{t}_1, \ldots, \vec{t}_n) = \sum_{k=1}^{n} \left[x^p \vec{t}_k - \langle \vec{t}_k, \vec{x} \rangle \epsilon_p \right]^m \circ e_k.$$

By (7), $F_m(\mathbf{x}, \vec{t}_1, \ldots, \vec{t}_n)$ are well defined mappings from S into V for any system $(\vec{t}_1, \ldots, \vec{t}_n) \in \underbrace{\mathbb{R}^{p-1} \times \cdots \times \mathbb{R}^{p-1}}_{n \text{ times}}$ for $m = 1, 2, \ldots$. They are polynomials of degrees $m = 1, 2, \ldots$, respectively, in the variable $\mathbf{x} \in S$.

LEMMA. $F_m(\mathbf{x}, \vec{t}_1, \ldots, \vec{t}_n)$, defined by (8), are regular mappings, i.e., they satisfy the F–H equation for $m = 1, 2, \ldots$.

PROOF. By induction we get

$$\partial^\alpha F_m = -m \sum_{k=1}^{n} t_k^\alpha \left[x^p \vec{t}_k - \langle \vec{t}_k, \vec{x} \rangle \epsilon_p \right]^{m-1} \circ e_k,$$

$$\partial^p F_m = m \sum_{k=1}^{n} \left[x^p \vec{t}_k - \langle \vec{t}_k, \vec{x} \rangle \epsilon_p \right]^{m-1} \vec{t}_k \circ e_k.$$

Hence

$$D^+ \circ F_m(\mathbf{x}, \vec{t}_1, \ldots, \vec{t}_n) = 0. \quad \square$$

By expanding the right-hand side of (8) in powers of t_k^α we obtain

$$(9) \qquad F_m(\mathbf{x}, \mathbf{t}) = \sum_{\sum_k m_k^\alpha = m} m! \, P_{m_1^\alpha \ldots m_k^\alpha}(\mathbf{x})(t_1^\alpha)^{m_1^\alpha} \cdots (t_k^\alpha)^{m_k^\alpha},$$

where

$$m_k^\alpha := m_k^1 + \cdots + m_k^{p-1}, \qquad k = 1, \ldots, n;$$

$$(t_l^\alpha)^{m_l^\alpha} := (t_l^1)^{m_l^1} \cdots (t_l^{p-1})^{m_l^{p-1}}, \qquad l = 1, \ldots, n,$$

and $P_{m_1^\alpha \ldots m_k^\alpha}(\mathbf{x})$ are polynomials of degree m in $x^1, x^2, \ldots x^p$ with values in V.

When applying D^+ to the left-hand side of (9) and putting $\vec{t}_1, \ldots \vec{t}_n$ as independent variables we see that $P_{m_1^\alpha \ldots m_k^\alpha}(\mathbf{x})$ are regular polynomials:

$$D^+ \circ P_{m_1^\alpha \ldots m_k^\alpha}(\mathbf{x}) = 0.$$

The exponential form $\exp[i(Ex_0 + \langle \vec{p}, \vec{x} \rangle)]$ plays an important role in the solutions of the wave equation. Analogously, we introduce the following mapping

$$\exp\left\{ i \left[\sum_{k=1}^n (x^p \vec{t}_k - \langle \vec{t}_k, \vec{x} \rangle \epsilon_p) \circ e_k \right] \right\}$$

$$(10) \qquad := \sum_{s=0}^\infty \frac{(i)^s}{s!} \left[\sum_{k=1}^n (x^p \vec{t}_k - \langle \vec{t}_k, \vec{x} \rangle \epsilon_p) \circ e_k \right]$$

$$= \sum_{s=0}^\infty \frac{(i)^s}{s!} F_s(\mathbf{x}, \mathbf{t}).$$

Of course, the function $\exp$ introduced above is a well defined mapping from S into $V^{\mathbb{C}} := V \otimes_{\mathbb{R}} \mathbb{C}$. It is also obvious that the exponential mapping is regular in the sense of Fueter–Hurwitz:

$$D^+ \circ \exp\left\{ i \left[\sum_{k=1}^n (x^p \vec{t}_k - \langle \vec{t}_k, \vec{x} \rangle \epsilon_p) \circ e_k \right] \right\} = 0,$$

because of the usual properties of the exponential map, since the individual terms $F_s(\mathbf{x}, \mathbf{t})$ are regular.

Fourier representation of regular mappings

The mapping (10) is a special solution of the F–H equation:

$$D^+ \circ F = 0, \qquad F : S \to V^{\mathbb{C}}.$$

A general solution of the F–H equation should be given by superposition of the special solutions (10) integrated over the parameters $\vec{t}_1, \ldots, \vec{t}_n$, $\vec{t}_k = (t_k^1, \ldots, t_k^{p-1})$, $k = 1, \ldots, n$:

$$(11)$$

$$\Phi(\mathbf{x}) = \int_{-\infty}^{+\infty} \cdots \int_{-\infty}^{+\infty} A(\vec{t}_1, \ldots, \vec{t}_n) \exp\left\{ i \left[\sum_{k=1}^n (x^p \vec{t}_k - \langle \vec{t}_k, \vec{x} \rangle \epsilon_p) \circ e_k \right] \right\} d\mathbf{t},$$

where

$$\mathbf{dt} := d\vec{t}_1 \cdots d\vec{t}_n := dt_1^1 \cdots dt_1^{p-1} dt_2^1 \cdots dt_2^{p-1} \cdots dt_n^1 \cdots dt_n^{p-1},$$

and A is a C^∞ mapping from $\underbrace{\mathbb{R}^{p-1} \times \cdots \times \mathbb{R}^{p-1}}_{n \text{ times}}$ into $\mathbb{R}$ satisfying the following condition:

$$\int_{-\infty}^{+\infty} \cdots \int_{-\infty}^{+\infty} |A(\vec{t}_1, \ldots, \vec{t}_n)| \, \mathbf{dt} < +\infty.$$

It is clear that (11) is a solution of the F–H equation.

As is seen from (11), $\Phi(\mathbf{x})$ is expressed by Fourier-type integral. By separating the factors containing x^p and $\vec{x}$, (11) can be written as

$$(12) \qquad \Phi(\mathbf{x}) = \int_{-\infty}^{+\infty} \cdots \int_{-\infty}^{+\infty} A(\vec{t}_1, \ldots, \vec{t}_n) e^{ix^p \sum_k \vec{t}_{k0} e_k} e^{-i \sum_k \langle \vec{t}_k, \vec{x} \rangle e_k} \, \mathbf{dt}.$$

Putting $x^p = 0$ into (12) the initial condition for $\Phi(\mathbf{x})$ is given by

$$(13) \qquad \Phi(\mathbf{x})\big|_{x^p=0} := G(\vec{x}) = \int_{-\infty}^{+\infty} \cdots \int_{-\infty}^{+\infty} A(\vec{t}_1, \ldots, \vec{t}_n) e^{-i \sum_k \langle \vec{t}_k, \vec{x} \rangle e_k} \, \mathbf{dt}.$$

It is clear that $G: S|_{x^p=0} \to V^{\mathbb{C}}$. The components $G^m(\vec{x})$ of $G(\vec{x}) = G^m(\vec{x}) e_m$ are as follows;

$$(14) \qquad G^m(\vec{x}) = \int_{-\infty}^{+\infty} \cdots \int_{-\infty}^{+\infty} A(\vec{t}_1, \ldots, \vec{t}_n) e^{-i \langle \vec{t}_m, \vec{x} \rangle} \, \mathbf{dt},$$

which are nothing but Fourier-type integrals. Indeed, by the definition (10) we have

$$e^{-i \sum_k \langle \vec{t}_k, \vec{x} \rangle e_k} := \sum_{s=0}^{+\infty} \frac{(-i)^s}{s!} \left[\sum_{k=1}^n \langle \vec{t}, \vec{x} \rangle^s e_k \right]$$

$$= \sum_{k=1}^n \left[\sum_{s=0}^{+\infty} \frac{(-i)^s}{s!} \langle \vec{t}_k, \vec{x} \rangle^s \right] e_k = \sum_{k=1}^n e^{-i \langle \vec{t}_k, \vec{x} \rangle} e_k.$$

Suppose that the initial condition $G(\vec{x})$ is given. We will try to determine $A(\vec{t}_1, \ldots, \vec{t}_n)$. To do this assume that

$$(15) \qquad \begin{cases} A(\vec{t}_1, \ldots, \vec{t}_n) = A_1(\vec{t}_1) \cdots A_n(\vec{t}_n), \\ \int_{-\infty}^{+\infty} A_k(\vec{t}_k) d\vec{t}_k = 1, \qquad k = 1, \ldots, n. \end{cases}$$

Then

$$(16) \qquad G^m(\vec{x}) = \int_{-\infty}^{+\infty} A_m(\vec{t}_m) e^{-i \langle \vec{t}_m, \vec{x} \rangle} d\vec{t}_m, \qquad m = 1, \ldots, n.$$

Applying the Fourier integral theorem to (16) we have

$$(17) \qquad A_m(\vec{t}_m) = \frac{1}{(2\pi)^{p-1}} \int_{-\infty}^{+\infty} \cdots \int_{-\infty}^{+\infty} G^m(\vec{y}_m) e^{i \langle \vec{t}_m, \vec{t}_m \rangle} d\vec{y}_m,$$

where
$$d\vec{y}_m = dy_m^1 \cdots dy_m^{p-1}.$$
Hence, we obtain

(18) $A(\vec{t}_1, \ldots, \vec{t}_n) = \dfrac{1}{(2\pi)^{n(p-1)}} \displaystyle\int_{-\infty}^{+\infty} \cdots \int_{-\infty}^{+\infty} G^1(\vec{y}_1) \cdots G^n(\vec{y}_n) \times$

$$\times \quad e^{i \sum_{k=1}^n \langle \vec{t}_k, \vec{y}_k \rangle} \, d\vec{y}_1 \cdots d\vec{y}_n.$$

Inserting $A(\vec{t}_1, \ldots, \vec{t}_n)$ given by (15), (17) and (18) into (12) we obtain

(19) $\Phi(\mathbf{x}) = \dfrac{1}{(2\pi)^{n(p-1)}} \displaystyle\int_{-\infty}^{+\infty} \cdots \int_{-\infty}^{+\infty} G^1(\vec{y}_1) \cdots G^n(\vec{y}_n) \times$

$$\times \quad e^{i \sum_{k=1}^n \langle \vec{t}_k, \vec{y}_k \rangle} e^{i \sum_{l=1}^n (x^p \vec{t}_l - \langle \vec{t}_l, \vec{x} \rangle) \circ e_l} \, \mathbf{dy} \, \mathbf{dt}.$$

THEOREM. *The* $\Phi(\mathbf{x})$, *given by (19), satisfies the F–H equation and the initial condition*

$$\Phi(\mathbf{x})\big|_{x^p=0} = \frac{1}{(2\pi)^{n(p-1)}} \int_{-\infty}^{+\infty} \cdots \int_{-\infty}^{+\infty} A(\vec{t}_1, \ldots, \vec{t}_n) e^{-i \sum_{k=1}^n \langle \vec{t}_k, \vec{x} \rangle e_k} \, \mathbf{dt} = G(\vec{x}),$$

provided that $A(\vec{t}_1, \ldots, \vec{t}_n)$ *satisfies condition (15).*

References.

1. A. Hurwitz, *Über die Komposition der quadratischen Formen von beliebig vielen Variablen*, Nachrichten von der Königlichen Gesellschaft der Wissenschaften zu Göttingen Math.-phys. Kl. (1898), 308–316, reprinted in: A. Hurwitz, Mathematische Werke II, Birkhäuser Verlag, Basel 1933, 565–571.

2. ———, *Über die Komposition der quadratischen Formen*, Math. Ann. **88** (1923), 1–25, reprinted in: A. Hurwitz, Mathematische Werke II, Birkhäuser Verlag, Basel 1933, 641–666.

3. K. Imaeda, *Quaternionic formulation of classical electrodynamics and theory of functions of a biquaternion variable*, Report of Fundamental Physics Laboratory, Department of Electronic Science, Faculty of Science, Okayama University of Science, Feb. 1983.

4. W. Królikowski and E. Ramírez de Arellano, *Fueter–Hurwitz regular mappings and an integral representation*, Clifford algebras and their applications in mathematical physics Proceedings, Montpellier 1989 (A. Micali, ed.), Kluwer Academic Publishers, Dordrecht (to appear).

5. J. Lawrynowicz and J. Rembieliński, *Hurwitz pairs equipped with complex structures*, Seminar on Deformations, Lódź-Warsaw 1982–84, Proceedings (Lecture Notes in Math. 1165) (J. Lawrynowicz, ed.), Springer-Verlag, Berlin, 1985, pp. 184–197.

6. ———, *Pseudo-Euclidean Hurwitz pairs and the Kałuża-Klein theories*, J. Phys. A: Math. Gen. **20** (1987), 5831–5848.

7. ———, *Supercomplex vector spaces and spontaneuos symmetry breaking*, Seminari di Geometria 1984, CNR (Università di Bologna, ed.), Bologna, 24 pp.

8. _____ , *Pseudo-Euclidean Hurwitz pairs and generalised Fueter equations*, Clifford algebras and their applications in mathematical physics, Proceedings, Canterbury 1985 (J. S. R. Chisholm and A. K. Common, eds.), D. Reidel, Dordrecht, 1986, pp. 39–48.

9. D. B. Shapiro, *Compositions of quadratic forms*, Book preprint.

INSTITUTE OF MATHEMATICS, POLISH ACADEMY OF SCIENCES, LÓDŹ BRANCH, NARUTO-WICZA 56, PL–90–136 LÓDŹ, POLAND

DEPARTMENT OF MATHEMATICS, CINVESTAV-I.P.N., APDO. 14–740, MÉXICO 07000 DF, MÉXICO

Contemporary Mathematics
Volume **137**, 1992

Corona Problem of Several Complex Variables

SONG-YING LI

This paper is dedicated to Dr. Walter Rudin.

ABSTRACT. Some distance formulas are obtained. As an application, we prove that $S_f : H^\infty(\Omega, C^N) \to H^\infty(\Omega)$ is onto if and only if $S_f : H^p(\Omega, C^N) \to H^p(\Omega)$ is onto for any $1 < p < \infty$ when Ω is a simply connected domain in the complex plane with $C^{1+\alpha}$ boundary for some $\alpha > 0$. As a Corollary, p=2, we give an elementary proof of a theorem of Arveson in [A] for the domain $\Omega = \Delta$, the unit disc in C. Moreover, for the higher dimension, $n > 1$, we prove that $S_f : H^p(\Omega, C^N) \to H^p(\Omega)$ is onto for all $1 < p < \infty$, if $f = (f_1, \cdots, f_n) \in H^\infty(\Omega, C^N)$ is a Corona data and Ω is the unit polydisc in C^n, where $S_f(u) = \sum_{j=1}^{N} f_j u_j$ and $u = (u_1, u_2, \cdots, u_N) \in H^p(\Omega, C^N)$.

1. Introduction.

Let Ω denote a bounded domain in C^n. Let $f_1, f_2, \cdots, f_N \in H^\infty(\Omega)$, the Banach algebra of all bounded holomorphic functions on Ω, satisfy the following condition

$$(1.1) \qquad 0 < \delta^2 \le \sum_{j=1}^{N} |f_j|^2 \le 1 \quad \text{on } \Omega.$$

Then the Corona problem consists of finding holomorphic functions $g_1, g_2, \cdots, g_N \in H^\infty(\Omega)$ such that

$$(1.2) \qquad \sum_{j=1}^{N} f_j g_j = 1, \quad \text{on } \Omega.$$

Let $H^p(\Omega)$ denote the usual Hardy spaces of holomorphic functions over Ω, $H^p(\Omega, C^N) = H^p(\Omega) \oplus H^p(\Omega) \oplus \cdots \oplus H^p(\Omega)$. Then for $f = (f_1, f_2, \cdots, f_N) \in$

1991 *Mathematics Subject Classification.* Primary 30D55. Secondary 32A35.

The author deeply thanks his thesis advisor, Professor F. Beatrous for his guides and encouragments such that this work becomes possible. Also he appreciates the valuable conversations he has had with Professors J. Burbea; K-S Lau; E. Ligocka and T. A. Metzger. Finally, author would like to thank the referee for his pointing out several misprints in his original manuscript.

This paper is in final form and no version of it will be submitted for publication elsewhere.

$H^\infty(\Omega, C^N)$, if we define S_f to be the operator from $H^p(\Omega, C^N)$ to $H^p(\Omega)$ by $S_f(g) = \sum_{j=1}^N f_j g_j$ for $g = (g_1, \cdots, g_N) \in H^p(\Omega, C^N)$, then the solvability of the Corona problem is equivalent to that $S_f : H^\infty(\Omega, C^N) \to H^\infty(\Omega)$ is onto.

For n=1, the Carleson's Corona theorem asserts that the Corona problem is solvable when Ω is the unit disc in C. It is also known that the Corona theorem remains true if Ω is a simply connected domain, Denjoy domain [GP] and a domain whose boundary lies on a $C^{1+\alpha}$ curve with $\alpha > 0$ [M]. There are many interesting results related to the Corona problem and some different proofs of Carleson's Corona theorem of one complex variable which can be found in the book of Garnett [G].

In [CS], L. Coburn and M. Schecter studied some problems about the joint spectra and interpolation of operators. In the end of their paper, they pointed out that as a consequence of the Carleson's Corona theorem, one has that the solvabilty of Corona problem is equivalent to that $S_f : H^2(\Omega, C^N) \to H^2(\Omega)$ is onto. In [A], Arveson proved the above intersting result: $S_f : H^\infty(\Omega, C^N) \to H^\infty(\Omega)$ is onto if and only if $S_f : H^2(\Omega, C^N) \to H^2(\Omega)$ is onto when $\Omega = \Delta$, the unit disc in C, by using the theory of the Nest algebras. Later on, the theorem of Arveson in [A] was also proved by Schubert in [Sc] by using the Lifting theorem of Sz.-Nagy and Foias.

For $n > 1$, there are Runge domains where the Corona problem has no solutions. More recently, N. Sibony [S] gave an example: There is a pseudoconvex domain Ω in C^3 with C^3 boundary such that (i) $\partial\Omega$ is stongly pseudoconvex with exception of one point $0 \in \partial\Omega$ and (ii) there are functions $f_j \in H^\infty(\Omega) \cap C^\infty(\overline{\Omega} - \{0\})$ satisfying (1.1) in Ω with $N = 3$, but the equation (1.2) has no solution $g_j \in H^\infty(\Omega), j = 1, 2, 3$. On the other hand, there are some positve partial results. For example, Varopoulos [V1] and [V2] proved that (1.2) has $\cap_{0<p<\infty} H^p(\Omega)$ solution when Ω is a strongly pseudoconvex domain and the unit polydisc in C^n, respectively. For the case, $\Omega = \Delta^n$, Chang [Ch] and Lin [Lin] gave the alternative proofs of the above results. Also, if we restrict Corona data $f_1, f_2, \cdots, f_N \in A(\Omega)$, the algebra of all functions which are holomorphic in Ω and continuous on $\overline{\Omega}$, then M. Hakim and N. Sibony [HS] proved that (1.2) has solutions $g_1, g_2, \cdots, g_N \in A(\Omega)$ when Ω is a pseudoconvex domain in C^n with smooth boundary. However, the solvability of the Corona problem for the most standard domain like the unit polydisc Δ^n and the unit ball B^n in C^n is still an open problem (see [R 2,3]).

In this paper, we shall prove some distance formulas. As an application, we shall give an elementary proof of the above theorem of Arveson in [A] for a simply connected domain Ω in C. Moreover, we shall prove that $S_f : H^\infty(\Omega, C^N) \to H^\infty(\Omega)$ is onto if and only if $S_f : H^p(\Omega, C^N) \to H^p(\Omega)$ is onto for any $1 < p < \infty$ when Ω is a simply connected domain in the complex plane with $C^{1+\alpha}$ boundary for some $\alpha > 0$. For the case of higher dimension $n > 1$, we generalize the theorem of Varopoulos [V2] by using the technique used in [Ch] and [Lin]. And we prove that $S_f : H^p(\Omega, C^N) \to H^p(\Omega)$ is onto for all $1 < p < \infty$, if

$f = (f_1, \cdots, f_n) \in H^\infty(\Omega, C^N)$ satisfies (1.1) and Ω is the unit polydisc in C^n. And the same result has been proved by M. Andersson [An] for $p = 2$ and E. Amar [Am] for $1 < p < \infty$ when $\Omega = B^n$. As a connection of the above theorem, we realize that if one can prove the above theorem of Arveson for $\Omega = \Delta^n$ or B^n, then the Corona problem is solvable when $\Omega = \Delta^n$ or B^n.

The paper is organized as follows. In Section 2, we shall give some definitions and the statements of our main theorems, Theorems 1–5. In Section 3, we prove some lemmas. In Section 4, we apply the lemmas proved in Section 3 to prove Theorems 1, 2 and 3. In Section 5, we collect some lemmas from [Lin]. In Section 6, we prove Theorems 4 and 5. Finally, in Section 7, we make some remarks, where we introduce a decomposition domain, and we prove that if Ω is a decomposition domain, then $S_f : H^\infty(\Omega, C^N) \to H^\infty(\Omega)$ is onto if and only if $S_f : H^2(\Omega, C^N) \to H^2(\Omega)$ is onto.

2. Notations and Main Results.

In this section, we shall give some notations, definitions and the statements of main results.

2.1. Notations and Definitions.

Let Ω denote a bounded domain in C^n with smooth boundary $\partial\Omega$. Let $L^p(\partial\Omega)$ denote the Lebesgue space over $\partial\Omega$ with respect to the normalized surface measure over $\partial\Omega$, and $H^p(\Omega)$ denote the usual Hardy space of holomorphic functions on Ω. For $1 \leq p \leq \infty$, we can identify $H^p(\Omega)$ as a subspace of $L^p(\partial\Omega)$ in the usual way by passing to the non-tangential limits on $\partial\Omega$. Also we let $L^p(\partial\Omega, l^2)$ denote the all vector valued functions

$$f = (f_1, f_2, \cdots, f_n, \cdots)$$

such that each component $f_n \in L^p(\partial\Omega)$, for all $n = 1, 2, \cdots$, and satisfying that $|||f|||_p < \infty$, where

$$(2.1) \qquad |||f|||_p = \left\{ \int_{\partial\Omega} \{ \sum_{j=1}^{\infty} |f_j(z)|^2 \}^{p/2} \right\}^{1/p}.$$

It is easy to prove that $L^p(\partial\Omega, l^2)$ is a Banach space with the norm $|||\cdot|||_p$ over C, for $1 \leq p \leq \infty$. And we let $H^p(\Omega, l^2)$ denote the holomorphic subspace of $L^p(\partial\Omega, l^2)$, i.e. $H^p(\Omega, l^2) = \{(f_1, f_2, \cdots, f_n, \cdots) \in L^p(\partial\Omega, l^2) : f_n \in H^p(\Omega), n = 1, 2, \cdots\}$. Also it is easy to see that $H^p(\Omega, l^2)$ is a closed subspace of $L^p(\partial\Omega, l^2)$. We shall define $\mathcal{H}^2(\partial\Omega) = \{\overline{f} : f \in H^2(\Omega)^\perp = L^2(\partial\Omega) \ominus H^2(\Omega)\}$. And for each $1 \leq p \leq \infty$, we let $\mathcal{H}^p(\partial\Omega)$ denote the closure of $L^p(\partial\Omega) \cap \mathcal{H}^2(\Omega)$ under the norm $||\cdot||_p$ over $\partial\Omega$. Also, we write $\mathcal{H}^p(\partial\Omega, l^2) = \{f = (f_1, \cdots, f_n, \cdots) \in L^p(\partial\Omega, l^2) : f_n \in \mathcal{H}^p(\partial\Omega), n = 1, 2, \cdots\}$. The spaces $\mathcal{H}^p(\partial\Omega, l^2)$ will play an important role in this paper. As an example of a view of these spaces, one can see that $\mathcal{H}^2(\partial\Omega) = H_0^2(\Omega)$ when $\Omega = \Delta$, the closed subspace of $H^2(\Delta)$ which vanishes at the origin $z = 0$. Moreover, it is not difficult to prove that $\mathcal{H}^p(\partial\Omega)$

and $\mathcal{H}^p(\partial\Omega, l^2)$ are Banach spaces for all $1 \le p \le \infty$. Let P be the orthogonal projection from $L^2(\partial\Omega) \to H^2(\Omega)$. Then for each function $f \in L^\infty(\partial\Omega)$, we use M_f to denote the multiplication operator, and we define the Toeplitz operator as $T_f = PM_f : L^p(\partial\Omega) \to H^p(\Omega)$ for $1 < p < \infty$. The big and small Hankel operator H_f and h_f with respect to the symbol f are defined as follows.

$$(2.2) \qquad H_f(u) = M_{\overline{f}}P(u) - PM_{\overline{f}}(u), \quad \text{and} \quad h_f(u) = P(f\overline{u})$$

for every $u \in L^p(\partial\Omega)$.

If $1 < p < \infty$, and $f = (f_1, f_2, \cdots, f_n, \cdots) \in L^\infty(\partial\Omega, l^2)$, we use M_f to denote the operator $(M_{f_1}, \cdots, M_{f_n}, \cdots) : L^p(\partial\Omega) \to L^p(\partial\Omega, l^2)$ with $M_f(u) = (M_{f_1}, \cdots, M_{f_n}, \cdots)$; also T_f and H_f have same interpretations.

2.2. Statements of Main Theorems.

Now we are ready to state our main theorems.

THEOREM 1. *Let $f \in L^\infty(\partial\Omega, l^2)$. Then we have the following statements hold*

(i)
$$\inf\left\{|||f - g|||_\infty : g \in H^\infty(\Omega, l^2)\right\}$$
$$= \sup\left\{|\int_{\partial\Omega} \langle u, f\rangle d\sigma(z)| : u \in \mathcal{H}^1(\partial\Omega, l^2) \text{ with } |||u|||_1 \le 1\right\}$$
$$\ge \sup\left\{|||H_f(u)|||_2 : u \in H^2(\Omega), \|u\|_2 \le 1\right\}$$
$$= |||H_f|||.$$

where $\langle u, v\rangle = \langle u, v\rangle_{l^2} = \sum_{j=1}^\infty u_j\overline{v}_j$ for any $u, v \in L^2(\partial\Omega, l^2)$.
(ii) If Ω is a simply connected domain in C, then we have

$$\inf\left\{|||f - g|||_\infty : g \in H^\infty(\Omega, l^2)\right\}$$
$$= \sup\left\{|||H_f(u)|||_2 : u \in H^2(\Omega), \|u\|_2 \le 1\right\}$$
$$= |||H_f|||.$$

As an application of Theorem 1, we shall prove the following theorem

THEOREM 2. *Let Ω be a bounded domain in C^n with C^2 boundary. Also we let $f \in H^\infty(\Omega, 2)$ and $|||f|||_\infty \le 1$. We consider the following statements:*
(a) There is a function $g \in H^\infty(\Omega, l^2)$ such that

$$(2.3) \qquad \sum_{j=1}^\infty f_j g_j = 1 \quad \text{in} \quad \Omega \quad \text{and} \quad |||g|||_\infty \le 1/\delta.$$

(b) $|f| \ge \delta$ on $\partial\Omega$, and

$$\inf\left\{|||\overline{f}/|f| - g|||_\infty : g \in H^\infty(\Omega, l^2)\right\} \le (1 - \delta^2)^{1/2}.$$

(c) The following inequality holds true

$$|||T_{\overline{f}}(u)|||_2 \ge \delta\|u\|_2, \quad \text{for } u \in H^2(\Omega).$$

Then we have

1) $(a) \Longrightarrow (b)$; $(b) \Longrightarrow (c)$. Moreover, (a) and (b) are equivalent if $|f| = 1$ on the boundary $\partial\Omega$.

2) If the statement (a) holds with δ replaced by δ^2, then the statement (b) implies (a).

3) If Ω is simply connected domain in C, then the statement (c), with δ replaced by $\delta^{1/2}$, implies (a).

THEOREM 3. Let Ω be a bounded, simply connected domain in C with $C^{1+\alpha}$ boundary for some $\alpha > 0$. Also we let $f \in H^\infty(\Omega, C^N)$. Then we have that $S_f : H^\infty(\Omega, C^N) \to H^\infty(\Omega)$ is onto if ond only if $S_f : H^p(\Omega, C^N) \to H^p(\Omega)$ is onto for some $1 < p < \infty$.

Let Δ^n denote the unit polydisc in C^n, T^n denotes the distingushed boundary of Δ^n. If we use T^n instead of $\partial\Omega$, then we are going to prove the following theorem.

THEOREM 4. Let N be positive integer. And let $f_1, f_2, \cdots, f_N \in H^\infty(\Delta^n)$ satisfy the following inequality

$$(2.4) \qquad \delta^2 \le \sum_{j=1}^{N} |f_j|^2 \le 1, \quad z \in \Delta^n.$$

Then for each $1 < p < \infty$, there are a bounded operator $A : H^p(\Delta^n) \to H^p(\Delta^n, C^N)$ and a constant $C = C(n, N, \delta, p) \in (0, \infty)$ such that

$$(2.5) \qquad \sum_{j=1}^{N} f_j A_j(u) = u$$

and

$$(2.6) \qquad \sum_{j=1}^{N} \|A_j(u)\|_p^2 \le C^2 \|u\|_p^2,$$

for all $u \in H^p(\Delta^n)$.

For the special case $p = 2$, we have the following more precise theorem.

THEOREM 5. Let $f_1, f_2, \cdots, f_n \in H^\infty(\Delta^n)$ satisfy condition (2.4). Then we have

(i) There is a constant $\epsilon = 1/C(n, N, \delta, 2) > 0$ such that

$$(2.7) \qquad \sum_{j=1}^{N} \|T_{\bar{f}_j}(u)\|_2^2 \ge \epsilon^2 \|u\|_2^2$$

for every $u \in H^2(\Delta^n)$. Where $C = C(n, N, \delta, 2)$ is the constant in Theorem 4.

(ii) There is a linear operator $A : H^2(\Delta^n) \to H^2(\Delta^n, C^N)$ such that $\|A\| \le 1/\epsilon$ and (2.5) holds true for all $u \in H^2(\Delta^n)$.

REMARK. *If one can generalize the theorem of Arveson to the polydisc Δ^n, then the Corona problem is solvable there by combining Theorem 2 with Theorem 5.*

3. Some Lemmas

In order to prove our theorems, we first prove some lemmas.

Let B denote a separable Banach space over C, we let $L^p(\partial\Omega, B)$ denote the Lebesgue space consisting of all vector valued functions u such that $u(z) \in B$ for almost all $z \in \partial\Omega$, and $|||u|||_{p,B} < \infty$, where

$$(3.1) \qquad |||u|||_{p,B} = \left\{ \int_{\partial\Omega} |u(z)|_B^p \, d\sigma(z) \right\}^{1/p}.$$

If $1 \leq p \leq \infty$, then it is easy to see that $L^p(\partial\Omega, B)$ is a Banach space over C. Thus we have the following duality property.

LEMMA 3.1. *Let H be a separable Hilbert space over C and $1 \leq p < \infty$. Then*

$$(3.2) \qquad L^p(\partial\Omega, H)^* = L^q(\partial\Omega, H)$$

with the paring $\langle f, g \rangle = \int_{\partial\Omega} \langle f, g \rangle_H d\sigma(z)$, $1/p + 1/q = 1$ and $\langle, \rangle_H$ is the inner product in H.

PROOF. It is easy to see that $L^q(\partial\Omega, B)$ can be embedded into $L^p(\partial\Omega, B)^*$ for any separable Banach space B over C, and that the embedding is contractive.

Now we are going to prove the converse. We first consider the case $1 \leq p \leq 2$. Let $f \in (L^p(\partial\Omega, H))^*$. Notice that $L^p(\partial\Omega, H)^*$ is subspace of $L^2(\partial\Omega, 2)^* = L^2(\partial\Omega, 2)$ since $d\sigma(\cdot)$ is bounded Borel measure over $\partial\Omega$, we have $f(z) \in H$ for almost all $z \in \partial\Omega$, and so we need only to prove that $f \in L^q(\partial\Omega, H)$ and the following inquality holds true,

$$(3.3) \qquad |||f|||_{q,H} \leq |||f|||_{L^p(\partial\Omega, H)^*}.$$

In fact, we choose

$$u(z) = \begin{cases} f(z)/|f(z)|_H, & \text{if } |f(z)|_H \neq 0; \\ e, & \text{if } |f(z)|_H = 0. \end{cases}$$

where e is an element in H such that $|e|_H = 1$.

Then, we have

$$|u(z)|_H = 1 \quad \text{on } \partial\Omega,$$

Since the dual space of $L^p(\partial\Omega)$ is $L^q(\partial\Omega)$, then one has

$$|||f|||_{q,H} = \sup\left\{ |\int_{\partial\Omega} |f|_H g(z) d\sigma(z)| : g \in L^p(\partial\Omega) \text{ with } ||g||_p \leq 1 \right\}$$

$$= \sup\left\{ |\int_{\partial\Omega} \langle f, gu \rangle_H d\sigma(z)| : g \in L^p(\partial\Omega) \text{ with } ||g||_p \leq 1 \right\}$$

$$\leq |||f|||_{L^p(\partial\Omega, H)^*}.$$

Hence by the symmetry, we have proved the inequality (3.3) holds true. Therefore the proof of Lemma 3.1 is complete.

As a special case, we have the following corollary

COROLLARY 3.2. *Let Ω as the same as it in Lemma 3.1. Then we have*

$$(3.4) \qquad L^1(\partial\Omega, l^2)^* = L^\infty(\partial\Omega, l^2).$$

As an application of the above Corollary, we have the following lemma

LEMMA 3.3. *Let Ω be bounded strongly pseudocovex domaian in C^n with smooth boundary. Then we have*

$$(3.7) \quad H^\infty(\Omega, 2) = \mathcal{H}^1(\partial\Omega, l^2)^\perp$$
$$= \left\{ f \in L^\infty(\partial\Omega, 2) : \int_{\partial\Omega} \langle u, f \rangle_{l^2} d\sigma(z) = 0, \text{ for all } u \in \mathcal{H}^1(\partial\Omega, l^2) \right\}.$$

Notice that $\mathcal{H}^2(\partial\Omega, l^2)$ is dense in $\mathcal{H}^1(\partial\Omega, l^2)$. Then it is easy to see that $H^\infty(\Omega, l^2)$ is a subspace of $\mathcal{H}^1(\partial\Omega, l^2)^\perp$. Also one can easily see that the converse is true.

Therefore, one has by Lemmas 3.2 and 3.3,

$$(3.8) \qquad L^\infty(\partial\Omega, l^2)/H^\infty(\Omega, l^2) \cong \mathcal{H}^1(\partial\Omega, l^2)^*.$$

With direct computation, one can easily prove the following lemma.

LEMMA 3.4. *Let $f \in L^\infty(\partial\Omega, l^2)$. Then*

$$(3.9) \qquad |||H_f(u)|||_2^2 = |||M_f(u)|||_2^2 - |||T_{\overline{f}}(u)|||_2^2.$$

In order to prove our main theorem, we need the following lemma

LEMMA 3.5. *Let $f \in L^\infty(\partial\Omega, 2)$. Then we have following inquality*

$$(3.10) \qquad \sup\left\{ |||H_f(u)|||_2 : u \in H^2(\Omega) \text{ and } ||u||_2 \leq 1 \right\} \leq$$
$$\inf\left\{ |||\overline{f} - g|||_\infty : g \in H^\infty(\Omega, l^2) \right\}.$$

PROOF. Since $f \in L^\infty(\partial\Omega, l^2)$, then we have

$$||H_f^*|| = ||H_f|| = \sup\left\{ |||H_f(u)|||_2 : u \in H^2(\Omega) \text{ and } ||u||_2 \leq 1 \right\}.$$

Since

$$(3.11) \qquad H_f^* : H^2(\Omega, l^2)^\perp \to H^2(\Omega).$$

We claim that

$$H_{f_j}^* = PM_{f_j} : H^2(\Omega)^\perp \to H^2(\Omega), \quad j = 1, 2, \cdots.$$

In fact, we let $u \in H^2(\Omega)^\perp$ and $v \in H^2(\Omega)$, then

$$\langle H_f^*(u), v \rangle = \langle u, H_{f_j}(v) \rangle$$
$$= \langle u, M_{\overline{f}_j} P(v) - P(\overline{f}_j v) \rangle$$
$$= \langle u, M_{f_j}(v) \rangle = \langle PM_{f_j}(u), v \rangle.$$

Therefore, we have proved our claim.

Moreover, we have

$$(3.12) \qquad H_f^* = (PM_{f_1}, PM_{f_2}, \cdots, PM_{f_n}, \cdots) = PM_f.$$

Thus

$$(3.13) \qquad \|H_f^*\| = \sup \left\{ \|PM_f(u)\|_2 : u \in H^2(\Omega, l^2)^\perp, \text{ and } \||u|\|_2 \le 1 \right\}.$$

and

$$(3.14) \qquad \||H_f|\| = \sup \left\{ \||PM_f(u)|\|_2 : u \in H^2(\Omega, l^2)^\perp \right\}$$
$$= \sup \left\{ |\langle M_f(u), v \rangle| : v \in H^2(\Omega), \|v\|_2 \le 1, u \in H^2(\Omega, l^2)^\perp, \text{ and } \||u|\|_2 \le 1 \right\}$$
$$= \sup \left\{ |\sum_{j=1}^\infty \int_{\partial\Omega} \overline{f}_j \overline{u}_j v d\sigma| : u \in H^2(\Omega, 2)^\perp, v \in H^2(\Omega); \||u|\|_2 \le 1, \|v\|_2 \le 1 \right\}$$
$$\le \sup \left\{ |\sum_{j=1}^\infty \int_{\partial\Omega} \overline{f}_j u_j d\sigma| : u \in \mathcal{H}^1(\partial\Omega, l^2) \text{ and } \||u|\|_1 \le 1 \right\}.$$

On the other hand, we have

$$(3.15) \inf \left\{ \||\overline{f} - g|\|_\infty : g \in H^\infty(\Omega, l^2) \right\}$$
$$= \sup \left\{ |\int_{\partial\Omega} \langle u, f \rangle d\sigma(z)| : u \in \mathcal{H}^1(\partial\Omega, l^2) \text{ with } \||u|\|_1 \le 1 \right\}$$
$$= \sup \left\{ |\int_{\partial\Omega} \sum_{j=1}^\infty \overline{f}_j u_j d\sigma(z)| : u \in \mathcal{H}^1(\partial\Omega, l^2) \text{ with } \||u|\|_1 \le 1 \right\}.$$

Therefore we have (3.10) holds true by (3.14) and (3.15). So we have completed the proof of Lemma 3.5. Now we are ready to prove Theorems 1, 2 and 3.

4. Proofs of Theorems 1, 2 and 3.

In this section, we are going to prove the main theorems 1, 2 and 3.

4.1. Proof of Theorem 1.

PROOF. By Lemma 3.5 and (3.8), we have (i) holds true.

So we need only to prove (ii).

From the conclusion of (i), we need only to prove

$$(4.1)$$
$$\sup \left\{ \||H_f(u)|\|_2 : u \in H^2(\Omega), \|u\|_2 \le 1 \right\} \ge \inf \left\{ \||\overline{f} - g|\|_\infty : g \in H^\infty(\Omega, l^2) \right\}.$$

Let $u \in \mathcal{H}^1(\partial\Omega, l^2)$ with $|||u|||_1 \leq 1$. For $0 < \epsilon < 1$, since Ω is simply connected, there is a holomorphic function $h \in H^2(\Omega)$ such that

$$\text{Re}h(z) = P_0(1/4\log(|u|^2 + \epsilon))(z),$$

where $P_0(H)$ denotes the poisson integral of H.

Thus, we define a function

$$H_1(z) = \exp(h(z)) \quad H_2(z) = \exp(-h(z)) \quad \text{in } \Omega.$$

So we have $H_1(z)H_2(z) = 1$ and $H_1, H_2 \in H^2(\Omega)$. Moreover, we have

$$|H_1(z)| = (|u|^2 + \epsilon)^{1/2}, \quad |H_2(z)| = (|u|^2 + \epsilon)^{-1/2}.$$

Therefore

$$\sup\left\{ |\int_{\partial\Omega} \sum_{j=1}^{\infty} \overline{f}_j u_j \, d\sigma(z)| : u \in \mathcal{H}^1(\partial\Omega, l^2) \text{ with } |||u|||_1 \leq 1 \right\}$$

$$= \sup\left\{ |\int_{\partial\Omega} \sum_{j=1}^{\infty} \overline{f}_j H_1 H_2 u_j \, d\sigma(z)| \right\}$$
$$\leq |||H_f^*(H_1)|||_2$$
$$\leq (1+\epsilon)\|H_f^*\|$$
$$= (1+\epsilon)\|H_f\|.$$

By letting $\epsilon \to 0^+$, we have (4.1) holds.

Therefore, we have completed the proof of Theorem 1.

As a direct consequence, we have

COROLLARY. $\|h_f\|_{H_0^2(\Delta)} = \|H_f\|_{H^2(\Delta)}$.

REMARK. *The above Corollary have been proved by S. Power [P] who used the Arveson distance formula in [A].*

4.2. Proof of Theorem 2.

Let us prove 1) first. By Lemma 3.5, we have that (b) implies (c). So we need only to prove that (a) implies (b). Applying Theorem 1 to the function $f/|f|$

and let $g \in H^{\infty}(\Omega, 2)$ such that (a) holds, then we have

$$\inf \left\{ |||\overline{f}/|f| - h|||_{\infty} : h \in H^{\infty}(\Omega, l^2) \right\}$$

$$= \sup \left\{ |\int_{\partial\Omega} \langle u, \overline{f}/|f| \rangle d\sigma(z)| : u \in \mathcal{H}^1(\partial\Omega, l^2) \text{ and } |||u|||_1 \leq 1 \right\}$$

$$= \sup \left\{ |\int_{\partial\Omega} \langle u, (\overline{f}/|f| - \delta^2 g) \rangle d\sigma(z)| : u \in \mathcal{H}^1(\partial\Omega, l^2) \text{ and } |||u|||_1 \leq 1 \right\}$$

$$\leq \sup \left\{ \int_{\partial\Omega} |||\overline{f}/|f| - \delta^2 g|||_{\infty} |||u|||_1 \right\}$$

$$= |||\overline{f}/|f| - \delta^2 g|||_{\infty}$$

$$= (|||f/|f||||_{\infty}^2 - 2\delta^2(1/|f|)\mathrm{Re}(\sum_{j=1}^{\infty} f_j g_j) + \delta^4 |||g|||_{\infty}^2)^{1/2}$$

$$\leq (1 - \delta^2)^{1/2}.$$

And one can easily see that the last inequality becomes equality if $|f| = 1$ on $\partial\Omega$ a.e..

Therefore we have completed the proof of 1).

Now we prove 2). Since

$$\inf \left\{ |||\overline{f}/|f| - g|||_{\infty} : g \in H^{\infty}(\Omega, l^2) \right\} \leq (1 - \delta^2)^{1/2},$$

then there is $g \in H^{\infty}(\Omega, l^2)$ such that the above inequality holds. So we have

$$1 - 2(1/|f|) \, \mathrm{Re}(\sum_{j=1}^{\infty} f_j g_j) + |||g|||_{\infty}^2 \leq 1 - \delta^2 \quad \text{on } \partial\Omega;$$

i.e.

$$2\mathrm{Re}(\sum_{j=1}^{\infty} f_j g_j) - |f||g|^2 \geq |f|\delta^2 \quad \text{on } \partial\Omega;$$

i.e.

$$2\mathrm{Re}(\sum_{j} f_j g_j) - \delta|g|^2 \geq \delta^3 \quad \text{on } \partial\Omega.$$

Since $2\mathrm{Re}(\sum_{j=1}^{\infty} f_j g_j) - \delta|g|^2$ is super-harmonic in Ω . So it attains its minimum on the boundary $\partial\Omega$. So we have

$$2\mathrm{Re}(\sum_{j=1}^{\infty} f_j g_j) - \delta|g|^2 \geq \delta^3 \quad \text{on } \overline{\Omega}.$$

Now we let

$$h_j = g_j/(\sum_{j=1}^{\infty} f_j g_j) \in H^{\infty}(\Omega), \quad h = (h_1, \cdots, h_n, \cdots) \quad \text{in } \Omega.$$

Then we have

$$\sum_{j=1}^{\infty} f_j h_j = 1 \quad \text{on } \overline{\Omega}.$$

and
$$|h|^2 \leq (4/\delta^2)(|g|^2/(|g|^2 + \delta^2))^2.$$

Notice the fact that function $r(x) = x^2/(\delta^2 + x^2)^2$ attains its maximum of $[0, \infty)$ at the point $x = \delta^2$ since that $r'(x) = 0 \iff x = \delta^2$.

So we have
$$|h|^2 \leq (4/\delta^2)(\delta^2/(2\delta^2)^2) = 1/\delta^4,$$

i.e. (a) holds. Therefore we have completed the proof of 2).

Finally, we prove 3). From the conclusion of 2), we need only to prove that (c) implies (b).

Applying Theorem 1 to the vector-valued function $f/|f| \in L^\infty(\partial\Omega, l^2)$, we have

$$\sup \left\{ | \int_{\partial\Omega} \sum_{j=1}^{\infty} \overline{f}_j/|f| u_j \, d\sigma(z)| : u \in \mathcal{H}^1(\partial\Omega, l^2) \text{ with } |||u|||_1 \leq 1 \right\}$$
$$= \|H_{f/|f|}\|$$
$$= (1 - \inf \{ \|T_{f/|f|}(u)\|_2^2 : u \in H^2(\Omega) \text{ and } \|u\|_2 \leq 1 \})^{1/2}$$
$$\leq (1 - \delta^2)^{1/2}.$$

Hence we have (b) holds.

Therefore, we have completed the proof of Theorem 2.

4.3. Proof of Theorem 3.

Now we start to prove Theorem 3.

PROOF. In order prove Theorem 3, we introduce a lemma from Rudin's book [R1] first

LEMMA 4.1 [R1]. *Let X and Y be two Banach spaces over C. $T \in \mathcal{B}(X, Y)$. Let U and V denote the unit balls of X and Y, respectively. Then we have*
  1) *$T : X \to Y$ is onto if and only if $\|T^*(y^*)\| \geq \delta\|y^*\|$ for some $\delta > 0$ and for every $y^* \in Y^*$.*
  2) *If $\delta\|y^*\| \leq \|T^*(y^*)\|$ for every $y^* \in Y^*$, then $\delta V \subset T(U)$.*

Now ready to prove Theorem 3. It is easy to see that $S_f : H^\infty(\Omega, C^N) \to H^\infty(\Omega)$ is onto $\implies S_f : H^p(\Omega, C^N) \to H^p(\Omega)$ is onto for all $0 < p < \infty$.

So we need only to prove the converse.

Let $1 < p < \infty$ be such that $S_f : H^p(\Omega, C^N) \to H^p(\Omega)$ is onto. Then, by Lemma 4.1, there is a constant $\delta > 0$ such that

$$|||S_f^*(u)|||_{H^p(\Omega, C^N)^*} \geq \delta\|u\|_{H^p(\Omega)^*}, \quad \text{for all } u \in H^p(\Omega)^*.$$

Notice that $H^p(\Omega)^* = L^q(\partial\Omega)/H^p(\partial\Omega) \approx H^q(\Omega)$, so we have

$$H^p(\Omega, C^N) \approx H^q(\Omega, C^N)^*, \quad \text{for } 1 < p < \infty; \ 1/p + 1/q = 1$$

since $P : L^q(\partial\Omega) \to H^q(\Omega)$ is bounded.

It is easy to prove $S_f^* = T_{\overline{f}}$. And also, one can prove that there is a constant $c_q > 0$ such that

$$1 \geq \sum_{j=1} |f_j(z)|^2 \geq c_q \delta^2, \quad \text{for all } z \in \Omega.$$

by choosing $\mathcal{K}_z(w) = K(w,z)K(z,z)^{1-q}$, $z \in \Omega$, $\|\mathcal{K}_z\|_q \approx c_q > 0$, and then applying $T_{\overline{f}}$ to $\mathcal{K}_z$ for $z \in \Omega$.

Applying Lemma 4.1 again, we have that for each $u \in H^p(\Omega)$, there is $v = (g_1(u), \cdots, g_N(u)) \in H^p(\Omega, C^N)$ such that $S_f(v) = u$ and $\||g(u)\||_p \leq 1/\delta \|u\|_p$.

Let $u \in \mathcal{H}^1(\partial\Omega, C^N)$ and $\||u\||_1 \leq 1$. Then for any $\epsilon > 0$, from the argument of the proof of theorem 1, there are $H_1(u) \in H^p(\Omega)$ and $H_2(u) \in H^\infty(\Omega)$ such that

$$\|H_1\|_p \leq (1+\epsilon), \||uH_2\||_q \leq 1, \quad H_1H_2 = 1, \text{ a.e. } z \in \partial\Omega.$$

Applying Theorem 1, we have

$$\inf\left\{\||f/|f| - g\||_\infty : g \in H^\infty(\Omega, C^N)\right\}$$

$$= \sup\left\{|\int_{\partial\Omega} \sum_{j=1}^N \overline{f}_j/|f|u_j d\sigma| : u \in \mathcal{H}^1(\Omega, C^N) \text{ and } \||u\||_1 \leq 1\right\}$$

$$\leq \sup\left\{|\int_{\partial\Omega} \sum_{j=1}^N \overline{f}_j/|f|H_1H_2u_j d\sigma| : u \in \mathcal{H}^1(\Omega, C^N), \||u\||_1 \leq 1\right\}$$

$$= \sup\left\{|\int_{\partial\Omega} \sum_{j=1}^N (\overline{f}_j H_1/|f| - \gamma g_j(H_1))H_2u_j d\sigma| : u \in \mathcal{H}^1(\Omega, C^N) \text{ and } \||u\||_1 \leq 1\right\}$$

$$\leq \sup\left\{\int_{\partial\Omega} (\sum_{j=1}^N |\overline{f}_j H_1 - \gamma g_j(H_1)|^2)^{1/2}(\sum_{j=1}^N |H_2u_j|^2)^{1/2}d\sigma :\right.$$

$$\left. u \in \mathcal{H}^1(\Omega, C^N), \||u\||_1 \leq 1\right\}$$

$$\leq \sup\left\{(\int_{\partial\Omega} ((1-2\gamma)|H_1|^2 + \gamma^2 \sum_{j=1}^N |g_j(H_1)|^2)^{p/2}d\sigma)^{1/p} :\right.$$

$$\left. u \in \mathcal{H}^1(\Omega, C^N), \||u\||_1 \leq 1\right\}.$$

We choose $\gamma = (1/2)^{1/(p-1)}\delta^{\frac{2p}{p-1}}$ if $p \leq 2$, and $\gamma = 1/4\delta^2$ if $p > 2$, and notice the fact that $(1-\gamma)^{p/2} \leq 1-\gamma$ with $p \geq 1$ and $\gamma \in (0,1)$, then

$$(\int_{\partial\Omega} \{(1-2\gamma)|H_1|^2 + \gamma^2 \sum_{j=1}^{N} |g_j(H_1)|^2\}^{p/2} d\sigma)^{1/p}$$

$$\leq \begin{cases} (\int_{\partial\Omega}(1-2\gamma)^{p/2}|H_1|^p + \gamma^p(\sum_{j=1}^{N}|g_j(H_1)|^2)^{p/2}d\sigma)^{1/p}, & \text{if } 1 < p \leq 2, \\ \{(1-2\gamma)\|H_1\|_p^2 + \gamma^2\||g(u)\||_p^2\}^{1/2}, & \text{if } 2 < p < \infty \end{cases}$$

$$\leq \max\left\{((1-2\gamma)^{p/2} + \gamma^p\delta^{-p})^{1/2} \text{ if} p \leq 2, (1-2\gamma) + \gamma^2\delta^{-2})^{1/2}\right\}(1+\epsilon)$$

$$\leq (1-1/4\gamma)^{1/2}(1+\epsilon).$$

Therefore we have, by letting $\epsilon \to 0^+$.

$$\inf\left\{\|\|\overline{f}/|f| - g\|\|_\infty : g \in H^\infty(\Omega, C^N)\right\} \leq (1-1/4\gamma)^{1/2}.$$

Then by the argument of the proof of Theorem 2, there is $g \in H^\infty(\Omega, C^N)$ such that (1.2) holds and

$$\|\|g\|\|_\infty \leq \frac{4}{C_q\gamma}.$$

Therefore, we have completed the proof of Theorem 3.

It remains to prove Theorems 4 and 5. We separate it into the following two sections. We shall follow the arguments have been given by S-Y. A. Chang [Ch] and K.C. Lin [Lin] to prove the Corona problem in Δ^n has $H^p(\Delta^n)$ solutions.

5. The Koszul Complex in C^n and Some Known Results

We shall use the Koszul complex to reduce the proof of Theorem 4 to solving the set of $\overline{\partial}$-equations. For the convience, we restrict our proof in the special case N=n=3, and from the proof, reader can see that the argument still works for general n and N.

Let $u \in H^p(\Delta^n)$, and $f_j, j = 1, 2, \cdots, N$ in Theorem 4. Then we define

$$\varphi_j(z) = \overline{f}_j/(\sum_{j=1}^{N}|f_j|^2) \quad \text{in } \Delta^n, \quad j = 1, 2, \cdots, N;$$

and

$$W_l = \varphi_l, \quad l = 1, 2, \cdots, N;$$
$$W_{k,l} = \varphi_k\overline{\partial}\varphi_l - \varphi_l\overline{\partial}\varphi_k, \quad k, l = 1, 2, \cdots, N;$$
$$W_{j,k,l} = \varphi_j\overline{\partial}\varphi_k \wedge \overline{\partial}\varphi_l - \varphi_j\overline{\partial}\varphi_l \wedge \overline{\partial}\varphi_k + \varphi_l\overline{\partial}\varphi_j \wedge \overline{\partial}\varphi_k$$
$$- \varphi_l\overline{\partial}\varphi_k \wedge \varphi_j + \varphi_k\overline{\partial}\varphi_l \wedge \overline{\partial}\varphi_j - \varphi_k\overline{\partial}\varphi_j \wedge \overline{\partial}\varphi_l;$$

and

$$W_{i,j,k,l} = \varphi_i\overline{\partial}\varphi_j \wedge \overline{\partial}\varphi_k \wedge \overline{\partial}\varphi_l - \cdots (\text{ 24 terms}).$$

From the definitions of $W_l, W_{k,l}, W_{j,k,l}$ and $W_{i,j,k,l}$, one can easily see that they are defined to be alternating and the sign of each term is determined by the order of the permutation. Also one can easily see the following proposition holds true.

PROPOSITION 5.1 [LIN].

$$\overline{\partial} W_{j,k,l} = \sum_i f_i W_{i,j,k,l};$$

$$\overline{\partial} W_{k,l} = \sum_j f_j W_{j,k,l};$$

$$\overline{\partial} W_l = \sum_k f_k W_{k,l}.$$

Since $W_{i,j,k,l}$ is a (0,3) form on C^3, then $\overline{\partial} W_{i,j,k,l} = 0$. By the exactness of the $\overline{\partial}$ operator, there are (0,2) form $\psi_{i,j,k,l}$ which is alternating such that

$$\overline{\partial} \psi_{i,j,k,l} = W_{i,j,k,l}.$$

Since

$$\overline{\partial}(W_{i,j,kl} u - \sum_i f_i \psi_{i,j,k,l} u)$$

$$= (\overline{\partial} W_{j,k,l} - \sum_i f_i W_{i,j,k,l}) u$$

$$= 0;$$

and Proposition 5.1, there are alternating (0,1) form $\psi_{j,k,l}$ such that

$$\overline{\partial} \psi_{j,k,l} = W_{j,k,l} u - \sum_i f_i \psi_{i,j,k,l} u.$$

Also

$$\overline{\partial}(W_{k,l} u - \sum_j f_j \psi_{j,k,l})$$

$$= u \overline{\partial} W_{k,l} - \sum_j f_j \overline{\partial} \psi_{j,k,l}$$

$$= u(\overline{\partial} W_{k,l} - \sum_j f_j W_{j,k,l} + \sum_{i,j} f_i f_j \psi_{i,j,k,l})$$

$$= 0.$$

Because of Proposition 5.1, and the alternating properties of $\psi_{j,k,l}$, there are alternating functions $\psi_{k,l}$ so that

$$\overline{\partial} \psi_{k,l} = W_{k,l} u - \sum_j f_j \psi_{j,k,l}.$$

By a similar argument as before, one has

$$
\overline{\partial}(W_l u - \sum_k f_k \psi_{k,l})
$$
$$
= u\overline{\partial}W_l - \sum_k f_k \overline{\partial}\psi_{k,l}
$$
$$
= u(\overline{\partial}W_l - \sum_k f_k W_{k,l}) + \sum_{k,j} f_k f_j \psi_{j,k,l}
$$
$$
= 0.
$$

Moreover, we have

$$
\sum_l f_l (W_l u - \sum_k f_k \psi_{k,l}) = \sum_l f_l W_l u - \sum_{k,l} f_l f_k \psi_{k,l} = u.
$$

So, let us define

$$
(5.2) \qquad g_k(u) = W_k - \sum_l f_l \psi_{k,l}
$$

for all $u \in H^p(\Delta^n)$.

Then $g_k \in H(\Delta^n)$ satisfies the following equality

$$
\sum_j f_j g_j(u) = u
$$

for all $u \in H^p(\Delta^n)$.

Therefore, what we need to do is to choose a good $g_j(u)$ such that

$$
\sum_{j=1}^N \|g_j(u)\|_p^p \le C(n, N, \delta, p).
$$

We shall leave it to Section 6.

Now let's collect some results from the paper of K.C. Lin [Lin].

LEMMA 5.2 [LIN]. *Let* $f_1, f_2, \cdots f_N \in H^\infty(\Delta^n)$ *satisfy the condition (1.1) with* $N = n = 3$. *Then we have*

(i) $|\gamma(z)|^2 d\omega(z) = |\gamma(z)|^2 \log(1/|z_1|) \log(1/|z_2|) \log(1/|z_3|) dA(z_1) dA(z_2) dA(z_3)$

is a Carleson measure over Δ^n, *where* $dA(\lambda)$ *is the area measure on* Δ *in* C; *and*

$$
\gamma(\cdot) = \sum{}' \varphi_i \frac{\partial \varphi_j}{\partial \overline{z}_1} \frac{\partial \varphi_k}{\partial \overline{z}_2} \frac{\partial \varphi_l}{\partial \overline{z}_3}.
$$

(ii)
$$
\sum_{i \ne j, j \ne k, i \ne k} \int_{\Delta^3} |\frac{\partial^3 h}{\partial z_i \partial z_j} \frac{\partial \gamma}{\partial z_k}| d\omega \le C \int_{T^3} |h| d\sigma.
$$

for all $h \in H^1(\Delta^3)$, *where* $d\sigma$ *is the area measure over* T^3.

(iii)
$$
\sum_{i \ne j, j \ne k, i \ne k} \int_{\Delta^3} |\frac{\partial h}{\partial z_i}| |\frac{\partial^2 \gamma}{\partial z_j \partial z_k}| d\omega \le C \int_{T^3} |h| d\sigma.
$$

for all $h \in H^1(\Delta^3)$.

(iv) $\left|\frac{\partial^3 \gamma}{\partial z_i \partial z_2 \partial z_3}\right| d\omega$ *is a Carleson measure on* Δ^3 *with norm* $\leq C$.

(v) *If* γ *satisfies* (i)–(iv), *then the equation*

$$(5.3) \qquad\qquad\qquad \frac{\partial^3 v}{\partial \overline{z}_1 \partial \overline{z}_2 \partial \overline{z}_3} = \gamma$$

has a solution $v \in C^\infty(\Delta^3)$ *with*

$$(5.4) \qquad\qquad\qquad \|v\|_\infty \leq C = C(n, N, \delta) < \infty.$$

Moreover, we can see that $C = C(n, N, \delta) \leq C(n, N)\delta^{-4}$.

LEMMA 5.3 [LIN]. *Suppose* $\alpha = \alpha_1 d\overline{z}_1 + \alpha_2 d\overline{z}_2 + \alpha_3 d\overline{z}_3$ *is a* $C^\infty(\overline{\Delta}^3)$ $\overline{\partial}$-*closed* (0, 1) *form on* Δ^3 *and suppose that the equation*

$$(5.5) \qquad\qquad\qquad \frac{\partial v}{\partial \overline{z}_j} = \alpha_j \quad on \ \Delta^3$$

has a solution $u_j \in C^\infty(\overline{\Delta}^3)$, *then there is a* $u \in C^\infty(\overline{\Delta}^3)$ *so that*

$$(5.6) \qquad\qquad\qquad \overline{\partial} u = \alpha;$$

and

$$(5.7) \qquad\qquad\qquad \|u\|_{L^p(T^3)} \leq C_p \max_{1 \leq j \leq 3} \left\{ \|u_j\|_{L^p(T^3)} \right\}.$$

Also we need the following lemma.

LEMMA 5.4[LIN]. *Suppose that*

$$(5.8) \qquad\qquad \beta = \beta_{1,2} d\overline{z}_1 \wedge d\overline{z}_2 + \beta_{1,3} d\overline{z}_1 \wedge d\overline{z}_3 + \beta_{2,3} d\overline{z}_2 \wedge d\overline{z}_3$$

is a $C^\infty(\overline{\Delta}^3)$ $\overline{\partial}$-*closed forms on* Δ^3 *and suppose that the equation*

$$(5.9) \qquad\qquad\qquad \frac{\partial^2 v}{\partial \overline{z}_j \partial \overline{z}_k} = \beta_{j,k}, \quad 1 \leq j < k \leq 3.$$

has a solution $u_{j,k} \in C^\infty(\overline{\Delta}^3)$. *Then there are some solutions* $u'_{j,k} \in C^\infty(\overline{\Delta}^3)$ *to* (4.9) *so that*

$$(5.10) \qquad\qquad\qquad u'_{1,2} - u'_{1,3} + u'_{2,3} = 0;$$

and

$$(5.11) \qquad\qquad\qquad \|u'_{j,k}\|_{L^p(T^3)} \leq C_p \max_{1 \leq j < k \leq 3} \left\{ \|u_{j,k}\|_{L^p(T^3)} \right\}.$$

Now we are ready to prove Theorems 4 and 5.

6. The Proof of Theorems 4 and 5.

Without loss of the generality, we may assume that $f_j{}'$s are holomorphic across T^3 by the argument of the normal family and the properties of H^p space.

Now we turn to the $\bar{\partial}-$equations in Section 5.

$$\text{(6.1)} \qquad \bar{\partial}\psi_{i,j,k,l} = W_{i,j,k,l}$$

$$\text{(6.2)} \qquad \bar{\partial}\psi_{j,k,l} = W_{j,k,l}u - \sum_i f_i \psi_{i,j,k,l}u$$

$$\text{(6.3)} \qquad \bar{\partial}\psi_{k,l} = W_{k,l}u - \sum_j f_j \psi_{j,k,l}$$

and

$$\text{(6.4)} \qquad g_l(u) = W_l u - \sum_k f_k \psi_{k,l},$$

where $\psi_{i,j,k,l}, \psi_{j,k,l}$ and $\psi_{k,l}$ are $(0,2)$, $(0,1)$ and $(0,0)$ alternating forms respectively. From the definitoin φ_l, we have $\varphi_l \in C^\infty(\overline{\Delta}^3)$ and

$$\sum_{l=1}^N |\varphi_l| \le 1/\delta^2, \quad \text{and} \quad \sum_{j=1}^N |\varphi_j|^2 = 1.$$

We divide the proof of theorem 4 into the following steps.

Step 1. Solving equation (6.1) so that every $d\bar{z}_r d\bar{z}_s$-coefficient of $\psi_{i,j,k,l}$ is a $\bar{\partial}_r\bar{\partial}_s$-derivative of some function whose $L^\infty(T^3)$ norm is less than $C = C(n, N, \delta)$.

Since $\psi_{i,j,k,l}$ is a $(0,2)$ form, so it has the form

$$\text{(6.5)} \qquad \psi_{i,j,k,l} = \psi^{1,2}_{i,j,k,l}d\bar{z}_1 \wedge d\bar{z}_2 + \psi^{1,3}_{i,j,k,l}d\bar{z}_1 \wedge d\bar{z}_3 + \psi^{2,3}_{i,j,k,l}d\bar{z}_2 \wedge d\bar{z}_3.$$

We shall suppress that indices i, j, k, l. Then the equation (5.1) can be written as follows

$$\text{(6.6)} \qquad \frac{\partial\psi^{1,2}}{\partial\bar{z}_3} - \frac{\partial\psi^{1,3}}{\partial\bar{z}_2} + \frac{\partial\psi^{2,3}}{\partial\bar{z}_1} = \sum{}' \varphi_i \frac{\partial\varphi_j}{\partial\bar{z}_1} \frac{\partial\varphi_k}{\partial\bar{z}_2} \frac{\partial\varphi_l}{\partial\bar{z}_3},$$

where $\sum'$ denotes a sum over different permutation on indices i, j, k, and l.

We are hoping that

$$\text{(6.7)} \qquad \psi^{r,s} = \frac{\partial^2 \phi^{r,s}}{\partial\bar{z}_r \partial\bar{z}_s} \quad \text{with } \|\phi^{r,s}\|_{L^\infty(T^3)} \le C(n, N, \delta).$$

Applying Lemma 5.2, we have the equation (5.3) has a solution $v \in C^\infty(\overline{\Delta}^3)$ with

$$\text{(6.8)} \qquad \|v\|_{L^\infty(T^3)} \le C(n, N, \delta).$$

Thus by letting $\phi^{1,2} = v$, $\phi^{1,3} = 0$ and $\phi^{2,3} = 0$ and combining (6.5)–(6.8), we solve the equation (6.1) with desired estimates.

Step 2. Solving the equation (6.2) so that every $d\bar{z}_r$-coefficient of $\psi_{j,k,l}$ is a $\bar{\partial}_r$-derivative of some functions in $L^p(T^3)$ norm is $\le C(n, N, \delta, p)\|u\|_p$.

Suppose

(6.10) $\psi_{j,k,l} = \psi^1_{j,k,l} d\bar{z}_1 + \psi^2_{j,k,l} d\bar{z}_2 + \psi^3_{j,k,l} d\bar{z}_3,$

then (6.2) can be written as

(6.11) $\dfrac{\overline{\partial}\psi^s_{j,k,l}}{\partial \bar{z}_r} - \dfrac{\overline{\partial}\psi^r_{j,k,l}}{\partial \bar{z}_s} = \sum{}' \varphi_j \dfrac{\partial \varphi_k}{\partial \bar{z}_r} \dfrac{\partial \varphi_l}{\partial \bar{z}_s} u - \sum_i f_i \psi^{r,s}_{i,j,k,l} u.$

What we want is to have the followings

(6.12) $\psi^r_{j,k,l} = \dfrac{\partial b^r_{j,k,l}}{\partial \bar{z}_r}$ with $\|b^r_{j,k,l}\|_p \leq C(n,N,\delta)\|u\|_p.$

Substituting $(6,7) \times u$ and (6.12) into (6.11), we have

(6.13) $\dfrac{\partial^2 (b^s_{j,k,l} - b^r_{j,k,l})}{\partial \bar{z}_r \partial \bar{z}_s} = \sum{}' \varphi_j \dfrac{\partial \varphi_k}{\partial \bar{z}_r} \dfrac{\partial \varphi_l}{\partial \bar{z}_s} u - \sum_i f_i \dfrac{\partial^2 \phi^{r,s}_{i,j,k,l}}{\partial \bar{z}_r \partial \bar{z}_s} u.$

Fix r and s for moments. The fact that the equation

$$\dfrac{\partial^2 u}{\partial \bar{z}_r \partial \bar{z}_s} = \sum{}' \varphi_j \partial \varphi_k \partial \bar{z}_r \dfrac{\partial \varphi_l}{\partial \bar{z}_s}$$

has a solution $u_{r,s}$ with $\|u_{r,s}\|_\infty \leq C(n,N,\delta)$ (see Chang[Ch] for the details).
    In particular, the equation

$$\dfrac{\partial^2 u}{\partial \bar{z}_r \partial \bar{z}_s} = \sum{}' \varphi_j \dfrac{\partial_k}{\partial \bar{z}_r} \dfrac{\partial \varphi_l}{\bar{z}_s} u$$

has a solution $u^1_{r,s}$ with $\|u^1_{r,s}\|_{L^p(T^3)} \leq C(n,N,\delta,p)\|h\|_p$ and we know the equation

$$\dfrac{\partial^2 u}{\partial \bar{z}_r \partial \bar{z}_s} = \sum_i f_i \dfrac{\partial^2 \phi^{r,s}_{i,j,k,l}}{\partial \bar{z}_r \partial \bar{z}_s} u$$

also has a solution $u^2_{r,s} = \sum_i f_i \phi^{r,s}_{i,j,k,l} u$ with $\|u^2_{r,s}\|_{L^p(T^3)} \leq C(n,N,\delta,p)\|u\|_p.$
    Thus, for each fixed r and s, the equation

(6.14) $\dfrac{\partial^2 u}{\partial \bar{z}_r \partial \bar{z}_s} = \sum{}' \varphi_j \dfrac{\partial \varphi_k}{\partial \bar{z}_r} \dfrac{\partial \varphi_l}{\partial \bar{z}_s} u - \sum_i f_i \psi^{r,s}_{i,j,k,l} u$

has a solution $u^{r,s}_{j,k,l}$ with

(6.15) $\|u^{r,s}_{j,k,l}\|_{L^p(T^3)} \leq C(n,N,\delta,p)\|u\|_{L^p(T^3)}.$

Now we apply Lemma 5.3 to get $u'^{r,s}_{j,k,l}$ for (6.14) so that

(6.16) $u'^{1,2}_{j,k,l} - u'^{1,3}_{j,k,l} + u'^{2,3}_{j,k,l} = 0$

and

(6.17) $\|u'^{r,s}_{j,k,l}\|_{L^p(T^3)} \leq C(n,N,\delta,p)\|u\|_p.$

Now we set

$$b^1_{j,k,l} = 0, \quad b^2_{j,k,l} = u'^{1,2}_{j,k,l}, \quad \text{and } b^3_{j,k,l} = u'^{1,3}_{j,k,l}.$$

Then (6.13) is satisfied with desired $L^p(T^3)$ bounds in $b^r_{j,k,l}$.

Step 3. Solving equation (6.3) so that

$$(6.18) \qquad ||\psi_{k,l}||_{L^p(T^3)} \le C(n, N, \delta, p)||u||_p.$$

Collecting the coefficients of $d\bar{z}_r$, we obtain

$$(6.19) \qquad \frac{\partial \psi_{k,l}}{\partial \bar{z}_r} = \sum' \varphi_k \frac{\partial \varphi_l}{\partial \bar{z}_r} u - \sum_j f_j \psi^r_{j,k,l}.$$

As in Step 2, we fix r and look at the equation individualy. Wolff's classical one variable result to $\frac{\partial u}{\partial \bar{z}_r} = \sum_k \varphi_k \frac{\partial \varphi_l}{\partial \bar{z}_r}$ and then to $\frac{\partial u}{\partial \bar{z}_r} = \sum' \varphi_k \frac{\partial \varphi_l}{\partial \bar{z}_r} u$, one has a solution u with $L^p(T^3)$ norm$\le C(n, N, \delta, p)||u||_p$ and notice that $\psi^r_{j,k,l}$ is $\bar{\partial}$-derivative of some function $b^r_{j,k,l}$ with $||b^r_{j,k,l}||_{L^p(T^3)} \le C(n, N, \delta, p)||u||_p$. These imply that equation (6.19) has a solution $\psi^r_{k,l}$ with

$$||\psi^r_{k,l}||_{L^p(T^3)} \le C(n, N, \delta, p)||u||_p.$$

Now Lemma 5.3 gives a common solution of (6.17) with

$$(6.20) \qquad ||\psi_{k,l}||_{L^p(T^3)} \le C(n, N, \delta, p).$$

Finally, we define

$$(6.21) \qquad g_l(u) = W_l u - \sum_k f_k \psi_{k,l},$$

and

$$(6.22) \qquad A(u) = (A_1(u), A_2(u), \cdots, A_N(u)) = (g_1(u), g_2(u), \cdots, g_N(u)),$$

which is an operator from $H^p(\Delta^n) \to H^p(\Delta^n, C^N)$ with

$$(6.23) \qquad |||A(u)|||_p \le C(\sum_{j=1}^N ||A_j(u)||_p^2)^{1/2} \le C(n, N, \delta, p)||u||_p.$$

Therefore

$$||A||_p \le C(n, N, \delta, p) < \infty.$$

Therefore, we have completed the proof of Theorem 4.

Now we turn to prove Theorem 5.

Let $u \in H^2(\Delta^n)$. Then

$$(6.23) \qquad T_{\bar{f}_j}(u) = \bar{f}_j u - H_{f_j}(u)$$

and

$$(6.24) \qquad ||T_{\bar{f}_j}(u)||_2^2 = ||\bar{f}_j u||_2^2 - ||H_{f_j}(u)||_2^2.$$

Also

(6.25)
$$\|H_{f_j}(u)\|_2^2 = \|\overline{f}_j u - \gamma A_j(u) - P(\overline{f}_j u - \gamma A_j(u))\|_2^2$$
$$= \|\overline{f}_j u - \gamma A_j(u)\|_2^2 + \|P(\overline{f}_j u - \gamma A_j(u))\|_2^2$$
$$- \langle \overline{f}_j u - \gamma A_j(u), P(\overline{f}_j u - \gamma A_j(u)) \rangle$$
$$- \langle P(\overline{f}_j u - \gamma A_j(u)), \overline{f}_j u - \gamma A_j(u) \rangle$$
$$= \|(\overline{f}_j u - \gamma A_j(u))\|_2^2 - \|P(\overline{f}_j u - \gamma A_j(u))\|_2^2$$
$$\leq \|\overline{f}_j u - \gamma A_j(u)\|_2^2$$
$$\leq \|\overline{f}_j u\|_2^2 + \gamma^2 \|A_j(u)\|_2^2 - 2\gamma \mathrm{Re}\langle u, f_j A_j(u) \rangle.$$

So we have

(6.26)
$$\sum_{j=1}^N \|T_{\overline{f}_j}(u)\|_2^2$$
$$\geq 2\gamma \sum_{j=1}^N \mathrm{Re}\langle u, f_j A_j(u) \rangle - \gamma^2 \sum_{j=1}^N \|A_j(u)\|_2^2$$
$$= 2\gamma \mathrm{Re}\langle u, \sum_{j=1}^N f_j A_j(u) \rangle - \gamma^2 \sum_{j=1}^N \|A_j(u)\|_2^2$$
$$\geq 2\gamma \|u\|_2^2 - \gamma^2 C(n, N, \delta, 2)^2 \|u\|_2^2.$$

By taking

(6.27)
$$\gamma = (1/C(n, N, \delta, 2))^2,$$

one has

$$\sum_{j=1}^N \|T_{\overline{f}_j}(u)\|_2^2 \geq (1/C(n, N, \delta, 2))^2 \|u\|_2^2.$$

So we let

$$\epsilon = \epsilon(n, N, \delta) = 1/C(n, N, \delta, 2) \in (0, \infty)$$

such that

$$\sum_{j=1}^N \|T_{\overline{f}_j}(u)\|_2^2 \geq \epsilon^2 \|u\|_2^2.$$

Notice the fact that

$$\sum_{j=1}^N \|T_{\overline{f}_j}(u)\|_2^2 \leq \|u\|_2^2.$$

Therefore we have completed the proof of (i) of Theorem 5.

Now we prove (ii)

Since $S_f : H^2(\Delta^n, C^N) \to H^2(\Delta^n)$ and defined by $S_f(u) = \sum_{j=1}^N f_j u_j$, for $u = (u_1, u_2, \cdots, u_n) \in H_N^2(\Delta^n)$, then it is easy to see that $S_f^* = (T_{f_1}, T_{f_2}, \cdots, T_{f_n}) : H^2(\Delta^n) \to H^2(\Delta^n, C^N)$.

From the conclusion of (i), we have that $S_f^* : H^2(\Delta^n) \to$ Range of S_f^* is invertible. Let A' denote the inverse operator of S_f^*. Then we have $\|A'\| \le 1/\epsilon$. Let $A : H^2(\Delta^n, C^N) \to H^2(\Delta^n)$ be the norm-preserving extension of A'. Then we have

$$AS_f^* = A'S_f^* = I.$$

So

$$S_f A^* = (AS_f^*)^* = I^* = I.$$

Therefore the operator $A^* : H^2(\Delta^n) \to H^2(\Delta^n, C^N)$ is the desired operator. Therefore we have completed the proof of Theorem 5.

7. Some Remarks.

In this section, we shall give some remarks on the domain where the decomposition theorem for $\mathcal{H}^1(\partial\Omega)$ holds true.

Let Ω be a bounded domain in C^n with C^2 boundary. Then we say that Ω is a decomposition domain, for simplicity, we call $\mathcal{D}$-domain if there is $\beta \in (0, 1]$, such that for any $u \in \mathcal{H}^1(\partial\Omega)$, there are $u_1 \in \mathcal{H}^2(\partial\Omega)$ and $u_2 \in L^2(\partial\Omega)$ satisfy the following properties:

$$u = u_1 u_2, \quad \|u\|_1 = \|u_1\|_2\|u_2\|_2, \text{ and } \|P(u_2)\|_2 \ge \beta\|u_2\|_2.$$

From the proof of Theorems 1 and 2, one has the following theorem.

THEOREM 6. *Let Ω be a $\mathcal{D}$-domain in C^n and let $f_1, f_2, \cdots, f_N \in H^\infty(\Omega)$. Then $S_f : H^\infty(\Omega, C^N) \to H^\infty(\Omega)$ is onto if and only if $S_f : H^2(\Omega, C^N) \to H^2(\Omega)$ is onto.*

By the arguments of the proofs of Theorems 1 and 2, one has that any simply connected domains in the complex plane with C^2 boundary are $\mathcal{D}$-domains.

Therefore, we may pose the following question.

QUESTION. *What kinds of domains are $\mathcal{D}$-domains except for simply connected domains in the complex plane? Are Δ^n and B^n $\mathcal{D}$-domains?*

REMARK. *Associating Theorem 5, we have that the Corona problem is solvable on Δ^n if Δ^n is a $\mathcal{D}$-domain.*

Add to the proof: The author wishes to point out that K. C. Lin has, independently, duplicated some of the main results presented in this paper.

References.

[A] W. Arveson, *Interpolation problems in Nest Algebras*, J. Functional Analysis **20** (1975), 208–233.

[Am] E. Amar, *On the Corona Problem*, preprint.

[An] M. Andersson, *Values in the interior of L^2-minial solutions of the $\bar{\partial}$-equation in the unit ball of C^n*, Publ. Mat. **32** no. no 2 (1988), 179-189.

[C] L. Carleson, *Interpolations by bounded analytic functions and Corona problem*, Ann. of Math. **76** (1962), 547–559.

[Ch] S-Y. A. Chang, *Two remarks of H^1 and BMO on the bidisc*, Conference on harmonic analysis in honor of A. Zygmund **II**, 373–393.

[CS] L. Coburn and M. Schecter, *Joint spectra and interpolation of operators*, J. Functional Anal. **2** (1968), 226–237.

[G] J. B. Garnett, *Bounded analytic functions*, Academic Press, New York, London, Toronto, 1981.

[GP] J. B. Garnett and P. Jones, The Corona theorem for Denjoy domains, Report 6, Institute Mittage-Leffler, 1984.

[HS] M. Hakim and N. Sibony, *Sectre de $A(\overline{\Omega})$ pour les domaines faiblement pseudoconvexes*, J. Functional Analysis **37** (1980), 127–135.

[L] S-Y. Li, *Teoplitz operators on the Hardy spaces $H^p(S)$ with $0 < p \le 1$*, Integral equations and operator theory (to appear).

[Lin] K. C. Lin, H^p solutions for the Corona problem on the polydisc in C^n, Bull A.M.S. **110** (1986), 69–84.

[M] C. M. Moore, *The Corona theorem for domains whose boundary lies in a smooth curve*, Proceeding A.M.S. **100** (1987), 200–204.

[P] S. Power, *Analysis in nest algebras*, Surveys of some recent results in operator theory, vol 2, Pitman Research Notes in Mathematics series **192**, 189–235.

[R1] W. Rudin, *Functional Analysis*, Inc., New York, 1974.

[R2] ———, *Function theory in polydisc*, W. A. Benjamin, New York, 1969.

[R3] ———, *Function theory in the unit ball in C^n*, Springer-Verlag, New York, 1980.

[S] N. Sibony, *Problem de la Courone pour des domaines pseudoconvexes á bord lissc*, Annals of Math. **126** (1987), 675–682.

[Sc] C. F. Schubert, *The Corona theorem as an operator theorem*, Proceeding A.M.S. **69** (1978), 73–76.

[V1] N. Th. Varopoules, *BMO functions and the ∂- equation*, Pacific J. of Math. **71** (1977), 221–272.

[V2] ———, *Probabilistic approach to some problems in complex analysis*, T. **105**, Bull. S. C. Math. Pairs (1981), 181–224.

DEPARTMENT OF MATHEMATICS AND STATISTICS, UNIVERSITY OF PITTSBURGH, PITTSBURGH, PENNSYLVANIA 15260

Contemporary Mathematics
Volume **137**, 1992

Sharp Estimates of the Kobayashi Metric
Near Strongly Pseudoconvex Points

DAOWEI MA

ABSTRACT. We give sharp estimates for the Kobayashi metric on bounded domains near strongly pseudoconvex points. Namely, we prove that near strongly pseudoconvex points, the Kobayashi metric on a bounded domain is the same as that on the "osculating ellipsoids" up to a factor of $1+O(\sqrt{u})$, where u is the distance from the base point to the boundary. We also give an expression in terms of the Levi form which approximates the Kobayashi metric up to a factor of $1+O(\sqrt{u})$. An example is given to show that these estimates are best possible.

Introduction.

In this paper we will give estimates for the Kobayashi metric on bounded domains near C^3 strongly pseudoconvex points.

Let D be a bounded domain in $\mathbb{C}^n$. Let Δ be the unit disk in $\mathbb{C}$. For $z \in D$, let $\mathrm{Hol}(\Delta_0, D_z)$ denote the set of holomorphic mappings from Δ to D that send 0 to z. The Kobayashi metric on D (see [**15**]) is the function $F_D : D \times \mathbb{C}^n \to \mathbb{R}^+$ defined by

$$F_D(z,\xi) = \inf\{|v| : v \in \mathbb{C} \text{ and there is an} f \in \mathrm{Hol}(\Delta_0, D_z) \text{ with } df(0)v = \xi\}.$$

The Kobayashi metric has been an important tool for the study of holomorphic maps and function spaces. But the explicit form of the Kobayashi metric is unlikely obtainable except for the symmetric domains and some Thullen domains (explicit formulae for the Kobayashi metric on some Thullen domains will appear in [**4**]). Thus it is desirable to obtain estimates for the Kobayashi metric.

The first work in this direction is by I. Graham ([**8**]), who gave asymptotic formulae for the Kobayashi metric and the Carathéodory metric (see [**14**]) near the

1991 *Mathematics Subject Classification*. Primary 32H15.
I wish to thank Franc Forstneric for bringing the reference [6] to my attention.
This paper is in final form and no version of it will be submitted for publication elsewhere.

boundary of strongly pseudoconvex domains. Namely, he obtained the following formulae:

(1.1)
$$\lim_{z \to p} F_D(z, \xi)u(z) = (1/2)\|\xi_n(p)\|,$$
$$\lim_{z \to p} F_D(z, \xi_t(z))^2 u(z) = (1/2)L_p(\xi_t(p)).$$

Here $p \in \partial D$; $\xi_n(z)$, $\xi_t(z)$ are the normal and tangential components at z, respectively, of the vector ξ; $u(z)$ is the Euclidean distance to the boundary, and L_p is the Levi form at p with respect to any defining function ϕ for D with $\|\nabla\phi\| = 1$. The limits in these formulae are approached uniformly in $p \in \partial D$ and in vectors ξ of unit length. Another slightly stronger formulation of these formulae was also given in [8].

Subsequently, Aladro ([1]) gave the following estimate for the Kobayashi and the Carathéodory metrics:

(1.2)
$$C^{-1}\left(\frac{L_p(\xi_t)}{2u(z)} + \frac{\|\xi_n\|^2}{4u(z)^2}\right)^{1/2} \le F_D(z, \xi)$$
$$\le C\left(\frac{L_p(\xi_t)}{2u(z)} + \frac{\|\xi_n\|^2}{4u(z)^2}\right)^{1/2},$$

where ξ_n and ξ_t are the normal and tangential components of ξ at p, the point on the boundary nearest to z. Sibony [16] also gave an estimate of the Kobayashi metric on complex manifolds that support plurisubharmonic functions with certain properties. For strongly pseudoconvex domains his estimate reduces to (1.2). Catlin [5] gave estimates of the Kobayashi and Carathéodory metrics on pseudoconvex domains of finite type in $\mathbb{C}^2$.

In [12] we showed that for the Kobayashi metric the constant C in (1.2) can be taken arbitrarily close to 1, provided that $u(z)$ is sufficiently small. More precisely, we obtained an estimate of the form

(1.3)
$$\exp(-Cu(z)^\lambda)\left(\frac{L_p(\xi_t)}{2u(z)} + \frac{\|\xi_n\|^2}{4u(z)^2}\right)^{1/2} \le F_D(z, \xi)$$
$$\le \exp(Cu(z)^\lambda)\left(\frac{L_p(\xi_t)}{2u(z)} + \frac{\|\xi_n\|^2}{4u(z)^2}\right)^{1/2}.$$

In [12], we obtained (1.3) with $\lambda = 1/3 - \epsilon$, where ϵ is an arbitrarily small positive number. In [11] we obtained estimates of the form

(1.4) $$\exp(-Cu^\lambda)F_{E^p}(z, \xi) \le F_D(z, \xi) \le \exp(Cu^\lambda)F_{E^p}(z, \xi)$$

for the Kobayashi metric, the Carathéodory metric and the Eisenman volume forms on strongly pseudoconvex domain, with $\lambda = 1/5 - \epsilon$. Here E^p denotes the osculating ellipsoid of D at p, to be defined below.

In [1,8,12], a crucial step in the proof of the estimates is to use a localization lemma due to Royden (see [15, Lemma 2]). In [6] Forstneric and Rosay obtained a more precise localization result for the Kobayashi metric by a very clever use of Hadamard's three circles lemma. In this paper we shall use Forstneric/Rosay's localization theorem and a variation of the scaling method (see [2,3,7,10,13]) to obtain estimates for the Kobayashi metric of form (1.3) and of form (1.4) with $\lambda = 1/2$. We emphasize that the estimates in [11] for the *Carathéodory metric* were obtained through a different approach and cannot be improved by the techniques in the present paper.

The statement and proof of the results are in §2. In §3, we give an example to show that the estimates obtained are best possible in the sense that the exponent $\lambda = 1/2$ cannot be improved.

Results.

Consider a bounded domain $D \in \mathbb{C}^n$, not assumed to be pseudoconvex. Assume that M_0 is a relatively open subset of ∂D, that M_0 is a C^3 strongly pseudoconvex hypersurface, and that D is on the pseudoconvex side of M_0. Let M be a compact subset of M_0. For $\delta > 0$, let $Q_\delta = \{z \in \mathbb{C}^n : d(z, M) < \delta\}$ and $D_\delta = D \cap Q_\delta$, where $d(z, M)$ is the euclidean distance from z to M. Let δ_0 be a positive number such that the following conditions are satisfied:

a) $\partial D \cap Q_{\delta_0}$ is relatively compact in M_0;

b) There is a real valued strictly plurisubharmonic function $\phi \in C^3(Q_{\delta_0})$ such that $D_{\delta_0} = \{z \in Q_{\delta_0} : \phi(z) < 0\}$ and such that $\|\nabla\phi(p)\| = 1$ for $p \in Q_{\delta_0} \cap M$, where $\nabla\phi = (\partial\phi/\partial z_1, \ldots, \partial\phi/\partial z_n)$.

For the existence of the function ϕ in b), see [9, pp. 263–264]. We fix such a function ϕ once for all. Let $\delta < \delta_0$ be a positive number such that for each $z \in D_\delta$, there is a unique point on ∂D closest to z and this point $p = \pi(z)$ is in $Q_{\delta_0} \cap M_0$, and such that the set $M_1 := \pi(D_\delta)$ is relatively compact in $Q_{\delta_0} \cap M_0$.

We now fix some notation. Let $B(c, r)$ denote the Euclidean ball in $\mathbb{C}^n$ of center c and radius r and $B^n = B(0, 1)$. For $z \in D_\delta$, let $u(z) = d(z, \partial D)$. Let $T_z = T_z D$ denote the holomorphic tangent space at z. If $\xi \in T_z$, ξ_n and ξ_t denote the the complex normal and the complex tangential components of ξ at $p = \pi(z)$, respectively. Let

$$H_p(\xi, \eta) = \sum_{i,j=1}^{n} \frac{\partial^2 \phi}{\partial x_i \partial \overline{x}_j}(p)\xi_i \overline{\eta}_j$$

be the complex Hessian of ϕ at p. Let $L_p(\xi_t) := H_p(\xi_t, \xi_t)$ be the Levi form. One may verify that L_p is independent of the choice of ϕ as long as $\|\nabla\phi\| = 1$ on $Q_{\delta_0} \cap M_0$.

For $p \in \partial\Omega$, define an "osculating ellipsoid" $E^p = E(\Omega, \phi, p)$ by

$$E^p = \left\{ x \in \mathbb{C}^n : 2\mathrm{Re}\left(\sum_{j=1}^{n} \frac{\partial\phi}{\partial x_j}(p)(x_j - p_j)\right) + \right.$$

$$\left. \sum_{i,j=1}^{n} \frac{\partial^2 \phi}{\partial x_i \partial \overline{x}_j}(p)(x_i - p_i)(\overline{x}_j - \overline{p}_j) < 0 \right\}.$$

One may verify that the definition of E^p is independent of the choice of coordinates in the sense that if T is the composition of a translation and a unitary transformation, then

$$T(E(\Omega,\phi,p)) = E(T(\Omega),\phi \circ T^{-1}, T(p)).$$

Also E^p is biholomorphic to the ball (see, e.g., [8]).

Now we can state our theorems.

THEOREM A. *Under the above hypothesis, there is a positive constant C, depending only on D, M, ϕ and δ, such that for $z \in D_\delta$, the Kobayashi metric $F_D(z,\xi)$ satisfies*

$$(2.1) \qquad \exp(-C\sqrt{u})F_{E^p}(z,\xi) \le F_D(z,\xi) \le \exp(C\sqrt{u})F_{E^p}(z,\xi),$$

where $u = u(z)$ and $p = \pi(z)$.

THEOREM B. *Under the above hypothesis, there is a positive constant C, depending only on D, M and δ, such that for $z \in D_\delta$, the Kobayashi metric $F_D(z,\xi)$ satisfies*

$$(2.2) \qquad \begin{aligned} \exp(-C\sqrt{u})\Big(\frac{L_p(\xi_t)}{2u} &+ \frac{\|\xi_n\|^2}{4u^2}\Big)^{1/2} \le F_D(z,\xi) \\ &\le \exp(C\sqrt{u})\Big(\frac{L_p(\xi_t)}{2u} + \frac{\|\xi_n\|^2}{4u^2}\Big)^{1/2}, \end{aligned}$$

where $u = u(z)$ and $p = \pi(z)$.

PROOF OF THEOREM A. It is enough to prove (2.1) for z with $u(z)$ sufficiently small. Fix such a z and $p = \pi(z)$. Note that both (2.1) and (2.2) are invariant under unitarty transformations. By a translation followed by a unitary transformation we may take p to 0 and the vector $\nabla\phi$ to the negative real x_1 axis. We then take another unitary transformnation in the $x_2,\dots,x_n$ directions so that the complex Hessian $H_p(x,x)$ of ϕ is diagonalized in the $x_2,\dots,x_n$ directions. We have

$$\phi(x) = \operatorname{Re}\Big(-2x_1 + \sum_{i,j=1}^n c_{ij}x_i x_j\Big) + H_p(x,x) + O(\|x\|^3),$$

where $c_{ij} = c_{ji}$ and

$$H_p(x,x) = \sum_{j=1}^n a_j^2 |x_j|^2 + 2\operatorname{Re}\Big(\sum_{j=2}^n b_j x_1 \overline{x}_j\Big)$$

is a positive definite Hermitian form. Note that $z = (u,0,\dots,0)$. In this coordinate system, E^p has the form

$$E^p = \{x \in \mathbb{C}^n : -2\operatorname{Re}x_1 + H_p(x,x) < 0\}.$$

Define a map $\Phi : \mathbb{C}^n \to \mathbb{C}^n$ by

(2.3)
$$w_1 = x_1 - \frac{1}{2} \sum_{i,j=1}^{n} c_{ij} x_i x_j,$$

$$w_k = x_k, \qquad k = 2, \ldots, n.$$

Then Φ is biholomorphic in a neighborhood of 0 of fixed size (as p runs over M_1). In terms of the w coordinates ϕ has the form

(2.4)
$$\phi \circ \Phi^{-1}(w) = -2\mathrm{Re}w_1 + H_p(w, w) + O(\|w\|^3).$$

In the following C denotes a large positive constant whose value may vary from appearance to appearance, but is independent of p and independent of u. If w is small and if $\phi \circ \Phi^{-1}(w) \le 0$, then $\|w\|^2 \le C\mathrm{Re}w_1$ for some constant C. Since $d\Phi(0) = \mathrm{id}$, we have $B(0, \rho/2) \subset \Phi(B(0, \rho)) \subset B(0, 2\rho)$ for sufficiently small ρ. Thus there are small constant $r > 0$ and large constant $C > 0$, independent of $p \in M_1$, such that the following are satisfied:

i)   The set $B(0, 3r)$ is relatively compact in Q_{δ_0};
ii)   for each $\rho \in (0, 3r]$, the boundary of $B(0, \rho)$ intersects M_0 transversally;
iii) Φ is biholomorphic in $B(0, 3r)$;
iv) $B(0, r) \subset \Phi(B(0, 2r)) \subset B(0, 4r)$;
v) $|w|^2 \le C\mathrm{Re}w_1$ for $w \in \overline{\Phi(D \cap B(0, 2r))}$.

We assume that $u < r$. Now $\Phi(z) = \zeta$, where

$$\zeta = (v, 0, \ldots, 0), \quad v = u - (1/2)c_{11}u^2.$$

Since $|c_{11}|$ is bounded above (as p rans over M_1), for u sufficiently small we have

(2.5)
$$\mathrm{Im}v < Cu^2, \qquad \mathrm{Re}v > u/2.$$

Now we need to invoke a result in [6]. By Theorem 2.1 and Lemma 2.2 in [6], for each $q \in M_1$ there is a positive constant $K = K_q$ such that

(2.6)
$$F_D(x, \xi) \ge \exp(-Ku(x))F_{D \cap B(q, 2r)}(x, \xi), \qquad x \in D \cap B(q, r).$$

For our purpose we need that the constant K in (2.6) be independent of $q \in M_1$. To achieve this, we choose a finite number of points $\{q_k : k = 1, \ldots, N\}$ in M_1 such that for each $q \in M_1$ there is a q_k such that

(2.7)
$$B(q, r) \subset B(q_k, 4r/3) \subset B(q_k, 5r/3) \subset B(q, 2r).$$

By the result in [6] mentioned above, there is a $K > 0$ such that

(2.8)
$$F_D(x, \xi) \ge \exp(-Ku(x))F_{D \cap B(q_k, 5r/3)}(x, \xi),$$
$$x \in D \cap B(q_k, 4r/3), \ k = 1, \ldots, N.$$

Since the Kobayashi metric has the mononicity property (i.e., $U \subset V$ implies $F_U \ge F_V$), (2.7) and (2.8) imply that (2.6) holds for a constant K independent of q. In particular, for $p = \pi(z) = 0$ we have

(2.9)
$$F_{D \cap B(0, 2r)}(z, \xi) \ge F_D(z, \xi) \ge \exp(-Ku)F_{D \cap B(0, 2r)}(z, \xi).$$

Let $\Omega = \Phi(D \cap B(0, 2r))$ and define a holomorphic mapping $\Psi : \Omega \to \mathbb{C}^n$ by

(2.10)
$$y_1 = \frac{w_1 - v}{w_1 + v},$$
$$y_k = \frac{\sqrt{2v} a_k w_k}{w_1 + v}, \qquad k = 2, \dots, n,$$

where $\sqrt{v}$ is the square root of v that satisfies $\arg(\sqrt{v}) < \pi/4$ (recall that $\operatorname{Re} v > 0$). Since $\operatorname{Re} v > 0$ and $\operatorname{Re} w_1 \geq 0$ on $\overline{\Omega}$, the mapping Ψ is biholomorphic in a neighborhood of $\overline{\Omega}$. We would like to show that for some constant $C > 0$

(2.11)
$$\big| \|\Psi(w)\|^2 - 1 \big| < C\sqrt{u}, \qquad w \in \partial\Omega,$$

which would imply that

(2.12)
$$B(0, \exp(-C\sqrt{u})) \subset \Psi(\Omega) \subset B(0, \exp(C\sqrt{u})).$$

To prove (2.11), consider the expression

(2.13)
$$L(w) = |w_1 + v|^2 (\|\Psi(w)\|^2 - 1)$$
$$= 2|v|M(w) + 4(|v|\operatorname{Re} w_1 - \operatorname{Re}(\overline{v} w_1)),$$

where

(2.14)
$$M(w) = -2\operatorname{Re} w_1 + \sum_{j=2}^{n} a_j^2 |w_j|^2.$$

Since $\operatorname{Re} w_1 \geq 0$ on $\overline{\Omega}$, (2.5) implies that for some constant C we have

(2.15)
$$|w_1 + v|^2 \geq C^{-1}(|w_1|^2 + u^2) \qquad \text{for } w \in \overline{\Omega}$$

and

(2.16)
$$\big||v|\operatorname{Re} w_1 - \operatorname{Re}(\overline{v} w_1)\big| \leq Cu^2 |w_1| \qquad \text{for } w \in \overline{\Omega}.$$

Now $\partial\Omega = V_1 \cup V_2$, where

$$V_1 = \Psi(\partial B(0, 2r) \cap \overline{D}), \qquad V_2 = \Psi(B(0, 2r) \cap M_0).$$

If $w \in V_1$, then
$$|w_1| \geq \operatorname{Re} w_1 \geq C^{-1} \|w\|^2 \geq C^{-1} r^2$$

by iv) and v). Hence it follows from (2.13), (2.15) and (2.16) that for $w \in V_1$ we have

$$\big| \|\Psi(w)\|^2 - 1 \big| = \frac{|L(w)|}{|w_1 + v|^2} \leq \frac{Cu + Cu^2 |w_1|}{C^{-1}(C^{-2}r^4 + u^2)} \leq Cu \leq C\sqrt{u}.$$

For $w \in V_2$, we have $\phi \circ \Phi^{-1}(w) = 0$. By (2.4) and (2.14) we obtain

$$|M(w)| = |M(w) - \phi \circ \Phi(w)|$$
$$\leq C|w_1|^2 + C|w_1| \cdot \|w\| + C\|w\|^3$$
$$\leq C|w_1|^{3/2}.$$

Hence for $w \in V_2$ we have

$$
\begin{aligned}
\left| \|\Psi(w)\|^2 - 1 \right| &= \frac{|L(w)|}{|w_1 + v|^2} \\
&\leq \frac{Cu|M(w)| + Cu^2|w_1|}{C^{-1}(|w_1|^2 + u^2)} \\
&\leq C^3\sqrt{u} \cdot \frac{\sqrt{u}|w_1|^{3/2}}{|w_1|^2 + u^2} + C^2 u \cdot \frac{u|w_1|}{|w_1|^2 + u^2} \\
&\leq C^3\sqrt{u} \cdot (3^{3/4}/4) + C^2 u/2 \\
&\leq C\sqrt{u}.
\end{aligned}
$$

Thus (2.11) and (2.12) hold.

Now (2.12) and the Schwarz lemma imply that for any $\omega \in \mathbb{C}^n$,

$$
\exp(-C\sqrt{u})\|\omega\| \leq F_{\Psi(\Omega)}(0, \omega) \leq \exp(C\sqrt{u})\|\omega\|.
$$

Since $F_{D \cap B(0,2r)}(z, \xi) = F_{\Psi(\Omega)}\big(0, d(\Psi \circ \Phi)(z)\xi\big)$, we obtain

$$
\begin{aligned}
\exp(-C\sqrt{u})\|d(\Psi \circ \Phi)(z)\xi\| &\leq F_{D \cap B(0,2r)}(z, \xi) \\
&\leq \exp(C\sqrt{u})\|d(\Psi \circ \Phi)(z)\xi\|.
\end{aligned}
\tag{2.17}
$$

We now need to consider $F_{E^p}(z, \xi)$. Since E^p is biholomorphic to the ball, the explicit form for the Kobayashi metric on E^p is available (see, e.g., [8]). But for our purpose it is more convenient to obtain an estimate for $F_{E^p}(z, \xi)$ analogous to (2.17). Define a holomorphic mapping $\Psi_1 : E^p \to \mathbb{C}^n$ by

$$
\begin{aligned}
y_1 &= \frac{x_1 - u}{x_1 + u}, \\
y_k &= \frac{\sqrt{2u}a_k x_k}{x_1 + u}, \qquad k = 2, \ldots, n,
\end{aligned}
\tag{2.18}
$$

By arguments very similar to those leading to (2.17), we obtain that

$$
\exp(-C\sqrt{u})\|d\Psi_1(z)\xi\| \leq F_{E^p}(z, \xi) \leq \exp(C\sqrt{u})\|d\Psi_1(z)\xi\|.
\tag{2.19}
$$

It is now sufficient to prove that

$$
\begin{aligned}
\exp(-C\sqrt{u})\|d\Psi_1(z)\xi\| &\leq \|d\Psi(\zeta) \circ d\Phi(z)\xi\| \\
&\leq \exp(C\sqrt{u})\|d\Psi_1(z)\xi\|,
\end{aligned}
\tag{2.20}
$$

since (2.1) then would follow from (2.9), (2.17), (2.19) and (2.20). The linear transformations $d\Psi(\zeta)$ and $d\Psi_1(z)$ are represented by the following matrices:

$$(2.21) \qquad d\Psi(\zeta) = \begin{pmatrix} 1/(2v) & & & 0 \\ & a_2/\sqrt{2v} & & \\ & & \ddots & \\ 0 & & & a_n/\sqrt{2v} \end{pmatrix},$$

$$(2.22) \qquad d\Psi_1(z) = \begin{pmatrix} 1/(2u) & & & 0 \\ & a_2/\sqrt{2u} & & \\ & & \ddots & \\ 0 & & & a_n/\sqrt{2u} \end{pmatrix}.$$

Since $v = u - c_{11}u^2/2$, it is clear that $\|d\Psi(\zeta)\theta\| = \exp(O(u))\|d\Psi_1(z)\theta\|$ for any $\theta \in \mathbb{C}^n$. Thus (2.20) is equivalent to

$$(2.23) \qquad \begin{aligned} \exp(-C\sqrt{u})\|d\Psi_1(z)\xi\| &\leq \|d\Psi_1(z) \circ d\Phi(z)\xi\| \\ &\leq \exp(C\sqrt{u})\|d\Psi_1(z)\xi\|. \end{aligned}$$

Putting $\eta = d\Psi_1(z)\xi$, we rewrite (2.23) as

$$(2.24) \qquad \exp(-C\sqrt{u})\|\eta\| \leq \|L\eta\| \leq \exp(C\sqrt{u})\|\eta\|,$$

where $L = d\Psi_1(z) \circ d\Phi(z) \circ \left(d\Psi_1(z)\right)^{-1}$. The linear transformation L is represented by the matrix

$$(2.25) \qquad L = \begin{pmatrix} 1 - c_{11}u & -\frac{c_{12}\sqrt{u}}{\sqrt{2}a_2} & \cdots & -\frac{c_{1n}\sqrt{u}}{\sqrt{2}a_n} \\ & 1 & & 0 \\ & & \ddots & \\ 0 & & & 1 \end{pmatrix}.$$

Now (2.24) follows easily from (2.25). This proves (2.20).□

PROOF OF THEOREM B. We use the same notation as in the proof of Theorem A. By (2.22),

$$\|d\Psi_1(z)\xi\|^2 = \frac{|\xi_1|^2}{4u^2} + \sum_{j=2}^{n} \frac{a_j^2|\xi_j|^2}{2u} = \frac{\|\xi_n\|^2}{4u^2} + \frac{L_p(\xi_t)}{2u}.$$

Thus (2.19) is equivalent to

$$(2.26) \qquad \begin{aligned} \exp(-C\sqrt{u})\left(\frac{\|\xi_n\|^2}{4u^2} + \frac{L_p(\xi_t)}{2u}\right)^{1/2} &\leq F_{E^p}(z,\xi) \\ &\leq \exp(C\sqrt{u})\left(\frac{\|\xi_n\|^2}{4u^2} + \frac{L_p(\xi_t)}{2u}\right)^{1/2}. \end{aligned}$$

Now (2.2) follows from (2.1) and (2.26).

Example.

We now give an example to show that in Theorem A and Theorem B the factors $\exp(\pm C\sqrt{u})$ cannot be improved to $\exp(\pm Cu^{1/2+\epsilon})$.

Example. Consider the domain

$$D = \{x \in \mathbb{C}^2 : -2\mathrm{Re}(x_1 + x_1x_2/2) + |x_1 + x_1x_2/2|^2 + |x_2|^2 < 0\}.$$

Let $u > 0$ be small and let $z = (u, 0)$ and $\xi = (\sqrt{u}, 1)$. We have $p = \pi(z) = 0$ and

$$E^p = \{x \in \mathbb{C}^2 : -2\mathrm{Re}x_1 + |x_1|^2 + |x_2|^2 < 0\}.$$

Since both E^p and D are biholomorphic to the ball (the mapping $f : D \to B^n$ defined by $(x_1, x_2) \mapsto (x_1 + x_1x_2/2 - 1, x_2)$ is a biholomorphism), $F_{E^p}(z, \xi)$ and $F_D(z, \xi)$ can be easily calculated. We have

$$F_D(z, \xi) = \frac{\sqrt{3u + u^{3/2} - 3u^2/4}}{2u - u^2},$$

$$F_{E^p}(z, \xi) = \frac{\sqrt{3u - u^2}}{2u - u^2},$$

$$\frac{\|\xi_n\|^2}{4u^2} + \frac{L_p(\xi_t)}{2u} = \frac{3}{4u}.$$

It is easy to see that

$$F_D(z, \xi) = F_{E^p}(z, \xi) \cdot (1 + \sqrt{u}/6 + O(u)),$$

$$F_D(z, \xi) = \left(\frac{\|\xi_n\|^2}{4u^2} + \frac{L_p(\xi_t)}{2u}\right)^{1/2} \cdot (1 + \sqrt{u}/6 + O(u)).$$

Thus the factors $\exp(\pm C\sqrt{u})$ in (2.1) and (2.2) cannot be improved.

References

1. Gerardo J. Aladro, *Some Consequences of the Boundary Behavior of the Carathéodory and Kobayashi Metrics and Applications to Normal Holomorphic Functions*, PhD thesis, Pennsylvania State University, 1985.
2. Eric Bedford and Sergey Pinchuk, *Domains in $\mathbb{C}^n$ with noncompact automorphism group*, Preprint.
3. _____, *Domains in $\mathbb{C}^2$ with noncompact group of automorphisms*, Math. USSR-Sb. **63** (1989), 141–151.
4. Brian Blank, Dashan Fan, David Klein, Steven G. Krantz, Daowei Ma, and Myung-Yull Pang, *The Kobayashi metric of a complex ellipsoid in $\mathbb{C}^2$*, Preprint.
5. David W. Catlin, *Estimates of invariant metrics on pseudoconvex domains of dimension two*, Math. Z. **200** (1989), 429–466.
6. Franc Forstneric and Jean-Pierre Rosay, Math. Ann. **279** (1987), 239–252.
7. Sidney Frankel, *Complex geometry of convex domains that cover varieties*, Acta Math. **163** (1989), 109–149.

8. Ian Graham, *Boundary behavior of the Carathéodory and Kobayashi metrics on strongly pseudoconvex domains in C^n with smooth boundary*, Trans. Amer. Math. Soc. **207** (1975), 219–240.

9. Robert C. Gunning and Hugo Rossi, *Analytic Functions of Several Complex Variables*, Prentice-Hall, Inc., Englewood Cliffs, N.J., 1965.

10. Kang-Tae Kim, *Complete localization of domains with noncompact automorphism groups*, Trans. Amer. Math. Soc. **319** (1989), 139–153.

11. Daowei Ma, *Boundary behavior of invariant metrics and volume forms on strongly pseudoconvex domains*, Duke Math. J. **63** (1991), 67–697.

12. _____ , *On iterates of holomorphic maps*, Math. Z. **207** (1991), 417–428.

13. Sergey Pinchuk, *Holomorphic inequivalences of some classes of domains in $\mathbb{C}^n$*, Math. USSR-Sb. **39** (1981).

14. H. J. Reiffen, *Die Carathéodorysche distanz und ihre zugehörige differential-metrik*, Math. Ann. **161** (1965), 315–324.

15. H. L. Royden, *Remarks on the Kobayashi metric*, Proceedings of the Maryland Conference on Several Complex Variables, Springer, Berlin, 1970, Lecture Notes in Math., v. 185.

16. Nessim Sibony, *A class of hyperbolic manifolds*, Recent Developments in Several Complex Variables, John E. Fornæss, editor, Princeton University Press, 1981, pp. 357–372.

DEPARTMENT OF MATHEMATICS, THE UNIVERSITY OF CHICAGO, CHICAGO, IL 60637

Contemporary Mathematics
Volume **137**, 1992

Proper Maps, Complex Geodesics and Iterates of Holomorphic Maps on Convex Domains in $\mathbb{C}^n$

PETER R. MERCER

Introduction.

This paper is an announcement of results, the details of which will appear in the author's doctoral thesis and elsewhere ([Me1], [Me2]).

We begin with a generalization of the well-known Hopf lemma which requires only a cone condition on the domain in question rather than boundary smoothness. This enables us to study, for example, a large class of bounded convex domains in $\mathbb{C}^n$ – called the m-convex domains – which carries no boundary regularity assumption. In particular, we examine boundary regularity of (i) complex geodesics into m-convex domains, and (ii) proper holomorphic maps from bounded convex domains into m-convex domains.

The existence of (i) above with prescribed boundary data is then established. We apply this result to obtain new results from iteration theory of holomorphic self-maps of convex domains in $\mathbb{C}^n$.

The author is grateful to Ian Graham, his thesis advisor, not only for helpful discussions and suggestions concerning the content of this paper, but also for making it possible for him to attend the Symposium on Complex Analysis in honour of Professor Rudin. He would also like to thank the organizers for their support.

The Results

A domain $\Omega \subset \mathbb{C}^n$ satisfies a cone condition if there is an $r > 0$ and a $\theta \in (0, \pi)$ such that each $z \in \Omega$ near $\partial\Omega$ lies on the axis of an open cone $\Gamma_p \subset \Omega$ with vertex $p \in \partial\Omega$, height r, and aperture θ.

Denote by $d_\Omega(z)$ the distance from $z \in \Omega$ to $\partial\Omega$. The following is our version of the Hopf lemma, the proof of which is a modification of that of Proposition 12.2 of [FS]. It was recently brought to our attention (by the referee of [Me2]) that an even more general version is known ([O], [Mi1, Mi2]).

1991 *Mathematics Subject Classification*. Primary 32H25, Secondary 32H15, 32H50.
The final (detailed) version of this paper will be submitted for publication elsewhere.

THEOREM 1. *Let $\Omega \subset\subset \mathbb{C}^n$ satisfy a cone condition and let $\varphi \colon \Omega \to [-\infty, 0)$ be plurisubharmonic. There is a $c > 0$ and an $\alpha \geq 1$ such that*

$$\varphi(z) \leq -c d_\Omega^\alpha(z) \qquad \forall z \in \Omega \ . \qquad \square$$

Denote by k_Ω the Kobayashi distance on Ω and compare the following proposition with Theorem 1.4 of [A1], in which Ω is strongly pseudoconvex.

PROPOSITION 2. *Let $\Omega \subset\subset \mathbb{C}^n$ be convex with $z_0 \in \Omega$. There are constants $c_1, c_2 > 0$ and an $\alpha \geq 1$ such that*

$$c_1 - \frac{1}{2} \log d_\Omega(z) \ \leq \ k_\Omega(z_0, z) \ \leq \ c_2 - \frac{1}{2} \log d_\Omega^\alpha(z) \qquad \forall z \in \Omega \ . \qquad \square$$

This estimate quickly yields (compare with Proposition 12 of [L]):

COROLLARY 3. *Let $\Omega \subset\subset \mathbb{C}^n$ be convex and $f \in \mathrm{Hol}(\Delta, \Omega)$ a complex geodesic with $z_0 = f(0) \in \Omega$. There are constants $c_1, c_2 > 0$ and an $\alpha \geq 1$ such that*

$$c_1(1 - |\lambda|) \ \leq \ d_\Omega(f(\lambda)) \ \leq \ c_2(1 - |\lambda|)^{1/\alpha} \qquad \forall \lambda \in \Delta \ . \qquad \square$$

The next result is a consequence of Theorem 1 together with some techniques developed by Pinčuk in ([P]).

COROLLARY 4. *Let $\Omega_1, \Omega_2 \subset\subset \mathbb{C}^n$ be convex and $F \in \mathrm{Hol}(\Omega_1, \Omega_2)$ proper. There are constants $c_1, c_2 > 0$ and an $\alpha \geq 1$ such that*

$$c_1 d_{\Omega_1}^\alpha(z) \ \leq \ d_{\Omega_2}(F(z)) \ \leq \ c_2 d_{\Omega_1}^{1/\alpha}(z) \qquad \forall z \in \Omega_1. \qquad \square$$

For $\Omega \subset\subset \mathbb{C}^n$ convex, $z \in \Omega$ and $v \in \mathbb{C}^n$ a unit vector, set

$$r_\Omega(z; v) \ = \ \sup[|\lambda| : \ z + \lambda v \in \Omega] \ .$$

Let $m \in (0, \infty)$. We say that such an Ω is m-convex if there is a $c > 0$ such that for each unit vector $v \in \mathbb{C}^n$ we have

$$r_\Omega(z; v) \ \leq \ c d_\Omega^{1/m}(z) \qquad \forall z \in \Omega \ .$$

A strongly convex domain (i.e. C^2-bounded with positive definite real Hessian) is 2-convex.

By the previous two corollaries, complex geodesics into convex domains and proper holomorphic maps between convex domains each change the distance to the boundary of any point by at most some fixed power. In either case, if the target domain is in fact m-convex we can say even more, using the estimates for the Kobayashi metric on convex domains established in [G1, G2]:

PROPOSITION 5. *Let $\Omega, \Omega_1, \Omega_2 \subset\subset \mathbb{C}^n$ be convex with Ω and Ω_2 m-convex.*

(a) *Let $f \in \mathrm{Hol}(\Delta, \Omega)$ be a complex geodesic. Then f extends to a Hölder continuous map on $\overline{\Delta}$.*

(b) Let $F \in \text{Hol}(\Omega_1, \Omega_2)$ be proper. Then F extends to a Hölder continuous map on $\overline{\Omega}_1$. $\square$

Corollary 4 and Proposition 5(b) may be compared with some results cited in [F]. With Proposition 5(a) in place, the following lemma is obtained. It is a consequence of some methods developed in [CHL] (see also [L]).

LEMMA 6. *Let $\Omega \subset\subset \mathbb{C}^n$ be m-convex and let $z, w \in \overline{\Omega}$. There is a complex geodesic $f \in \text{Hol}(\Delta, \Omega)$ whose continuous extension contains $\{z, w\}$ in its image.* $\square$

The classical Denjoy-Wolff theorem (on the convergence of the sequence of iterates of a holomorphic, fixed point free, self map of the unit disk) was generalized to the context of bounded strongly convex domains by Abate ([A1], Theorem 0.6). An important tool was the correct generalization to $\mathbb{C}^n$ of the classical horocycle: For $\Omega \subset \mathbb{C}^n$, the big horosphere with pole $z_0 \in \Omega$, centre $x \in \partial\Omega$ and radius $R > 0$ is given by

$$F_x(z_0, R) \; = \; \{z \in \Omega : \; \liminf_{w \to x}[k_\Omega(z, w) - k_\Omega(z_0, w)] < \frac{1}{2}\log R\} \; .$$

The small horosphere is defined similarly using lim sup.

The ideas of Lemma 6 yield (compare with Theorems 1.7 and 0.6 of [A1]):

LEMMA 7. *Let $\Omega \subset\subset \mathbb{C}^n$ be m-convex with $z_0 \in \Omega$ and $x \in \partial\Omega$. For each $R > 0$ we have*

$$\overline{F}_x(z_0, R) \cap \partial\Omega \; = \; \{x\} \; .$$ $\square$

THEOREM 8. *Let $\Omega \subset\subset \mathbb{C}^n$ be m-convex. Let $F \in \text{Hol}(\Omega, \Omega)$ with $\{z \in \Omega : F(z) = z\} = \phi$. There is an $x \in \partial\Omega$ such that the sequence $\{F^j\}$ of iterates of F converges (uniformly on compact subsets of Ω) to x.* $\square$

This result does not hold if Ω is merely convex and bounded ([A1]). Related results concerning the case where Ω is (weakly) convex and bounded with C^2 boundary appear in [A2]. Theorem 8 is currently the most general version of the Denjoy-Wolff theorem in the sense that the m-convex domains is the largest subclass of the bounded convex domains for which such a result is known to hold.

References.

[A1] M. Abate, *Horospheres and iterates of holomorphic maps*, Math. Z. **198** (1988), 225–238.

[A2] ——, *Iteration theory of holomorphic maps on taut manifolds*, Mediterranean Press, Rende, Cosenza, 1989.

[CHL] C.H. Chang, M.C. Hu, H.P. Lee, *Extremal analytic discs with prescribed boundary data*, Trans. Am. Math. Soc. **310** no. no. 1 (1988), 355-369.

[FS] J.E. Fornaess and B. Stensones, *Lectures on counterexamples in several complex variables*, Math. Notes no. 33, Princeton University Press, 1987.

[F] F. Forstnerič, *Proper holomorphic mappings: a survey*, Preprint Series, Dept. Math. University E.K. Ljubljana.

[G1] I. Graham, *Distortion theorems for holomorphic maps between convex domains in* $\mathbb{C}^n$, Complex Variables Theory Appl. **15** no. no. 1 (1990), 37–42.

[G2] _____, *Sharp constants for the Koebe theorem and for estimates of intrinsic metrics on convex domains*, Proc. Symp. Pure Math. **52** no. part 2 (1991), 233–238.

[L] L. Lempert, *Intrinsic distances and holomorphic retracts*, Complex Analysis and Applications, 1981, Bulgar. Acad. Sci., Sofia, 1984, pp. 341–364.

[Me1] P.R. Mercer, *Complex geodesics and iterates of holomorphic maps on convex domains in* $\mathbb{C}^n$, Trans. Am. Math. Soc. (to appear).

[Me2] _____, *A general Hopf lemma and proper holomorphic mappings between convex domains in* $\mathbb{C}^n$, Proc. Am. Math. Soc. (to appear).

[Mi1] K. Miller, *Barriers on cones for uniformly elliptic operators*, Ann. Mat. Pura Appl. **76** (1967), 93–105.

[Mi2] _____, *Extremal barriers on cones with Phragmen-Lindelöf theorems and other applications*, Ann. Mat. Pura Appl. **90** (1971), 297–329.

[O] J.K. Oddson, *On the boundary point principle for elliptic equations in the plane*, Bull. Am. Math. Soc. **74** (1968), 666–670.

[P] S.I. Pinčuk, *On proper holomorphic mappings of strictly pseudoconvex domains*, Sib. Math. J. **15** no. no. 4 (1974), 644–649.

DEPARTMENT OF MATHEMATICS, UNIVERSITY OF TORONTO, TORONTO, CANADA M5S 1A1

Contemporary Mathematics
Volume **137**, 1992

On The Polynomial Hull Of Two Balls

CARL MUELLER

The main topic of this note (part **A**) is the discussion of the polynomial hull of two intersecting balls in $\mathbf{C}^2$. In part **B**, which is independent of part **A**, we study the hull of the union of three disjoint complex ellipsoids.

A. A partial description of the hull of two intersecting balls

Let B_1 and B_2 be the closed balls in $\mathbf{C}^2$ with radius $\sqrt{2}$ and centers $(-1, 0)$ and $(1, 0)$;

$$B_1 = \{(z, w) \in \mathbf{C}^2 : |z+1|^2 + |w|^2 \le 2\}, \quad B_2 = \{(z, w) \in \mathbf{C}^2 : |z-1|^2 + |w|^2 \le 2\}.$$

Let $\Omega = B_1 \cup B_2$, and let $\widehat{\Omega}$ denote the polynomial hull of Ω. The precise boundary of $\widehat{\Omega}$ is unknown thus far, but partial results are known. These results fall into three categories:

i) Points in the boundary of Ω which are in the interior of $\widehat{\Omega}$ due to the edge-of-the-wedge theorem (preliminary remark below),

ii) Points of $\widehat{\Omega}$ which have been found by embedding analytic discs, and

iii) Points outside $\widehat{\Omega}$ which have been determined by examining a family of analytic surfaces missing Ω.

Each of these types of points will be discussed in turn. Additionally, some approximate results attained with the help of a computer are discussed.

Note: If K is a compact set in $\mathbf{C}^2$ defined as $K = \{(z, w) : |w| \le f(z)\}$, it follows that the hull $\widehat{K}$ is of the form $\widehat{K} = \{(z, w) : |w| \le F(z)\}$ where $F \ge f$. If we let z_0 be so that $F(z_0) > f(z_0)$, then, at least if we assume that F is smooth in a neighborhood of z_0, it is clear that $-\log F$ will be harmonic in a neighborhood of z_0. Indeed the hypersurface $|w| = F(z)$ must be Levi flat there. (If the Levi form of $|w| = F(z)$ at (z_0, w_0) is greater than zero, we have a local maximum which is ruled out by Rossi's local maximum principle [4]. If

---

1991 *Mathematics Subject Classification*. Primary 32E20.

The author is indebted to both Walter Rudin and Jean-Pierre Rosay for some of the ideas expressed in this note.

This paper is in final form and no version of it will be submitted for publication elsewhere.

the Levi form is less than zero, the point is in the interior of the polynomial hull by extension of holomorphic functions.) As this hypersurface is Levi flat, $-\log F(z)$ is harmonic. So, in principle, it is possible to determine $-\log F(z)$ by taking the pointwise supremum (in fact the maximum) of all the subharmonic functions which are less than or equal to $-\log f(z)$ at all points in the domain. In practice, this is difficult to do precisely, but, using a computer, it is possible to get some approximate results.

Note that $(0, w) \in \widehat{\Omega}$ if and only if $(0, |w|) \in \widehat{\Omega}$. We therefore sometimes refer to $\widehat{\Omega}$ with coordinates from $\mathbf{C} \times \mathbf{R}^+$.

I: Preliminary Remark

If we consider the union of two convex domains, then only exceptional points of their intersection fail to be in the hull of the union. This result follows immediately from the edge-of-the-wedge theorem as explained below for the present example.

The boundaries of B_1 and B_2 intersect in the set $S = \{(z, w) \in \mathbf{C}^2 : x = 0, y^2 + |w|^2 = 1\}$, $(z = x + yi)$. All of S with the exception of the points $(\pm i, 0)$ is contained in the interior of $\widehat{\Omega}$. It is clear that the intersection of $\widehat{\Omega}$ with the z plane ($\{w = 0\}$) is simply the set $\omega = \{(z, 0) : |z - 1|^2 \leq 2, |z + 1|^2 \leq 2\}$, the union of two discs. Thus, the points $(\pm i, 0)$ are in the boundary of $\widehat{\Omega}$. Let (z_0, w_0) be a point in $S \setminus \{(\pm i, 0)\}$. Then, in a neighborhood of (z_0, w_0), S is a totally real surface of dimension 2. Indeed, a defining function for B_1 is $\rho_1(z, w) = 2 - |z + 1|^2 - |w|^2$ and one for B_2 is $\rho_2(z, w) = 2 - |z - 1|^2 - |w|^2$. We compute $\partial \rho_1 \wedge \partial \rho_2$:

$$-\partial \rho_1 = (\bar{z} + 1)dz + \bar{w}dw \qquad -\partial \rho_2 = (\bar{z} - 1)dz + \bar{w}dw$$

$$\partial \rho_1 \wedge \partial \rho_2 = ((\bar{z} + 1)\bar{w} - (\bar{z} - 1)\bar{w})dz \wedge dw = 2\bar{w}dz \wedge dw$$

Hence $\partial \rho_1 \wedge \partial \rho_2 \neq 0$ (and hence the edge is totally real) for $w \neq 0$.

Let (z_0, w_0) be a point in $S \setminus \{(\pm i, 0)\}$. We can apply the edge-of-the-wedge theorem [3] to wedges with edge S in a neighborhood of (z_0, w_0). We then get a neighborhood U of (z_0, w_0) so that every function holomorphic on Ω has a unique holomorphic extension to U. If f is holomorphic and bounded on Ω then $\sup_U |f| \leq \sup_\Omega |f|$ since homomorphisms of $H^\infty(\Omega)$ have norm 1. So, U is included in the polynomial hull of Ω.

In **II** and **III** we will establish the following proposition.

PROPOSITION. $(0, h)$ is in $\widehat{\Omega}$ for $h \leq \sqrt{(1 + \sqrt{2})/2}$ and is not in $\widehat{\Omega}$ for $h > (1 + \sqrt{2})/2$.

II: Embedded Analytic Discs

We wish to define a map Φ from the closed unit disc to $\mathbf{C}^2$ so that the image of the unit circle lies in Ω. This would imply that the image of the disc lies in $\widehat{\Omega}$. The map will be defined by shrinking the unit disc, "lifting" it so that the images of $\pm i$ are on the boundary of Ω, and finally "bending" it until the images of ± 1 are on the boundary of Ω. Thus the image of the unit circle will touch the boundary of Ω at four points. The remainder of the unit circle will be mapped into the interior of Ω. We define Φ_1 as:

$$\Phi_1(\lambda) = (\delta\lambda, \sqrt{1-\delta^2}),$$

where $\delta < 1$ will be chosen later. This map shrinks the unit disk by the factor δ and lifts it by an amount $\sqrt{1-\delta^2}$, so that the images of $\pm i$ are on the boundary of Ω. Now, we define Φ as:

$$\Phi(\lambda) = (\delta\lambda, \sqrt{1-\delta^2} + \epsilon(1+\lambda^2)).$$

This keeps the images of $\pm i$ fixed and bends the disc so that the images of ± 1 are lifted toward the boundary of Ω. We choose ϵ and δ so that
(1)
$$2\epsilon(\sqrt{1-\delta^2} + \epsilon) \le \delta.$$
This keeps the image of the unit circle under Φ inside Ω. This is not hard to show. We will show that $\Phi(\cos\theta + i\sin\theta)$ lies in B_2 for $-\pi/2 < \theta < \pi/2$. A similar argument shows that the rest of the unit circle is mapped into B_1.

$$\Phi(\cos\theta + i\sin\theta) = (\delta(\cos\theta + i\sin\theta), \sqrt{1-\delta^2} + \epsilon(1 + (\cos\theta + i\sin\theta)^2)) = (z_\theta, w_\theta).$$

To show that (z_θ, w_θ) is in B_2, we need only show that $|z_\theta - 1|^2 + |w_\theta|^2 - 2 \le 0$.

$$|z_\theta - 1|^2 + |w_\theta|^2 - 2 = (\delta\cos\theta - 1)^2 + (\delta\sin\theta)^2$$
$$+ (\sqrt{1-\delta^2} + 2\epsilon\cos^2\theta)^2 + (2\epsilon\sin\theta\cos\theta)^2 - 2$$
$$= -2\delta\cos\theta + 4\epsilon\cos^2\theta\sqrt{1-\delta^2} + 4\epsilon^2\cos^2\theta \overset{?}{\le} 0.$$

Rearranging terms a bit gives:

$$2\epsilon(\sqrt{1-\delta^2} + \epsilon)\cos\theta \overset{?}{\le} \delta.$$

This last inequality follows from (1) because $0 < \cos\theta \le 1$ for the range of θ in question. So, it has been shown that the unit circle is mapped into Ω. Thus $(0, \sqrt{1-\delta^2} + \epsilon)$ lies in $\widehat{\Omega}$ as do all the other points in the image of the unit disc.

We wish to choose ϵ and δ so that $\sqrt{1-\delta^2} + \epsilon$ is maximized. By choosing ϵ so that equality holds in (1), we get that $\sqrt{1-\delta^2} + \epsilon = (\sqrt{1+2\delta-\delta^2} + \sqrt{1-\delta^2})/2$ which is maximized when $\delta = \sqrt{2} - 1$. This maximum value is $\sqrt{1-\delta^2} + \epsilon = \sqrt{(1+\sqrt{2})/2}$. Thus $\left(0, \sqrt{(1+\sqrt{2})/2}\right)$ is in $\widehat{\Omega}$.

III: Analytic Surfaces Missing Ω

As Ω is "topologically simple," we might expect that every polynomial P that does not vanish on Ω also does not vanish on $\widehat{\Omega}$. But, this would require a proof, and [1] contains an example for which this "principle" fails. However, if we restrict our attention to polynomials of the form $Q(z, w) = w - P(z)$, we have the following lemma.

LEMMA. *If the zero set of $Q(z, w) = w - P(z)$ misses Ω then the zero set also misses $\widehat{\Omega}$.*

For the convenience of the reader we give a simple argument (as in proving Runge's theorem), although one could simply refer to "Oka's principle."

PROOF. We denote by $\mathbf{P}(\Omega)$ the closed subalgebra of $\mathcal{C}(\overline{\Omega})$ generated by polynomials in (z, w). Define $Q_t(z, w) = tw - P(z)$ for $0 \leq t \leq 1$, and let $\Sigma_t = \{(z, w) : Q_t = 0\}$. The hypothesis is that Σ_1 misses Ω. It follows that Σ_t does as well for $t < 1$ (for fixed z, $|w|$ increases as t decreases so the surfaces move away from Ω). Q_0 is invertible in $\mathbf{P}(\Omega)$ by Runge's theorem. That is, $1/Q_0$ is the uniform limit of polynomials on Ω. Furthermore, the set of t for which Q_t is invertible in $\mathbf{P}(\Omega)$ is both closed (since $\mathbf{P}(\Omega)$ is closed) and open (since the set of invertible elements in a Banach algebra is open). Thus Q_t is invertible for all t, $0 \leq t \leq 1$. This implies that Σ_t misses $\widehat{\Omega}$.

We now define $P(z) = az^2 + b$ where a and b are positive real numbers and are chosen so that the surface $\Sigma_1 = \{(z, w) : w - az^2 - b = 0\}$ defined in the proof of the lemma misses Ω. The lemma implies that Σ_1 misses $\widehat{\Omega}$ and in fact the proof shows that Σ_t misses $\widehat{\Omega}$ for $0 \leq t \leq 1$. In particular, the point $(0, b/t)$ is outside $\widehat{\Omega}$ for $t < 1$. If we choose

$$b = \frac{4a^2 + 1 + \sqrt{(4a^2 + 1)^2 + 16a^2}}{8a}$$

then the surface Σ_1 is tangent to Ω at four points and is otherwise outside Ω. So, Σ_t misses Ω entirely for $t < 1$. Setting $a = 1/2$ yields $b = (1 + \sqrt{2})/2$, the smallest value of b for which Σ_1 is tangent to Ω. Hence, $(0, (1 + \sqrt{2})/2t)$ lies outside $\widehat{\Omega}$ for $t < 1$.

It may (or may not) be helpful to note that the z coordinates ($z = x + iy$) of the four points of tangency mentioned above are given in terms of a and b as

$$x^2 = \frac{1}{16a^2b^2} \quad \text{and} \quad y^2 = \frac{16ab^3 - 8b^2 - 1}{16a^2b^2}.$$

When b is defined as above and $.447488 < a < \infty$, these values of $z = x + iy$ form the boundary of a compact region in the z plane. The region is roughly elliptical in shape and passes through $z = \pm i$ and $z \approx \pm.4608$. The points on the boundary of Ω which project onto the boundary of this region are also points on the boundary of $\widehat{\Omega}$ (since points immediately "above" them are outside $\widehat{\Omega}$). Furthermore, it is not difficult to see that $\widehat{\Omega} \setminus \Omega$ is connected. Thus $\widehat{\Omega} \setminus \Omega$ projects entirely inside this region in the z plane.

The proposition has been established. Thus, it is known that $(0, h)$ is in $\widehat{\Omega}$ for $h \leq \sqrt{(1 + \sqrt{2})/2} \approx 1.099$ and is not in $\widehat{\Omega}$ for $h > (1 + \sqrt{2})/2 \approx 1.207$. This leaves a range of values for h about which little is known. It should be noted that the results in **II** and **III** are not the best possible. The maps used in **II** take the unit circle into the interior of Ω with the exception of four points which are taken to the boundary. We can perturb these maps slightly so that the image of the unit circle remains in Ω and so that the image of the origin is slightly farther from Ω. Thus the result in **II** is not sharp. In a similar way, the surfaces from **III** can be perturbed slightly so that they continue to miss Ω and so that they dip slightly closer to Ω above the origin. So the result in **III** is not sharp.

IV: Computer Results

Computer aided computations seem to indicate that the boundary of the hull is near $(0, 1.189)$. These computer approximations were carried out as detailed below.

Define $\omega = \{z : |z - 1|^2 \leq 2, |z + 1|^2 \leq 2\}$ and let $f(z): \omega \to (\mathbf{R}^+ \cup \{0\})$ be the function which describes the boundary of Ω, $|w| = f(z)$. Similarly, we let $F(z)$ be the function that describes the boundary of $\widehat{\Omega}$. We define $g(z) = -\log f(z)$ and $G(z) = -\log F(z)$. Then, as discussed in the note earlier, $G(z)$ is subharmonic and is the maximum of all the subharmonic functions defined on the interior of ω which are no larger than $g(z)$. A computer can be used to approximate $G(z)$ as follows. The approximation starts with $G(z)$ set equal to $g(z)$. Successive approximations are attained by computing the average of $G(z)$ over small discs and updating the value of $G(z)$ at the center of the circle with this average if it is smaller than the old value of G at the center. In this way, the function $G(z)$ is continually updated until a limiting function is approached. Thus $F(z) = \exp{-G(z)}$, the function that describes the boundary of $\widehat{\Omega}$, is found. Problems with infinities near the points $(0, \pm i)$ and the finite memory capacities of computers make the results nothing more than good approximations, but these approximations indicate that the boundary of the hull is near $(0, 1.189)$. That is, $F(0) \approx 1.189$. The set ω was broken into a grid of 256 points per unit length to make these computations, and the discs over which the function was averaged were of radius $16/256$ (and thus contained roughly 800 grid points). One iteration in the approximation consisted of computing these averages for circles centered at each grid point for which $0 \leq y \leq 236/256$ and $0 \leq x \leq 125/256$. (Symmetry allows us to consider only non-negative values of x and y.) The value $F(0) = 1.189$ (to this accuracy) appeared after 293 iterations. After 1000 iterations the value was $1.188677899...$ and a limit could be extrapolated from the data. This limit (to three decimal places) is 1.189. Of course the computer tells us much more, giving a reasonably good picture of the entire hull (away from the points $z = \pm i$). This picture indicates that $\widehat{\Omega} \setminus \Omega$ projects onto a roughly elliptical region somewhat smaller than the one derived in **III** above.

Graphs of $g(x) = \log(1/f(x))$ and of $G(x) = \log(1/F(x))$

$f(x) =$ distance from $(x, 0)$ to the boundary of Ω in the w variable

$F(x) =$ distance from $(x, 0)$ to the boundary of $\widehat{\Omega}$ in the w variable

$$-1 - \sqrt{2} \leq x \leq 1 + \sqrt{2}$$

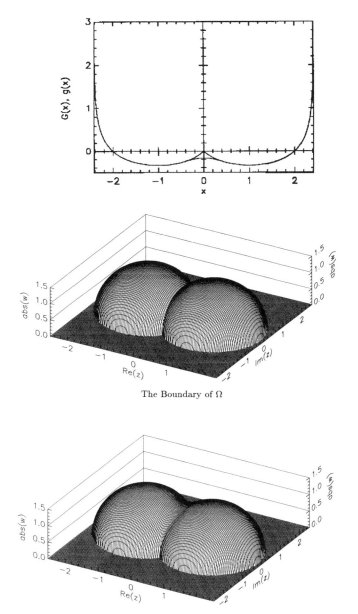

The Boundary of Ω

The (Approximate) Boundary of $\widehat{\Omega}$

B. An example of disjoint complex ellipsoids with a hull

It is known that the union of three disjoint balls in $\mathbf{C}^n$ is polynomially convex. It is also known that there exist three disjoint polydiscs in $\mathbf{C}^n (n > 2)$ whose union fails to be polynomially convex. These results are due to Eva Kallin ([2]) and may also be found in [5]. It is the purpose of the remainder of this note to exhibit three disjoint (and congruent) complex ellipsoids whose union fails to be polynomially convex. The development below closely parallels the treatment in Stout's book ([5] page 389).

Following Kallin's example, we define V to be the variety defined by $z_1 z_2 = 1$ and $z_3(1 - z_1) = 1$, and we define D to be the planar domain $\mathbf{C} \setminus \{0, 1\}$. We define $\Phi : D \to \mathbf{C}^3$ by

$$\Phi(\zeta) = \left(\zeta, \frac{1}{\zeta}, \frac{1}{1 - \zeta}\right).$$

The image of D under Φ is V. We define Δ to be the image under Φ of the set

$$\{\zeta \in D : |\zeta| < M, |\zeta| > M^{-1}, |\zeta - 1| > M^{-1}\}.$$

Kallin set $M > 2$, whereas the current example will require a somewhat larger M. The boundary of Δ in V is made up of the three curves

$$\gamma_1 = \left\{\left(\frac{e^{i\theta}}{M}, Me^{-i\theta}, \frac{M}{M - e^{i\theta}}\right) : \theta \in \mathbf{R}\right\},$$

$$\gamma_2 = \left\{\left(1 + \frac{e^{i\theta}}{M}, \frac{M}{M + e^{i\theta}}, -Me^{-i\theta}\right) : \theta \in \mathbf{R}\right\},$$

$$\gamma_3 = \left\{\left(Me^{i\theta}, \frac{e^{-i\theta}}{M}, \frac{1}{1 - Me^{i\theta}}\right) : \theta \in \mathbf{R}\right\}.$$

The maximum modulus theorem implies that the polynomial hull of any set containing $\gamma_1 \cup \gamma_2 \cup \gamma_3$ must contain all of Δ.

Kallin's example gave three disjoint polydiscs each containing one of the γ_i. Thus the polynomial hull of the union of the polydiscs contains all of Δ and hence points outside the polydiscs. Kallin's polydiscs are replaced here with complex ellipsoids E_1, E_2, and E_3 defined as follows:

$$E_1 = \left\{(z_1, z_2, z_3) : \frac{|z_1|^2}{(1/2)^2} + \frac{|z_2|^2}{(kM)^2} + \frac{|z_3 - 1|^2}{(1/2)^2} \leq \alpha\right\},$$

$$E_2 = \left\{(z_1, z_2, z_3) : \frac{|z_1 - 1|^2}{(1/2)^2} + \frac{|z_2 - 1|^2}{(1/2)^2} + \frac{|z_3|^2}{(kM)^2} \leq \alpha\right\},$$

$$E_3 = \left\{(z_1, z_2, z_3) : \frac{|z_1|^2}{(kM)^2} + \frac{|z_2|^2}{(1/2)^2} + \frac{|z_3|^2}{(1/2)^2} \leq \alpha\right\},$$

where k and M are to be determined and α is slightly less than 1 (to make the ellipsoids disjoint). We wish to show that $\gamma_i \subset E_i$ for $i = 1, 2, 3$.

$\gamma_1 \subset E_1$:

$$4|\frac{e^{i\theta}}{M}|^2 + \frac{|Me^{-i\theta}|^2}{(kM)^2} + 4|\frac{M}{M-e^{i\theta}} - 1|^2 = \frac{4}{M^2} + \frac{M^2}{(kM)^2} + \frac{4}{|M-e^{i\theta}|^2}$$

$$\leq \frac{4}{M^2} + \frac{1}{k^2} + \frac{4}{(M-1)^2} \overset{?}{<} \alpha$$

$\gamma_2 \subset E_2$:

$$4|\frac{e^{i\theta}}{M}|^2 + 4|\frac{M}{M+e^{i\theta}} - 1|^2 + \frac{|-Me^{-i\theta}|^2}{(kM)^2} = \frac{4}{M^2} + \frac{4}{|M+e^{i\theta}|^2} + \frac{M^2}{(kM)^2}$$

$$\leq \frac{4}{M^2} + \frac{4}{(M-1)^2} + \frac{1}{k^2} \overset{?}{<} \alpha$$

$\gamma_3 \subset E_3$:

$$\frac{|Me^{i\theta}|^2}{(kM)^2} + 4|\frac{e^{-i\theta}}{M}|^2 + 4|\frac{1}{1-Me^{i\theta}}|^2 = \frac{M^2}{(kM)^2} + \frac{4}{M^2} + \frac{4}{|M-e^{i\theta}|^2}$$

$$\leq \frac{1}{k^2} + \frac{4}{M^2} + \frac{4}{(M-1)^2} \overset{?}{<} \alpha$$

If we set $k = 2$ and $M = 4$ then $\frac{1}{k^2} + \frac{4}{M^2} + \frac{4}{(M-1)^2} = \frac{1}{4} + \frac{1}{4} + \frac{4}{9} = \frac{17}{18}$. So, if $\frac{17}{18} < \alpha < 1$, $\gamma_i \subset E_i$ for $i = 1, 2, 3$.

The question of how far these ellipsoids are from balls is a natural one. The ratio of the longest axis to the shortest axis for these ellipsoids is 16. Altering the values of k and M slightly can push this ratio down a bit to about 15.84. It is unknown how small this ratio may be and still allow the ellipsoids to be arranged in such a way that their union is not polynomially convex.

References.

1. J. Duval, *Convexité Rationelle Des Surfaces Lagrangiennes*, preprint, 1990.

2. E. Kallin, *Polynomial Convexity: The Three Spheres Problem*, Proceedings of the Conference on Complex Analysis, Minneapolis, 1964, (A. Appeli, E. Calabi, and H. Röhrl, Eds.), New York, 1965.

3. S. Pinčuk, *Bogoljubov's theorem on the "edge of the wedge" for generic manifolds*, Math. USSR-Sb. **23** (1974), 441-455.

4. H. Rossi, *The Local Maximum Modulus Principle*, Ann. of Math. **72** (1960), 1-11.

5. E. L. Stout, *The Theory of Uniform Algebras*, Bogden & Quigley, Inc., 1971.

Contemporary Mathematics
Volume **137**, 1992

Analytic automorphisms of $\mathbf{C}^2$
which preserve the coordinate axes

YASUICHIRO NISHIMURA

1. Introduction.

(1.1) Let us denote by IX the set of all injective holomorphic maps F : $\mathbf{C}^2 \longrightarrow \mathbf{C}^2$ which preserve the coordinate axes. Then, $F \in IX$ is of the form $F = (f, g)$ with $f, g \in \mathcal{O}(\mathbf{C}^2)$ satisfying $f(0, y) = 0$ and $g(x, 0) = 0$, where $\mathcal{O}(\mathbf{C}^2)$ denotes the algebra of all holomorphic functions on $\mathbf{C}^2$. Since the order of zero $\{x = 0\}$ of f is equal to 1 and $f(x, y)/x$ takes no zero in $\mathbf{C}^2$, $f(x, y) = xe^{u(x,y)}$ with $u \in \mathcal{O}(\mathbf{C}^2)$. Similarly, $g(x, y) = ye^{v(x,y)}$ with $v \in \mathcal{O}(\mathbf{C}^2)$.

Denote by $\mathrm{Aut}(\mathbf{C}^2)$ the group of all holomorphic automorphisms of $\mathbf{C}^2$, and by AX the subgroup of $\mathrm{Aut}(\mathbf{C}^2)$ consisting of all $T \in \mathrm{Aut}(\mathbf{C}^2)$ which preserve the coordinate axes. Then, since $AX \subset IX$, we can say that

$$AX = \{T(x, y) = (xe^{u(x,y)}, ye^{v(x,y)}); u, v \in \mathcal{O}(\mathbf{C}^2), T \in \mathrm{Aut}(\mathbf{C}^2)\}.$$

In fact, we have $AX \subsetneq IX$, since we can find examples of maps in $IX \setminus AX$ in [3] and [5].

(1.2) In [3], we called a domain D in $\mathbf{C}^n$ to be of type **C** if there is no nonnegative integrable plurisubharmonic function on D except for the constant zero function. Being of type **C** is invariant under biholomorphic maps with constant Jacobian. Using this notion, we proved the following theorem.

Let F be a map in IX. Suppose that the Jacobian JF of F satisfies $JF \equiv c$ (constant). Then, there is an entire holomorphic function χ of one variable such that $F(x, y) = (xe^{\chi(xy)}, cye^{-\chi(xy)})$. So, F must be an automorphism of $\mathbf{C}^2$.

In the present paper, using the same method as in [3], we will deal with an injective holomorphic map F of a domain $D \subset \mathbf{C}^2$ of type **C** into $\mathbf{C}^2$ when JF is not necessarily constant but $|JF|^{-1}$ is bounded in a certain subdomain (See Theorem (4.4)). As a result, we will obtain a theorem concerning AX.

1991 *Mathematics Subject Classification*. Primary 32H02.

Research partially supported by Grant-in-Aid for Scientific Research (C) (No.01540163),The Ministry of Education, Science and Culture.

This paper is in final form and no version of it will be submitted for publication elsewhere.

To state this theorem more precisely, we need to introduce some notation. Let I be an interval in $[0, \infty]$. Let

$$\Lambda_I = \{(a, b) \in \mathbf{Z}_+ \times \mathbf{Z}_+; b = ra \quad \text{with} \quad r \in I\},$$

where $\mathbf{Z}_+$ denotes the set of nonnegative integers. Using the Taylor series expansion of $h \in \mathcal{O}(\mathbf{C}^2)$ about the origin, we define a subalgebra $\mathcal{O}_I$ of $\mathcal{O}(\mathbf{C}^2)$ as

$$\mathcal{O}_I = \{h(x, y) = \sum h_{ab} x^a y^b \in \mathcal{O}(\mathbf{C}^2); h_{ab} = 0 \text{ for every } (a, b) \notin \Lambda_I\}.$$

THEOREM. *Let I be an interval in $[0, \infty]$ such that $1 \in I$. Let $T(x, y) = (xe^{u(x,y)}, ye^{v(x,y)}) \in AX$, where $u, v \in \mathcal{O}(\mathbf{C}^2)$. Suppose that the Jacobian JT satisfies $JT \in \mathcal{O}_I$. Then, we have $u, v \in \mathcal{O}_I$ (Theorem (5.2)).*

Let us denote by $\mathcal{O}(\mathbf{C})$ the algebra of all holomorphic functions on $\mathbf{C}$. Then, as a corollary, we have:

Let $\alpha \in \mathcal{O}(\mathbf{C})$ and $T \in AX$. Suppose that $JT(x, y) = e^{\alpha(xy)}$. Then α is constant and T is of the form $T(x, y) = (xe^{\chi(xy)}, cye^{-\chi(xy)})$, where $c = e^\alpha$ and $\chi \in \mathcal{O}(\mathbf{C})$ (Corollary (5.3)).

(1.3) Let us consider the meromorphic 2-form ω on $\mathbf{C}^2$ defined by $\omega = dx \wedge dy/(xy)$. Define a subgroup AP of AX by

$$\begin{aligned} AP &= \{T \in AX; T^*\omega = \omega\} \\ &= \{T(x, y) = (xe^{u(x,y)}, ye^{v(x,y)}) \in AX; JT = e^{u+v}\}. \end{aligned}$$

In [4], E. Peschl claimed that $AP = AX$, and in [3], we referred to [4] as a motivation of our study. But, it turned out that the proof in [4] contains a gap. So, it is an open question whether $AP = AX$ or not (See [5]).

Let $G \subset \mathrm{Aut}(\mathbf{C}^n)$ be a group and let $S \subset G$ be a subset. In [1], E. Andersén gives a method to prove an assertion of the following type: "the group generated by S is dense in G", where "dense" is in the sense of uniform convergence on compact sets in $\mathbf{C}^n$.

In the final section of this paper, we will apply this method to prove the following theorem.

THEOREM. *Consider the subgroup $G = AP$ of AX and the subset $S = \cup_{r \in [0, \infty]} AX_r$ of AP (See (2.7) for the set AX_r). Then the group generated by S is dense in G.*

(1.4) In §2, we collect some fundamental properties of subalgebras $\mathcal{O}_I$ and introduce subgroups AX_I of the group AX. In §3, for a closed interval I, we give a condition for $h \in \mathcal{O}(\mathbf{C}^2)$ to belong to $\mathcal{O}_I$ in terms of the boundedness of $|h|$ on a certain subset in $\mathbf{C}^2$. In §4, using the notion of a domain of type $\mathbf{C}$, we prove a theorem about exceptional sets of a certain injective holomorphic map. In §5, we prove the theorem in (1.2) when I is closed. The assumption of the closedness of I will be removed in §6. Finally, in §7, we deal with the topic which we explained in (1.3).

2. The automorphism group AX and its subgroups.

(2.1) We will employ the following notation throughout:
$\mathbf{R}_+ = \{r \in \mathbf{R}; r \geqq 0\}, \mathbf{Q}_+ = \mathbf{Q} \cap \mathbf{R}_+, \mathbf{Z}_+ = \mathbf{Z} \cap \mathbf{R}_+$. $\mathbf{N}$ denotes the set of all positive integers, so $\mathbf{N} = \mathbf{Z}_+ \setminus \{0\}$. Put $\widehat{\mathbf{R}}_+ = \mathbf{R}_+ \cup \{\infty\}$. Similarly, $\widehat{\mathbf{Q}}_+$ and $\widehat{\mathbf{Z}}_+$ are defined. By an interval I in $\widehat{\mathbf{R}}_+$, we mean either an interval I in $\mathbf{R}_+$ (including the case $I = \{r\}$), $(r, \infty]$, $[r, \infty]$ or $\{\infty\}$. Put $\Lambda = \mathbf{Z}_+ \times \mathbf{Z}_+$. For an interval I in $\widehat{\mathbf{R}}_+$, we define a set

$$\Lambda_I = \{(a, b) \in \Lambda \; ; \; b = ra, r \in I\}.$$

Here, we mean

$$\Lambda_{[r,\infty]} = \{(a, b) \in \Lambda; ar \leqq b\}, \quad \Lambda_{\{\infty\}} = \{(0, b); b \in \mathbf{Z}_+\}, \quad \text{etc.}$$

It follows immediately that $O = (0, 0) \in \Lambda_I$ and $\Lambda_I \subset \Lambda_J$ if $I \subset J$. If $I = \{r\}$ with $r \in \widehat{\mathbf{Q}}_+$, then, we sometimes write Λ_r instead of Λ_I.

(2.2) For an interval I in $\widehat{\mathbf{R}}_+$, by using the Taylor series expansion $h(x, y) = \sum h_{ab} x^a y^b$ of $h \in \mathcal{O}(\mathbf{C}^2)$ about the origin $O \in \mathbf{C}^2$, we will define a subalgrebra $\mathcal{O}_I$ of $\mathcal{O}(\mathbf{C}^2)$ as

$$\mathcal{O}_I = \{h(x, y) = \sum h_{ab} x^a y^b \in \mathcal{O}(\mathbf{C}^2); h_{ab} = 0 \text{ for every } (a, b) \notin \Lambda_I\}.$$

Then, we have $\mathcal{O}_I \subset \mathcal{O}_J$ if $I \subset J$. These $\mathcal{O}_I$ contain the field $\mathbf{C}$ of constant functions as a subalgebra. If $I = \{r\}$ with $r \in \widehat{\mathbf{Q}}_+$, then, we sometimes write $\mathcal{O}_r$ instead of $\mathcal{O}_I$.

For $r \in \mathbf{Q}_+ \setminus \{0\}$, we take $k, l \in \mathbf{N}$ with $r = l/k$ and $(k, l) = 1$. For $r = 0$, we put $(k, l) = (1, 0)$ and for $r = \infty$, $(k, l) = (0, 1)$. Then, we have
$\mathcal{O}_r = \{h \in \mathcal{O}(\mathbf{C}^2); h(x, y) = \alpha(x^k y^l), \alpha \in \mathcal{O}(\mathbf{C})\}$.
We also have $\mathcal{O}_I = \mathbf{C}$ for $I = \{r\}$ with $r \in \widehat{\mathbf{R}}_+ \setminus \widehat{\mathbf{Q}}_+$.

(2.3) LEMMA. *Let $h \in \mathcal{O}_I$ and let $\varphi \in \mathcal{O}(\mathbf{C})$. Then, $\varphi \circ h \in \mathcal{O}_I$.*

(2.4) LEMMA. *Let $h \in \mathcal{O}(\mathbf{C}^2)$. Then, $h \in \mathcal{O}_I$ iff $e^h \in \mathcal{O}_I$.*

(2.5) Remember the group AX which we have introduced in (1.1).
Definition. Let $I \subset \widehat{\mathbf{R}}_+$ be an interval. We put

$$AX_I = \{T(x, y) = (x e^{u(x,y)}, y e^{v(x,y)}) \in AX; u, v \in \mathcal{O}_I\}.$$

(2.6) PROPOSITION. *AX_I is a subgroup of AX.*

PROOF. We will show that $T \in AX_I$ implies $T^{-1} \in AX_I$. Without loss of generality, we assume $dT(O) = id$. Then, we put $T = (f, g)$ with

$$f = x(1 + \sum p_{ab} x^a y^b) \quad \text{and} \quad g = y(1 + \sum q_{ab} x^a y^b).$$

We also put $T^{-1} = (F, G)$ with

$$F = x(1 + \sum P_{ab} x^a y^b) \quad \text{and} \quad G = y(1 + \sum Q_{ab} x^a y^b).$$

Here, the summations extend over $\Lambda \setminus \{O\}$. By $T^{-1} \circ T = id$,

$$
\begin{aligned}
x &= f(x,y)(1 + \sum P_{mn} f(x,y)^m g(x,y)^n) \\
\text{(2.6.1)} \quad &= x(1 + \sum p_{ab} x^a y^b) \\
&\quad + x \sum P_{mn} x^m y^n (1 + \sum p_{ab} x^a y^b)^{m+1}(1 + \sum q_{ab} x^a y^b)^n.
\end{aligned}
$$

By assumption,

$$
\text{(2.6.2)} \qquad p_{ab} = 0 \quad \text{and} \quad q_{ab} = 0 \quad \text{if} \quad (a,b) \notin \Lambda_I.
$$

We will prove, by induction on c+d, that $P_{cd} = 0$ if $(c,d) \notin \Lambda_I$. Similarly, $Q_{cd} = 0$ for $(c,d) \notin \Lambda_I$ will be proved.

If $c + d = 1$ and $(c,d) \notin \Lambda_I$, then, comparing the coefficients of $x^{c+1} y^d$ in (2.6.1), we have $0 = p_{cd} + P_{cd}$. By (2.6.2), we have $p_{cd} = 0$, hence $P_{cd} = 0$. Next, let $c + d \geqq 2$ with $(c,d) \notin \Lambda_I$ and suppose that $P_{mn} = 0$ if $(m,n) \notin \Lambda_I$ and $m + n < c + d$. Comparing the coefficients of $x^{c+1} y^d$ in (2.6.1), we have $0 = p_{cd} + P_{cd} + \gamma$, where γ is the coefficient of $x^c y^d$ of

$$
\sum_{(m,n) \in \Lambda_I, \, m+n < c+d} P_{mn} x^m y^n (1 + \sum p_{ab} x^a y^b)^{m+1}(1 + \sum q_{ab} x^a y^b)^n.
$$

Note that, if $(m,n) \in \Lambda_I$, then

$$
x^m y^n (1 + \sum p_{ab} x^a y^b)^{m+1}(1 + \sum q_{ab} x^a y^b)^n \in \mathcal{O}_I.
$$

Since $(c,d) \notin \Lambda_I$, the coefficient γ of $x^c y^d$ is 0. By (2.6.2), we have $p_{cd} = 0$. So, we obtain $P_{cd} = 0$.

Similarly, we can prove that $T, S \in AX_I$ implies $S \circ T \in AX_I$.   q.e.d.

(2.7) If $I = \{r\}$ with $r \in \widehat{\mathbf{Q}}_+$, then, we sometimes write AX_r instead of AX_I. As usual, we write $\mathbf{C}^* = \mathbf{C} \setminus \{0\}$.

LEMMA.
(a) $AX_0 = \{(cx, ye^{\chi(x)}); \chi \in \mathcal{O}(\mathbf{C}), c \in \mathbf{C}^*\}$.
(b) Let $k, l \in \mathbf{N}$ with $(k, l) = 1$ and let $r = l/k$.
Then, $AX_r = \{(xe^{l\chi(x^k y^l)}, cye^{-k\chi(x^k y^l)}); \chi \in \mathcal{O}(\mathbf{C}), c \in \mathbf{C}^*\}$.
(c) $AX_\infty = \{(xe^{\chi(y)}, cy); \chi \in \mathcal{O}(\mathbf{C}), c \in \mathbf{C}^*\}$.
(d) For $r \in \widehat{\mathbf{Q}}_+$, we have $AX_r \subset AP$, where the group AP is defined in (1.3).

PROOF. (a) It is obvious that $(cx, ye^{\chi(x)}) \in AX_0$. Conversely, let $T \in AX_0$. Then (2.2) shows $T(x,y) = (xe^{\alpha(x)}, ye^{\chi(x)})$. By restricting T to $\{y = 0\} \simeq \mathbf{C}$, we have an automorphism $x \to xe^{\alpha(x)}$ of $\mathbf{C}$. So, α must be a constant.

Similarly, we can prove (c).

Next, we will prove (b). We have

$$
T(x,y) = (xe^{l\chi(x^k y^l)}, cye^{-k\chi(x^k y^l)}) \in AX_r,
$$

because we can see that

$$T^{-1}(x, y) = (xe^{-l\chi(c^{-1}x^k y^l)}, c^{-1}ye^{k\chi(c^{-1}x^k y^l)}).$$

Conversely, let $T \in AX_r$. Then (2.2) shows $T(x, y) = (xe^{\alpha(x^k y^l)}, ye^{\beta(x^k y^l)})$ by some $\alpha, \beta \in \mathcal{O}(\mathbf{C})$. Consider $h(x, y) = x^k y^l \in \mathcal{O}(\mathbf{C}^2)$ and $h \circ T \in \mathcal{O}(\mathbf{C}^2)$. Then,

$$h \circ T(x, y) = x^k y^l e^{k\alpha(x^k y^l) + l\beta(x^k y^l)}.$$

Put $\gamma(z) = ze^{k\alpha(z) + l\beta(z)} \in \mathcal{O}(\mathbf{C})$ so that $h \circ T(x, y) = \gamma \circ h(x, y)$. We will show that $\gamma(z) = tz$ with $t \in \mathbf{C}^*$. If this is not the case, then by the little Picard theorem, $\{z; \gamma(z) = 1\}$ consists of infinitely many points. So, the analytic set in $\mathbf{C}^2$ defined by

$$A = \{(x, y); (h \circ T)(x, y) = 1\} = \{(x, y); \gamma \circ h(x, y) = 1\}$$

has infinitely many components. This is a contradiction, because

$$T(A) = \{h(x, y) = 1\} = \{x^k y^l = 1\}$$

has only one component and T is an automorphism. So, $t = e^{k\alpha(z) + l\beta(z)} \in \mathbf{C}^*$. Put $\chi(z) = \alpha(z)/l \in \mathcal{O}(\mathbf{C})$. Then, $e^{l\beta(z)} = te^{-kl\chi(z)}$, so $e^{\beta(z)} = ce^{-k\chi(z)}$ by some $c \in \mathbf{C}^*$. This proves $T(x, y) = (xe^{l\chi(x^k y^l)}, cye^{-k\chi(x^k y^l)})$.

Finally, (d) is proved by direct computations using (a)-(c). q.e.d.

(2.8) Let us denote by $\mathcal{O}^*(\mathbf{C}^2)$ the set of all nowhere vanishing holomorphic functions on $\mathbf{C}^2$, and let $\mathcal{O}_I^* = \mathcal{O}_I \cap \mathcal{O}^*(\mathbf{C}^2)$. By computation, we see that

$$JT(x, y) = \{1 + xu_x + yv_y + xy(u_x v_y - u_y v_x)\}e^{u+v}$$

for $T(x, y) = (xe^{u(x,y)}, ye^{v(x,y)}) \in AX$. So, it holds that $JT \in \mathcal{O}_I^*$ if $T \in AX_I$. Later, in §5 and §6, we will prove that, when $1 \in I$, it holds that $T \in AX_I$ if $JT \in \mathcal{O}_I^*$.

(2.9) If I has the following special form, then AX_I is naturally isomorphic to AX. Take $k, l, m, n \in \mathbf{Z}_+$ so that $kn - lm = 1$. Put $I = [\frac{l}{k}, \frac{n}{m}]$. Then, $L : \Lambda_I \longrightarrow \Lambda$ defined by $L(a, b) = (na - mb, -la + kb)$ is a bijection. Consider the automorphism G of $\mathbf{C}^{*2}$ defined by $G(x, y) = (x^k y^l, x^m y^n)$ with $G^{-1}(x, y) = (x^n y^{-l}, x^{-m} y^k)$.

PROPOSITION. *In the above situation, the following hold.*
(a) $M : \mathcal{O}_I \longrightarrow \mathcal{O}(\mathbf{C}^2)$ *defined by* $M(h) = h \circ G^{-1}$ *is an isomorphism.*
(b) $N : AX_I \longrightarrow AX$ *defined by* $N(S) = G \circ S \circ G^{-1}$ *is an isomorphism.*
(c) *Put* $S(x, y) = (xe^{u(x,y)}, ye^{v(x,y)}) \in AX_I, U = (ku + lv) \circ G^{-1}, V = (mu + nv) \circ G^{-1}$ *and* $T = N(S)$. *Then,* $T(x, y) = (xe^{U(x,y)}, ye^{V(x,y)})$.
(d) *Put* $h = e^{(m+k-1)u+(n+l-1)v}$. *Then,* $JT = (h \circ G^{-1}) \cdot (JS \circ G^{-1})$.

(2.10) We will use the next lemma in (5.2).

LEMMA. *Let $I \subsetneq \widehat{\mathbf{R}}_+$ be an interval. Let $k, l \in \mathbf{Z}_+$ and let $v \in \mathcal{O}(\mathbf{C}^2)$. If $x^k y^l e^{v(x,y)} \in \mathcal{O}_I$, then $v \in \mathcal{O}_I$.*

PROOF. It is enough to prove the lemma when $v(0) = 0$ and $I = [r, \infty]$ or $I = (r, \infty]$ with $r > 0$. Since the coefficient of $x^k y^l$ of the Taylor expansion of $x^k y^l e^{v(x,y)}$ is equal to 1, we have $(k, l) \in \Lambda_I$. Put $A = \{(a, b) \in \Lambda; (a, b) + (k, l) \in \Lambda_I\}$. Put $v(x, y) = \sum v_{ab} x^a y^b$ and $e^{v(x,y)} = \sum p_{ab} x^a y^b$. Then,

$$(2.10.1) \qquad p_{ab} = 0 \text{ for } (a, b) \in \Lambda \setminus A.$$

Put $B = \{(a, b) \in (\Lambda \setminus \Lambda_I); p_{ab} \neq 0\}$. Suppose that $B \neq \phi$. Take $(c, d) \in B$ such that $d/c = \min\{b/a; (a, b) \in B\}$. Put $C = \{(a, b); \frac{d}{c} > \frac{b}{a}, v_{ab} \neq 0\}$. Then, we have

$$(2.10.2) \qquad p_{ab} = 0 \text{ for } (a, b) \in C.$$

We will show that $C = \phi$. In fact, if $C \neq \phi$, take $(m, n) \in C$ with $m + n = \min\{a + b; (a, b) \in C\}$. Then $p_{mn} = v_{mn} \neq 0$, which is a contradiction to $(2.10.2)$. So, we can write

$$v(x, y) = \sum{}' v_{ab} x^a y^b + \sum{}'' v_{ab} x^a y^b,$$

where the summation $\sum'$ extends over all (a, b) with $\frac{b}{a} = \frac{d}{c}$ and the summation $\sum''$ extends over all (a, b) with $\frac{b}{a} > \frac{d}{c}$. Put $v'(x, y) = \sum' v_{ab} x^a y^b$. Then,

$$e^{v(x,y)} = e^{v'(x,y)} + \sum{}'' p_{ab} x^a y^b.$$

Since $e^{v'}$ is a transcendental function in $\mathcal{O}(\mathbf{C}^2)$, there is $(a, b) \notin A$ such that $\frac{b}{a} = \frac{d}{c}$ and $p_{ab} \neq 0$. This is a contradiction to $(2.10.1)$. So, we have $B = \phi$, which implies $e^v \in \mathcal{O}_I$. So $v \in \mathcal{O}_I$.   q.e.d.

3. A geometric characterization of $\mathcal{O}_I$ when I is closed.

(3.1) Definition. Let p, q be positive numbers. Put
$$A(p, q, r) = \{(x, y) \in \mathbf{C}^2; |y| \leq q, |x||y|^r \leq pq^r\} \text{ for } r \in (\widehat{\mathbf{R}}_+ \setminus \{0\}),$$
$$B(p, q, s) = \{(x, y) \in \mathbf{C}^2; |x| \leq p, |x||y|^s \leq pq^s\} \text{ for } s \in \mathbf{R}_+.$$
We have $A(p, q, r) \subset A(p, q, r')$ if $r < r'$ and $B(p, q, s') \subset B(p, q, s)$ if $s < s'$.

For a closed interval $I \subsetneq \widehat{\mathbf{R}}_+$, we define
$$C_I(p, q) = \begin{cases} A(p, q, r) & \text{if } I = [r, \infty], \\ B(p, q, s) & \text{if } I = [0, s], \\ A(p, q, r) \cup B(p, q, s) & \text{if } I = [r, s] \text{ with } 0 < r \leq s < \infty. \end{cases}$$

(3.2) In the next lemma, we consider $k, l, m, n \in \mathbf{Z}_+$ and G as in (2.9). Set $r = l/k$. For $p, q > 0$, put
$$\widetilde{A}(p, q, r) = \mathbf{C}^{*2} \cap \text{Int}(A(p, q, r))$$
$$= \{(x, y); 0 < |x|^k |y|^l < p^k q^l, \, 0 < |y| < q\}.$$
Then,
$$G(\widetilde{A}(p, q, r)) = \{(x, y); 0 < |x| < p^k q^l, 0 < |y|^k < q|x|^m\}.$$

LEMMA. *Let* $h \in \mathcal{O}(\mathbf{C}^2)$. *Put* $H = h \circ G^{-1} \in \mathcal{O}(\mathbf{C}^{*2})$. *Then the following are equivalent.*

(a) $H(x, y)$ *is continued holomorphically to* $\{y = 0\}$.

(b) H *is bounded in* $G(\widetilde{A}(p, q, r))$ *for some* $p, q > 0$.

(c) H *is bounded in* $G(\widetilde{A}(p, q, r))$ *for any* $p, q > 0$.

PROOF. (b) $\Rightarrow$ (a) follows from the Riemann's continuation theorem.

(a) $\Rightarrow$ (c). Take $p, q > 0$. Put

$$h(x, y) = \sum h_{ab} x^a y^b \quad \text{and} \quad H(x, y) = \sum H_{ab} x^a y^b.$$

If $\begin{pmatrix} a \\ b \end{pmatrix} = \begin{pmatrix} k & m \\ l & n \end{pmatrix} \begin{pmatrix} c \\ d \end{pmatrix}$, then $h_{ab} = H_{cd}$. Since $H(x, y)$ is continued to $\{y = 0\}$, $H_{cd} = 0$ for $d < 0$. For $d \geqq 0$ and for $(x, y) \in G(\widetilde{A}(p, q, r))$,

$$|H_{cd} x^c y^d| \leqq q^{d/k} |H_{cd}||x|^{c + dm/k} = |h_{ab}| q^{d/k} |x|^{a/k} \leqq |h_{ab}| p^a q^b.$$

So, $|H(x, y)| \leqq \sum |h_{ab}| p^a q^b$ for $(x, y) \in G(\widetilde{A}(p, q, r))$. q.e.d.

(3.3) LEMMA. *Let* $k, l \in \mathbf{N}$ *with* $(k, l) = 1$. *Put* $r = l/k$ *and* $I = [r, \infty]$. *Let* $h \in \mathcal{O}(\mathbf{C}^2)$. *Then the following are equivalent.*

(a) $h \in \mathcal{O}_I$.

(b) h *is bounded in* $A(p, q, r)$ *for some* $p, q > 0$.

(c) h *is bounded in* $A(p, q, r)$ *for any* $p, q > 0$.

PROOF. Take $m, n \in \mathbf{N}$ such that $kn - lm = 1$. Use the notation as in (3.2). Then, h is bounded in $A(p, q, r)$ iff H is bounded in $G(\widetilde{A}(p, q, r))$. Since $d = -la + kb < 0$ iff $b < ra$, H is continued to $\{y = 0\}$ iff $h \in \mathcal{O}_I$. q.e.d.

(3.4) PROPOSITION. *Let* $I \subsetneqq \widehat{\mathbf{R}}_+$ *be a closed interval. Then, for* $h \in \mathcal{O}(\mathbf{C}^2)$, *the following are equivalent.*

(a) $h \in \mathcal{O}_I$.

(b) h *is bounded in* $C_I(p, q)$ *for some* $p, q > 0$.

(c) h *is bounded in* $C_I(p, q)$ *for any* $p, q > 0$.

PROOF. It is enough to prove the proposition when $I = [r, \infty]$ with $r > 0$. Put $h(x, y) = \sum h_{ab} x^a y^b$.

(b) $\Rightarrow$ (a). Take (a, b) with $b < ra$. Take $k, l \in \mathbf{N}$ with $(k, l) = 1$ and $\frac{b}{a} < \frac{l}{k} < r$. Since $|h|$ is bounded in $A(p, q, r)$ and $A(p, q, \frac{l}{k}) \subset A(p, q, r)$, $|h|$ is bounded in $A(p, q, \frac{l}{k})$. By the lemma (3.3), $h_{ab} = 0$.

(a) $\Rightarrow$ (c). Let $(x, y) \in A(p, q, r)$. Then,

$$|h(x, y)| \leqq \sum |h_{ab}||x|^a |y|^b$$
$$= \sum |h_{ab}|(|x||y|^r)^a |y|^{b - ra} \leqq \sum |h_{ab}| p^a q^b.$$

q.e.d.

4. Domains of type C.

(4.1) Definition. Let D be a domain in $\mathbf{C}^n$. Let us denote by $P_+(D)$ the set of all non-negative plurisubharmonic functions in D and by $L^1(D)$ the space of complex valued functions which are integrable with respect to the Lebesgue measure. We call D a domain of type **C**, if $P_+(D) \cap L^1(D) = \{0\}$, where 0 means the constant zero function. When n=1, a domain D in **C** is of type **C** iff $D = \mathbf{C}$. In this paper, we use the fact that, for a domain G in $\mathbf{C}$, $G \times \mathbf{C}$ is a domain of type **C** in $\mathbf{C}^2$. Some information about domains of type **C** may be found in [3].

(4.2) We will define two subsets in $\mathbf{C}^2$ which will appear in the next lemma. Let p, q, r be positive numbers. Put

$$A'(p,q,r) = \{0 < |x||y|^r < pq^r, |y| < q\}$$

and

$$E(p,q) = \{|x| \leqq p, |y| = q\} \cup \{x = 0, |y| \leqq q\} \cup \{y = 0\}.$$

So, we have

$$E(p,q) \subset \partial A'(p,q,r) \quad \text{and} \quad A'(p,q,r) \subset A(p,q,r),$$

where $A(p,q,r)$ is as in (3.1).

(4.3) LEMMA. *Let $p, q > 0$ and let $G = \mathbf{C}^2 \setminus E(p,q)$. Suppose that $0 < r < 1$ and define a function $\Phi(x,y)$ in G by*

$$\Phi(x,y) = \begin{cases} \dfrac{1}{|x||y|^r} - \dfrac{1}{pq^r} & in \quad A'(p,q,r), \\ 0 & in \quad G \setminus A'(p,q,r). \end{cases}$$

Then, $\Phi \in P_+(G) \cap L^1(G)$ and $\{\Phi > 0\} = A'(p,q,r)$.

PROOF. Put $\Phi_1(x,y) = \frac{1}{|x||y|^r} - \frac{1}{pq^r}$ in a neighbourhood U_1 of $A'(p,q,r)$ in $G \setminus G \cap \{x = 0\}$ and put $\Phi_2(x,y) = 0$ in a neighbourhood U_2 of $G \setminus A'(p,q,r)$ in G. Then, Φ_i is plurisubharmonic in U_i ($i = 1,2$). On $U_1 \cap U_2$, we have $\Phi = max(\Phi_1, \Phi_2)$. So, $\Phi \in P_+(G)$. A direct computation gives $\int_G \Phi dV = \pi^2 p q^{2-r}/(1-r)$.

(4.4) THEOREM. *Let $p, q > 0$ and let $0 < r \leqq 1$. Let D be a domain of type **C** in $\mathbf{C}^2$. Let $S : D \longrightarrow \mathbf{C}^2$ be an injective holomorphic map such that $S(D) \cap E(p,q) = \phi$. Suppose that there is a positive number L which satisfies $|JS(x,y)| \geqq L$ whenever $S(x,y) \in A'(p,q,r)$. Then, in reality, we have $S(D) \cap A(p,q,r) = \phi$.*

PROOF. First assume that $r < 1$ and consider the function Φ in G defined in (4.3). Then,

$$\int_{S(D)} \Phi dV \leqq \int_G \Phi dV < +\infty.$$

Put $\Psi = S^*\Phi$. Then,

$$\int_{S(D)} \Phi dV = \int_{S(D)\cap A'(p,q,r)} \Phi dV = \int_{D\cap S^{-1}(A'(p,q,r))} \Psi |JS|^2 dV$$

$$\geqq L^2 \int_{D\cap S^{-1}(A'(p,q,r))} \Psi dV = L^2 \int_D \Psi dV.$$

This implies $\Psi \in P_+(D) \cap L^1(D)$. Since D is of type $\mathbf{C}$, we have $\Psi = 0$ in D, i.e., $S(D) \cap A(p,q,r) = \phi$. This proves the theorem when $r < 1$. When $r = 1$, since $S(D) \cap A(p,q,r) = \phi$ for any $r < 1$, we have $S(D) \cap A(p,q,1) = \phi$. q.e.d.

   (4.5) Example. As we have shown in [2] and [3], there is an injective holomorphic map $S : \mathbf{C}^2 \longrightarrow G$ such that $S(\mathbf{C}^2) \cap A'(p,q,r) \neq \phi$. Since $\mathbf{C}^2$ is of type $\mathbf{C}$, for any $\epsilon > 0$, there is a point $(x,y) \in S^{-1}(A'(p,q,r))$ such that $|JS(x,y)| < \epsilon$.

5. Main theorem.

   (5.1) LEMMA.. *Let $I \subsetneqq \widehat{\mathbf{R}}_+$ be an interval.*
If $T(x,y) = (xe^{u(x,y)}, ye^{v(x,y)}) \in AX$ satisfies $JT \in \mathcal{O}_I^*$ and $v \in \mathcal{O}_I$, then we have $T \in AX_I$.

   PROOF. Without loss of generality, we suppose $dT(O) = id$. So, $T = (f,g)$ is of the form

$$f(x,y) = x(1 + \sum p_{ab} x^a y^b) \quad \text{and} \quad g(x,y) = y(1 + \sum q_{ab} x^a y^b).$$

Put $JT(x,y) = 1 + \sum J_{ab} x^a y^b$. Here the summations extend over $\Lambda \setminus \{O\}$. A simple computation shows

$$JT(x,y)$$
$$= f_x g_y - f_y g_x$$
(5.1.1)
$$= 1 + \sum (a+1) p_{ab} x^a y^b + \sum (b+1) q_{ab} x^a y^b$$
$$+ \left(\sum (a+1) p_{ab} x^a y^b\right)\left(\sum (b+1) q_{ab} x^a y^b\right)$$
$$- \left(\sum b p_{ab} x^a y^b\right)\left(\sum a q_{ab} x^a y^b\right).$$

By hypothesis, it holds that

(5.1.2)
$$J_{ab} = q_{ab} = 0 \quad \text{if} \quad (a,b) \notin \Lambda_I.$$

We will show, by induction on $c+d$, that $p_{cd} = 0$ if $(c,d) \notin \Lambda_I$. If $c+d = 1$ and $(c,d) \notin \Lambda_I$, comparing the coefficients of $x^c y^d$ in (5.1.1), we have

$$J_{cd} = (c+1)p_{cd} + (d+1)_{cd}.$$

By (5.1.2), we have $p_{cd} = 0$. Next, let $c+d \geqq 2$ with $(c,d) \notin \Lambda_I$ and suppose that $p_{ab} = 0$ if $(a,b) \notin \Lambda_I$ and $a+b < c+d$. Then, comparing the coefficients of $x^c y^d$ in (5.1.1), we have

(5.1.3) $\qquad J_{cd} = (c+1)p_{cd} + (d+1)q_{cd} + \sum((k+1)(n+1) - lm)p_{kl}q_{mn},$

where the last summation extends over k, l, m, n with $k + m = c$ and $l + n = d$. Since $(c, d) \notin \Lambda_I$, we have either $(k, l) \notin \Lambda_I$ or $(m, n) \notin \Lambda_I$, which implies $p_{kl}q_{mn} = 0$. By (5.1.2) and (5.1.3), we have $p_{cd} = 0$.   q.e.d.

(5.2) THEOREM. *Let* $I \subsetneq \widehat{\mathbf{R}}_+$ *be an interval such that* $1 \in I$. *If an automorphism* T *in* AX *satisfies* $JT \in \mathcal{O}_I^*$, *then we have* $T \in AX_I$.

PROOF. Here, we will prove the theorem under the additional assumption that I is a closed interval, which we need when we apply the propoition (3.4). This assumption will be removed in §6. Under this assumption, it is enough to prove the theorem when $I = [r, \infty]$ with $0 < r \leq 1$. Put $S = T^{-1}$. Take $p, q > 0$ arbitrarily and consider $E(p, q)$ and $A(p, q, r)$ as in (4.2). By $T \circ S = id$, we have $(JT \circ S) \cdot JS = 1$. By hypothesis and by the proposition (3.4), there is $M > 0$ such that

$$|JT(x, y)| \leq M \quad \text{for} \quad (x, y) \in A(p, q, r).$$

This implies that

$$|JS(x, y)| \geq 1/M \quad \text{if} \quad S(x, y) \in A(p, q, r).$$

On the other hand take $Q > 0$ so that $D = \{(x, y) \in \mathbf{C}^2; |y| > Q\}$ satisfies $S(D) \cap E(p, q) = \phi$. This is possible because $\overline{E(p, q) \setminus \{y = 0\}}$ is compact. Note that the domain D is of type $\mathbf{C}$. Applying the theorem (4.4) for the map $S|_D : D \to \mathbf{C}^2 \setminus E(p, q)$, we have $S(D) \cap A(p, q, r) = \phi$. So, putting $T(x, y) = (xe^{u(x,y)}, ye^{v(x,y)})$, we have $|ye^{v(x,y)}| \leq Q$ on $A(p, q, r)$. By the lemmas (2.10), (5.1) and by the proposition (3.4), it follows that $T \in AX_I$.   q.e.d.

(5.3) COROLLARY. *Let* $\alpha \in \mathcal{O}(\mathbf{C})$ *and* $T \in AX$. *Suppose that* $JT(x, y) = e^{\alpha(xy)}$. *Then* α *is constant and* T *is of the form*

$$T(x, y) = (xe^{\chi(xy)}, cye^{-\chi(xy)}), \quad \text{where} \quad c = e^\alpha \quad \text{and} \quad \chi \in \mathcal{O}(\mathbf{C}).$$

PROOF. By hypothesis, $JT \in \mathcal{O}_1$. Applying the theorem for $I = \{1\}$, we have $T \in AX_1$. As we have seen in (2.7), there are $\chi \in \mathcal{O}(\mathbf{C})$ and $c \in \mathbf{C}^*$ such that $T(x, y) = (xe^{\chi(xy)}, cye^{-\chi(xy)})$. By computation, $JT = c$. So, α is constant. q.e.d.

(5.4) Remark.1.   The condition $1 \in I$ is necessary. For example, consider the case $I = \{r\}$ with $r \neq 1$. Let $T(x, y) = (xe^{\chi(xy)}, cye^{-\chi(xy)})$, where $\chi \in \mathcal{O}(\mathbf{C})$ is a nonconstant function. Then $JT = c$. So, we have $JT \in \mathcal{O}_I^*$ and $T \notin AX_I$.

(5.5) Remark.2.   The corollary is a generalization of the theorem in [3] (See (1.2).) in the sense that $\mathbf{C}^*$ is generalized to the case $\mathcal{O}_1^*$. But, though we assumed merely $T \in IX$ in [3], we must assume $T \in AX$ in the corollary. The author does not know whether there is $F \in IX$ such that $JF(x, y) = e^{\alpha(xy)}$ with a nonconstant $\alpha \in \mathcal{O}(\mathbf{C})$.

(5.6) Remark.3.   Let us continue to consider $F \in IX$ with $JF \in \mathcal{O}_r^*$, where $r \in \widehat{\mathbf{Q}}_+$. In (5.3) and (5.5), we have already seen the case when $r = 1$. In contrast with this case, the situation is completely different when $r \neq 1$. To see this, first, we state the following proposition.

PROPOSITION. *Let $F \in IX$ and let $T \in AX$. Suppose that $JT \equiv JF$ on $\mathbf{C}^2$. Then, there is $S \in AX_1$ with $JS \equiv 1$ such that $F = S \circ T$. So, $F \in AX$.*

PROOF. By (2.7) and a simple computation of the Jacobian, we have

$$\{S \in AX_1; JS \equiv 1\} = \{(xe^{\chi(xy)}, ye^{-\chi(xy)})\}.$$

So, the proposition follows from the theorem in [3] (See (1.2).) applied to $S = F \circ T^{-1}$. q.e.d.

Let $r \in \widehat{\mathbf{Q}}_+$ with $r \neq 1$. For simplicity, assume $r = l/k$ by $k, l \in \mathbf{N}$ with $(k, l) = 1$. For $\alpha \in \mathcal{O}(\mathbf{C})$, put $\beta = \frac{\alpha}{l-k}$ and

$$T(x, y) = (xe^{l\beta(x^k y^l)}, ye^{-k\beta(x^k y^l)}) \quad \text{so that} \quad JT(x, y) = e^{\alpha(x^k y^l)}.$$

Then, in view of this proposition, we have

$$\{F \in IX; JF(x, y) = e^{\alpha(x^k y^l)}\} = \{S \circ T; S \in AX_1, JS \equiv 1\} \subset AX.$$

6. Proof of theorem (5.2) for general interval.

(6.1) In §5, we have proved the theorem (5.2) when the interval I is closed. In order to prove this theorem for general I, it is enough to prove it when $I = (r, \infty]$ with $r \in \mathbf{R}_+$. If $r \notin \mathbf{Q}_+$, then we have $\mathcal{O}_{(r,\infty]} = \mathcal{O}_{[r,\infty]}$ and $AX_{(r,\infty]} = AX_{[r,\infty]}$. So, this case reduces to the case $I = [r, \infty]$. Therefore, it remains to deal with the case where $I = (r, \infty]$ with $r \in \mathbf{Q}_+$.

(6.2) Definition. In this section, we will employ the following notation. Let $r \in \mathbf{Q}_+$, $I = (r, \infty]$ and put $K = [r, \infty]$. For $h = \sum h_{ab} x^a y^b \in \mathcal{O}_K$, we set

$$h' = \sum{}' h_{ab} x^a y^b \in \mathcal{O}_r \quad \text{and} \quad h'' = \sum{}'' h_{ab} x^a y^b \in \mathcal{O}_I,$$

where $\sum'$ extends over all (a, b) in Λ_r and $\sum''$ in $\Lambda_K \setminus \Lambda_r$. The constant terms of h and h' are equal and every $(a, b) \in \Lambda_K \setminus \Lambda_r$ satisfies $a + b \geq 1$. This is an important remark which will be used subsequently.

(6.3) LEMMA. *Let $h \in \mathcal{O}_K$. Then $(e^h)' = e^{h'}$.*

PROOF. Observe that

$$e^h = e^{h'+h''} = e^{h'}\{1 + (e^{h''} - 1)\}$$
$$= e^{h'} + e^{h'}(e^{h''} - 1).$$

Then, we can see that $(e^h)' = e^{h'}$ and $(e^h)'' = e^{h'}(e^{h''} - 1)$. q.e.d.

(6.4) Let $T = (xe^u, ye^v) \in AX_K$. Using the decomposition $u = u' + u''$ and $v = v' + v''$, we define $T' = (xe^{u'}, ye^{v'})$.

LEMMA. *If $T \in AX_K$, then $T' \in AX_r$.*

PROOF. Without loss of generality, we may assume that $dT(O) = id$. We set

$$T(x, y) = (f(x, y), g(x, y)) = (xp(x, y), yq(x, y))$$

and

$$T^{-1}(x,y) = (F(x,y), G(x,y)) = (xP(x,y), yQ(x,y)).$$

We write as $P(x,y) = 1 + \sum P_{ab}x^a y^b$, etc., where $\sum$ extends over $\Lambda_K \setminus \{O\}$. By the lemma (6.3), denoting $S = T^{-1}$, we have $T' = (xp', yq')$ and $S' = (xP', yQ')$. In view of $S \circ T = id$, we have

$$1 = p + \sum P_{ab}x^a y^b p^{a+1} q^b.$$

By considering the '-part of the decomposition of functions of both sides, we have

$$1 = p' + \sum{}' P_{ab}x^a y^b (p')^{a+1}(q')^b.$$

By the same argument, we will have

$$1 = q' + \sum{}' Q_{ab}x^a y^b (p')^a (q')^{b+1}.$$

These identities are equivalent to $S' \circ T' = id$. Similarly, we can prove $T' \circ S' = id$. q.e.d.

(6.5) LEMMA. *Let* $T = (xe^u, ye^v) \in AX_K$. *Then* $(T')^{-1} \circ T \in AX_I$.

PROOF. By the same notation as in the proof of the lemma (6.4), we have

$$(T')^{-1} \circ T = S' \circ T = (x(p + \sum{}' P_{ab}x^a y^b p^{a+1} q^b), y(q + \sum{}' Q_{ab}x^a y^b p^a q^{b+1})).$$

Therefore, we have

$$p + \sum{}' P_{ab}x^a y^b p^{a+1} q^b = p' + \sum{}' P_{ab}x^a y^b (p')^{a+1}(q')^b + R = 1 + R,$$

where $R \in \mathcal{O}_I$. So, $p + \sum{}' P_{ab}x^a y^b p^{a+1} q^b \in \mathcal{O}_I$. This and the similar result for the second component mean that $(T')^{-1} \circ T \in AX_I$. q.e.d.

(6.6) LEMMA. *Let* $h \in \mathcal{O}_r$. *Let* $T \in AX_I$ *such that* $dT(O) = id$. *Then,* $(h \circ T)' = h$.

PROOF. By the notation of (2.2), there is $\alpha \in \mathcal{O}(\mathbf{C})$ such that $h(x,y) = \alpha(x^k y^l)$. We put $\alpha(z) = \alpha_0 + \sum \alpha_m z^m$, hence

$$h(x,y) = \alpha_0 + \sum \alpha_m x^{mk} y^{ml}.$$

Set $T = (xp, yq)$. Then, by assumption, we have $p' = 1$ and $q' = 1$. So,

$$h \circ T = \alpha_0 + \sum \alpha_m x^{mk} y^{ml} (1 + p'')^{mk} (1 + q'')^{ml}.$$

Therefore, $(h \circ T)' = \alpha(x^k y^l) = h(x,y)$. q.e.d.

(6.7) Proof of Theorem (5.2). As we have seen in (6.1), we may assume that $I = (r, \infty]$ with $r \in \mathbf{Q}_+$. We also assume that $dT(O) = id$. From this and from $JT \in AX_I$, we have

(6.7.1) $(JT)' = 1.$

Since $I \subset K$, we have $JT \in \mathcal{O}_K$. By the result of §5, we have $T \in AX_K$. Take $T' \in AX_r$. Put $R = (T')^{-1} \circ T$ so that $T' \circ R = T$. Then we have $dT'(O) = id, dR(O) = id$ and

(6.7.2) $$JT = (JT' \circ R)JR.$$

By the lemma (6.5), $R \in AX_I$. Hence we have

(6.7.3) $$(JR)' = 1.$$

By the lemma (6.6),

(6.7.4) $$(JT' \circ R)' = JT'.$$

Combining (6.7.1)-(6.7.4), we have

(6.7.5) $$(JT') = (JT)' = 1.$$

By the notation of (k, l) as in (2.2) and by the lemma (2.7), we have $T'(x, y) = (xe^{l\chi(x^k y^l)}, cye^{-k\chi(x^k y^l)})$. A computation shows $JT'(x, y) = ce^{(l-k)\chi(x^k y^l)}$. Since $r < 1$, we have $k \neq l$. By (6.7.5), $\chi = constant = \chi(0)$. So, $T'(x, y) = (e^{l\chi(0)}x, ce^{-k\chi(0)}y)$. Since $R \in AX_I$, we have $T = T' \circ R \in AX_I$.   q.e.d.

7. The subgroup AP and its dense subgroup.

(7.1) Remember the subgroup AP of AX (See (1.3)) and the subgroups AX_r of AP (See (2.7)). Then, we can state the following theorem.

THEOREM. *Let $I \subset \widehat{\mathbf{R}}_+$ be an interval. Then, the group generated by $\bigcup_{r \in I} AX_r$ is dense in $AP \cap AX_I$.*

Specially, the group generated by $\bigcup_{r \in \widehat{\mathbf{Q}}_+} AX_r$ is dense in AP.

Here, "dense" is in the sense of uniform convergence on compact sets in $\mathbf{C}^2$.

Since our proof pursues the same route as that in [1], we will give only a brief explanation of the change that we need.

In this section, putting $z = (x, y)$, we write $f(z) = f(x, y)$ for $f \in \mathcal{O}(\mathbf{C}^2)$. For $F = (f, g)$ with $f, g \in \mathcal{O}(\mathbf{C}^2)$ and for $z = (x, y) \in \mathbf{C}^2$, we put $z \cdot F(z) = xf(z) + yg(z)$. For simplicity, instead of writing as $T(z) = (xe^{u(z)}, ye^{v(z)})$, we will write as $T(z) = < u(z), v(z) >$.

(7.2) After E. Andersén [1], we use the following notation :

Let $h(z, \varepsilon)$ be a function such that, for each ε, $h(\cdot, \varepsilon) \in \mathcal{O}(\mathbf{C}^2)$. Then, $h(z, \varepsilon) = o_z(\varepsilon)$ means that $h(z, \varepsilon)/\varepsilon \longrightarrow 0$ as $\varepsilon \longrightarrow 0$, uniformly on compact sets. Similarly, for a holomorphic map $H(\cdot, \varepsilon) : \mathbf{C}^2 \longrightarrow \mathbf{C}^2$, the symbol $H(z, \varepsilon) = o_z(\varepsilon)$ is used.

(7.3) LEMMA. *Let $\{T_\varepsilon(\cdot)\} \subset AX$ and $f, g \in \mathcal{O}(\mathbf{C}^2)$. Put $F = (f, g)$. Then,*

$$T_\varepsilon(z) = < \varepsilon f(z) + o_z(\varepsilon), \varepsilon g(z) + o_z(\varepsilon) >$$

iff

$$T_\varepsilon(z) = z + \varepsilon z \cdot F(z) + o_z(\varepsilon).$$

(7.4) LEMMA. *Let*

$$T_{i,\varepsilon}(z) = <\varepsilon f_i(z) + o_z(\varepsilon), \varepsilon g_i(z) + o_z(\varepsilon)> \quad for \quad i \in \{1,2\}.$$

Then,

$$T_{1,\varepsilon} \circ T_{2,\varepsilon}(z) = <\varepsilon(f_1(z) + f_2(z)) + o_z(\varepsilon), \varepsilon(g_1(z) + g_2(z)) + o_z(\varepsilon)>.$$

(7.5) Definition. For $F = (f,g)$ with $f,g \in \mathcal{O}(\mathbf{C}^2)$, put $\mathcal{D}F = xf_x + yg_y$.

(7.6) LEMMA. *Let* $\quad T_\varepsilon(z) = z + \varepsilon z \cdot F(z) + o_z(\varepsilon)$. *If* $\quad T_\varepsilon(\cdot) \in AP$ *for each* ε, *then it holds that* $\mathcal{D}F = 0$.

PROOF. By computation,

$$JT_\varepsilon(z) = 1 + \varepsilon(f + g + xf_x + yg_y) + o_z(\varepsilon).$$

On the other hand, since $T_\varepsilon(\cdot) \in AP$,

$$JT_\varepsilon(z) = \exp(\varepsilon(f(z) + g(z)) + o_z(\varepsilon)) = 1 + \varepsilon(f + g) + o_z(\varepsilon).$$

So, we have $\mathcal{D}F = 0$.   q.e.d.

(7.7) LEMMA. *Let* $a \geq 1$. *Let* $f(z)$ *and* $g(z)$ *be homogeneous polynomials of degree* a *in* x *and* y. *Put* $F(z) = (f(z), g(z))$. *If* F *satisfies* $\mathcal{D}F = 0$, *then there is* $\{T_\varepsilon\} \subset AP$ *such that*
(a) $T_\varepsilon(z) = z + \varepsilon z \cdot F(z) + o_z(\varepsilon)$,
(b) T_ε *is a finite composition of maps in* $\bigcup_{r \in \widehat{\mathbf{Q}}_+} AX_r$.

PROOF. Put

$$f(z) = \sum_m f_m x^m y^{a-m}, \qquad g(z) = \sum_m g_m x^m y^{a-m}.$$

For each $m \in \{0, 1, \cdots, a\}$, set

$$T_{m,\varepsilon}(z) = <\varepsilon f_m x^m y^{a-m}, \varepsilon g_m x^m y^{a-m}>.$$

Then, by $\mathcal{D}F = 0$, it follows that $mf_m + (a-m)g_m = 0$, which implies $T_{m,\varepsilon} \in AX_r$ with $a - m = rm$. Using the lemmas (7.3) and (7.4), it can easily be shown that $T_\varepsilon(z) = T_{0,\varepsilon} \circ \cdots \circ T_{a,\varepsilon}$ has the desired properties.   q.e.d.

(7.8) Remark. Let $I \subsetneq \widehat{\mathbf{R}}_+$ be an interval. In the lemma (7.7), if we add an assumption that the homogeneous polynomials f,g belong to $\mathcal{O}_I$, then we have the following (b') instead of (b) in (7.7):
(b') T_ε is a finite composition of maps in $\bigcup_{r \in I} AX_r$.

(7.9) With the lemmas (7.6) and (7.7), by the same procedure as that of [1], we can prove the theorem in (7.1).

Acknowledgement.
    The author wishes to thank Professor Walter Rudin for giving him valuable comments and for letting him know about the paper [1].

References.

[1] E. Andersén, *Volume-preserving automorphisms of* $\mathbf{C}^n$, Complex Variables **14** (1990), 223–235.

[2] Y. Nishimura, *Applications holomorphes injectives de* $\mathbf{C}^2$ *dans lui-même qui exceptent une droite complexe*, J. Math. Kyoto Univ. **24** (1984), 755–761.

[3] Y. Nishimura, *Applications holomorphes injectives à jacobien constant de deux variables*, J. Math. Kyoto Univ. **26** (1986), 697–709.

[4] E. Peschl, *Automorphismes holomorphes de l'espace à n dimensions complexes*, C. R. Acad. Sci. Paris **242** (1956), 1836-1838.

[5] J. P. Rosay and W. Rudin, *Holomorphic maps from* $\mathbf{C}^n$ *to* $\mathbf{C}^n$, Trans. Amer. Math. Soc. **310** (1988), 47–86.

DEPARTMENT OF MATHEMATICS, OSAKA MEDICAL COLLEGE

Contemporary Mathematics
Volume 137, 1992

Analytic Transversality and Nullstellensatz in Bergman Space

MIHAI PUTINAR AND NORBERTO SALINAS

1. Introduction.

Our purpose in this paper, is to study the structure of certain analytically invariant subspaces of the Bergman space $L_a^2(\Omega)$ over a (bounded) pseudoconvex domain Ω in $\mathbf{C}^n$. We use analytic transversality and the notion of privileged domains to show that the "Hilbert's Nullstellensatz" holds for these subspaces.

We recall that the Bergman space $L_a^2(\Omega)$ is the space of holomorphic functions on Ω which are square integrable with respect to the "volume" Lebesgue measure on Ω. Hereafter, Ω is assumed to be a bounded domain in $\mathbf{C}^n$. A (closed) subspace $\mathcal{S}$ of $L_a^2(\Omega)$ is called analytically invariant if $f.\mathcal{S} \subseteq \mathcal{S}$ for every $f \in \mathcal{O}(\overline{\Omega})$. As usual, we denote (hereafter) by $\mathcal{O}(\overline{\Omega})$ the algebra of holomorphic functions defined on neighborhoods of the closure of Ω.

Even in the case that Ω is the unit disk $\mathbf{D}$ in $\mathbf{C}$, the lattice of analytically invariant subspaces of $L_a^2(\mathbf{D})$ is huge and its structure is very poorly understood (see [4]). Of course, the structure of these subspaces become more complex as one considers (bounded) domains in several variables. The only progress made so far has been on the very special case of invariant subspaces of finite codimension (see [3], [1], [2], and [23]), and on analitically invariant subspaces whose zeros satisfy a rather severe restriction (see [11]).

In this paper, we study a natural class of analytically invariant subspaces. In order to discuss these subspaces we need to introduce the following terminology.
  (a) Given a closed subset F of Ω, we let $\mathcal{S}(F)$ be the (analytically invariant) subspace of $L_a^2(\Omega)$ consisting of those functions that vanish on F.
  (b) Given an analytically invariant subspace $\mathcal{S}$ we let $\mathcal{I}(\mathcal{S})$ be the ideal $\mathcal{I}(\mathcal{S}) = \mathcal{S} \cap \mathcal{O}(\overline{\Omega})$. Likewise, given a closed subset F of $\overline{\Omega}$, we let $\mathcal{I}(F)$ be the ideal $\mathcal{I}(F) = \mathcal{I}(\mathcal{S}(F))(= \mathcal{S}(F) \cap \mathcal{O}(\overline{\Omega}))$.

1991 *Mathematics Subject Classification*. Primary 47A15. Secondary 46E20.
Supported by NSF grant DMS 9002958.
This paper is in final form and no version of it will be submitted for publication elsewhere.

(c) Given an ideal $\mathcal{J}$ of $\mathcal{O}(\overline{\Omega})$, we let $\mathcal{J}^{(2)}$ be the (non necessarily closed analytically invariant subspace and) extended ideal $\mathcal{J}^{(2)} = \mathcal{J}.L_a^2(\Omega)$.

(d) Given a subset of functions $\mathcal{A}$ in $\mathcal{O}(\Omega)$, we let $V(\mathcal{A})$ be the set of common zeros of functions in $\mathcal{A}$.

We recall that (when Ω is pseudoconvex and one works with the algebra $\mathcal{O}(\Omega)$) every ideal $\mathcal{J}$ is locally finitely generated, and hence, for each compact subset K of Ω, there are finitely many functions $f_1, \ldots, f_m$ in $\mathcal{O}(\Omega)$ such that $V \cap K = V(\mathcal{J}) \cap K = \{z \in K : f_j(z) = 0, 1 \le j \le m\}$. In particular, $V(\mathcal{I}(V)) = V$. On the other hand, we also recall that a radical ideal $\mathcal{J}$ in $\mathcal{O}(\Omega)$ is characterized by the property $\mathcal{J} = \mathcal{I}(V(\mathcal{J}))$. An important classical fact (see [13]) is that $\mathcal{J}$ is a radical ideal if and only if $\mathcal{J}.\mathcal{O}$ is a subsheaf of radical ideals of the structural sheaf $\mathcal{O}$ over Ω. These statements are the globalized version of the "Hilbert's Nullstellensatz" which can be obtained via Cartan's Theorem (B) (see [13]). Our intention, in the present work, is to examine the corresponding assertions for analytically invariant subspaces, where we substitute the standard structural sheaf $\mathcal{O}$ by the structural Bergman sheaf over $\overline{\Omega}$ (i.e., the sheaf generated by local Bergman spaces on neighborhoods of points of $\overline{\Omega}$). Our point of view is to consider $L_a^2(\Omega)$ as a set of functions in $\mathcal{O}(\Omega)$ subject to a growth condition at $\partial\Omega$.

There are several interesting questions within this circle of ideas. Most of such problems originated naturally in the study of Hilbert modules over function algebras (see [10] and [25, Remark 3.2]). We are indebted to Professor Ronald Douglas who called our attention to some of the following questions:

(A) Given an ideal $\mathcal{J}$ of $\mathcal{O}(\overline{\Omega})$, when is the extended ideal $\mathcal{J}^{(2)}$ closed in $L_a^2(\Omega)$?

(B) Given an analytically invariant subspace $\mathcal{S}$, when is $\mathcal{I}(\mathcal{S})(= \mathcal{S} \cap \mathcal{O}(\overline{\Omega}))$ dense in $\mathcal{S}$? When is $(\mathcal{I}(\mathcal{S}))^{(2)} = \mathcal{S}$?

(C) Given a proper ideal $\mathcal{J}$ of $\mathcal{O}(\overline{\Omega})$, when is $\mathcal{J}$ dense in $L_a^2(\Omega)$?

(D) Given a radical ideal $\mathcal{J}$, when is $\mathcal{J}(= \mathcal{I}(V(\mathcal{J})))$ dense in $\mathcal{S}(V(\mathcal{J}))$? When is $\mathcal{J}^{(2)} = \mathcal{S}(V(\mathcal{J}))$?

Remark 1.1 (a) Problems related to the above questions were already considered by A. Douady and his school, in the context of extended ideals in Banach spaces of analytic functions defined by global conditions such as the Bergman space (see [9], [21], and [22]).

(b) Notice that an affirmative answer to the second part of question (D) will give a positive answer to the second part of (B) (and hence to question (A)) in the particular case of radical ideals. On the other hand, a positive answer to the first part of (D) will give an affirmative answer to (B), in the special case $\mathcal{S} = \mathcal{S}(V(\mathcal{J}))$.

In the present paper we give a complete answer to question (C), in the case that Ω has a "nice" boundary. We also give a complete answer to question (D), in the case that $V(\mathcal{J})$ is the union of a sequence of smooth k-dimensional manifolds V_k that intersect transversally $\partial\Omega$, $0 \le k \le n - 1$. In particular, question (A) has an affirmative answer in this case, so that the 0-th-homology of the derived functor $\mathcal{J}\widehat{\otimes}_{\mathcal{O}(\mathbf{C}^n)}L_a^2(\Omega)$ is Hausdorff, and we also prove that the higher homology is trivial. This means that $\mathcal{J}$ and $L_a^2(\Omega)$ are analytically transversal (or, equivalently, that Ω is privileged with respect to $\mathcal{J}$). This is the main theme of our discussion. In fact, we conjecture that analytic transversality and separation of the corresponding extended ideal are equivalent notions. We prove this conjecture in the case of radical ideals $\mathcal{J}$ of $\mathcal{O}(\overline{\Omega})$ such that $V(\mathcal{J})$ is a smooth manifold. We shall also discuss the relation between the notion of analytic transversality and the notion that a domain (i.e. Ω) be privileged with respect to a coherent sheaf (i.e. the one induced by $\mathcal{J}$) in the sense of Douady (see [8]). There are a series of other Nullstellensatz phenomena for spaces of analytic functions with growth conditions on the boundary obtained by integral representation formuli, sharp estimates for the $\overline{\partial}$-operator or the theory of currents. They are supplementary to our approach and we will not mention them in this paper. An excellent current account of all these topics can be found in [29].

Finally, but not last, we would like to thank professor T. Ohsawa for having clarified and explained to us results from his work [19].

2. Preliminaries.

In order to proceed with our discussion about the answer to question (C) above (which we complete in the next section), we shall need to make some natural working assumptions that will be satisfied by any (bounded) strongly pseudoconvex domain with smooth boundary. These are the following:

(α_1) $\mathcal{O}(\overline{\Omega})$ is dense in $L_a^2(\Omega)$.

(α_2) $\overline{\Omega}$ is Stein, i.e. $\overline{\Omega}$ has a neighborhood basis consisting of pseudoconvex domains.

(α_3) For every $w \in \partial\Omega$ there exists a sequence $\{k_m\} \subseteq \mathcal{O}(\overline{\Omega})$ such that $\|k_m\|_{2,\Omega} \to 0$ and $\inf|k_m(w)| > 0$.

(α_4) If A is a closed analytic subset of a neighborhood of $\overline{\Omega}$ such that $A \cap \Omega = \emptyset$, then $A \cap \partial\Omega$ is finite.

A domain Ω enjoying the above properties α wil be called an α-domain.

Remark 2.1 (a) Condition α_3 was first introduced in [2, Theorem 3.1]. it is also pointed out in [2, Example 4.3] that α_3 is valid for every (bounded) complete pseudoconvex Reinhardt domain. Notice that conditions α_1 and α_2 are also obviously valid for these domains.

(b) Observe that condition α_3 is equivalent to:

1. there exists no bounded point evaluation of $\mathcal{O}(\overline{\Omega})$ in the $L_a^2(\Omega)$-norm on $\partial\Omega$, i.e., given any $w \in \partial\Omega$, there exists no $C > 0$ such that $|f(w)| \le C\|f\|_2$, for every $f \in L_a^2(\Omega)$ (see [23]).

It readily follows that a sufficient condition for a domain Ω to be an α-domain is the existence of a peaking function in $\mathcal{O}(\overline{\Omega})$ for every point in $\partial\Omega$ (see [24]).

Now we turn our attention to the notion of analytic transversality.

Remark 2.2 Let M and N be two analytic submanifolds of $\mathbf{C}^n$ passing through zero. It is a well known fact that M and N are geometrically transversal at zero,

i.e. their tangent spaces are transversal at zero (the dimension of their intersection is minimal) if and only if the corresponding local rings are algebraically
transversal, i.e. they satisfy

$$\mathcal{O}_{M,0} \otimes_{\mathcal{O}_{\mathbf{C}^n,0}} \mathcal{O}_{N,0} \approx \mathcal{O}_{M \cap N,0}, \text{and}$$
$$\operatorname{Tor}_q^{\mathcal{O}_{\mathbf{C}^n,0}}(\mathcal{O}_{M,0}, \mathcal{O}_{N,0}) = 0, \ q \geq 1.$$

The above algebraic transversality relations are the basis of intersection theory
in local algebra. In passing from the germs at a point to a Stein manifold X,
one first take any complete locally convex tensor product of the given modules
(we don't need to specify the topology on this complete tensor product because
we shall only work with nuclear spaces), and then one needs to substitute the
algebraic operations by those induced by the module tensor product over the
algebra $\mathcal{O}(X)$, which we shall denote by $.\widehat{\otimes}_{\mathcal{O}(X)}..$ Likewise, the corresponding
derived functor will be denoted by $\widehat{\operatorname{Tor}}$.

Definition 2.3 Let X be a Stein manifold, and let $\mathbf{M}$ and $\mathbf{N}$ be Frechet $\mathcal{O}(X)$
modules. We call $\mathbf{M}$ and $\mathbf{N}$ analytically transversal over $\mathcal{O}(X)$, if $\mathbf{M}\widehat{\otimes}_{\mathcal{O}(X)}\mathbf{N}$ is
Hausdorff as a locally convex space, and $\widehat{\operatorname{Tor}}_q^{\mathcal{O}(X)}(\mathbf{M}, \mathbf{N}) = 0$ for every $q \geq 1$.
(The above example satisfies this definition, where X is a Stein neighborhood
of zero, and $\mathbf{M} = \mathcal{O}(M)$, $\mathbf{N} = \mathcal{O}(N)$.) A Frechet $\mathcal{O}(X)$ module $\mathbf{M}$ is called
quasi-coherent if $\mathbf{M}$ and $\mathcal{O}(U)$ are analytically transversal over $\mathcal{O}(X)$ for every
Stein subdomain U of X.

Remark 2.4 (a) Given a quasi-coherent module $\mathbf{M}$ over X, the sheaf induced
by $U \to \mathcal{O}(U)\widehat{\otimes}_{\mathcal{O}(X)}\mathbf{M}$ (where U ranges over Stein neighborhoods of points of
X) represents a canonical analytic localization of the global object consisting of
the module $\mathbf{M}$ and it is also called quasi-coherent. Of course, every coherent
analytic sheaf is quasi-coherent in the above sense (the global object being the
space of global sections with the canonical Frechet structure).

(b) A typical example of a quasi-coherent $\mathcal{O}(\mathbf{C}^n)$ module is the Bergman space
$L_a^2(\Omega)$ over a (bounded) pseudoconvex domain $\Omega \subseteq \mathbf{C}^n$. Its associated analytic
sheaf is induced by $\mathcal{F}(U) = \{f \in \mathcal{O}(U \cap \Omega) : \|f\|_{2,K \cap \Omega} < \infty, K \subseteq\subseteq U\}$, where
U is an arbitrary pseudoconvex domain in $\mathbf{C}^n$ (see [21]). This sheaf was already
used in [23] to characterize invariant subspaces of finite codimension in $L_a^2(\Omega)$.

Definition 2.5 As before, let Ω be a (bounded) pseudoconvex domain in $\mathbf{C}^n$, and
let $\mathcal{F}$ be a coherent analytic sheaf defined on a pseudoconvex domain U containing $\overline{\Omega}$. Following A. Douady's program, we say that Ω is privileged with respect
to $\mathcal{F}$ and with respect to the Bergman hermitian structure over Ω (or briefly,
$\mathcal{F}$-privileged), if $\mathcal{F}(U)$ and $L_a^2(\Omega)$ are (analytically) transversal over $\mathcal{O}(U)$.

Remark 2.6 (a) The above definition is equivalent to the following more track-
able condition. By shrinking U (if necessary) to a smaller pseudoconvex domain
containing $\overline{\Omega}$, one may suppose that there exists a finite free resolution of $\mathcal{F}$ of
the form

$$0 \to \mathcal{O}_U^{k_m} \to \ldots \to \mathcal{O}_U^{k_1} \to \mathcal{O}_U^{k_0} \to \mathcal{F} \to 0.$$

By Cartan's Theorem B, this sequence remains exact after passing to the space of global sections on U. Thus, the domain Ω is privileged with respect to the sheaf $\mathcal{F}$ (i.e. $\mathcal{F}$-privileged) if the induced complex

$$0 \to L_a^2(\Omega)^{k_m} \to \ldots \to L_a^2(\Omega)^{k_1} \to L_a^2(\Omega)^{k_0} \to 0$$

is exact in positive degrees and has Hausdorff zero-th homology, i.e. the range of the last non-trivial map in the above complex is closed. It is worth mentioning the fact (proved by Douady in his thesis, see [8]) that given a coherent analytic sheaf $\mathcal{F}$ on the Stein domain U and a fixed point λ in U, there is a fundamental system of $\mathcal{F}$-privileged Stein neighborhoods of λ.

(b) Actually, one can obtain equivalent notions of privileged domains by substituting, in the above definition, the Bergman space $L_a^2(\Omega)$ by other global objects such as the analog of the disk algebra, or the Hardy space, etc (see [20]).

(c) The simplest example of non privileged phenomenon is obtained by considering the open unit disk $\mathbf{D}$ and the principal ideal $[z-1]$ in $\mathcal{O}_{\mathbf{C}}$ generated by the coordinate function on $\mathbf{D}$. Then $\mathbf{D}$ is not privileged with respect to the (coherent) sheaf $\mathcal{O}_{\mathbf{C}}/[z-1]$ because the operator $z-1 : L_a^2(\mathbf{D}) \to L_a^2(\mathbf{D})$ induced by the canonical resolution $0 \to \mathcal{O} \to \mathcal{O} \to \mathcal{O}/[z-1] \to 0$ does not have closed range.

(d) There are very few geometric criteria for domains in $\mathbf{C}^n$ to be privileged with respect to a standard coherent sheaf, say the ideal sheaf of an analytic variety in $\mathbf{C}^n$. A very interesting criterion for the special case of polydomains, is due to G. Pourcin (see [20]). We recall the essential features of this criterion because it will serve as a good testing device. Given a boundary point λ in the polydomain $\Omega = \Omega_1 \times \ldots \times \Omega_n$ in $\mathbf{C}^n$, one lets $I_\lambda = \{i \in \{1, \ldots, n\} : \lambda_i \in \partial\Omega_i\}$, and one defines the corresponding projection as $\pi_\lambda : \overline{\Omega} \to \prod_{i \in I_\lambda} \mathbf{C}$. We also, let $\overline{\Omega}^p = \{\lambda \in \overline{\Omega} : \operatorname{Card}(I_\lambda) \geq p\}$. It is obvious that $\overline{\Omega}^1 = \partial\Omega$, and that $\overline{\Omega}^n$ coincides with the distinguished boundary of Ω. The last piece of notation we need in order to state Pourcin's result is the following: let $\mathcal{F}$ be a coherent sheaf of $\mathcal{O}_U$ modules, where U is a given pseudoconvex domain in $\mathbf{C}^n$; we recall the stratification of the support of $\mathcal{F}$ by closed analytic sets: $S_p(\mathcal{F}) = \{\lambda \in U : \operatorname{prof}_\lambda(\mathcal{F}) \leq p, p \geq 0\}$. Here, $\operatorname{prof}_\lambda(\mathcal{F})$ denotes the profounder of $\mathcal{F}$ at λ (see [28]). It follows that $\operatorname{supp}(\mathcal{F}) = \cup_{p \geq 0} S_p(\mathcal{F})$, $\dim(S_p(\mathcal{F})) \leq p, p \geq 0$.

THEOREM 2.7. (Pourcin) Let Ω be a polydomain in $\mathbf{C}^n$ with $\Omega = \operatorname{int}(\overline{\Omega})$. Let U be a Stein open neighborhood of $\overline{\Omega}$, and let $\mathcal{F}$ be a coherent $\mathcal{O}_U$ module. Then the following conditions are equivalent:

(a) Ω is $\mathcal{F}$-privileged.

(b) For any integer p, $0 \leq p \leq n$, $S_p(\mathcal{F}) \cap \overline{\Omega}^{p+1} = \emptyset$.

(c) For any $\lambda \in \overline{\Omega}$, $\mathcal{F}$ is π_λ-flat at λ.

(d) The space $\mathcal{F}(U))\widehat{\otimes}_{\mathcal{O}(U)} L_a^2(\Omega)$ is Hausdorff.

(e) The canonical restriction map $\mathcal{F}(U)\widehat{\otimes}_{\mathcal{O}(U)} L_a^2(\Omega) \to \mathcal{F}(\Omega)$ is continuous and injective.

Remark 2.8 The remarkable feature of the above theorem is the fact that it gives a complete solution to the global privilege problem for polydomains. In order to exhibit a simple example in $\mathbf{C}^2$, consider the (radical) ideal $\mathcal{J}$ of a hypersurface $H \subseteq \mathbf{C}^2$. Then, the unit polydisk $P = \mathbf{D} \times \mathbf{D}$ is $\mathcal{O}_H$-privileged if and only if $H \cap (\partial \mathbf{D} \times \partial \mathbf{D}) = \emptyset$. Indeed, in this case, $S_1(\mathcal{O}/\mathcal{J}) = H$, and $\overline{P}^2 = \partial \mathbf{D} \times \partial \mathbf{D}$.

Another direct application of Pourcin's Theorem is contained in the following result.

COROLLARY 2.9. *Let Ω be a polydomain, as above, and let V be a smooth p-dimensional submanifold of an open neighborhood of $\overline{\Omega}$. Then the domain Ω is privileged with respect to the structural sheaf $\mathcal{O}_V$ if and only if $V \cap \overline{\Omega}^{p+1} = \emptyset$.*

PROOF. It readily follows from the fact that $S_r(\mathcal{O}_V) = \emptyset$ unless $r = p$, in which case $S_p(\mathcal{O}_V) = V$ (see [28]). $\square$

The same conclusion of the above corollary remains valid if we substitute V by a singular variety which is locally complete intersection.

3. Density of ideals in Bergman Space.

We begin our discussion with the extreme case in which the zero set of an ideal does not intersect the domain that carries the Bergman space. The following result gives a complete answer to question (C) of the introduction for α-domains.

THEOREM 3.1. *Let Ω be a pseudoconvex domain in $\mathbf{C}^n$ that satisfies the α-conditions (see section 2), and let $\mathcal{J}$ be an ideal in $\mathcal{O}(\overline{\Omega})$. Then $\mathcal{J}$ is dense in $L_a^2(\Omega)$ if and only if $V(\mathcal{J}) \cap \Omega = \emptyset$.*

PROOF. If $V(\mathcal{J}) \cap \Omega \neq \emptyset$, then the linear subspace $\mathcal{J}^{(2)}(= \mathcal{J} . L_a^2(\Omega))$ cannot be dense in $L_a^2(\Omega)$ because all its elements vanish on $V(\mathcal{J})$. Conversely, assume that $V(\mathcal{J}) \cap \Omega = \emptyset$, and, by way of contradiction, suppose that $\mathcal{J}$ is not dense in $L_a^2(\Omega)$. Let $g \in L_a^2(\Omega)$ be a unit vector in the orthogonal complement of $\mathcal{J}$. Then, g induces the linear functional $\langle ., g \rangle$ on $\mathcal{O}(\overline{\Omega})$, which produces a continuous linear functional $L : \mathcal{O}(\overline{\Omega})/\mathcal{J} \to \mathbf{C}$. By Malgrange's (flatness and separation) Theorem on ideals of smooth functions, [17], the extended ideal $\mathcal{J} . C^\infty(\overline{\Omega})$ is closed in the locally convex topology of the space of germs of smooth functions on $\overline{\Omega}$. Moreover, $(\mathcal{J} . C^\infty(\overline{\Omega})) \cap \mathcal{O}(\overline{\Omega}) = \mathcal{J}$. This enables us to extend L (using Hahn-Banach's Theorem) to a continuous functional $\widetilde{L} : C^\infty(\overline{\Omega})/(\mathcal{J} . C^\infty(\overline{\Omega})) \to \mathbf{C}$. By assumption, $A = V(\mathcal{J}) \cap \overline{\Omega} = V(\mathcal{J}) \cap \partial\Omega$ is necessarily finite. Thus, since the distribution $\widetilde{L}$ is supported by A, it follows that $\widetilde{L}$ is the finite sum of derivatives of Dirac measures at points of A. Consequently, the functional $\widetilde{L}$ (and, afortiori, L) vanishes on (the image of) some k-th power $\mathcal{I}^k(A)$ of the ideal $\mathcal{I}(A)$ in $\mathcal{O}(\overline{\Omega})$ associated with A. Denoting by $\mathcal{M}_\lambda$ the maximal ideal of $\mathcal{O}(\overline{\Omega})$ at $\lambda \in \partial\Omega$, we observe that α_3 in the α-conditions of section 2 is equivalent to the density of $\mathcal{M}_\lambda$ in $L_a^2(\Omega)$. In other words, $\sum_{j=1}^n (z_j - \lambda_j) . L_a^2(\Omega) = \mathcal{M}_\lambda . L_a^2(\Omega) = \mathcal{M}_\lambda^{(2)}$ is a dense (linear) subspace of $L_a^2(\Omega)$. This shows that $\sum_{j=1}^n (z_j - \lambda_j)[\mathcal{M}_\lambda . L_a^2(\Omega)] =$

$\mathcal{M}_\lambda^2 . L_a^2(\Omega)$ is also dense in $L_a^2(\Omega)$. Continuing this argument, we deduce that $\mathcal{M}_\lambda^k$ is still dense in $L_a^2(\Omega)$, for every $\lambda \in A$. Furthermore, since $\sum_{\lambda \in A} \mathcal{M}^k(\lambda) \subseteq \mathcal{I}^k(A)$ is dense in $L_a^2(\Omega)$, we therefore conclude L vanishes in a dense set so that $g = 0$, reaching a contradiction. □

Remark 3.2 We point out that the proof of the above theorem does not make a crucial use of the Hilbert structure of $L_a^2(\Omega)$. One can take first any non-zero continuous linear functional that vanish on $\mathcal{J}$ and then define the functional L as above. Thus, we see that the same proof works for other Banach spaces of analytic functions. For example, the next corollary can be obtained via such a proof.

COROLLARY 3.3. *Let Ω be a domain with strongly pseudoconvex smooth boundary, and let $p \in \mathbf{R}, 1 \le p < \infty$. Then, an ideal $\mathcal{J}$ of $\mathcal{O}(\overline{\Omega})$ is dense in $L_a^p(\Omega)$ if and only if $V(\mathcal{J}) \cap \Omega = \emptyset$.*

4. Geometric versus analytic transversality.

Pourcin's result (mentioned in section 2) strongly suggest that the relative position of the closed sets $V(\mathcal{J})$ and $\partial\Omega$ is the basic geometric criterion that determines whether or not $\mathcal{J}^{(2)}(= \mathcal{J}.L_a^2(\Omega))$ is closed, for a given ideal $\mathcal{J}$ of $\mathcal{O}(\overline{\Omega})$. Thus, this could be also a criterion to study a Nullstellensatz phenomenon in the Bergman space $L_a^2(\Omega)$.

In this section, we investigate a series of geometric conditions on $V(\mathcal{J})$ and $\partial\Omega$ which imply the analytic transversality of $\mathcal{O}/\mathcal{J}$ and $L_a^2(\Omega)$.

We shall see that, at least in the smooth case, analytic transversality is equivalent to geometric transversality. As in the preceding sections, we shall suggest rather than develop the most general conditions under which our results are valid.

THEOREM 4.1. *Let Ω be a pseudoconvex domain in $\mathbf{C}^n$, and let Ω' be a pseudoconvex neighborhood of $\overline{\Omega}$. Let V be a smooth complex submanifold of Ω'. If $\partial\Omega$ is smooth at any point of intersection between V and $\partial\Omega$, and if $\partial\Omega$ and V are geometrically transversal at these points, then Ω is privileged with respect to the reduced structural sheaf $\mathcal{O}_V$ of V.*

PROOF. Let $\mathcal{J}$ be the radical ideal in $\mathcal{O}(\Omega')$ associated with V. Since Ω' is Stein, $\mathcal{J}$ has finite free resolutions on $\overline{\Omega}$. Therefore, $\mathcal{O}_{\overline{\Omega}}/\mathcal{J}$ also has finite free resolutions. Let $\mathcal{L}.$ be any of such resolutions. We must prove that $\mathcal{L}(\overline{\Omega}) \widehat{\otimes}_{\mathcal{O}(\overline{\Omega})} L_a^2(\Omega)$ is an exact complex in positive degrees, and that it has Hausdorff zero-th homology. Let $\mathcal{F}$ be the analytic sheaf induced by (the quasi-coherent module) $L_a^2(\Omega)$ (see definition quasi). The localization of the preceding complex with copies of $L_a^2(\Omega)$ is given by $\mathcal{L}.\widehat{\otimes}_{\mathcal{O}_{\overline{\Omega}}}\mathcal{F}$. Using the acyclicity of quasi-coherent modules over Stain manifolds, it is sufficient to prove that $\mathcal{L}.\widehat{\otimes}_{\mathcal{O}_{\overline{\Omega}}}\mathcal{F}$ is an exact complex in positive degrees, and its zero-th homology is the space of global sections of a sheaf of Frechet spaces. Assume first that λ is a fixed point of Ω. Since $\mathcal{F}|_\Omega = \mathcal{O}_\Omega$ (see [13]), $(\mathcal{L}.\widehat{\otimes}_{\mathcal{O}_{\overline{\Omega}}}\mathcal{F})_\lambda \approx \mathcal{L}_{.,\lambda}$ and the above conditions are obviously satisfied. Now, if $\lambda \notin \overline{\Omega}$ then $\mathcal{F}_\lambda = 0$ and hence $(\mathcal{L}.\widehat{\otimes}_{\mathcal{O}_{\overline{\Omega}}}\mathcal{F})_\lambda = 0$. Next assume that $\lambda \in \overline{\Omega} \setminus V$. Then $\mathcal{I}(V)_\lambda = \mathcal{O}_\lambda$. Therefore, $\mathcal{L}_{k,\lambda}$ is a $\mathcal{O}_\lambda$ split complex, so that the same is true for $(\mathcal{L}.\widehat{\otimes}_{\mathcal{O}_{\overline{\Omega}}}\mathcal{F})_\lambda$. So, only the case of a point $\lambda \in V \cap \partial\Omega$ remains to be analyzed. Since by assumption, $\partial\Omega$ is smooth in a neighborhood of λ, we can perform a local holomorphic change of coordinates around λ, and we may assume that, for a small polydisk U with center zero, we have: $\lambda = 0$, $V \cap U = \{z \in U : z_1 = \ldots = z_p = 0\}$ for some $1 \leq p \leq n$, and the normal of the real hypersurface $\partial\Omega$ at λ is contained in V. A resolution of $\mathcal{O}_U/\mathcal{J}|_U$ is given by the Koszul complex $\mathcal{K}.(z', \mathcal{O}|_U)$, where $z' = (z_1, \ldots, z_p)$, $z'' = (z_{p+1}, \ldots, z_n)$, so that $z = (z', z'')$. Thus, it suffices to show that $\mathcal{K}.(z', \mathcal{F}(U))$ is exact for positive degrees, and has Hausdorff zero-th homology. In other words, for a fixed q, $0 \leq q \leq p$ and any sequence $\{\eta_m\} \subseteq \mathcal{K}_q(z', \mathcal{F}(U))$ satisfying $\partial\eta_m \to 0$ (here ∂ denoes the boundary operator in the Koszul complex), we must prove that there exists a sequence $\{\xi_m\} \subseteq \mathcal{K}_{q+1}(z', \mathcal{F}(U))$ with the property $\eta_m - \partial\xi_m \to 0$. The last assertion is obviously true for $\mathcal{O}(U \cap \Omega)$ in place of $\mathcal{F}(U)$. We shall exploit this fact, keeping in mind that $\mathcal{F}|_{U\cap\Omega} \approx \mathcal{O}|_{U\cap\Omega}$, via a deformation of the domain $U \cap \Omega$. Indeed, fix q with $0 \leq q \leq p$, and let $\{\eta_m\} \subseteq \mathcal{K}_q(z', \mathcal{F}(U))$ be a sequence suchh that $\partial\eta_m \to 0$. We translate the domain $U\cap\Omega$ in the direction of the exterior normal τ at $0 \in \partial\Omega$. Then, by shrinking the radius of the polydisk U (if necessary), we see that the hypersurfaces $U\cap\partial\Omega$ and $(U\cap\partial\Omega)+\epsilon\tau$ do not intersect each other for a fixed small $\epsilon > 0$. Then, $\partial_p(z')\eta_m(z', z''+\epsilon\tau)$ converges to zero in the new domain $(U\cap\Omega)+\epsilon\tau$. Therefore, there exists a sequence $\{\xi_m^\epsilon\} \subseteq \mathcal{K}_{q+1}(z', \mathcal{O}[(U\cap\Omega)+\epsilon\tau])$ such that $\lim_{m\to\infty}[\eta_m(z', z''+\epsilon\tau) - \partial(z')\xi_m^\epsilon(z', z'')] = 0$ in the Frechet topology of $\mathcal{K}_q(z', \mathcal{O}[(U\cap\Omega)+\epsilon\tau])$. This proves that for every compact subset K of U the restriction $\eta_m|_{K\cap\Omega}$ can be approximated in the $(2, K\cap\Omega)$-norm by elements of $\partial\mathcal{K}_{q+1}(z', \mathcal{F}(K))$. The next step is to take an exhaustion $K_j \to U$ by compact subpolydisks of U, in order to find a sequence $\{\xi_m\} \subseteq \mathcal{K}_{q+1}(z', \mathcal{F}(U))$ with the property $(\partial\xi_m) - \eta_m \to 0$, as desired. $\square$

Remark 4.2 The same argument of the last proof enables us to identify the zero-th homology space of the Koszul complex $\mathcal{K}.(z', \mathcal{F}(U))$. Observe first that the image $\mathrm{Im}(\partial_1)$ of $\partial_1 : \mathcal{K}_1(z', \mathcal{F}(U)) \to \mathcal{K}_0(z', \mathcal{F}(U))$ is contained in $\{f \in \mathcal{F}(U) : f|_{U\cap V\cap\Omega} = 0\}$. In order to prove the opposite inclusion, let $f \in \mathcal{F}(U)$ such that $f|_{U\cap V\cap\Omega} = 0$. Then, the translated function $f(z', z''+\epsilon\tau)$ is in $\mathcal{J}|_{(U\cap\Omega)+\epsilon\tau}$ because of the standard Nullstellensatz. By passing to the limit as $\epsilon \to 0$, we deduce that f is in the closure of $\mathrm{Im}(\partial_1)$. But, we already know that $\mathrm{Im}(\partial_1)$ is closed. Thus, we proved the following result, which gives a partial answer to questions (A) and (D) of the introduction.

COROLLARY 4.3. *Under the same assumptions of Theorem 4.1, we have:*
$$\mathcal{I}(V).L_a^2(\Omega) = \mathcal{S}(V).$$

Remark 4.4 (a) We observe that, in Theorem 4.1 (and hence in the last corollary), the ideal $\mathcal{J}$ is finitely generated in $\mathcal{O}(\overline{\Omega})$ because Ω' is a pseudoconvex neighborhood of $\overline{\Omega}$. Let $f_1, \dots, f_k \in \mathcal{O}(\overline{\Omega})$ be a system of generators of $\mathcal{J}$. Corollary 4.3 then asserts that every function in $L_a^2(\Omega)$ vanishing on $V \cap \Omega$ is a linear combination of $f_1, \dots, f_k$ with coefficients in $L_a^2(\Omega)$.

(b) A natural generalization of Theorem 4.1 is obtained by considering the ideal $\mathcal{J}$ in $\mathcal{O}(\Omega')$ with an ascending filtration of ideals of the form $\mathcal{J} = \mathcal{J}_N \subseteq \mathcal{J}_{N-1} \subseteq \dots \subseteq \mathcal{J}_1 \subseteq \mathcal{J}_0 = \mathcal{O}(\Omega')$ where the consecutive quotients satisfy $\mathcal{J}_{j-1}/\mathcal{J}_j \approx \mathcal{O}(V_j)$ and V_j is a smooth manifold transversal to $\partial\Omega$, as in Theorem 4.1, $1 \le j \le N$. Indeed, a repeated application of an argument involving a long exact sequence of "tors" shows that the domain Ω is privileged with respect to (the coherent sheaf induced by) the ideal $\mathcal{J}$. Moreover, if $\mathcal{J}$ is radical, then the conclusion of Corollary 4.3 is also valid. We leave the details to the interested reader (see also our discussion in the next section).

(c) Another mild generalization of Theorem 4.1, is obtained by observing that the smoothness of V was only needed near points of $V \cap \partial\Omega$. Thus, we may allow the variety V in Theorem 4.1 to have isolated singularities in Ω.

The following is one of the simplest consequences of Theorem 4.1.

PROPOSITION 4.5. *Let Ω be a strongly convex domain with smooth boundary in $\mathbf{C}^n$, and let L be a linear variety in $\mathbf{C}^n$ such that $L \cap \overline{\Omega} \ne \emptyset$. Then the following assertions are equivalent:*

(a) Ω is $\mathcal{O}_L$-privileged.

(b) L intersects transversally $\partial\Omega$.

(c) $\mathcal{I}(L)^{(2)}(= \mathcal{I}(L).L_a^2(\Omega))$ is a closed subspace of $L_a^2(\Omega)$.

(d) $\mathcal{I}(L)^{(2)} = \mathcal{S}(L)$.

PROOF. In considering the relative position of L and $\partial\Omega$, there are only two possible alternatives:

(*) L is tangent to $\partial\Omega$.

(**) L intersects transversally $\partial\Omega$. In the first case, $\mathcal{I}(L)$ is dense in $L_a^2(\Omega)$ by Theorem 3.1, so that assertions (a), (3), and (d) are not true. In case (**), all assertions are valid by Theorem 4.1 and Corollary 4.3. Thus, (a), (b), (c), and (d) are all equivalent statements. $\square$

The following results give some other partial answers to the questions raised in the introduction.

PROPOSITION 4.6. *Let Ω be a pseudoconvex domain in $\mathbf{C}^n$ and let H be a (non-necessarily smooth) analytic hypersurface defined in a pseudoconvex domain Ω' that contains $\overline{\Omega}$. Then, the following are equivalent conditions:*

(a) Ω is $\mathcal{O}_H$-privileged.

(b) $\mathcal{I}(H)^{(2)}$ is a closed subspace.

PROOF. Let $\mathcal{F}$ be the sheaf induced by $L_a^2(\Omega)$ (as in the proof of Theorem 4.1). We recall that Ω is $\mathcal{O}_H$-privileged if and only if, at the sheaf level, we have: $\mathcal{O}_H \widehat{\otimes}_{\mathcal{O}_{\overline{\Omega}}} \mathcal{F}$ is Hausdorff, and

$$\mathrm{Tor}_q^{\mathcal{O}_{\overline{\Omega}}}(\mathcal{O}_H, \mathcal{F}) = 0, \forall q \ge 1.$$

But, locally, H has a single (complex-valued) defining analytic function f, so, locally, $\mathcal{I}(H)$ has the free resolution $0 \to \mathcal{O} \to^f \mathcal{I}(H) \to 0$. Therefore, the corresponding "Tors" (mentioned above) are trivial. Since the separation of the zero-th homology is clear, the proof is complete. $\square$

PROPOSITION 4.7. *Let Ω be a domain in $\mathbf{C}^2$ with smooth strongly pseudo-convex boundary, and let V be a closed analytic variety in a neighborhood of $\overline{\Omega}$. Then the following assertions are equivalent:*

(a) Ω is $\mathcal{O}_V$-privileged.

(b) $\mathcal{I}(V)^{(2)}$ is a closed subspace.

PROOF. Let $V = V_0 \cup V_1$ be the disjoint union decomposition of V in purely dimensional subspaces of dimension 0 and 1, respectively. If $V_0 = \emptyset$, then the previous proposition can be applied. Since $\mathcal{O}_V = \mathcal{O}_{V_0} \oplus \mathcal{O}_{V_1} = \mathcal{O}_{V_1} \oplus \bigoplus_{\lambda \in V_0} \mathcal{O}_\lambda / \mathcal{M}_\lambda$, it follows, accordingly, that $L_a^2(\Omega)/(\mathcal{I}(V)^{(2)}) \approx L_a^2(\Omega)/(\mathcal{I}(V_1)^{(2)}) \oplus \bigoplus_{\lambda \in V_0} L_a^2(\Omega)/\mathcal{M}_\lambda^{(2)}$. If at least one of the points in V_0 is in $\partial\Omega$, then (by Theorem 3.1) the corresponding quotient $L_a^2(\Omega)/\mathcal{M}_\lambda^{(2)}$ is not Hausdorff, i.e., $\mathcal{M}_\lambda$ is dense in $L_a^2(\Omega)$. Therefore, if $V_0 \subseteq \Omega$, then $L_a^2(\Omega)/(\mathcal{I}(V_0)^{(2)})$ is a Hausdorff space by [2] and [23]. Thus, we have reduced the statement to the preceding proposition. $\square$

In view of Pourcin's Theorem (see [20]) and the results in sections 3 and 4, we formulate the following:

CONJECTURE. *Let Ω be a pseudoconvex domain in $\mathbf{C}^n$ and let $\mathcal{J}$ be an ideal in $\mathcal{O}(\overline{\Omega})$. The domain Ω is $\mathcal{O}_{\overline{\Omega}}/\mathcal{J}$-privileged if and only if $\mathcal{J}^{(2)}(= \mathcal{J}.L_a^2(\Omega))$ is a closed subspace of $L_a^2(\Omega)$.*

Under some additional smoothness assumptions on V and $\partial\Omega$, we expect that analytic transversality be equivalent to geometric transversality, as in Proposition 4.5.

Remark 4.8 (a) Besides the evidence shown in this paper in support of the above conjecture, we point out that it is easily seen to be true for domains $\Omega \subseteq \mathbf{C}$ with piecewise smooth boundaries. Indeed, in this case, every ideal $\mathcal{J} \subseteq \mathcal{O}(\overline{\Omega})$ is supported by finitely many points. If one of this points lies in $\partial\Omega$, then one applies Theorem 3.1 to infer that Ω is not $\mathcal{O}_{\overline{\Omega}}/\mathcal{J}$-privileged. If, on the other hand, $\operatorname{supp}(\mathcal{J}) \subseteq \Omega$, then one appeals to the main result of [2] and [23] to conclude that Ω is $\mathcal{O}_{\overline{\Omega}}/\mathcal{J}$-privileged, and this is the only case in which $\mathcal{J}^{(2)}$ is closed.

(b) It is worth noticing that the corresponding version of Theorem 4.1 and its consequences where $L_a^2(\Omega)$ is replaced by $L_a^p(\Omega)$, with $1 \le p < \infty$ are also valid, provided that Ω has a smooth strongly pseudoconvex boundary. Indeed, the existence of L^p-estimates for the $\overline{\partial}$-operator on Ω (see [15]) insures the quasi-coherence of $L_a^p(\Omega)$. This is the crucial fact needed in the proofs of the above results.

5. The spectral picture of the Bergman n-tuple.

In this section, we use the results of the preceding sections to study the spectral properties of the compressions of the so-called Bergman n-tuple (see [26], [23], and [7]) to (orthogonal) complements of the analytically invariant subspaces considered above. We recall that the Bergman n-tuple B_Ω is the n-tuple of multiplication operators by the standard coordinate functions on $L_a^2(\Omega)$, where Ω is any (bounded) domain in $\mathbf{C}^n$. The spectral and C^*-algebraic properties of B_Ω are fairly well understood (see [26], [27], [21], and [7]). However, the situation is much less clear, when dealing with compressions to complements of invariant subspaces.

Each analytically invariant subspace $\mathcal{S}$ defines an uppertriangular 2×2-matricial representation of B_Ω, where the diagonal elements are, respectively, $(B_\Omega)|_{\mathcal{S}}$ (the restriction of B_Ω to $\mathcal{S}$) and $[(B_\Omega^*)|_{\mathcal{S}^\perp}]^*$ (the compression of B_Ω to the complement of $\mathcal{S}$).

In [19], T. Ohsawa proved that when Ω is a pseudoconvex domain in $\mathbf{C}^n$ and V is a smooth submanifold of Ω, every holomorphic function on $V \cap \Omega$ which is square integrable with respect to a natural "surface area" measure on V, has an extension to a function in $L_a^2(\Omega)$. It is not known whether the restriction map from $L_a^2(\Omega)$ into $L_a^2(V)$ is well defined. We suspect that this is the case, when V is a smooth submanifold of a neighborhood of $\overline{\Omega}$ that intersects transversally $\partial\Omega$. Thus, we denote by $L_a^2(\Omega, V)$ the range of the restriction map. Ohsawa's result tells us that $L_a^2(V) \subseteq L_a^2(\Omega, V)$. Therefore, we may identify the space $\mathcal{S}(V)^\perp (\equiv L_a^2(\Omega)/\mathcal{S}(V))$ with the space $L_a^2(\Omega, V)$. Thus, it is natural to ask questions about the $(2,2)$-diagonal entry of B_Ω in the above upper-triangular matricial representation on $L_a^2(\Omega) = \mathcal{S}(V) \oplus \mathcal{S}(V)^\perp$ (which is the one related to $\mathcal{S}(V)^\perp$). These n-tuples are natural examples of elements in the Cowen-Douglas class over the manifold $V \cap \Omega$. Some of such questions are not easy to answer (like those related to almost normality and essential normality, see [10]). On the other hand, some of the spectral properties of such compressions are more tractable, as we shall see below.

Henceforth, Ω stands for a bounded pseudoconvex domain in $\mathbf{C}^n$ with smooth boundary.

Remark 5.1 As in section 4, we shall consider an ideal $\mathcal{J}$ of analytic functions defined on a pseudoconvex domain Ω' containing $\overline{\Omega}$, such that there exists an ascending filtration of ideals

$$\mathcal{J} = \mathcal{J}_N \subseteq \mathcal{J}_{N-1} \subseteq \cdots \subseteq \mathcal{J}_1 \subseteq \mathcal{J}_0 = \mathcal{O}(\Omega'), \tag{1}$$

having the consecutive quotients of the form $\mathcal{J}_{j-1}/\mathcal{J}_j \approx \mathcal{O}(V_j)$, where V_j are closed complex analytic submanifolds of Ω', $1 \leq j \leq N$ (see also Remark 4.4 (b)). If each of the submanifolds V_j is transversal to $\partial\Omega$, then the domain Ω is $\mathcal{O}_{\Omega'}/\mathcal{J}$-privileged, in fact it is $\mathcal{O}_{\Omega'}/\mathcal{J}_j$-privileged, $0 \leq j \leq N-1$. Notice that $\mathcal{J}_j^{(2)}$ is closed, $0 \leq j \leq N$, by analytic transversality.

LEMMA 5.2. *With the above assumptions, we have:* $\mathcal{J}_{j-1}^{(2)}/\mathcal{J}_j^{(2)} \approx L_a^2(\Omega, V_j \cap \Omega)$, $1 \leq j \leq N$.

PROOF. Let j be fixed, $1 \leq j \leq N$. According to Corollary 4.3, the radical ideal $\mathcal{I}(V_j)$ satisfies

$$\mathcal{I}(V_j)^{(2)}(= \mathcal{I}(V_j).L_a^2(\Omega)) =$$
$$\mathcal{S}(V_j)(= \{f \in L_a^2(\Omega) : f|_{V_j} = 0\}).$$

On the other hand, we have the natural identification $L_a^2(\Omega)/\mathcal{S}(V_j) \approx L_a^2(\Omega, V_j \cap \Omega)$. Therefore,

$$\mathcal{J}_{j-1}^{(2)}/\mathcal{J}_j^{(2)} \approx (\mathcal{J}_{j-1}/\mathcal{J}_j)\widehat{\otimes}_{\mathcal{O}(\Omega')}L_a^2(\Omega) \approx$$
$$(\mathcal{O}(\Omega')/\mathcal{I}(V_j))\widehat{\otimes}_{\mathcal{O}(\Omega')}L_a^2(\Omega) \approx L_a^2(\Omega)/\mathcal{I}(V_j)^{(2)} \approx$$
$$L_a^2(\Omega)/\mathcal{S}(V_j) \approx L_a^2(\Omega, V_j \cap \Omega),$$

as desired. □

LEMMA 5.3. *With the above assumptions, the compression of B_Ω to the complement of $\mathcal{J}^{(2)}$, on the decomposition*

$$L_a^2(\Omega) \ominus \mathcal{J}^{(2)} = \bigoplus_{j=1}^N \mathcal{J}_{j-1}^{(2)} \ominus \mathcal{J}_j^{(2)},$$

has the following upper-triangular matricial representation

$$[B_\Omega^*|_{(\mathcal{J}^{(2)})^\perp}]^* = \begin{bmatrix} B_{V_1 \cap \Omega} & * & \cdots & * \\ 0 & B_{V_2 \cap \Omega} & \cdots & * \\ \vdots & \vdots & \ddots & \vdots \\ 0 & 0 & \cdots & B_{V_N \cap \Omega} \end{bmatrix}.$$

Here, $B_{V_j \cap \Omega}$ denotes the n-tuple of multiplication operators by the restrictions of the coordinate functions $z_k : \Omega \to \mathbf{C}, 1 \leq k \leq n$ to V_j, acting on $L_a^2(\Omega, V_j \cap \Omega)$, $1 \leq j \leq N$.

PROOF. It is an immediate consequence of Lemma 5.2 and the remarks at the beginning of the present section. □

Notice that some of the varieties V_j in the pervious lemma can occur with non-trivial multiplicity in the corresponding representation.

THEOREM 5.4. *Let $\mathcal{J}$ be an ideal of analytic functions defined on a pseudoconvex domain $\Omega' \subseteq \mathbf{C}^n$ containing $\overline{\Omega}$ such that $V(\mathcal{J})$ is the union of smooth manifolds transversal to $\partial\Omega$. Then the (Taylor) joint spectrum $\sigma(B_\Omega/(\mathcal{J}^{(2)}))$ and the (Taylor) joint essential spectrum $\sigma_e(B_\Omega/(\mathcal{J}^{(2)}))$ (see [7]) of $B_\Omega/(\mathcal{J}^{(2)}) := [B_\Omega^*|_{(\mathcal{J}^{(2)})^\perp}]^*$ are given by $\sigma(B_\Omega/(\mathcal{J}^{(2)})) = V \cap \overline{\Omega}$, $\sigma_e(B_\Omega/(\mathcal{J}^{(2)})) = V \cap \partial\Omega$.*

PROOF. We point out that the notation $B_\Omega/(\mathcal{J}^{(2)})$ for the compression of B_Ω to $(\mathcal{J}^{(2)})^\perp$ is due to the fact that such an n-tuple is unitarily equivalent to the n-tuple induced by B_Ω in the quotient $L^2_a(\Omega)/(\mathcal{J}^{(2)})$. The present theorem is a direct consequence of Lemma 5.3 and the spectral pictures of each diagonal entry in the corresponding matricial representation. A second proof may be obtained by first localizing $L^2_a(\Omega)/(\mathcal{J}^{(2)})$, and then by using the sheaf-model theory of [22]. We briefly outline the latter proof. By Theorem 4.1, the sheaf $\mathcal{O}_{\Omega'}/\mathcal{J}$ is transversal to the Bergman space $L^2_a(\Omega)$ over $\mathcal{O}(\Omega')$. This implies that $\mathcal{J}^{(2)} = \mathcal{J}\widehat{\otimes}_{\mathcal{O}(\Omega')}L^2_a(\Omega)$ and $L^2_a(\Omega)/\mathcal{J}^{(2)}$ are quasi-coherent $\mathcal{O}(\Omega')$-modules. Let $\mathcal{F}$, $\mathcal{F}'$, and $\mathcal{F}''$ be the sheaves induced, respectively, by $L^2_a(\Omega)$, $\mathcal{J}^{(2)}$, and $L^2_a(\Omega)/\mathcal{J}^{(2)}$. It follows that $\mathcal{F}/\mathcal{F}' \approx \mathcal{F}''$, and obviously $\mathrm{supp}(\mathcal{F}'') \subseteq V\cap\overline{\Omega}$. Since $\sigma(B_\Omega/\mathcal{J}^{(2)}) = \mathrm{supp}(\mathcal{F}'')$ (see [22]) and clearly $V\cap\Omega \subseteq \sigma(B_\Omega/\mathcal{J}^{(2)})$, the first part of the statement follows. For the second part, it suffices to observe that $V\cap\partial\Omega$ is certainly contained in $\sigma_e(B_\Omega/\mathcal{J}^{(2)})$. To prove the opposite inclusion, we further note that $\mathcal{F}$, $\mathcal{F}'$, and $\mathcal{F}''$ are actually coherent sheaves on Ω. In view of [22], Ω is then disjoint from $\sigma_e(B_\Omega/\mathcal{J}^{(2)})$. This completes the proof of the theorem. $\square$

Remark 5.5 (a) Since the quotient $L^2_a(\Omega)/\mathcal{J}^{(2)}$, under the action of the Bergman n-tuple B_Ω, is a quasi-coherent $\mathcal{O}(\mathbf{C}^n)$-module, it follows that the above spectra coincide, in fact, with the corresponding versions of right spectra. Due to the fact that $\sigma(B_\Omega/\mathcal{J}^{(2)}) = V\cap\overline{\Omega}$, it follows that $\sigma(B_\Omega/\mathcal{J}^{(2)})$ is thin in Ω, and hence the usual index (see [7]) of the Fredholm n-tuple $(B_\Omega - \lambda)/\mathcal{J}^{(2)}$ vanishes for $\lambda \in \Omega$. However, a more refined index data is non-trivial. For instance, the sheaf $\mathcal{F}''$ corresponding to $L^2_a(\Omega)/\mathcal{J}^{(2)}$ has the property $\mathcal{F}''|_\Omega \approx \mathcal{O}_\Omega/\mathcal{J}$, and, therefore, Levy's analytic index can be readily computed as $\mathrm{ind}_a(B_\Omega/\mathcal{J}^{(2)}) = \sum_{j=1}^N [\mathcal{O}_{V_j}]$, where $[\mathcal{O}_{V_j}]$ is the class in the Grotendieck-group $K^{an}(\Omega)$ induced by the coherent $\mathcal{O}_\Omega$-module $\mathcal{O}_{V_j}$, $1 \le j \le N$ (see [16]). Of course, each component V_j of V appears with a certain multiplicity, which is precisely the algebraic multiplicity of V_j as irreducible component of $\mathcal{J}$. This is the right substitute of the global Fredholm index. Again, as in section 4, we point out that V_j can be an analytic variety with isolated singularities in Ω, $1 \le j \le N$.

(b) If, under the above assumptions, V is an algebraic variety of dimension at most 1, then, according to the main result of [11], all commutators $[T_j, T_k^*]$ are trace class, where T_j is the j-th component of the n-tuple $B_\Omega/\mathcal{J}^{(2)}$, $1 \le j, k \le n$. In this case, the principal current on V associated with $B_\Omega/\mathcal{J}^{(2)}$ (according to [5]) can be expressed as a cyclic cocycle on the algebra $C^\infty(\mathbf{C}^n)$ of smooth functions on $\mathbf{C}^n$. For higher-dimensional subvarieties, there is no satisfactory analogue of this result.

References.

1. O. Agrawal, D. Clark, and R. Douglas, *Invariant Subspaces in the Polydisk*, Pacific J. Math. **121** no. no. 1 (1986), 1-11.

2. O. Agrawal, and N. Salinas, *Sharp Kernels and Canonical Subspaces*, Amer. J. Math. **109** (1987), 23-48.

3. P. Ahern, and D. Clark, *Invariant subspaces and Analytic Continuation in*

Several Variables, J. Math. Mech. **19** (1970), 963-969.

4. H. Bercovici, C. Foias, and C. Pearcy, *Dual Algebras with Applications to Invariant Subspaces and Dilation Theory*, MBMS Regional Conference Series in Math., 56, A.M.S., Providence, Rhode Island, 1985.

5. R. Carey and J. Pincus, *Principal Currents*, Integral Equations Operator Theory **8** (1985), 614-640.

6. M. Cowen and R. Douglas, *On Operators Possessing an Open Set of Eigenvalues*, Proc. Fejer-Riesz Conference, Budapest, 1980.

7. R. Curto, *Applications of Several Complex Variables to Multi-parameter Spectral Theory*, Survey of Recent Results in Operator Theory, Part I, edited by J. Conway, and B. Morrel, Pitman Research Notes in Math,171 (1989).

8. A. Douady, *Le Probleme des Modules*, Ann. Inst. Fourier **16** no. 1 (1966), 1-95.

9. A. Douady and J. Verdier, *Seminaire de Geometrie Analytique*, Asterisque, 16 Soc. Math. France (1974).

10. R. Douglas and V. Paulsen, *Hilbert Modules Over Function Algebras*, Pitman Research Notes in Math. **217** (1989).

11. R. Douglas, V. Paulsen, and K. Yan, *Operator Theory and Algebraic Geometry*, Bull. Amer. Math. Soc. **20** (1989), 67-71.

12. R. Douglas and K. Yan, *A Multivariable Berger-Shaw Theorem*, preprint 1990.

13. H. Grauert and R. Remmert, *Theory of Stein Spaces*, Springer, Berlin et all, 1979.

14. G. Henkin and J. Leiterer, *Theory of Functions on Complex Manifolds*, Akademie Verlag, Berlin, 1984.

15. N. Kerzman, *Holder and L^p-Estimates for solutions of $\overline{\partial}u = f$ on Strongly Pseudoconvex Domains*, Comm. Pure Appl. Math. **24** (1971), 301-380.

16. R. Levy, *Algebraic and Topological K-Functors of Commuting N-Tuples of Operators*, J. of Operator Theory **21** (1989), 219-253.

17. B. Malgrange, *Ideals of Differentiable Functions*, Oxford University Press, Oxford, 1966.

18. B. Mityagin and G. Henkin, *Linear Problems of Complex Analysis*, Uspehi Math. Nauk **26** (1971), 93-152.Russian Math. Surveys (1972), 99-164.

19. T. Ohsawa, *On the Extension of L^2-Holomorphic Functions. II*, Publ. RIMS, Kyoto Univ. **24** (1988), 265-275.

20. G. Pourcin, *Sous-espaces Priviligies d'un Polycylindre*, Ann. Inst. Fourier **25** no. 1 (1975), 151-193.

21. M. Putinar, *Spectral Theory and Sheaf Theory. II*, Math. Z. **192** (1986), 473-490.

22. _____ , *Spectral Theory and Sheaf Theory. IV*, Proc. Symposia Pure Math, AMS, Providence, Rhode Island, **51**, 2 (1990) pp. 273-293.

23. _____ , *On Invariant Subspaces of Several Variables Bergman Spaces*, Pacific J. of Math. **147** no. 2 (1991), 355-364.

24. W. Rudin, *Function Theory in the Unit Ball of* $\mathbf{C}^n$, Springer Verlag, Berlin et all, 1980.

25. N. Salinas, *Products of Kernel Functions and Module Tensor Products*, Oper. Theory: Adv. Appl., Birkhauser, Basel-Boston, 32, 1988, pp. 219-241.

26. _____ , *The $\bar{\partial}$-Formalism and the C^*-Algebra of the Bergman n-tuple*, J. of Operator Theory **22** (1989), 325-343.

27. _____ , *Toeplitz Operators and Weighted Wiener-Hopf Operators, Pseudoconvex Reinhardt and Tube Domains*, to appear in Trans. of the Amer. Math. Soc..

28. J. P. Serre, *Algebre Locale. Multiplicites*, Lectures Notes in Math., 11, Springer Verlag, Berlin et all, 1965.

29. A. G. Vitushkin (Ed.), *Several complex Variables. I*, Encyclopaedia of Mathematical Sciences, 7, Springer Verlag, Berlin et all, 1990.

DEPARTMENT OF MATHEMATICS, UNIVERSITY OF KANSAS, LAWRENCE, KANSAS 66045-2142

DEPARTMENT OF MATHEMATICS, UNIVERSITY OF CALIFORNIA, RIVERSIDE, CALIFORNIA 92521

Contemporary Mathematics
Volume **137**, 1992

On the Radial Maximal Function and the
Hardy Littlewood Maximal Function in Wedges

JEAN PIERRE ROSAY

This work has been motivated by a work of A. Boggess and A. Nagel [3]. The author is greatly indebted to A. Nagel for very helpful discussions.

The title is intentionally misleading. We will spend most of the paper discussing the existence of so-called "admissible limits" in wedges. It happens that this matter is pretty simple but quite long to discuss (gathering well-known results). The arguments given in Part I involve the control of the radial or nontangential maximal function by a rather unnatural type of maximal function on the edge. This raised the question of the control of nontangential maximal functions by a more natural type of maximal function on the edge.

In Part II, we show that indeed holomorphic functions can be controlled by the Hardy Littlewood (or isotropic) maximal function. This differs from what is classically done in the polydisk, for which iterated maximal functions, or the so-called strong maximal function (averages on rectangles with sides parallel to the axis, but in unrestricted ratios) have rather been used. Part II justifies the title.

Recently F. Forstnerič has studied restricted admissible limits in wedges for bounded holomorphic functions [9]. In I we simply go back to the technique developed by Stein in [15]. For standard facts about the Hardy Littlewood maximal functions there are many references, let us mention [12], [14], [16] and [17].

The problems under consideration are of local nature. Even if not explicitly stated, at any step one may have to restrict to smaller neighborhood etc. ... .

In I.2 the result to be used later is Proposition 1, which is nothing very original but is quite long to discuss in detail. It is mainly an adaptation of well-known results in the polydisk (see I.1) using almost analytic change of variables (totally real case) together with addition of variables (for the generic case). The

1991 *Mathematics Subject Classification.* Primary 32A40. Secondary 32A35.

Research partially supported by a N.S.F grant at the University of Wisconsin.

This paper is in final form and no version of it will be submitted for publication elsewhere.

reader accepting Proposition 1 can proceed directly to I.3 for the discussion of admissible limits.

An effort has been made to make Part II readable independently of Part I (except II.1). One can start reading II.2 where notations are defined again. For the simplicity of the exposition we will consider only smooth ($\mathcal{C}^\infty$) edges. The unit disk in $\mathbb{C}$ will always be denoted by U.

Part I. Admissible limits.

I.1 The "radial" maximal function in the polydisk U^n.

We denote by $\mathbb{T}^n$ the n torus (n integer ≥ 1), $\mathbb{T}^n = (\mathbb{R}/2\pi\mathbb{Z})^n$. If $\varphi \in L(^1\mathbb{T}^n)$ and $\varphi \geq 0$, we will denote by $\overline{\mathcal{M}}(\varphi)$ the *strong maximal functions* of φ defined by:

> $\overline{\mathcal{M}}(\varphi)(w)$ is the supremum of the average of φ over all the "rectangles" centered at w, with sides parallel to the axis. Notice that there is no restriction on the ratios of the sides, possibly $\overline{\mathcal{M}}(\varphi)(w) = +\infty$.

The Hardy Littlewood maximal function is bounded on $L^p(\mathbb{T})$ if $1 < p \leq \infty$. By iteration one sees that if $\varphi \in L^p(\mathbb{T}^n)$ $1 < p \leq +\infty$, then $\overline{\mathcal{M}}(\varphi) \in L^p(\mathbb{T}^n)$, and $\|\overline{\mathcal{M}}\varphi\|_{L^p(\mathbb{T}^n)} \leq C_p \|\varphi\|_{L^p(\mathbb{T}^n)}$, for some constant C_p depending only on $p \in (1 \ \infty]$, and n.

The following theorem is proved in Rudin's book [13] (see the "basic estimate", page 29).

THEOREM. *Let $\varphi \in L^1(\mathbb{T}^n)$, $\varphi \geq 0$. Let $\tilde{\varphi}$ be its n harmonic extension to the open unit polydisk U^n given by the Poisson kernel. For any $w \in \mathbb{R}^n$, set $\varphi^*(w) = \sup\limits_{0 \leq r < 1} \tilde{\varphi}(rw)$. Then $\varphi^* \leq C\overline{\mathcal{M}}(\varphi)$ (with $C = (4\pi)^n$).*

Remarks. 1) In [10], or [18] chapter 17 (proof of theorem 2.14), and in [16] page 117, a somewhat weaker statement appears, with $\overline{\mathcal{M}}(\varphi)$ replaced by the n-fold iterate of the one dimensional Hardy Littlewood maximal function. In fact, the "basic estimate" in [13] page 29, gives an even stronger result than the one stated above (with a sum instead of a sup), that we will not need for dealing only with holomorphic functions. This basic estimate is much tied with the notion of n harmonicity which has no natural generalization to twisted edges. Part II adds to our motivation for keeping only in mind the theorem as stated.

2) The definition of $\overline{\mathcal{M}}(\varphi)$ will not be useful. We only need to know that $\varphi \mapsto \overline{\mathcal{M}}(\varphi)$ is a bounded (non linear) operator from $L^p(\mathbb{T}^n)$ into itself ($1 < p < \infty$), and that $\varphi^* \leq C \overline{\mathcal{M}}(\varphi)$.

3) So far we have only considered radial behavior. Let us remind the reader that it is trivial to extend the results to nontangential approach. If Γ_q is a truncated cone in $\mathbb{C}^n$, with vertex at $q \in \mathbb{T}^n$ (now considered as the set of points $(e^{i\theta_1}, \ldots, e^{i\theta_n})$ in $\mathbb{C}^n$), so that $\overline{\Gamma}_q \subset U^n \cup \{q\}$, then there exists $C > 0$, so that for all $\varphi \in L^1(\mathbb{T}^n)$ $\varphi \geq 0$, $\sup\limits_{\Gamma_q} \tilde{\varphi} \leq C\varphi^*(q)$, with the notations as above. Indeed,

this is just a matter of trivial comparison of the Poisson kernels of points in Γ_q and points along the radius $[o \ q]$.

I.2. The "nontangential" maximal function in wedges.

I.2.1. Definitions, Notations.

Let E be a smooth generic (i.e at each point of E, the $\mathbb{C}$-linear span of the tangent space to E is $\mathbb{C}^n$) submanifold of $\mathbb{C}^n$, $0 \in E$, and Γ be an open convex cone in $\mathbb{C}^n$ with vertex at 0. Let $\delta > 0$. The set $W = \{z = e + \gamma, e \in E, \gamma \in \Gamma, |e| < \delta, |\gamma| < \delta\}$ is called a wedge with edge E. Let Γ_0 be an open sub cone of Γ, with $\overline{\Gamma}_0 \subset \Gamma \cup \{0\}$. We say that Γ_0 is a proper sub cone of Γ. If g is a nonnegative function defined on W, we will define g^* on E, the nontangential maximal function (or to be more precise the Γ_0 nontangential maximal function), by setting

$$g^*(e) = \sup_{\substack{\gamma \in \Gamma_0 \\ |\gamma| < \delta/2}} g(e + \gamma).$$

We hope that no confusion will arise from the fact that this notation does not mention explicitly the choices: E, Γ_0, δ.

I.2.2. L^p boundary values..

As well known, there are several ways of defining H^p spaces. Here we have to choose an approach, we felt guided by the presently bigger role played by "functions with polynomial growths", (after e.g. the work of Baouendi-Chang-Treves [1]). The setting is the following: Let W be a wedge (given as in I.2.1 by E, Γ, δ). A holomorphic function h defined on E is said to have *polynomial growth* if there exists $k \in \mathbb{N}$ and $C > 0$ such that for every $z \in W$,

$$|h(z)| \leq C \ \text{dist}(z, E)^{-k}.$$

Such a function has a boundary value on E, which is a distribution which we denote by h^b. We set:

$$S_k(h) = \sup(|h(z)|\text{dist}(z, E)^k).$$

If the distribution h^b is given by a L^p function ($1 \leq p < +\infty$) we say that h is in the Hardy space H^p (local). (for H^1 see the note at the end of I.2.3).

It will be convenient to consider not only the usual Banach space L^p, but also the set $\mathcal{L}_+^p$ of measurable functions (not equivalence classes of functions) with value in $[0 \ +\infty]$ and with p^{th} power integrable (bounded if $p = +\infty$).

The quantity $S_k(h)$ has to be used because we are dealing with a local situation (e.g. the maximum principle does not apply: $\sup_W \neq \sup_E$). It will basically come into "error terms".

Now we can state the proposition which is the translation, in this setting, of the theorem mentioned above.

PROPOSITION 1. *Let W be a wedge in $\mathbb{C}^n$ with smooth edge E, a generic manifold, and given by a cone Γ and $\delta > 0$. Let Γ_0 be a proper sub cone of Γ, and $.^*$ denotes the Γ_0 nontangential maximal function. For every $p \in [1 \ +\infty]$, there exists M_p a bounded nonlinear operator from $L^p(E)$ into $\mathcal{L}^p_+(E)$ with the following property: If h is any holomorphic function on W, with polynomial growth and L^1 boundary value, and E_0 compact $\subset E \cap \{|z| < \delta\}$ one has $|h|^* \in L^p(E_0)$ if $h^b \in L^p(E)$ and more precisely; for any $k \in \mathbb{N}$, and $e \in E_0$:*

$$|h|^*(e) \leq C_{p,k,E_0}(M_p(h^b)(e) + S_k(h)),$$

with C_{p,k,E_0} a constant which does not depend on h.

Remark. The proposition could be stated in a simpler way by just saying that the L^p norm of $|h|^*$ is bounded in terms of the L^p norm of h^b and $S_k(h)$. On the other hand we like to think of M_p as an operator explicitly defined on the edge, a kind of maximal function.

It is an easy consequence of Proposition 1, using the mean value property, that if h has L^1 boundary values then in any strictly smaller wedge one has the growth estimate:

$$|h(z)| = \mathcal{O}[(\operatorname{dist}(z, E))^{-2n}],$$

which is not sharp.

The proof of the proposition will occupy the next three paragraphs. First we will prove it in the straight edge case: $E = \mathbb{R}^n \subset \mathbb{C}^n$. Then we will reduce the case of totally real edge to the straight case. Finally the case of the generic edge itself will be reduced to the totally real one.

I.2.3. The straight case.

Let P^n denote the "standard wedge"

$$P^n = \{z \in \mathbb{C}^n; \ \mathcal{I}mz_j > 0 \quad j = 1, \ldots, n\}.$$

Here we consider the case where W is the intersection of P^n with a neighborhood of 0 in $\mathbb{C}^n$.

As well known, the general case of the straight edge case ($E = \mathbb{R}^n$) can be reduced to this case. Let us remind why it is so. In the setting of $E = \mathbb{R}^n$ it is more usual to define a wedge W by giving an open convex cone Λ in $\mathbb{R}^n$, and considering this set of $z = e + i\gamma$, $e \in \mathbb{R}^n$, $\gamma \in \Lambda$ ($|e| < \delta$, $|\gamma| < \delta$). This is equivalent to the definition given above (except for the trivial case when Γ contains a vector in $\mathbb{R}^n$ and then W is a neighborhood of $\mathbb{R}^n$). If Λ is the interior of the convex hull of n linearly independent vectors $v_1, \ldots, v_n$, then a linear transformation which maps $v_1, \ldots, v_n$ to the canonical basis of $\mathbb{R}^n$ will establish a correspondence between W and P^n. Let $v \in \mathbb{R}^n - \{0\}$. One can choose $v_1, \ldots, v_n$ linearly independent, close to V in $\mathbb{R}^n$ so that the cone Λ defined as above is a "small" conic neighborhood of v. Finally any cone in $\mathbb{R}^n$ can be covered by cones of this special type.

This allows, in case of the straight edge $E = \mathbb{R}^n$, to reduce proposition 1 to the following:

PROPOSITION 1'. *Let $\delta > 0$, $p \in [1, +\infty]$, $k \in \mathbb{N}$. Let Λ be a convex cone in $\mathbb{R}^n$ containing $(\mathbb{R}^+)^n$ in its interior. Let h be a holomorphic function defined on $\mathbb{R}^n + i\Lambda \cap \{|Re\ z| \leq \delta, |\mathcal{I}mz| \leq \delta\}$. Assume that $|h(z)| = \mathcal{O}(|\mathcal{I}mz|^{-k})$, and that along $\mathbb{R}^n$ h has a boundary value h^b in L^p (in the ball of radius δ). Let $A > 0$. Then for every $x_0 \in \mathbb{R}^n$, $|x_0| \leq \delta/4$:*

$$\sup_{\substack{(x,y) \in \mathbb{R}^n \times (\mathbb{R}^+)^n \\ |x|, |y| \leq \delta/2 \\ |x - x_0| \leq A|y|}} |h(x + iy)| \leq C[M(h^b)(x_0) + S_k(h)].$$

(the conditions on the left simply describe a nontangential approach to x_0). In the right hand side: C is a constant; M is a bounded operator from L^p to $\mathcal{L}^p_+$, M and C both dependent on δ, p, k, A and Λ, not on h neither on x_0, and $S_k(h) = \sup |h(z)(\mathcal{I}mz)^k|$.

PROOF. To the function h one can associate a function H defined on the intersection of the polydisk U^n with a neighborhood V of $(1, \ldots, 1)$ by setting

$$H(z, \ldots, z_n) = h\left(i\frac{1 - z_1}{1 + z_1}, \ldots, i\frac{1 - z_n}{1 + z_n}\right).$$

The function H is smooth on $V \cap (\overline{U}^n - \mathbb{T}^n)$ and when approaching $\mathbb{T}^n$ satisfies a growth condition:

$$H(z) = \mathcal{O}(\text{dist}(z, \mathbb{T}^n)^{-\ell}), \quad \text{for some } \ell \in \mathbb{N}.$$

By Cauchy estimates, and using the fact that $(\mathbb{R}^+)^n$ is a proper sub cone of Λ one sees that the same is true (with different ℓ's) for any derivative of H. And these estimates can be made in terms of $S_k(h)$ (where k is arbitrary, to be taken such that $S_k(h) < +\infty$).

Similarly, set $H^b(e^{i\theta_1}, \ldots, e^{i\theta_n}) = h^b(i\frac{1 - e^{i\theta_1}}{1 + e^{i\theta_1}}, \ldots, i\frac{1 - e^{i\theta_n}}{1 + e^{i\theta_n}})$. Then H^b is the boundary value of H.

Let χ be a $\mathcal{C}^\infty$ function on $\mathbb{C}$, supported in a "small" neighborhood of 1, and identically 1 in a neighborhood of 1 so that $\bar{\partial}\chi$ vanishes to infinite order along the unit circle. Set $H^\# = H(z_1, \ldots, z_n)\chi(z_1) \cdots \chi(z_n)$. It is not hard to see that $\bar{\partial}H^\#$ is smooth on $\overline{U}^n$ (0 on $\mathbb{T}^n$) and that $\bar{\partial}H^\#$ and each of its derivative can be estimated (uniformly on U^n) in terms of the quantity $S_k(h)$. There exists α a continuous function (in fact smooth) on $\overline{U}^n$ so that $\bar{\partial}H^\# = \bar{\partial}\alpha$ and $|\alpha| \leq CS_k(h)$ ([8]). By the very definition of boundary value, $H^\# - \alpha$ is the Poisson extension of the function $H^b(e^{i\theta_1}, \ldots, e^{i\theta_n}) \chi(e^{i\theta_1}) \cdots \chi(e^{i\theta_n}) - \alpha(e^{i\theta_1}, \ldots, e^{i\theta_n})$. On U^n: $|H^\# - \alpha| \leq \tilde{\varphi}$ where $\tilde{\varphi}$ is the Poisson extension of the function φ defined on $\mathbb{T}^n$ by:

$$\varphi(e^{i\theta_1}, \ldots, e^{i\theta_n}) = |H^b(\ldots)\chi(e^{i\theta_1}) \cdots \chi(e^{i\theta_n}) - \alpha(\quad)|.$$

We can apply the theorem. For $p > 1$, this gives the Proposition 1' which $M(h^b)$ given by the strong maximal function $\overline{\mathcal{M}}$ applied to $|h^b|$ (pulled back to $\mathbb{T}^n$).

The "error terms" are absorbed in S_k. As usual the case $p = 1$ deserves a special argument (due to the fact that the Hardy Littlewood function is not bounded on L^1). The function $H^\# - \alpha$ is holomorphic in U^n. And the estimate $|H^\# - \alpha| \leq \tilde{\varphi}$ shows that the restrictions of $H^\# - \alpha$ to the tori $r\mathbb{T}^n$ ($0 < r < 1$) are uniformly bounded in L^1 norm. From this point we propose two solutions to the reader.

i) With no desire of giving a self-contained proof, one is now in a position to apply well known (AND stated) results on the (classical) space $H^1(U^n)$. Denote by $(H^\# - \alpha)_r$ the function on $\mathbb{T}^n$ defined by $(H^\# - \alpha)_r(z) = (H^\# - \alpha)rz$. Then $(H^\# - \alpha)_r$ has a limit in L^1 norm as $r \to 1$ (see e.g. Theorem 3.4.3 in [13]), this limit has to be the weak limit $H^b\chi - \alpha$. Therefore $|(H^\# - \alpha)_r|^{1/2}$ converges in L^2 norm, as r tends to 1, to $\varphi^{1/2}$, with, as above, $\varphi(e^{i\theta_1}, \dots, e^{i\theta_n}) = |H^b(\ )\chi(e^{i\theta_1}) \cdots \chi(e^{i\theta_n}) - \alpha(\ )|$. By n subharmonicity $|H^\# - \alpha|^{1/2}$ is bounded from above by the Poisson extension of $\varphi^{1/2}$. This leads to proposition 1 for $p = 1$ with $\overline{\mathcal{M}}(|h^b|)$ replaced by $(\overline{\mathcal{M}}|h^b|^{1/2})^2$.

ii) For the reader who wishes a more self-contained proof, we outline the following alternate solution. We have that $|H^\# - \alpha|^{1/2}$ is uniformly bounded in L^2 norm on the tori $r\mathbb{T}^n$. By weak compactness and n subharmonicity, there exists $\psi \in L^2(\mathbb{T}^n)$ so that $|H^\# - \alpha|^{1/2} \leq \tilde{\psi}$ (Poisson extension). We then get Proposition 1' with $M(h^b)$ replaced by $(\overline{\mathcal{M}}(\psi))^2$. This is enough to show that the nontangential maximal function is in L^1, with norm bounded in terms of $S_k(h)$ and the L^1 norm of h^b. This is all that is needed in order to prove Proposition 2, to be stated shortly. Once Proposition 2 will be proved we are back to the situation considered above in i).

NOTE. We have presented the proof in such a way that the reader, who wishes to do so, can see that we are only using the fact that h has a boundary value which is a measure (not *a priori* a L^1 function) to show that the nontangential maximal function of h is in fact in L^1, from which it follows easily that the boundary value must be in L^1.

I.2.4. The Totally Real case (E generic, $\dim E = n$).

This case can be reduced to the straight case by an almost analytic change of variables.

Consider (with the notation of Proposition 1) τ a local diffeomorphism of $\mathbb{C}^n$ defined in a neighborhood of 0, so that $\tau(0) = 0$, $\tau(\mathbb{R}^n) \subset E$ and $\overline{\partial}\tau$ vanishes to infinite order along $\mathbb{R}^n$. The function $h \circ \tau$ is defined on a "wedge-like" region $\tau^{-1}(W)$ with edge $\mathbb{R}^n$, but is not holomorphic. However if W_1 is a strictly smaller wedge, corresponding to a cone Γ_1 smaller than Γ, then on $\tau^{-1}(W_1)$ $|\overline{\partial}(h \circ \tau)| \leq C_k S_k(h)$ where C_k depends only on the "geometry" (W, W_1) and k but not on h. And similar estimates hold for any derivative of $\overline{\partial}(h \circ \tau)$, with different constants C_k. Indeed, when applying chain rule, on one hand the derivatives of h have polynomial growth by Cauchy estimate, on the other hand $\overline{\partial}\tau$ vanishes to infinite order.

We can assume that the tangent space to E at 0 is $\mathbb{R}^n$ and $d\tau(0)$ is the identity. Let Γ' be an open convex cone with vertex at 0 so that $\overline{\Gamma}_0 \subset \Gamma' \cup \{0\}$, $\overline{\Gamma}' \subset \Gamma_1 \cup \{0\}$. Then for some $\epsilon > 0$, $\tau^{-1}(W_1)$ contains the intersection of the ball of radius ϵ with the wedge $\mathbb{R}^n + \Gamma'$. On this intersection we can solve the equation $\overline{\partial} u = \overline{\partial}(h \circ \tau)$ with bounds $|u| \leq C_k' S_k(h)$, e.g. by applying the results in [8] for solving $\overline{\partial}$ on convex sets (We also refer the reader who prefers integral kernels or who is interested in $\mathcal{C}^k$ results to [5],[6],[7]). Finally we can apply Proposition 1' to $h \circ \tau - u$, introducing thus another "error term" coming from u and bounded in terms of any of the quantities $S_k(h)$, $(k \in \mathbb{N})$.

I.2.5. The Generic Case.

Now the edge E is a smooth generic manifold in $\mathbb{C}^n$. We can assume that the tangent space to E at 0 is $\mathbb{C}^d \times \mathbb{R}^{n-d}$ (d the CR dimension of E, $n - d$ the real codimension). Let Φ be the map from $\mathbb{C}^n$ to $\mathbb{C}^{n+d}$ defined by $\Phi(z_1, \ldots, z_n) = (\zeta_1, \ldots, \zeta_{n+d})$, with

$$\begin{cases} \zeta_j = z_j & \text{if } 1 \leq j \leq n \\ \zeta_{j+n} = \overline{z}_j & \text{if } 1 \leq j \leq d. \end{cases}$$

Set $\tilde{E} = \Phi(E)$, this is a smooth totally real manifold of dimension $n + d$ in $\mathbb{C}^{n+d}$ (therefore generic). It will be convenient to denote points in $\mathbb{C}^{n+d}$ by $\tilde{z} = (z, z')$ with $z \in \mathbb{C}^n$, $z' \in \mathbb{C}^d$. And we denote by π the projection of $\mathbb{C}^{n+d}$ on $\mathbb{C}^n$, $\tilde{z} \mapsto z = \pi(\tilde{z})$. Fix, totally arbitrarily, a cone Ω in $\mathbb{C}^d$. Let $\tilde{W} = \{\tilde{z} \in \mathbb{C}^{n+d}, \tilde{z} = \tilde{e} + (\gamma, \gamma'), \tilde{e} \in \tilde{E}, \gamma \in \Gamma, \gamma' \in \Omega, |\tilde{e}| \leq \delta, |\gamma|^2 + |\gamma'|^2 \leq \delta^2\}$. Obviously $\pi(\tilde{W}) \subset W$. If h is any function holomorphic on W, h can be lifted to $\tilde{h}$ a holomorphic function on $\tilde{W}$, $\tilde{h} = h \circ \pi$, independent of $z_{n+1}, \ldots, z_{n+d}$; etc. .... . By applying Proposition 1 to the totally real edge $\tilde{E}$ in $\mathbb{C}^{n+d}$, one gets Proposition 1 for the generic edge E in $\mathbb{C}^n$.

There is no claim that this trick of adding variables is original. On the contrary it has been used so often that we gave up any attempt of research of the origin (let us just mention [11] as a recent example).

1.2.6. Remark.

The proof of Proposition 1 is entirely based on the control of the radial behavior of a function on the polydisk by means of the strong maximal function. This strong maximal function is naturally tied with the product structure of the polydisk. It has already lost any natural character when transported to a totally real edge by an almost analytic change of variable.

And it looks even more artificial when transported to the generic edge by addition of variables. This is the justification for part II.

I.2.7 Nontangential Limits, L^p Convergence ($1 \leq p < +\infty$).

Same notations as in I.2.2 Proposition 1. There is only a small gap in between proving estimates for nontangential maximal functions and proving existence of nontangential limits almost everywhere (i.e. for almost all $e \in E \cap \overline{W}$ the limit of $f(e + \gamma)$ exists as $\gamma \to 0$, $\gamma \in \Gamma_0$). One only needs to show that any function h (holomorphic in the wedge, with L^1 boundary value) can be approximated, on the intersection of W (possibly shrinked) with some neighborhood of 0, by a sequence of holomorphic functions h_j continuous up to the edge. Then one controls the oscillation of h by the maximal function of $|h - h_j|$.

Such an argument, by approximation, is very standard. In textbooks, one finds it to show the existence of Lebesgue points in differentiation theory (see [12] Theorem 7.7), then the existence of limits a.e for classes of harmonic functions is easily derived (ibid. 11.23). Here, as in [15] page 38-9, we have to follow a very similar scheme, but staying in the setting of holomorphic functions.

To be precise, we need an approximation in the following sense. In some neighborhood G of 0 we want the following: on $G \cap E$, h_j tends to h^b in L^1 norm, and for some integer k

$$\sup_{G \cap W} \left| (h(z) - h_j(z)) \text{dist}(z, E)^k \right| \longrightarrow_{j \to \infty} 0 \, .$$

This is easy to do in the case of the straight edge $E = \mathbb{R}^n$ by convolution, in the real direction, with an approximate identity kernel.

$$\left(h_j(x + iy) = \int h(x - x' + iy) \psi_j(x') dx' \dots, \ \psi_j \text{ the kernel} \right) \, .$$

One takes $k > k'$, where k' is so that $S_{k'} < +\infty$, in order to have $h(z) \text{dist}(z, E)^k$ $\to 0$ as z approaches E. By the same arguments as developed earlier, it is enough to prove the existence of nontangential limits in the straight edge case. The details will not be repeated. The existence of nontangential limits almost everywhere is thus established.

Let us however notice that the approximation above (of h by h_j) can be done in the totally real case (possibly even the generic case?) by use of the Baouendi Treves approximation formula ([2]). Notice also that, as already pointed out, we could not just decompose h^b into an arbitrary sum of a continuous function and a small L^1 function (as done in [13] page 30), because having used a less sophisticated version of the estimate in [13] we restricted ourself to holomorphic functions.

Finally notice that the nontangential convergence almost everywhere is a dominated convergence in the sense of Lebesgue (dominated by the nontangential maximal functions) due to Proposition 1. Let us state the following useful consequence:

PROPOSITION 2. *Let h be as in Proposition 1. Fix $\gamma_0 \in \Gamma$. For $t > 0$, t small enough let h_t be the function defined on E_0 by $h_t(e) = h(e + t\gamma_0)$. If $h^b \in L^p(E)$ $(1 \le p < +\infty)$, then h_t converges to h^b on E_0 in L^p norm.*

This, of course, corresponds to a more classical definition of H^p. Similar results can be found in [4] (Section 6.5).

I.3 Admissible Limits.

Although a better definition in the spirit of the definition of approach region page 32 of [15] would be possible, we choose to be a little bit more restrictive in the hope of making the proof easier to read.

We consider the case of a smooth generic edge E of dimension $n+d$ ($0 \leq d < n$) in $\mathbb{C}^n$, and a wedge given by a cone Γ (and some δ) as in I.2.1. We assume $0 \in E$ and that $\mathbb{C}^d \times \mathbb{R}^{n-d}$ is the tangent space to E at 0. We fix a unit vector ν inside the cone Γ, transverse to E at 0.

I.3.1. Approach Region at 0.

Denote by π_1, and π_2 respectively the projections of $\mathbb{C}^n$ ($= \mathbb{C}^d \times \mathbb{C}^{n-d}$) onto $\mathbb{C}^d$ and $\mathbb{C}^{n-d}$, i.e. the complex tangent to E at 0 and its orthogonal complement. If $q \in \mathbb{C}^n$ and $s > 0$ we will set

$$P(q,s) = \{q' \in \mathbb{C}^n, \pi_1(q) - \pi_1(q') \in \sqrt{s}U^d, \pi_2(q) - \pi_2(q') \in sU^{n-d}\}.$$

That is $P(q,s)$ is a polydisk centered at p with polyradii $\sqrt{s}$ in the first d components and s in the last $n-d$ components.

There exist $t_0 > 0$ and a constant A (related to the aperture of the cone and the "curvature" of the edge ...) so that for all $t \in (0, t_0)$ the polydisk $P(t\nu, 2At)$ is included in the wedge W. Fix A.

We will consider the approach region at 0 to be the set $\mathcal{A}_0$ defined by:

$$\mathcal{A}_0 = \bigcup_{0 < t < t_0} P(t\nu, At).$$

Although as pointed out earlier this is not optimal, we keep the essential *non-isotropic* character (the square root in the complex directions).

I.3.2. Approach Regions at Other Points.

They are obtained similarly replacing the decomposition of $\mathbb{C}^n$ into $\mathbb{C}^d \times \mathbb{C}^{n-d}$ by the decomposition of $\mathbb{C}^n$ into the product of the complex tangent at a point $e \in E$ and its orthogonal. The approach region thus obtained will be denoted by $\mathcal{A}_e$. One may have to restrict t_0 and A considered in I.3.1 for the construction to be valid not only at 0 but in some neighborhood of 0 in E that we denote by E_0.

I.3.3. The Nonisotropic Maximal Function on the Edge E.

If φ is a nonnegative function on E, we denote by $\mathcal{M}_2(\varphi)$ the maximal function related to a family of "balls" with dimension $\sqrt{d}$ in the complex tangential directions and d in the other directions. See [15].

If $\varphi \in L^p(E)$, $p > 1$, then $\mathcal{M}_2(\varphi) \in L^p(E)$. If $\varphi \in L^1(E)$ then $\mathcal{M}_2(\varphi)$ is in weak L^1.

I.3.4. Limits.

With the notations as above we have:

PROPOSITION 3. *If h is a holomorphic function in the wedge W in the Hardy class H^1 (see I.2.2) then at almost every point $e \in E_0$ the limit of $h(z)$ as z approaches e within the approach region $\mathcal{A}_e$ exists.*

The proof will follow [15] pages 32-37.

PROOF. Fix k so that $h(z)\mathrm{dist}(z, E)^k$ (is not only bounded, but) tends to 0 as z approaches E. Then we can write h as a sum $\sum_{j=0}^{\infty} h_j$, where:

$$\begin{cases} \text{each } h_j \text{ is smooth on the closure of the wedge } W \\ \qquad \text{(possibly shrinked)} \\ \\ \sum_{j=0}^{\infty} 2^j S_k(h_j) < +\infty \\ \\ \sum_j 2^j \|h_j\|_{L^1(E)} < +\infty . \end{cases}$$

It is enough indeed to use translates of h. For $t > 0$ set $h_t(z) = h(z - t\nu)$ (ν is a fixed vector in Γ transverse to E). One has $S_k(h - h_t) \to 0$ as $t \to 0$ and $h_t - h\big|_E$ tends to L^1 in $L^1(E)$ (Proposition 2). Take (t_k) a sequence converging fast enough to 0 $t_0 > t_1 > t_2 \ldots$, one can write

$$h = h_{t_0} + (h_{t_1} - h_{t_0}) + (h_{t_2} - h_{t_1}) + \cdots .$$

One takes $h_0 = h_{t_0}$, $h_j = h_{t_j} - h_{t_{j-1}}$ $j \geq 1$. For $e \in E$ and $\ell \in \mathbb{N}$ set

$$\mathcal{O}_\ell(e) = \sum_{j=\ell}^{+\infty} \sup_{\mathcal{A}_e} |h_j| .$$

It is enough to prove that for almost every $e \in E$, $\mathcal{O}_\ell(e)$ tends to 0 as ℓ tends to $+\infty$. This can be achieved by proving that for any $\epsilon > 0$, the set of e in E so that $\sup_{\mathcal{A}_e} |h_j| > \dfrac{\epsilon}{2^j}$ has measure at most

$$\frac{c}{\epsilon}\left(2^j \|h_j\|_{L^1(E)} + 2^j S_k(h_j)\right) .$$

The crucial fact is that, with the notations of I.3.1, the value of a holomorphic function u on a polydisk $P(t\nu, At)$ is bounded by the average of its modulus on the bigger polydisk $P(t\nu, \frac{3}{2}At)$, multiplied by some fixed constant.

If q is a point in the polydisk $P(t\nu, \frac{3}{2}At)$ one can bound $|u(q)|$ by $|u|^*(\tilde{q})$ where $*$ denotes an appropriate nontangential maximal function and say $\tilde{q}$ is the point nearest to q in E. Then the value of u on the smaller polydisk can be bounded by the average of $|u|^*$ on the "projection" of $P(t\nu, \frac{3}{2}At)$ on E (multiplied by some constant). This leads to an inequality:

$$\sup_{\mathcal{A}_e} |u| \leq C\mathcal{M}_2(|u|^*)(e) \qquad (e \in E) .$$

One applies this inequality to each h_j.

Finally, we get that, as desired, the set of $e \in E$ so that

$$\sup_{\mathcal{A}_e} |h_j| > \frac{\epsilon}{2^j}$$

has measures less than

$$\frac{C}{\epsilon} \left(2^j \|h_j\|_{L^1(E)} + 2^j S_k(h_j) \right) ,$$

using Proposition 1 and the fact that $\mathcal{M}_2$ maps L^1 to weak L^1. As in [15] the function h has been controlled in $\mathcal{A}_e$ by the composition of two maximal functions: "the" nontangential one and the nonisotropic one on the edge, with additional "errors terms" absorbed in $S_k(\cdot)$. Even for H^∞ functions, the proof uses H^p with $p < +\infty$.

II. The Hardy Littlewood maximal function.

II.1.

Motivations for this part of the paper have been explained in I.2.6. If E is a generic smooth submanifold of $\mathbb{C}^n$, $0 \in E$. Locally, near 0, we can identify E with its tangent space at 0, that we can assume to be $\mathbb{C}^d \times \mathbb{R}^{n-d}$. We call Hardy-Littlewood maximal function on E the operator (defined on functions defined in a fixed neighborhood of 0 in E – extended by 0) corresponding to the classical Hardy-Littlewood maximal function on $\mathbb{C}^d \times \mathbb{R}^{n-d}$ ($\simeq \mathbb{R}^{n+d}$), defined by the supremum of the average on Euclidean balls. (It does not matter that there is some arbitrary character in this definition). We state rather informally the result of this section.

PROPOSITION 4. *In the inequality*

$$|h|^*(e) \le C(M_p(h^b)(e) + S_k(h))$$

in Proposition 1, one can replace $M_p(h^b)(e)$ by the value at e of the Hardy-Littlewood maximal function of the function $|h^b|$. (for $p = 1$ this gives an upper bound not in L^1 but only in weak L^1).

As in I the essential work can be done in the setting of the straight edge $\mathbb{R}^n$, the reduction to this case is similar to what has been done in detail in I. And we will just expose in detail the case of the straight edge. In the hope of making this paper more readable we will set up the notations again.

II.2. Integral Kernel H (for bounded holomorphic functions).

Let P^n be the standard wedge: $P^n = \{z \in \mathbb{C}^n ; \mathcal{I}m z_j > 0\}$. From now on, we will fix an arbitrary (slightly) larger wedge P^n_+ defined by $P^n_+ = \{z \in \mathbb{C}^n ; \mathcal{I}m z \in \Lambda\}$, where Λ is a conic neighborhood of $\overline{(\mathbb{R}^+)^n} - \{0\}$ in $\mathbb{R}^n$. If h is a bounded holomorphic function defined in P^n, h has a boundary value on $\mathbb{R}^n$ which is a L^∞ function that we denote by h^b. Poisson's representation formula is

$$h(z) = \int_{\mathbb{R}^n} h^b(t) K(z,t) dt$$

with $z = (x_1 + iy_1, \ldots, x_n + iy_n)$, $y_j > 0$, and

$$K(z,t) = \frac{1}{\pi^n} \prod_{j=1}^{n} \frac{y_j}{(x_j - t_j)^2 + y_j^2} \, .$$

Let us look in detail at what happens at the point $(i, i, \ldots, i)$. We set $a = (i, i, \ldots, i)$.

$$K(a,t) = \frac{1}{\pi^n} \prod_{j=1}^{n} \left(\frac{1}{t_j^2 + 1} \right) \, .$$

This kernel has a nonisotropic behavior: if $t = (t_1, 0, \ldots, 0)$ then $K(a,t) = \frac{1}{\pi^n} \frac{1}{1 + |t|^2}$, while if $t_1 = \cdots = t_n$ then $K(a,t) = \mathcal{O}\left(\frac{1}{(1+|t|^2)^n} \right)$. We want to replace this kernel by a more isotropic kernel. But this will be possible only for functions holomorphic and bounded (and not only n-harmonic) on the larger wedge P_+^n. Here is the basic trick: Let R be a linear transformation of $\mathbb{R}^n$ (acting also an $\mathbb{C}^n$) with determinant 1, fixing the point $(1, \ldots, 1)$, and close enough to the identity so that $R(P^n) \subset P_+^n$. We have:

$$h(a) = h(R^{-1}(a)) = \int_{\mathbb{R}^n} h^b \circ R^{-1}(t) K(a,t) dt$$

$$= \int_{\mathbb{R}^n} h^b(t) K(a, R(t)) dt \, .$$

The interesting fact is that the kernel does not have the same directions of weak decay. Now we define a kernel H. Let G be the group of invertible linear transformations of $\mathbb{R}^n$, which fix the point $(1, 1, \ldots, 1)$ and with determinant 1. Let Ω be a compact neighborhood of the identity in G small enough so that for any $R \in \Omega$, $R(P^n) \subset P_+^n$. Let μ be a smooth measure of total mass 1 on Ω (e.g. the normalized restriction of Haar measure).

Let

$$H(a,t) = \int_\Omega K(a, R(t)) d\mu(R) \, .$$

We trivially have.

PROPOSITION 4. *If h is a bounded holomorphic function on P_+^n, then $h(a) = \int_{\mathbb{R}^n} h^b(t) H(a,t) dt$.*

II.3 Estimate of H.

PROPOSITION 5. *As $|t| \to +\infty$, $(t \in \mathbb{R}^n)$,*

$$H(a,t) = \mathcal{O}(|t|^{-n-1}) \, .$$

PROOF. Of course in some directions (in particular in the direction of $(1, 1, \ldots, 1)$) we have a better decay in $\mathcal{O}(|t|^{-2n})$. Let S_{n-1} be the unit sphere in $\mathbb{R}^n$, let F be a neighborhood of the points $\pm\left(\frac{1}{\sqrt{n}}, \ldots, \frac{1}{\sqrt{n}} \right)$. There exists a constant C so that for every $t \in \mathbb{R}^n - \{0\}$ and any $E \subset S_{n-1} - F$ the set $\Omega(t, E) = \left\{ R \in \Omega, \frac{R(t)}{|R(t)|} \in E \right\}$ satisfies $\mu(\Omega(t, E)) \leq C|E|$, where $|E|$ denotes the area measure of E. This is due to the fact that the map $R \rightarrow R(t)$ has maximum rank. This leads to an estimate, where $r = |t|$:

$$H(a,t) \leq C \int_{S^{n-1}} \frac{d\sigma(\xi)}{(1 + r^2 \xi_1^2) \cdots (1 + r^2 \xi_n^2)} \, .$$

This integral is of the same size as

$$I = \int_{1 < |\xi| < 2} \frac{d\xi_1 \cdots d\xi_n}{(1 + r^2 \xi_1^2) \cdots (1 + r^2 \xi_n^2)} \, .$$

If $|\xi| > 1$ then at least one component of ξ is larger than $\frac{1}{\sqrt{n}}$. This gives (by decomposing $\{|\xi| > 1\}$ into n regions):

$$I \leq n \frac{1}{1 + \frac{r^2}{n^2}} \int_{\mathbb{R}^{n-1}} \frac{d\xi_1 \cdots d\xi_{n-1}}{(1 + r^2 \xi_1^2) \cdots (1 + r^2 \xi_{n-1}^2)} \, ,$$

so

$$I \leq \frac{n^3}{n^2 + r^2} \frac{\pi^{n-1}}{r^{n-1}} = \mathcal{O}(r^{-n-1})$$

as desired.

II.4 The "radial" maximal function and Hardy-Littlewood maximal function.

DEFINITION. *If g is a nonnegative function defined on P^n, for every $x \in \mathbb{R}^n$ we set*

$$\hat{g}(x) = \sup_{\lambda > 0} g(x + \lambda(i, \ldots, i)) \, .$$

$\hat{g}$ is the radial maximal function.

If φ is a nonnegative locally measurable function on $\mathbb{R}^n$, for every $x \in \mathbb{R}^n$ we set

$$\mathcal{M}_0 \varphi(x) = \sup_{r > 0} \frac{1}{r^n} \int_{|y - x| < r} \varphi(y) dy \, .$$

This is the Hardy-Littlewood maximal function. For $\rho > 0$ we set

$$\mathcal{M}_0^\rho \varphi(x) = \sup_{r > \rho} \frac{1}{r^n} \int_{|y - x| < r} \varphi(y) dy \, .$$

PROPOSITION 6. *There exists a constant $C > 0$, so that for every bounded holomorphic function h on P_+^n (with boundary value $h^b \in L^\infty(\mathbb{R}^n)$), for every $x \in \mathbb{R}^n$:*

$$\widehat{|h|}(x) \leq C \mathcal{M}_0(|h^b|)(x) \, .$$

More precisely

$$|h(x + \lambda(i, i, \ldots, i))| \leq C \mathcal{M}_0^\lambda(|h^b|)(x) \, .$$

PROOF. By translation it is enough to prove the proposition for $x = 0$. By rescaling it is enough to prove

$$|h(a)| \leq C \mathcal{M}_0^1(|h^b|)(0) \, .$$

By Propositions 4 and 5 this reduces to the simple Lemma:

LEMMA. *There exists a constant $A > 0$ so that for every nonnegative $\varphi \in L^\infty(\mathbb{R}^n)$*

$$\int_{\mathbb{R}^n} \frac{\varphi(x)}{1 + |x|^{n+1}} dx \leq A \mathcal{M}_0^1 \varphi(0) .$$

A well-known and easy fact! For a precise reference see [14], pages 62-63 (Theorem 2(a), just take $\epsilon = 1$), and take advantage of the fact that the kernel is bounded.

Remark. It is not so easy to describe what we have done in the setting of the polydisk. Starting with the measure $\frac{d\theta}{2\pi} \cdots \frac{d\theta_n}{2\pi}$ to represent the origin (and then use conformal mappings) is wrong! The homogeneity together with isotropy is easier to use in the setting of P^n.

Also the larger cone P_+^n is very natural.

II.5 Nontangential Limits.

There is no need for repeating the details (replace a by some other vector ...) we just state the result:

PROPOSITION 6'. *Let Γ_0 be an open cone in $\mathbb{C}^n$ with vertex at 0 so that $\overline{\Gamma}_0 \subset P^n \cup \{0\}$. There exists $C_1 > 0$ so that for every bounded holomorphic function h on P_+^n and for every $x \in \mathbb{R}^n$*

$$\sup_{z - x \in \Gamma_0} |h(z)| \leq C_1 \mathcal{M}_0 |h^b|(x) .$$

more precisely if $\gamma \in \Gamma_0$

$$|h(z + \gamma)| \leq C_1 \mathcal{M}_0^{|\gamma|} |h^b|(x) .$$

II.6. The End!

We now state the special case of Proposition 4, for the straight edge $\mathbb{R}^n$ (from which, as we said earlier, Proposition 4 follows following the same arguments as in I).

PROPOSITION 7. *Let Γ_0 be a cone in $\mathbb{C}^n$ with vertex as 0 so that $\overline{\Gamma}_0 \subset P^n \cup \{0\}$. Let h be a holomorphic function on $P_+^n \cap \{|z| < 3\}$. Assume that*

   i) h has polynomial growth: $h(z) = \mathcal{O}(|\mathcal{I}mz|^{-k})$

   ii) h has a boundary value along $\mathbb{R}^n$ which belongs to L^1 (in the ball of radius 3), denoted by h^b.

Then for every $x \in \mathbb{R}^n$ so that $|x| \leq 1$ and $\gamma \in \Gamma_0$, $|\gamma| \leq 1$ we have:

$$|h(x + \gamma)| \leq C(\mathcal{M}_0^{|\gamma|}(|h^b|)(x) + S_k(h))$$
$$\leq C(\mathcal{M}_0(|h^b|)(x) + S_k(h)) ,$$

where C is a positive constant independent of h and $S_k(h) = \sup |h(z)(\mathcal{I}mz)^{-k}|$. (For defining $\mathcal{M}_0(|h^b|)$ extend h^b by 0 off the ball of radius 3.)

PROOF. If we do not intend to give a proof independent of the remarks made in I, we can restrict ourself to the case when h is smooth up to the boundary (use translation and homotheties to approximate). This argument of approximation is legitimate because, in computing $\mathcal{M}_0^{|\gamma|}$, one has only to use balls of radius bounded away from 0. Let $\chi \in \mathcal{C}_0^\infty(\mathbb{C}^n)$ a cut off function $\chi \equiv 1$ on the ball or radius 2, and with support in the ball of radius 3. Take χ so that $\overline{\partial}\chi$ vanishes to infinite order along $\mathbb{R}^n$. Then $\overline{\partial}(h\chi)$ and all its derivating can be uniformly bounded, in terms of the quantity $S_k(h)$. Shrinking P_+^n if necessary we can assume that P_+^n is the image of P^n under the action of a linear map. To solve the equation $\overline{\partial}\alpha = \overline{\partial}(h\chi)$ with bounds one can therefore go back to the setting of the polydisk. The conclusion is that there exists α (smooth on $\overline{P}_+^n$) so that $|\alpha| \leq CS_k(h)$ and $(\chi h - \alpha)$ is a bounded holomorphic function on P_n^+. Proposition 6' applied to the function $(\chi h - \alpha)$ yields Proposition 7.

References.

1. M. S. Baouendi, C. H. Chang, F. Treves, *Microlocal hypoanalyticity and extension of CR functions*, J. Diff. Geom. **18** (1983), 331–391.

2. M. S. Baouendi, F. Treves, *A property of the functions and distributions annihilated by a locally integrable system of complex vector fields*, Ann. of Math **113** (1981), 387–421.

3. A. Boggess, A. Nagel, Proc. of the Fourier Analysis Conference in honor of E. M. Stein on his 60th birthday, Princeton.

4. R. Carmichael, D. Mitrovic, *Distributions and analytic functions*, Pitnam Research notes in Math., vol. 206, 1989.

5. J. Chaumat, A. M. Chollet, *Noyaux pour résoudre l' équation $\overline{\partial}u = v$ dans des classes indéfiniment différentiables*, C.R.A.S **306** (1988), 585-588.

6. J. Chaumat, A. M. Chollet, *Noyaux pour résoudre l' équation $\overline{\partial}$ dans des classes ultradifférentiables sur des compacts irréguliers de C^n "*, Proceedings of the special year on several complex variables at the Institute Mittag Leffler (to appear).

7. J. Chaumat, A. M Chollet, *Estimations Höldériennes pour les équations de Cauchy-Riemann dans les convexes compacts de C^n "*, Math. Zeit. (to appear).

8. A. Dufresnoy, *Sur l'operateur d'' et les fonctions différentiables au sens de Whitney*, Ann. Inst. Fourier **29** (1979), 229–238.

9. F. Forstnerič,, *Admissible boundary values of bounded holomorphic functions in wedges*, Preprint.

10. B. Jessen, J. Marcinkiewicz, A. Zygmund, *Note on the differentiability of multiple integrals*, Fund.Math. **25** (1935), 217–34.

11. M. Marson, *Wedge extendability for hypo-analytic structures*, Preprint.

12. W. Rudin, *Real and Complex Analysis*, 3rd edition, McGraw-Hill, 1987.

13. W. Rudin, *Function Theory in polydisks*, Benjamin, 1969.

14. E. M. Stein, *Singular integrals and differentiability properties of functions*, Princeton University Press, 1970.

15. E. M. Stein, *Boundary behavior of holomorphic functions of several complex variables*, Math. Notes, Princeton University Press, 1972.

16. E. M. Stein, G. Weiss, *Introduction to Fourier Analysis on Euclidean space*, Princeton University Press, 1971.

17. A. Torchinsky, *Real variable methods in harmonic analysis*, Academic Press, 1986.

18. A. Zygmund, *Trigonometric Series*, 2nd edition, Cambridge University Press, 1959.

DEPARTMENT OF MATHEMATICS, UNIVERSITY OF WISCONSIN, MADISON WI 53706

Contemporary Mathematics
Volume **137**, 1992

Local and Semi-Global Existence Theorems for $\bar{\partial}_b$ on CR Manifolds

MEI-CHI SHAW

In this paper we discuss the recent development of the local and semi-global existence theorems for the tangential Cauchy-Riemann complex on a CR manifold in $\mathbb{C}^n$. Let M be a hypersurface in $\mathbb{C}^n$, $n > 2$. We consider the solvability of the tangential Cauchy-Riemann equations

$$(1) \qquad \bar{\partial}_b u = \alpha$$

in an open subset ω of M, where α is a (p, q) form on ω, $1 \leq p \leq n$ and $0 \leq q \leq n - 1$. Since $\bar{\partial}_b$ is a complex, we must assume that α satisfies the compatibility condition

$$(2) \qquad \bar{\partial}_b \alpha = 0 \qquad \text{in} \quad \omega$$

when $1 \leq q < n - 1$. When $q = n - 1$, it is related to the Lewy's nonsolvable example. When $1 \leq q < n - 1$ and M is strongly pseudo-convex, the local solvability of $\bar{\partial}_b$ has been studied by numerous people. In particular, we shall mention the work of Andreotti-Hill [1], Henkin [12], Treves [25], and Rosay [19] and Boggess-Shaw [5], Shaw [21,22] and Webster[26]. Recently the local and semi-global existence theorems of $\bar{\partial}_b$ on pseudo-convex CR manifolds have been obtained in Shaw [23,24]. By local existence we mean that given a (p, q) form α satisfying (2) in ω, can one find a solution u satisfying Equation (1) in an open set $\omega' \subset \omega$. While semi-global existence means the solution u exists on ω without shrinking. Naturally the semi-global existence results will depend on the shape of the boundary of ω. We are also interested in the regularity of the solutions.

1991 *Mathematics Subject Classification*. Primary 32F40, 35N10.
Partially supported by NSF grant DMS 91-01161.
The final (detailed) version of this paper will be submitted for publication elsewhere.

The solvability results for the case when $1 \leq q < n - 2$ and the case when $q = n - 2$ are quite different. In the first case (see Theorems 1 and 1'), we present the existence theorems on ω for any α satisfies (2) when M is of finite type (in the sense of D'Angelo). Theorem 2 gives the local regularity of the solutions and the closed range property of $\bar{\partial}_b$ in the Frechet spaces. We also show that near a point of finite 1-type, one can construct a neighborhood basis such that its boundary is holomorphically flat using the work of Catlin [6,7]. This gives the local solvability results near points of finite type as well. In the case when $q = n - 2$, condition (2) is no longer sufficient for Eq.(1) to be solvable in general. Another compatibility condition arises in this case. This new condition together with condition (2) are the necessary and sufficient conditions for Eq.(1) to be solvable when $q = n - 2$ (see Theorems 3 and 3'). The author would like to thank Professor Catlin for many discussions concerning the proof of Theorem 2'.

1. Solvability for $\bar{\partial}_b$ when $1 \leq q < n - 2$

Let M be the boundary of a weakly pseudo-convex domain D in $\mathbb{C}^n, n \geq 2$ and ρ be a defining function for D, i.e., $D = \{z\epsilon\ \mathbb{C}^n | \rho(z) < 0\}$ and $M = \{z\epsilon\ \mathbb{C}^n | \rho(z) = 0\}, |d\rho| \neq 0$ on M. We shall assume M is of finite type in the sense of D'Angelo. A hypersurface M_o is called flat if it is defined by a pluriharmonic function $r(z), M_o = \{r(z) = 0\}$ where r is pluriharmonic in a neighborhood of M_o and $dr \neq 0$ on M_o. M_o is called Levi-flat if the Levi form vanishes completely on M_o. When M is a compact weakly pseudo-convex CR manifolds, the global solvability has been studied by many authors (see Rosay [18], Shaw [20], Boas-Shaw [4], and Kohn [16]). In particular, it is established in [20,4,16] that $\bar{\partial}_b$ is solvable in the L^2 space on any compact weakly pseudo-convex CR manifolds under compatibility conditions. This implies that the range of $\bar{\partial}_b$ is closed for all degrees including the top degree (i.e., when $q = n - 1$). The global solvability results will yield local solvability for $\bar{\partial}_b$ on ω for any $\bar{\partial}_b$-closed (p, q) form with compact support in ω when $1 \leq q < n - 1$. However, they do not give any local solvability results for $\bar{\partial}_b$-closed forms which are not compactly supported. We first state the local solvability results when $1 \leq q < n - 2$.

THEOREM 1. *Let $D \subset \mathbb{C}^n$ be a bounded pseudo-convex domain with smooth boundary M of finite type in the degree of $min(q, n - q - 1)$, $1 \leq q < n - 2$ and $n \geq 4$. Let $\omega \subset M$ be a connected CR manifold with smooth boundary $\partial\omega$ such that $\partial\omega$ is the transversal intersection of M with a simply connected Levi-flat hypersurface M_0. We assume that the set $\overline{M_0 \cap D} \equiv \overline{\omega}_0$ has a Stein neighborhood basis. Let $\omega' \subset\subset \omega$ be any relatively compact subset of ω. For any $\alpha \in C_{(p,q)}^{\infty}(\omega)$, where $0 \leq p \leq n$ and $1 \leq q < n - 2$, such that*

$$\bar{\partial}_b\alpha = 0 \quad in \quad \omega,$$

there exists a $u \in C_{(p,q-1)}^{\infty}(\omega')$ such that

$$\bar{\partial}_b u = \alpha \quad in \quad \omega'.$$

If we assume that ω can be exhausted by subsets whose boundaries lie in Levi-flat hypersurfaces, then we have the following semi-global existence results.

THEOREM 1'. *Let M and ω be the same as in Theorem 1. Furthermore we assume $\omega = \cup \omega_i$ such that $\omega_i \subset\subset \omega_{i+1} \subset\subset \omega$ and $\partial \omega_i$ lies in a Levi-flat hypersurface for each i. For any $\alpha \in C^\infty_{(p,q)}(\omega)$, where $0 \leq p \leq n$ and $1 \leq q < n - 2$, such that*

$$\bar\partial_b \alpha = 0 \quad in \quad \omega,$$

there exists a $u \in C^\infty_{(p,q-1)}(\omega)$ such that

$$\bar\partial_b u = \alpha \quad in \quad \omega.$$

COROLLARY 1.1. *Under the same assumption as in Theorem 1', the range of $\bar\partial_b$ is closed in the $C^\infty_{(p,q)}(\omega)$ space.*

COROLLARY 1.2. *If M_0 is simply connected and defined by a pluriharmonic function, then the assertions in Theorem 1' holds. In particular, if M_0 is a holomorphically flat, then the assertions in Theorem 1' hold*

In fact, we have the following more general result which gives precise information on the interior regularity of the solution. Let $W^s(\omega, loc)$ denote the Frechet space of functions which have locally Sobolev spaces W^s coefficients.

THEOREM 2. *Let ω be the same as in Theorem 1'. Let s be any nonnegative number. For any $\alpha \in W^s_{(p,q)}(\omega, loc)$, $1 \leq q < n - 2$, such that*

$$\bar\partial_b \alpha = 0 \quad in \quad \omega,$$

there exists a $u \in W^{s+\epsilon}_{(p,q-1)}(\omega, loc)$ for some $\epsilon > 0$ such that

$$\bar\partial_b u = \alpha \quad in \quad \omega,$$

where $0 < \epsilon \leq \frac{1}{2}$ is independent of α and is dependent of the finite type only.

COROLLARY 2.1. *Under the same assumption as in Theorem 2, the range of $\bar\partial_b$ is closed in the $L^2_{(p,q)}(\omega, loc)$ space under the Frechet norms.*

COROLLARY 2.2. *Under the same assumption as in Theorem 2, the $\bar\partial_b$ closed $C^\infty_{(p,q)}(\omega, loc)$ forms are dense in $\bar\partial_b$-closed $L^2_{(p,q)}(\omega, loc)$ forms in the Frechet norm. This is also true for $q = 0$. In particular, for any CR function $f \in L^2(\omega, loc)$, there exists a sequence of CR functions $f_n \in C^\infty(\omega, loc)$ such that $f_n \to f$ in the Frechet norm $L^2(\omega, loc)$.*

We note that the assumption that M is of finite type in the degree of $\min(q, n - q - 1)$ implies that the $\bar\partial$-Neumann problem is subelliptic in the degrees q and $n - q - 1$. The ϵ obtained in Theorem 2 is the minimum of the "gain" of the subellipticty for the $\bar\partial$-Neumann problem in the degrees q and $n - q - 1$. Theorems 1,1' and 2 can be applied to any real-analytic pseudo-convex boundary M since real analytic domains are domains of finite type (see [8,15]). When M

is strongly pseudo-convex, we can choose $\epsilon = \frac{1}{2}$ since M is of finite type 2 and the $\bar{\partial}$-Neumann problem is subelliptic with exponent $\frac{1}{2}$ in all degrees following the results of Kohn (see Folland-Kohn [11]). The gain of $\frac{1}{2}$ derivative in the regularity of the solution u in this case is best possible.

We note that Levi-flat hypersurfaces can not always be flattened, even locally. In fact, there are Levi-flat hypersurfaces which can not be flattened locally even from one side (see Bedford-De Bartolomeis [2]). Thus Theorem 1 can be applied to a much wider class of ω's. We also note that Bedford-Fornaess [3] has given an example of a weakly pseudo-convex domain with smooth boundary which does not have a Stein neighborhood basis (see also Diederich-Fornaess [10]. If we assume that z_0 is a point of finite 1- type, we can have the following local solvability results.

THEOREM 2'. *Let M be a smooth be a smooth pseudo-convex hypersurface in $\mathbb{C}^n$, $n \geq 3$ and $z_0 \in M$. Suppose z_0 is a point of finite 1- type, then there exists a local neighborhood basis $\{\omega_\epsilon\}_{\epsilon > 0}$ of z_0 for M such that $\partial\omega_\epsilon$ lies in a holomorphically flat hypersurface for every $\epsilon > 0$. Thus Theorem 1' and 2 hold on neighborhood ω_ϵ for every $\epsilon > 0$.*

The proof of Theorem 2' uses the work of Catlin [6,7].

2. Solvability for $\bar{\partial}_b$ when $q = n - 2$

When $q = n - 2$, it was observed by Rosay [19] that condition (2) is not sufficient for Eq.(1) to be solvable semi-globally even when ω lies in a strongly pseudo-convex boundary and satisfies the condition of Theorems 1 and 2. In this case there is another campatibility condition that α must satisfy if Eq.(1) is solvable. Without loss of generality, we may assume that $p = n$.

THEOREM 3. *Let M be the boundary of a smooth pseudo-convex domain in $\mathbb{C}^n$, $n \geq 3$ and M is of finite type. Let $\omega \subset M$ be a connected subset such that the boundary $\partial\omega$ is the transversal intersection of M with a simply connected Levi-flat hypersurface M_0 which has a Stein neighborhood basis. Let ω' be any relatively compact subset of ω. For any $\alpha \in C^\infty_{(n,n-2)}(\bar{\omega})$ such that $\bar{\partial}_b\alpha = 0$ in ω and α satisfies the following campatibility condition*

(A) $\displaystyle\int_{\partial\omega} \alpha \wedge g = 0$ *for all $g \in \mathcal{O}(\partial\omega)$.*

where $\mathcal{O}(\partial\omega)$ denote the space of functions which are defined and holomorphic in a neiborhood of $\partial\omega$, then there exists a $u \in C^\infty_{(n,n-3)}(\omega')$ such that $\bar{\partial}_b u = \alpha$ in ω'.

If one assumes that ω can be exhausted by subsets whose boundaries lie in Levi-flat hypersurface, then we have the following semi-global existence result similar to Theorem 1'.

THEOREM 3'. *Let M and ω be the same as in Theorem 1'. For any $\alpha \in C^\infty_{(n,n-2)}(\bar{\omega})$ such that $\bar{\partial}_b \alpha = 0$ in ω and α satisfies the condition (A), there exists a $u \in C^\infty_{(n,n-3)}(\omega)$ such that $\bar{\partial}_b u = \alpha$ in ω.*

It is easy to see that condition (A) is necessary for Eq.(1) to be solvable when α is a $(n, n-2)$ form by Stoke's Theorem. The following proposition characterizes all the domains ω such that condition (2) will imply condition (A).

PROPOSITION. *If $\mathcal{O}(\bar{\omega})$ is dense in $\mathcal{O}(\partial\omega)$ (in the $C(\partial\omega)$ norm), for any α satisfying condition (2), α satisfies condition (A). In particular, if polynomials are dense in $\mathcal{O}(\omega)$, then condition (2) implies condition (A).*

Thus we can have the following corollary.

COROLLARY 3.1. *Let M and ω be the same as in Theorem 3'. If we assume that $\mathcal{O}(\bar{\omega})$ is dense in $\mathcal{O}(\partial\omega)$ in the sup norm in $\partial\omega$, then for any $\alpha \in C^\infty_{(n,n-2)}(\bar{\omega})$ such that $\bar{\partial}_b \alpha = 0$ in ω, there exists a $u \in C^\infty_{(n,n-3)}(\omega)$ such that $\bar{\partial}_b u = \alpha$ in ω.*

We shall give an example of a $\bar{\partial}_b$-closed $(n, n-2)$ form which does not satisfy condition (A) (see Rosay [19]). Let S_n be the unit sphere in $\mathbb{C}^n$, $n \geq 3$ and $\Sigma_1 = S_n \cap \{|z_1|^2 < \frac{1}{2}\}$, $\Sigma_2 = S_n \cap \{|z_1|^2 > \frac{1}{2}\}$. It is proved in [19] that one can solve Eq.(0.1) for any (p, q) form α satisfying condition (0.2) in Σ_2 for all $1 \leq q \leq n-2$. While on Σ_1, this is only true when $1 \leq q < n-2$. We note that the boundary of Σ_1 and Σ_2 lie in the Levi flat hypersurface $M_0 = \{|z_1|^2 = \frac{1}{2}\}$ which has a Levi-flat Stein neighborhood basis. Let $n = 3$ and $\zeta(z_3)$ be a cut-off function such that $\zeta(z_3) = 0$ when $|z_3|^2 \geq \frac{1}{4}$ and $\zeta(z_3) > 0$ when $|z_3|^2 < \frac{1}{4}$. We define

$$f = \frac{\zeta(z_3)}{z_2} dz_1 \wedge dz_2 \wedge dz_3 \wedge d\bar{z}_3$$

and

$$\alpha = \tau f$$

where τ is the projection operator from $(3, 1)$ form in $\mathbb{C}^3$ to Σ_1. It is easy to see that α is a smooth $(3,1)$ form on Σ_1, since for $z \in S_3$, $|z_2|^2 = 1 - |z_1|^2 - |z_3|^2 \geq \frac{1}{2} - |z_3|^2 > 0$ on the support of ζ. Also $\bar{\partial} f = 0$ on Σ_1 which implies that $\bar{\partial}_b \alpha = 0$ on Σ_1. If we set $h(z) = \frac{1}{z_1}$, then $h \in A(\partial\Sigma_1)$ and

$$\int_{\partial\Sigma_1} \alpha \wedge h = \int_{\partial\Sigma_1} \frac{1}{z_1} \frac{\zeta(z_3)}{z_2} dz_1 \wedge dz_2 \wedge dz_3 \wedge d\bar{z}_3$$

$$= \iint_{|z_3|^2 \leq \frac{1}{2}} \zeta(z_3) \int_{|z_2|^2 = \frac{1}{2} - |z_3|^2} \int_{|z_1|^2 = \frac{1}{2}} \frac{1}{z_1 z_2} dz_1 \wedge dz_2 \wedge dz_3 \wedge d\bar{z}_3$$

$$= (2\pi i)^2 \iint_{|z_3| \leq \frac{1}{2}} \zeta(z_3) dz_3 \wedge d\bar{z}_3$$

$$\neq 0$$

Thus α does not satisfy the necessary condition (A) and thus can not be solved on Σ_1 (or on arbitrarily large subset of Σ_1). We note that α is not smooth on

Σ_2. On the other hand, since $A(\Sigma_2)$ is dense in $A(\partial\Sigma_2)$, Any $\bar{\partial}_b$-closed (3,1) form can be solved on Σ_2 from Theorem 5. We also note that the boundary of Σ_2 is not Runge.

The proof of these results can be found in [23,24].

References.

1. A. Andreotti and C.D. Hill, *Convexity and the H. Lewy problem. I and II.*, Ann. Scuola Norm. Sup. Pisa **26** (1972), 325-363, ibid,747-806.

2. E. Bedford and P. de Bartolomeis, *Levi flat hypersurfaces which are not holomprphically flat*, Pro. A. M. S. (1981), 575-578.

3. E. Bedford and J. E. Fornaess, *Domains with pseudoconvex neiborhood systems*, Invent. Math. **47** (1978), 1-27.

4. H. Boas and M.-C. Shaw, *Sobolev estimates for the Lewy operator on weakly pseudo-convex boundaries*, Math. Ann. **274** (1986), 221-231.

5. A. Boggess and M.-C. Shaw, *Kernels for the tangential Cauchy-Riemann equations*, Trans. Am. Math. Soc. **289** (1985), 643-659.

6. D. Catlin, *Boundary behavior of holomorphic functions on pseudoconvex domains*, J. of differential Geometry **15** (1980), 605-625.

7. D. Catlin, *Global regularity of the $\bar{\partial}$-Neumann problem*, Pro. Symp. pure Math. **41** (1984), 39-49.

8. D. Catlin, *Subelliptic estimates for the $\bar{\partial}$-Neumann problem on pseudo-convex domains*, Ann. of Math. **126** (1987), 131-191.

9. J. P. D'Angelo, *Real hypersurfaces, orders of contact, and applications*, Ann. of Math. **115** (1982), 615-637.

10. K. Diederich and J.E. Fornaess, *Pseudoconvex domains: An example with nontrivial Nebenhülle*, Math. Ann. **225** (1977), 275-292.

11. G.B. Folland and J. J. Kohn, *The Neumann problem for the CauchyRiemann complex*, vol. 75, Ann. of Math. Studies, Princeton Univ. Press, Princeton, N.J., 1972.

12. G. M. Henkin, *The Lewy equation and analysis on pseudo-convex manifolds*, Uspheki Mat. Nauk **32** (1977), 57-118 English transl. in Russ. Math. Surv., 32, 1977, 59-130.

13. L. Hörmander, *Linear partial differential operators*, Springer-Verlag, New York, 1963.

14. H. Lewy, *On the local character of the solution of an atypical differential equation in three variables and related proble for regular functions of two complex variables*, Ann. Math. **64** (1956), 514-522.

15. J. J. Kohn, *Subellipticity of the $\bar{\partial}$-Neumann problem on pseudoconvex domains: Sufficient conditions*, Acta Math. **142** (1979), 79-122.

16. J. J. Kohn, *The range of the tangential Cauchy-Riemann operator*, Duke Math. Journ. **53** (1986), 525-545.

17. J. J. Kohn and H. Rossi, *On the extension of holomorphic functions from the boundary of a complex manifold*, Ann. Math. **81** (1965), 451-472.

18. J.-P. Rosay, *Equation de Lewy-résolubilite globale de l'équation $\partial_b u = f$ sur la frontiére de domaines faiblement pseudo-convexes de $\mathbb{C}^2$ (ou $\mathbb{C}^n$)*, Duke Math. J. **49** (1982), 121-128.

19. J.-P. Rosay, *Some applications of Cauchy-Fantappie forms to (local) problems on $\bar{\partial}_b$*, Ann. Scuola Normale Sup. Pisa **13** (1986), 225-243.

20. M.-C. Shaw, *L^2 estimates and existence theorems for the tangential Cauchy-Riemann complex*, Invent. Math. **82** (1985), 133-150.

21. M.-C. Shaw, *L^p estimates for local solutions of $\bar{\partial}_b$ on strongly pseudo-convex CR manifolds*, Math. Ann. **288** (1990), 35-62.

22. M.-C. Shaw, *L^2 existence theorems for the $\bar{\partial}_b$-Neumann problem on strongly pseudo-convex CR manifolds*, Jour. Geometric Analysis **1** (1991), 139-163.

23. M.-C. Shaw, *Local existence theorems with estimates for $\bar{\partial}_b$ on weakly pseudo-convex CR manifolds*, to appear.

24. M.-C. Shaw, *Semi-Global Existence Theorems for $\bar{\partial}_b$ for (o,n-2) forms on pseudo-convex boundaries in $\mathbb{C}^n$*, to appear.

25. F. Treves, *Homotopy formulas in the tangential Cauchy-Riemann complex*, Memoirs of the Amer. Math. Society, Providence, Rhode Island.

26. S. M. Webster, *On the local solution of the tangential CauchyRiemann equations*, Ann. Inst Henri Poicaré **6** (1989), 167-182.

DEPARTMENT OF MATHEMATICS, UNIVERSITY OF NOTRE DAME, NOTRE DAME, INDIANA 46556

Contemporary Mathematics
Volume **137**, 1992

Making an Outer Function
from Two Inner Functions

DONALD SARASON

1. Introduction.

Can the sum of two inner functions be an outer function? Anyone who has read the well-known paper [6] of Karel de Leeuw and Walter Rudin realizes that the answer is "yes". It is proved in that paper that a holomorphic function in the unit disk, D, is an outer function if its real part is positive. Hence, the sum of the constant inner function 1 and any inner function other than the constant function -1 is an outer function.

This note will examine cases in which the sum of two nonconstant inner functions is an outer function, mainly when one of the summands is a finite Blaschke product. Motivation for the study comes from a conjecture about the structure of the rigid functions in the Hardy space H^1 (on D). A function in H^1 is called rigid if the only functions in H^1 with the same argument as it almost everywhere on ∂D are the positive multiples of itself. The notion goes back to the paper of de Leeuw and Rudin, although the term "rigid function" was not used there. A nonconstant inner function is nonrigid, for, if u is such a function, then its argument on ∂D agrees with that of $(1+u)^2$. The following consequences are immediate.

1. Rigid functions are outer functions. (For, if the H^1 function f is divisible by the nonconstant inner function u, then the argument of f agrees with that of the H^1 function $(1+u)^2 f/u$ on ∂D.)

2. If the outer H^1 function f is divisible in H^1 by $(1+u)^2$, where u is a nonconstant inner function, then f is nonrigid. (For then the argument of f agrees with that of the H^1 function $uf/(1+u)^2$ on ∂D.)

The conjecture referred to above is that the converse of the last statement holds, namely, that if the outer H^1 function f is nonrigid then there is a nonconstant inner function u such that $f/(1+u)^2$ is in H^1. This is part of a more elaborate conjecture that is discussed and partly established in [8], [9], [10].

1991 *Mathematics Subject Classification*. Primary 30D55.

This paper is in final form and no version of it will be submitted for publication elsewhere.

If v_1 and v_2 are nonconstant inner functions then $(v_1+v_2)^2$ is nonrigid because its argument on ∂D agrees with that of $v_1 v_2$. As will be clear shortly, the sum $v_1 + v_2$ is often an outer function, tempting one to test the conjecture on functions of the form $(v_1 + v_2)^2$. That special case of the conjecture is just as open as the general case, and possibly it is no simpler. The latter is suggested by a result of H. Helson [4]: If the outer H^1 function f is nonrigid, then there are inner functions v_1 and v_2, not both constant, such that $v_1 + v_2$ is an outer function and $f/(v_1 + v_2)^2$ is in H^1.

In Section 5 the conjecture will be verified for outer functions of the form $(v_1 + v_2)^2$ with v_1 a finite Blaschke product. More generally, it will be shown that if v_1 is a finite Blaschke product and v_2 is any inner function, then the conjecture holds for the outer part of $(v_1 + v_2)^2$: that outer function either is rigid or is divisible in H^1 by $(1 + u)^2$ for some nonconstant inner function u.

Given two relatively prime inner functions, is some linear combination of them an outer function? This is a slight variant of the originally posed question. A simple argument given in Section 2 shows that a linear combination of two inner functions will fail to be outer if the two coefficients have different absolute values. Hence, it is equivalent to ask whether, given two relatively prime inner functions v_1 and v_2, there is a complex number α on ∂D such that $v_1 - \alpha v_2$ is an outer function. The answer turns out to be sometimes "yes" and sometimes "no". An extremely elementary case is treated in Section 3, that where v_1 is the function z and v_2 is a Blaschke factor. The answer here is "yes". It can be "no" however when $v_1 = z^2$ and v_2 is a Blaschke product of order 2. That also is shown in Section 3. Section 4 deals with the case where v_1 is z and Section 5 the case where v_1 is a finite Blaschke product. Section 2 contains various preliminaries.

I am especially pleased to be able to include this note in a volume honoring Walter Rudin, whose paper with de Leeuw initiated the trend it follows. The present paper is a minisample of Rudin's pervasive influence in modern function theory.

2. Preliminaries.

LEMMA 1. *If v is a nonconstant inner function and h is a function in H^∞ such that $\|h\|_\infty < 1$, then $v - h$ is not an outer function.*

The proof will employ a few simple properties of Toeplitz operators on H^2 [1]. For φ a function in L^∞ of ∂D, the Toeplitz operator on H^2 with symbol φ will be denoted, as usual, by T_φ.

Suppose $v - h$ is an outer function. Then it is invertible in H^∞, since it is bounded away from 0 on ∂D. The Toeplitz operator T_{v-h} is therefore invertible, and since $v - h = (1 - \bar{v}h)v$ (on ∂D), it factors as $T_{v-h} = T_{1-\bar{v}h}T_v$. Because $T_{1-\bar{v}h} = 1 - T_{\bar{v}h}$ and $\|T_{\bar{v}h}\| = \|\bar{v}h\|_\infty < 1$, the operator $T_{1-\bar{v}h}$ is also invertible. It follows that T_v is invertible, a contradiction.

COROLLARY. *If v_1 and v_2 are nonconstant inner functions and α is a complex* *number such that $v_1 - \alpha v_2$ is an outer function, then $|\alpha| = 1$.*

LEMMA 2. *Let v_1 and v_2 be holomorphic self-maps of D and let f be a non-constant function in H^1. Then the inner factor of $v_1 - v_2$ divides the inner factor of $f(v_1) - f(v_2)$.*

Let M be the invariant subspace of the shift operator on H^1 generated by $v_1 - v_2$ (in other words, $M = vH^1$ where v is the inner factor of $v_1 - v_2$). It will suffice to show that $f(v_1) - f(v_2)$ is in M.

Since M is invariant under multiplication by functions in H^∞, the function $v_1^n - v_2^n$ is in M for every positive integer n. Hence $p(v_1) - p(v_2)$ is in M for every polynomial p. Take a sequence $(p_k)_1^\infty$ of polynomials converging to f in H^1 norm. Since the composition operator induced by v_1 is bounded on H^1 [**11, p. 220**], we have $p_k(v_1) \to f(v_1)$ in H^1 norm. Similarly $p_k(v_2) \to f(v_2)$ in H^1 norm. Hence $f(v_1) - f(v_2)$ is in M, as desired.

Carathéodory's theorem on angular derivatives and the Denjoy–Wolff fixed point theorem will be needed and so will be stated here. A holomorphic self-map v of D is said to have an angular derivative in the sense of Carathéodory at the point λ of ∂D if v has a unimodular nontangential limit, $v(\lambda)$, at λ, and the difference quotient $\frac{v(z) - v(\lambda)}{z - \lambda}$ has a limit (denoted $v'(\lambda)$) as z tends nontangentially to λ.

CARATHÉODORY'S THEOREM. *The holomorphic self-map v of D has an angular derivative in the sense of Carathéodory at the point λ of ∂D if and only if the number*

$$c = \liminf_{z \to \lambda} \frac{1 - |v(z)|}{1 - |z|}$$

is finite. In that case $v'(\lambda) = c\bar\lambda v(\lambda)$, and $\frac{1 - |v(z)|}{1 - |z|}$ tends to c as z tends nontangentially to λ. The number c is never 0.

ADDENDUM. *If v is an inner function, then it has an angular derivative in the sense of Carathéodory at λ if and only if it has a unimodular nontangential limit at λ and the quotient $\frac{v(z) - v(\lambda)}{z - \lambda}$ belongs to H^2.*

It is an elementary consequence of Schwarz's lemma that if the holomorphic self-map v of D has the point λ of D as a fixed point, then $|v'(\lambda)| \le 1$, and the inequality is strict unless v is an elliptic Möbius transformation. A point λ of $\bar D$ is called a Denjoy–Wolff point of v if λ is in D and $v(\lambda) = \lambda$, or if λ is in ∂D, and v has λ as its nontangential limit at λ, and v has an angular derivative at λ satisfying $v'(\lambda) \le 1$.

DENJOY–WOLFF THEOREM. *Any holomorphic self-map v of D, other than the identity map, has a unique Denjoy–Wolff point. If λ is that point, and v is not an elliptic Möbius transformation, then the iterates of v, defined by $v_1 = v$, $v_n = v(v_{n-1})$ for $n = 2, 3, \ldots$, tend locally uniformly in D to λ as $n \to \infty$.*

Proofs of the theorems of Carathéodory and Denjoy–Wolff can be found in [**10**], which also contains references to the original papers.

3. Simplest Case.

PROPOSITION 1. *Let a be a point in $D\backslash\{0\}$ and ρ a point in ∂D. Then the function $\rho z + \bar{\rho}\left(\frac{z-a}{1-\bar{a}z}\right)$ is an outer function if and only if $|\mathrm{Re}\rho| \leq |a|$.*

This is an elementary calculation. The function in question, being rational, is an outer function if and only if it is free of zeros in D. Its zeros are easily found to be the numbers

$$\frac{\mathrm{Re}\rho \pm \sqrt{(\mathrm{Re}\rho)^2 - |a|^2}}{\bar{a}\rho}.$$

If $|\mathrm{Re}\rho| > |a|$ one of these numbers is in D, but if $|\mathrm{Re}\rho| \leq |a|$ both are on ∂D.

COROLLARY. *If $0 < |a| < 2^{-1/2}$, then the function $\rho^2 z^2 + \bar{\rho}^2\left(\frac{z-a}{1-\bar{a}z}\right)^2$ fails to be an outer function for every ρ in ∂D.*

In fact, the function in question factors as

$$-i\left[\rho z + \bar{\rho}\left(\frac{z-a}{1-\bar{a}z}\right)\right]\left[i\rho + \overline{(i\rho)}\left(\frac{z-a}{1-\bar{a}z}\right)\right].$$

By Proposition 1, the first factor is nonouter unless $|\mathrm{Re}\rho| \leq |a|$ and the second factor is nonouter unless $|\mathrm{Im}\rho| \leq |a|$. If $|a| < 2^{-1/2}$, one of the preceding inequalities fails (since $|\rho| = 1$).

4. The Case $z - v$.

PROPOSITION 2. *Let v be a nonconstant inner function, not the function z, and let λ be its Denjoy–Wolff point. If λ is in ∂D, the function $z - v$ is an outer function, and if λ is in D, its inner part is the Blaschke factor vanishing at λ.*

For the proof, it will be assumed that v is not an elliptic Möbius transformation, the result being obvious in that case. Let M be the invariant subspace of the shift operator on H^2 generated by the function $z - v$. It will suffice to prove that M contains the function $z - \lambda$.

By Lemma 2, M contains the function $f - f(v)$ for every f in H^∞. Let $v_1, v_2, \ldots$ be the iterates of v, defined as in the statement of the Denjoy–Wolff theorem. Applying the last conclusion with $f = v_n$, we see that M contains $v_n - v_{n+1}$ for all n. Hence M contains $z - v_n$ for all n. By the Denjoy–Wolff theorem, $v_n \to \lambda$ locally uniformly in D and hence in the weak topology of H^2. Therefore, M contains $z - \lambda$, as desired.

COROLLARY 1. *Let v be a nonconstant inner function, with $v(0) \neq 0$. There is a complex number α such that $z - \alpha v$ is an outer function if and only if there is a point λ in ∂D where v has an angular derivative in the sense of Carathéodory satisfying $|v'(\lambda)| \leq 1$.*

This is an immediate consequence of Proposition 2. If λ satisfies the preceding conditions, then the corresponding α for which $z - \alpha v$ is an outer function is $\lambda\overline{v(\lambda)}$.

In particular, if v is a finite Blaschke product with $v(0) \neq 0$, then $z - \alpha v$ fails to be an outer function for every α if and only if $|v'| > 1$ everywhere on ∂D. It is easy to see that there are Blaschke products of order 2 satisfying the last condition.

COROLLARY 2. *The function* $z - \alpha e^{\frac{z+1}{z-1}}$ *is an outer function if and only if* $\alpha = e^{i\gamma}$ *with* γ *real and* $\frac{\pi}{2} + 1 \leq |\gamma| \leq \pi$.

In fact, a computation shows that at the point $z = e^{i\beta}$ $(-\pi \leq \beta \leq \pi)$ the absolute value of the derivative of the function $e^{\frac{z+1}{z-1}}$ equals $1/2\sin^2(\beta/2)$, which is less than or equal to 1 exactly when $\frac{\pi}{2} \leq |\beta| \leq \pi$. It follows that the Denjoy–Wolff point of the function $\alpha e^{\frac{z+1}{z-1}}$ lies on ∂D if and only if $\bar\alpha = e^{-i\beta} \exp\left(\frac{e^{i\beta}+1}{e^{i\beta}-1}\right)$ with $\frac{\pi}{2} \leq |\beta| \leq \pi$. Since $\frac{e^{i\beta}+1}{e^{i\beta}-1} = -i\cot\frac{\beta}{2}$, the latter happens if and only if $\alpha = e^{i\gamma}$ with $\gamma = \beta + \cot\frac{\beta}{2}$ and β as above. As one easily sees, the function $\beta + \cot\frac{\beta}{2}$ is increasing on the interval $\frac{\pi}{2} \leq \beta \leq \pi$; its range on that interval is the interval $\left[\frac{\pi}{2} + 1, \pi\right]$. The desired conclusion now follows by Proposition 2.

5. The Case $b - v$

We consider now the difference between two nonconstant inner functions one of which is a finite Blaschke product. The following lemma from a paper of Helson and the author [5] is needed.

LEMMA 3. *Let* g *be a meromorphic function in* D *with only finitely many poles. Assume that the product of* g *with the Blaschke product whose zero set coincides with the set of poles of* g *belongs to the space* $H^{1/2}$, *and that the boundary function of* g *is positive almost everywhere on* ∂D. *Then* g *can be continued holomorphically across* ∂D.

This is actually a mild strengthening of the lemma from [5], obtainable by the same argument. Of course, from the Schwarz reflection principle one sees that the function g of the lemma must in fact be a rational function. In particular, if it has no poles then it is constant (a result of J. Neuwirth and D. J. Newman [7]).

PROPOSITION 3. *Let* b *be a finite Blaschke product of order* m (> 0) *and let* v *be a nonconstant inner function, different from* b. *Then the inner factor of* $b - v$ *is a Blaschke product of order at most* m.

That $b - v$ has at most m zeros in D is elementary. Namely, for $0 < r < 1$, the function $b - rv$ has exactly m zeros in D, by Rouché's theorem. Letting r tend to 1 and applying Hurwitz's theorem, one obtains the desired conclusion.

It remains to show that $b - v$ has no singular factor. The argument for this involves a standard procedure that appears for example (and perhaps originally) in the proof of the well-known Helson–Szegö theorem [2]. The function $1 - \frac{v}{b}$ has a positive real part on ∂D. Let s be the branch of $\arg\left(1 - \frac{v}{b}\right)$ on ∂D with values in the interval $\left(-\frac{\pi}{2}, \frac{\pi}{2}\right)$, and let $\tilde{s}$ be the conjugate function of s. The function $e^{\tilde{s}-is}$ is then an outer function with a positive real part, so it is in H^p for $p < 1$ [**3, p. 116**]; in particular, it is in $H^{1/2}$. The boundary function of the function $e^{\tilde{s}-is}\left(1 - \frac{v}{b}\right)$ is positive, and the product of this function with b is in $H^{1/2}$. By Lemma 3, the function can be continued holomorphically across ∂D. The same is therefore true of $e^{\tilde{s}-is}(b - v)$, so the latter function has no singular factor. Therefore the function $b - v$ also has no singular factor, as desired.

PROPOSITION 4. *Let b be a finite Blaschke product of order m (> 0), and let v be a nonconstant inner function, different from b. If $b - v$ has fewer than m zeros in D, then there is a point λ of ∂D where v has an angular derivative in the sense of Carathéodory and where $b(\lambda) = v(\lambda)$.*

As in the preceding proof, for $0 < r < 1$ the function $b - rv$ has m zeros in D; let λ_r be one of largest modulus. Assuming $b - v$ has fewer than m zeroes, we must have $|\lambda_r| \to 1$ as $r \to 1$. Let the point λ of ∂D be a cluster point of the net $(\lambda_r)_{0<r<1}$. We have

$$1 - \left|\frac{b(\lambda_r) + v(\lambda_r)}{2}\right| = 1 - \left(\frac{1 + \frac{1}{r}}{2}\right)|b(\lambda_r)|$$
$$\leq 1 - |b(\lambda_r)|.$$

Since b is a finite Blaschke product, the ratio $\frac{1-|b(z)|}{1-|z|}$ is bounded in D. We can conclude that

$$\liminf_{z \to \lambda} \frac{1 - \left|\frac{b(z)+v(z)}{2}\right|}{1 - |z|} < \infty,$$

so, by Carathéodory's theorem, the function $\frac{b+v}{2}$ has an angular derivative in the sense of Carathéodory at λ. In particular, $\left|\frac{b(\lambda)+v(\lambda)}{2}\right| = 1$, implying that $b(\lambda) = v(\lambda)$. Finally, because b, being a finite Blaschke product, has an angular derivative at λ, so also must v.

PROPOSITION 5. *Let b be a finite Blaschke product of order m (> 0), and let v be an inner function, relatively prime to b, whose order is at least m. Let w be the inner part and f the outer part of $b - v$. If v and w both have order m then f^2 is rigid; otherwise f^2 is nonrigid.*

Consider first the case where v and w are both Blaschke products of order m. Then $b - v$ has m zeros in D. It follows that $b - v$ also has m zeros outside of $\bar{D}$, for at a point z outside of $\bar{D}$, the value of b is $1/\overline{b(1/\bar{z})}$, and similarly for v. Since $b - v$ is a rational function of order $2m$, it can have no zeros on ∂D. Hence f in this case is a rational function in H^1 without zeros in $\bar{D}$, and f^2 is the same sort of function. It is elementary that such a function is rigid.

By Proposition 3, the function w is a finite Blaschke product of order at most m. It thus remains only to consider the case where either the order of v is larger than m or the order of w is less than m. We factor b as $b = -b_1 b_2$ and v as $v = v_1 v_2$, where b_1 and b_2 are Blaschke products, the order of b_1 is the same as that of w, and v_1 and v_2 are inner functions, with the order of v_1 at least as large as that of w, and v_2 nonconstant. Since $\arg(b - v)^2 = \arg(-bv)$, we have

$$\arg f^2 = \arg\left(\frac{-bv}{w^2}\right) = \arg\left(\frac{b_1 b_2 v_1 v_2}{w^2}\right).$$

The function b_1/w is continuous and unimodular on ∂D, and its winding number on ∂D about the origin is 0; the argument of this function therefore has a continuous branch, say q, on ∂D. The function $e^{-\tilde{q}+iq}$ is an outer function which belongs with its reciprocal to H^p for all finite p [**3, p. 117**], and its argument agrees with that of b_1/w on ∂D. Consider the function v_1/w; it is asserted that the L^∞ distance of this function from H^∞ is less than 1. In fact, the Toeplitz operator $T_{v_1\bar{w}}$ factors as $T_{\bar{w}}T_{v_1}$. The operator $T_{\bar{w}}$ is Fredholm with index equal to the order of w, and the operator T_{v_1} is left-Fredholm with index equal to the negative of the order of v_1. Hence $T_{v_1\bar{w}}$ is left-Fredholm with a nonpositive index. Because a nonzero Toeplitz operator on H^2 has either a trivial kernel or a trivial cokernel [**1, p. 185**], the kernel of $T_{v_1\bar{w}}$ must be trivial, so $T_{v_1\bar{w}}$ is in fact left-invertible. That implies the desired inequality, $\mathrm{dist}(v_1\bar{w}, H^\infty) < 1$ [**1, p. 187**]. We now follow (once again) the reasoning in the proof of the Helson–Szegő theorem. Let h be a function in H^∞ satisfying $\|v_1\bar{w} - h\|_\infty < 1$. Then also $\|1 - \bar{v}_1 wh\|_\infty < 1$, implying that $|\arg(\bar{v}_1 wh)| < \frac{\pi}{2} - \delta \pmod{2\pi}$ on ∂D for some $\delta > 0$. Let s be the branch of $\arg(v_1\bar{w}h)$ on ∂D taking values in the interval $\left(-\frac{\pi}{2} + \delta, \frac{\pi}{2} - \delta\right)$. The function $e^{-\tilde{s}+is}$ is then an outer function, and it belongs to H^p for $p < 1/\left(1 - \frac{2\delta}{\pi}\right)$ [**3, p. 116**]. We have $\arg\left(\frac{v_1}{w}\right) = \arg(he^{-\tilde{s}+is})$. All together, then,

$$\arg f^2 = \arg(b_2 v_2 h e^{-\tilde{q}+iq} e^{-\tilde{s}+is}).$$

The function on the right side is in H^1, and it is not a multiple of f^2 since it is not an outer function. Hence f^2 is nonrigid, as desired.

PROPOSITION 6. *Let b, v and f be as in Proposition 5. If f^2 is not rigid, then there is a nonconstant inner function u such that $f/(1 - u)$ is in H^2.*

The proof will be by induction on the order of b. The induction starts with the trivial case where b has order 0, that is, where b is a constant.

When b has positive order there are two possibilities. One is that $b - v$ has no zeros in D, which means, by Proposition 3, that $f = b - v$. Then, by Proposition 4, there is a point λ in ∂D where v has an angular derivative in the sense of Carathéodory and where $b(\lambda) = v(\lambda)$. In that case, by the addendum to Carathéodory's theorem, the function $\frac{b-v}{z-\lambda}$ is in H^2, so the inner function $u = \bar{\lambda}z$ has the required property. (This argument can in fact be used whenever $b - v$ has fewer than m zeros in D.) The other possibility is that $b - v$ vanishes at a point z_0 of D. Let φ be a Möbius transformation of D onto D that maps $b(z_0)$ to 0 and ψ a Möbius transformation that maps z_0 to 0. Then ψ divides both

$\varphi(b)$ and $\varphi(v)$, say $\varphi(b) = \psi b_1$ and $\varphi(v) = \psi v_1$. We have

$$b - v = \psi(b_1 - v_1) \left(\frac{b - v}{\varphi(b) - \varphi(v)} \right).$$

The last factor on the right side is an invertible function in H^∞. The product of that function with the outer factor of $b_1 - v_1$ is the outer factor of $b - v$. The case of the proposition where b has order m is thus reduced to the case where it has order $m - 1$, which completes the proof.

The preceding argument, if slightly augmented, also gives an alternative proof of Proposition 5.

References.

1. R. G. Douglas, *Banach Algebra Techniques in Operator Theory*, Academic Press, New York and London, 1972.

2. J. B. Garnett, *Bounded Analytic Functions*, Academic Press, New York and London, 1981.

3. H. Helson, *Harmonic Analysis*, Addison Wesley, Reading, Massachusetts, 1983.

4. H. Helson, *Large analytic functions*, II. Analysis and Partial Differential Equations, C. Sadosky (Ed.), Marcel Dekker, New York and Basel (1990), 217–220.

5. H. Helson and D. Sarason, *Past and future*, Math. Scand. **21** (1967), 5–16.

6. K. de Leeuw and W. Rudin, *Extreme points and extremum problems in H_1*, Pacific J. Math. **8** (1958), 467–485.

7. J. Neuwirth and D. J. Newman, *Positive $H^{1/2}$ functions are constants*, Proc. Amer. Math. Soc. **18** (1967), 958.

8. D. Sarason, *Exposed points in H^1*, I. Operator Theory: Advances and Applications **41** (1989), 485–496.

9. D. Sarason, *Exposed points in H^1*, II. Operator Theory: Advances and Applications **48** (1990), 333–347.

10. D. Sarason, *Function Theory in the Unit Disk from a Hilbert Space Perspective*, University of Arkansas Lecture Series in the Mathematical Sciences, Vol. 10, Wiley, New York, 1992.

11. K. Zhu, *Operator Theory in Function Spaces*, Marcel Dekker, New York and Basel, 1990.

DEPARTMENT OF MATHEMATICS, UNIVERSITY OF CALIFORNIA, BERKELEY, CA 94720

Contemporary Mathematics
Volume **137**, 1992

Arclength formulas in conformal mapping

ALEXANDER STANOYEVITCH

Introduction.

Let $\Omega \subseteq \mathbb{C}$ be a Jordan domain (i.e., a bounded planar domain whose boundary is a Jordan curve) and let $\mathbb{D}$ denote the unit disk. Throughout we let "ℓ" denote arclength or more specifically the Hausdorff one-dimensional measure. We consider a homeomorphism $\varphi : \overline{\mathbb{D}} \longrightarrow \overline{\Omega}$ which is C^1 on $\mathbb{D}$. Given an arc $J \subseteq \partial\mathbb{D}$, we seek to express the length of the image arc: $\ell(\varphi(J))$ in terms of integral formulas involving (first-order) derivatives of φ. In this general situation, it is always true and straightfoward to prove that arclength is lower semicontinuous, i.e.,

$$(1) \qquad \ell(\varphi(J)) \leq \varliminf_{r\uparrow 1} \ell(\varphi(rJ)) = \lim_{r\uparrow 1} \int_J \left| \frac{\partial}{\partial\theta} \varphi(re^{i\theta}) \right| d\theta$$

(here rJ denotes the dilation of $J : \{ re^{i\theta} : e^{i\theta} \in J \}$).

The first progress on this problem when φ is conformal on $\mathbb{D}$ dates back to 1916 when F. & M. Riesz [**Rz-16**] proved that in case $\partial\Omega$ is rectifiable we then have the following formula valid for each arc $J \subseteq \partial\mathbb{D}$:

$$(2) \qquad \ell(\varphi(J)) = \int_J \lim_{r\uparrow 1} \left| \varphi'(re^{i\theta}) \right| d\theta \quad,$$

The question arises whether the hypothesis of rectifiability of $\partial\Omega$ in the Riesz result (2) can be weakened. By a result of Gehring and Hayman [**Ge&Ha–62**] if the image arc $\varphi(J)$ is rectifiable then so will be the image of the hyperbolic geodesic in $\mathbb{D}$ which joins the endpoints of J. The resulting Jordan subdomain of $\varphi(\mathbb{D})$ has rectifiable boundary and with an appropriate change of variables the Riesz result (2) can be shown to remain valid if only $\ell(\varphi(J)) < \infty$. With a bit

1991 *Mathematics Subject Classification*. Primary 30C35, 28A75. Secondary 30C85, 30D40.

Key words and phrases. arclength, conformal mapping, fractal sets, hyperbolic geometry.

This paper is in final form and no version of it will be submitted for publication elsewhere.

more work, making use of the nontangential maximal function, one can go on to show that

$$(3) \qquad \ell(\varphi(J)) = \lim_{r \uparrow 1} \ell(\varphi(rJ)) = \lim_{r \uparrow 1} \int_J |\varphi'(re^{i\theta})| \, d\theta \quad .$$

By (1), (3) is certainly true when $\ell(\varphi(J)) = \infty$ hence (3) is always valid, i.e., when φ is conformal, arclength is continuous.

In attempting to extend (1) to the remaining case $\ell(\varphi(J)) = \infty$, one problem is that the limits in (1) can fail to exist almost everywhere on J. In fact, as was shown by Lohwater, Piranian & Rudin [**LP&R–55**] the behavior of φ can be so extreme that for a.e. θ

$$\varlimsup_{r \uparrow 1} |\varphi'(re^{i\theta})| = \infty \quad , \quad \varliminf_{r \uparrow 1} |\varphi'(re^{i\theta})| = 0 \; ,$$

(also $\varlimsup\limits_{r \uparrow 1} \arg \varphi'(re^{i\theta}) = \infty$, $\varliminf\limits_{r \uparrow 1} \arg \varphi'(re^{i\theta}) = -\infty$). It is also true that any univalent function φ on $\mathbb{D}$ satisfies $\varlimsup_{r \uparrow 1} |\varphi'(re^{i\theta})| > 0$ a.e. Hence it seems plausible that if we replace the limit in (2) with a limit supremum, the resulting formula:

$$(4) \qquad \ell(\varphi(J)) = \int_J \varlimsup_{r \uparrow 1} |\varphi'(re^{i\theta})| \, d\theta$$

might always be valid. The main purpose of this note is to give an example of a domain for which the formula (4) fails. The boundary of the domain is of fractal type and our methods rely on the recent pioneering work of Makarov [**Mak–85**] on boundary behavior of conformal mappings.

The author would like to express his thanks to Chris Bishop who had suggested to him such an approach to an example. Previously the author had a less pathological example constructed directly by its series expansion.

One can construct examples to show that with the exception of (1), none of our formulas have valid analogues in case $\varphi \in C^\infty(\mathbb{D})$ (with or without the rectifiability of $\partial\Omega$). We close this introduction with a question.

QUESTION. *Which of our formulas have analogues in case φ is quasi-conformal on $\mathbb{D}$?*

The Construction.

In this section we construct a Jordan domain $\tilde{\Omega}$ and a conformal map $h : \mathbb{D} \longrightarrow \tilde{\Omega}$ such that $\ell(\partial\tilde{\Omega}) = \infty$ but

$$\int_{\partial\mathbb{D}} \varlimsup_{\substack{z \to \zeta \\ z \in \mathbb{D}}} |h'(z)| \, |d\zeta| < \infty$$

This integral differs from the one in (4) in that the limit supremum is unrestricted as opposed to radial and thus the above map is *a fortiori* a counterexample to (4). For the requisite definitions of Hausdorff measures and dimension we cite

[**Fa–85**]. The domain $\tilde{\Omega}$ will be obtained from the familiar (Van Koch) snowflake domain Ω which is constucted as follows: Let Ω^0 be an equilateral triangle of side length 1. To obtain Ω^1, we build 3 new equilateral triangles exterior to Ω^0 each having side length $\frac{1}{3}$ and sharing the middle $\frac{1}{3}$ segments of the different sides of Ω^0. The domain Ω^1 is defined to be the union of Ω^0 with these 3 new equilateral triangles. Note that Ω^1 has $3 \cdot 4$ sides each of length 3^{-1}. We continue this construction so that the nth generation we have a polygon Ω^n with $3 \cdot 4^n$ sides of length 3^{-n} each. The snowflake domain Ω is defined to be $\lim \Omega^n = \cup \Omega^n$.

The snowflake Ω has recently been the subject of several investigations. In 1982, Kaufman and Wu proved [**Ka&Wu–85**] that the harmonic measure of Ω is supported on a set of Hausdorff dimension strictly smaller than that ($\log 4 / \log 3$) of the boundary curve. Subsequently, Carleson [**Ca–85**] showed that the harmonic measure of Ω lives on a set of Hausdorff dimension ≤ 1. Finally Makarov [**Ma–P**] showed that it lives on a set of length ($\equiv$ Hausdorff one–dimensional measure) zero. The techniques used to obtain these latter two results relied heavily on information and ergodic theory. We give an independent proof of the Makarov result based entirely on complex analysis which yields another property of the snowflake which we will use. Recall that a <u>Plessner point</u> of a meromorphic function $g : \mathbb{D} \longrightarrow \mathbb{C} \cup \{\infty\}$ is a point $\zeta \in \partial\mathbb{D}$ such that $\{g(z) : r < |z| < 1, z \in A_\zeta\}$ is dense in $\mathbb{C} \cup \{\infty\}$ for any $r < 1$ and Stolz angle A_ζ at ζ. Plessner proved [**Pl–27**] that the nontangential limit $\lim_{\substack{z \longrightarrow \zeta \\ z \in A_\zeta}} g(z)$ exists in $\mathbb{C}$ at a.e. non-Plessner points $\zeta \in \partial\mathbb{D}$.

LEMMA. *The harmonic measure of the snowflake Ω is supported on a set of length zero on $\partial\Omega$. Moreover, if $f : \mathbb{D} \longrightarrow \Omega$ is a Riemann map then a.e. $\zeta \in \partial\mathbb{D}$ is a Plessner point of f'.*

PROOF. A theorem of Makarov states that for any Riemann map $g : \mathbb{D} \longrightarrow D$ which is so pathological that the nontangential (or radial) limits of g' exist in $\mathbb{C}$ almost nowhere, the harmonic measure of D will be supported on a set of zero length (i.e. there exists $A_0 \subseteq \partial\mathbb{D}$ such that $\ell(A_0) = 2\pi$ but $\ell(g(A_0)) = 0$; see [**Ma–84**] Theorem 3 or [**Po–86**] Corollary 1). We define $A_T = \{\zeta \in \partial\mathbb{D} : \lim_{r\uparrow 1} f'(r\zeta)$ exists in $\mathbb{C}\}$. The function f will necessarily be conformal at each point $\zeta \in A_T$ (see Theorem 10.5 and the accompanying discussion in [**Po–75**]). Define a <u>cone</u> with vertex p and angle α to be a set of the form

$$\{p + re^{i\theta} : 0 < r < r_0, \quad \theta_0 < \theta < \theta_0 + \alpha\}.$$

The conformality condition means, in particular, that for each $\epsilon > 0$ and each image point $f(\zeta)$ ($\zeta \in A_T$) on the boundary $\partial\Omega$ of the snowflake, there will be a cone with vertex $f(\zeta)$ and angle $\pi - \epsilon$ which is entirely contained in Ω. For a subset $A \subset A_T$ it is known that

$$\ell(A) = 0 \Longleftrightarrow \ell(f(A)) = 0$$

(see [Po–75] Theorem 10.16). These facts together with the aforementioned theorem of Plessner show that the lemma will be proved as soon as we show that the set T of those points in $\partial\Omega$ which possess the above cone property is of zero length. Letting,

$$T_n = \{p \in \partial\Omega| \text{ there exists a cone in } \Omega \text{ with vertex } p,$$

$$\text{angle } \frac{99}{100}\pi \text{ and diameter } \geq 10 \cdot 3^{-n}\}$$

it is clear that $T \subseteq \cup T_n$ so we need only show that $\ell(T_n) = 0$. We consider first the nth generation Ω^n of the snowflake. The boundary $\partial\Omega^n$ is made up of congruent arcs which consist of 4 segments each of length 3^{-n} such that the outer two segments are colinear and the inner two segments form two sides of an equilateral triangle. For a given such configuration, points on the outer two segments might be on T_n but points on the inner two segments (or points of later generations arising from these segments) cannot (because there is no room for the cone). Similarly when we pass from Ω^n to Ω^{n+1}, each of the outer two segments gets replaced by 4 segments and the 2 middle segments of each must not contain points of T_n. We conclude that for any $m \geq n$,

$$\ell(T_n) \leq \ell(T_n \cap \partial\Omega^m)$$

$$\leq \ell(\partial\Omega^n)\left(\frac{2}{3}\right)^{m-n}$$

Hence $\ell(T_n) = 0$, as desired.

Aside: This argument shows that the Hausdorff dimension of T is $\leq \log 2/\log 3$.

We are now prepared to define the domain $\tilde{\Omega}$. For each Plessner point $\zeta \in \partial\mathbb{D}$ for f' there exists a sequence $z_n = z_n(\zeta)$ such that $z_n \longrightarrow \zeta$ in some fixed Stolz angle and $f'(z_n) \longrightarrow 0$. For each $z = z_n$ we construct a hyperbolic tent T_z with hyperbolic vertex z and base I_z. This means that T_z is the (smaller) subdomain of $\mathbb{D}$ formed by 2 circular arcs which meet $\partial\mathbb{D}$ at right angles and have z as a common end point, the base I_z is defined as $\partial\overline{T_z} \cap \partial\mathbb{D}$. By applying (the conformally invariant analogue of) Theorem 10.8 in [Po–75], T_z can be made to have the following properties: ("$\approx$" means comparable and "$\underset{\sim}{<}$" shall mean "$\leq$" with a constant)

(i) $\zeta \in I_z$,

(ii) $\ell(\partial T_z) \approx \ell(I_z)$

(iii) $\displaystyle\int_{\partial T_z \cap \mathbb{D}} |f'(w)|\,|dw| \underset{\sim}{<} |f'(z)|(1 - |z|) \underset{\sim}{<} \ell(I_z)$.

Letting $F \subseteq \partial\mathbb{D}$ be a compact null set whose image $f(F)$ has Hausdorff dimension greater than 1 (the existence of F is guaranteed by the lemma since $\partial\Omega$ has Hausdorff dimension > 1), we note that for a Plessner point $\zeta \varepsilon \partial\mathbb{D} \sim F$, and for $z_n(\zeta)$ close enough to ζ we have $I_{z_n} \cap F = \varnothing$. Hence the collection $\langle I_{z_n}(\zeta)\rangle$ covers $\partial\mathbb{D} \sim F$ a.e. in the sense of Vitali so we can find a disjoint collection of such intervals $\langle Ij\rangle$ such that

(iv) $Ij \cap F = \varnothing$ for all j and

(v) $\ell(\cup I_j) = 2\pi$

Let $\tilde{D} = \mathbb{D} \sim \overline{\cup T_j}$ and $\tilde{\Omega} = f(\tilde{D})$ then $\dim(\partial \mathbb{D}) > 1$ since it contains $f(F)$. On the other hand, f' is continuous at each point of $\partial \tilde{D} \sim \partial \mathbb{D}$ and

$$\int_{\partial \tilde{D}} |f'(w)| \, |dw| = \sum \int_{\partial T_j} |f'(w)| \, |dw|$$

$$\lesssim \sum |f'(z_j)|(1 - |z_j|)$$

$$\lesssim \sum \ell(I_j)$$

$$= 2\pi < \infty \ .$$

The desired Riemann map is now readily obtained by composing f with any Riemann map $k : \mathbb{D} \longrightarrow \tilde{D}$.

References.

[Ca–85] Carleson, L., *On the support of harmonic measure for sets of Cantor type*, Ann. Acad. Sci. Fenn. Ser. A I Math **10** (1985), 113–123.

[Fa–85] Falconer, K. J., *The Geometry of Fractal Sets*, Cambridge University Press (1985).

[Ge&Ha–62] Gehring, F. W. and Hayman, W. K., *An inequality in the theory of conformal mappings*, J. Math. Pures Appl. **44** (1962), 353–361.

[Ka&Wu–85] Kaufman, R. and Wu, J.-M., *On the snowflake domain*, Arkiv för Mat. **23** (1985), 177–183.

[LP&R–55] Lohwater, A. H., Piranian, G., and Rudin, W., *The derivative of a schicht function*, Math. Scand. **3** (1955), 103–106.

[Ma–84] Makarov, N. G., *Defining subsets, the support of harmonic measure, and perturbations of the spectra fo operators in Hilbert space*, Soviet Math. Dokl. **29** (1984), 103–106.

[Ma–P] _____, *On the harmonic measure of a snowflake*, Lomi Preprint (1986), Leningrad.

[Pl–27] Plessner, A. I., *Uber das Verhalten analytischer Functionen am Rande ihres Definitionsberreichs*, J. Reine Angew. Math **158** (1927), 219–227.

[Po–75] Pommerenke, C., *Univalent Functions*, Vandenhoeck & Ruprecht.

[Po–86] _____, *On conformal mapping and linear measure*, J. Analyse Math. **46** (1986), 231–238.

[Rz–16] Riesz, F. & M., *Über die Randwerte einer analytischen Funktion*, Quatrième congrès des math. scand. (1916), 27–44.

DEPARTMENT OF MATHEMATICS, UNIVERSITY OF HAWAII, HONOLULU, HAWAII 96822

Contemporary Mathematics
Volume **137**, 1992

Sharp Geometric Estimates
of the Distance to VMOA

DAVID A. STEGENGA AND KENNETH STEPHENSON

Dedicated to Walter Rudin on the occasion of his retirement.

1. Introduction

We assume familiarity with the space BMOA of analytic functions on the unit disc **D** having bounded mean oscillation and its subspace VMOA. For general background, see [G, Chp. VI]. The distance of a function $f \in$ BMOA from VMOA is defined as

$$\|f + \text{VMOA}\|_* = \inf_{g \in \text{VMOA}} \|f - g\|_*.$$

Since VMOA is the closure in BMOA of the disc algebra, this may be viewed as a measure of how close f is to being continuous. Our interest is in geometric estimates of this distance based on the boundary behavior of f.

For a compact set K in the plane, $C(K)$ denotes the Banach algebra of continuous functions on K under the supremum norm, and $d(\,\cdot\,,\,\cdot\,)$ represents distance in $C(K)$. The subspace $R(K)$ is the closure of the space of functions which extend to rational functions and $h(K)$ is the closure of the space of functions which extend to be harmonic in a neighborhood of K.

Assume $f \in H^\infty$; for $\zeta \in \mathbf{T} = \partial\mathbf{D}$, write K_ζ for the cluster set $\text{Cl}(f,\zeta)$. Our topic started with Axler and Shapiro [AS], who proved

$$\|f + \text{VMOA}\|_* \lesssim \limsup_{\zeta \in \mathbf{T}} \sqrt{area(K_\zeta)} \ .$$

(The symbol "$\lesssim$" means the left is bounded by a universal constant times the right.) This was improved in Carmona and Cufí [CC] to

$$\|f + \text{VMOA}\|_* \lesssim \limsup_{\zeta \in \mathbf{T}} d(\overline{z}, R(K_\zeta)) \ .$$

1991 *Mathematics Subject Classification.* Primary 30D50. Secondary 46E15.

The second author gratefully acknowledges support of the National Science Foundation and the Tennessee Science Alliance.

This paper is in final form and no version of it will be submitted for publication elsewhere.

Here we establish the inequality

$$(1) \qquad\qquad \|f + \text{VMOA}\|_* \lesssim \limsup_{\zeta \in \mathbf{T}} \sqrt{d(|z|^2, h(K_\zeta))} \ .$$

Each of these is strictly stronger than its predecessor, and we show via a construction that (1) is best possible in the sense requested by Carmona and Cufí — namely, we answer a question which they posed in [CC] and which motivated this research:

THEOREM A. *Let K be a compact subset of the plane. The condition $h(K) = C(K)$ is necessary and sufficient to conclude that $f \in$ VMOA for all functions $f \in H^\infty \cap C(\overline{\mathbf{D}} \backslash \{1\})$ having $Cl(f, 1) \subseteq K$.*

The condition $h(K) = C(K)$ will be shown to be equivalent to $d(|z|^2, h(K)) = 0$, in which case one says that the **harmonic content** of K is zero.

We establish Theorem A and other results by studying a geometric quantity associated with the image surface of a function f and showing that it is comparable to the distance from f to VMOA. The relevant ideas and techniques were introduced by the authors in [SS1]. One consequence of the study is an additional method for measuring the size of compact sets; this gives an estimate of distance to VMOA which is at least as strong as (1) and which can be shown to be sharp within the context of cluster sets.

If G is an open set in the plane, $a \in G$, and E is a closed set, then $\omega(a, G, E)$ denotes the harmonic measure of E at a with respect to $G \backslash E$. Throughout the paper, $D(a, t)$ denotes the open disc of radius t centered at a. For U open and $0 < \delta < 1$, $r(U, \delta) = \sup\{r : \omega(a, D(a, r), U^c) \le \delta, \text{ some } a \in U\}$. For a compact set K, $r(K, \delta) = \inf\{r(U, \delta) : K \subset U, U \text{ open}\}$. A small value for $r(K, \delta)$ suggests that K is "small" in the local sense that a certain small disc, when centered at any point of K, will hit the complement of K with harmonic impact at least δ. The following is the sharp geometric estimate alluded to in the title:

THEOREM B. *Fix $0 < \delta < 1$. Then for $f \in H^\infty$,*

$$\|f + \text{VMOA}\|_* \lesssim \sup_{\zeta \in \mathbf{T}} r(K_\zeta, \delta) \ .$$

Moreover, there is a constant C_δ such that for any compact connected set K in the plane, there exists a function $f \in H^\infty \cap C(\overline{\mathbf{D}} \backslash \{1\})$ with $Cl(f, 1) \subseteq K$ for which

$$\|f + \text{VMOA}\|_* \ge C_\delta r(K, \delta) \ .$$

We wish to thank J. J. Carmona, J. Cufí, and Carl Sundberg for valuable conversations on the material presented here. We also express our deep gratitude to Walter Rudin for his teaching, his friendship, and his encouragement over the years, and we extend best wishes to him and to Mary Ellen on their retirement.

2. A Sharp Estimate

The space BMOA consists of functions $f \in H^1(\mathbf{D})$ whose boundary values are in BMO($\mathbf{T}$) and $\| \cdot \|_*$ denotes the usual norm as introduced by John and Nirenberg (see [G]). There are several equivalent norms; in particular, for $f \in$ BMOA and $a \in \mathbf{D}$, define $f_a(z) = f((z+a)/(1+\bar{a}z)) - f(a)$. Then (see [B])

$$(2) \qquad \|f\|_* \approx \sup_{a \in \mathbf{D}} \|f_a\|_1.$$

The subspace VMOA, the space of functions with vanishing mean oscillation, consists of functions whose oscillations go to zero on intervals whose lengths go to zero. VMOA is the closure of the disc algebra in BMOA.

Let f be analytic in $\mathbf{D}$ and fix $0 < \delta < 1$. For $a \in \mathbf{D}$ and $t > 0$, define $\Omega_f(a,t)$ to be the component of $f^{-1}(D(f(a),t))$ containing a and $\Gamma_f(a,t) = \partial\Omega_f(a,t) \cap \mathbf{T}$. The maximum principle shows that $\Omega_f(a,t)$ is simply connected, and for large t, $\Gamma_f(a,t) \neq \emptyset$. Define

$$r_f(a,\delta) = \sup\{r : \omega(a, \Omega_f(a,r), \Gamma_f(a,r)) \le \delta\}.$$

The following theorem is in the authors' paper [SS1] (see Theorem 3, with $\rho \equiv \tilde{\rho} \equiv 1$, in conjunction with (2) above):

THEOREM 1 [SS1]. *If $0 < \delta < 1$ is fixed and f is analytic in $\mathbf{D}$, then* $\|f\|_* \approx \sup_{a \in \mathbf{D}} r_f(a,\delta)$.

This result is a "geometric" statement in the context of the image Riemann surface $\mathcal{R}_f$ for f; it says there is a disc of a fixed radius which, when placed anywhere in $\mathcal{R}_f$, will hit $\partial\mathcal{R}_f$ with a uniformly large harmonic impact. It may be helpful to the reader to refer to [SS1], where this viewpoint was first exploited. That paper also suggests a Brownian motion interpretation, which the authors find particularly intuitive.

The following sharp estimate for distance to VMOA is geometric in this image-surface sense and is the main result of this section:

THEOREM 2. *If $0 < \delta < 1$, then for f analytic in $\mathbf{D}$,*

$$(3) \qquad \|f + \text{VMOA}\|_* \approx \limsup_{|a| \to 1} r_f(a,\delta).$$

Note that f is not assumed to be bounded; that condition is needed later only to ensure that the cluster sets are compact. We begin with several preliminary lemmas.

LEMMA 1. *Let U be open, $r, \delta > 0$. The condition $\omega(a, D(a,r), U^c) > \delta, \forall a \in U$, implies for $n \ge 1$ that $\omega(a, D(a,nr), U^c) > 1 - (1-\delta)^n, \forall a \in U$.*

PROOF. Proof is by induction; assume the conclusion holds for some $n \ge 1$. Fix $a \in U$ and let f be the harmonic function in $U \cap D(a,(n+1)r)$ with boundary value 1 on $\partial U \cap D(a,(n+1)r)$ and 0 on $U \cap \partial D(a,(n+1)r)$. If $|z-a| = r$ and $z \in \partial(U \cap D(a,r))$ then either $z \in U$ and by the induction hypothesis, $f(z) >$

$1 - (1 - \delta)^n$, or $z \in \partial U$ and $f(z) = 1 > 1 - (1 - \delta)^n$. Let $\eta = \omega(a, D(a, r), U^c)$, so $\eta > \delta$. By the observations above,

$$
\begin{aligned}
\omega(a, D(a, (n+1)r), U^c) = f(a) &> \eta + (1 - \eta)[1 - (1 - \delta)^n] \\
&= \eta(1 - \delta)^n + 1 - (1 - \delta)^n \\
&> \delta(1 - \delta)^n + 1 - (1 - \delta)^n \\
&= 1 - (1 - \delta)^{n+1}. \quad \square
\end{aligned}
$$

For $\zeta \in \mathbf{T}$, $t > 0$, we write $\Delta(\zeta, t)$ for the set $D(\zeta, t) \cap \mathbf{D}$. For $a \in \mathbf{D}, a \neq 0$, I_a denotes the interval of $\mathbf{T}$ centered at $a/|a|$ and of length $|I_a| = 1 - |a|$.

LEMMA 2. *Let f and ϕ be analytic on $\mathbf{D}$, $\phi(\mathbf{D}) \subseteq \mathbf{D}$. Then for $b \in \mathbf{D}$ and $0 < \delta < 1$, the inequality $r_{f \circ \phi}(b, \delta) \leq r_f(\phi(b), \delta)$ holds.*

PROOF. Let $a = \phi(b)$; for $t > 0$, write $\Omega = \Omega_f(a, t)$ and $\Lambda = \Omega_{f \circ \phi}(b, t)$. It is immediate that $\phi(\Lambda) \subseteq \Omega$, so one may define a harmonic function h on Λ by $h(z) = \omega(\phi(z), \Omega, \partial \Omega \cap \mathbf{T})$. The key observation is that $\phi(\partial \Lambda \cap \mathbf{D}) \subset \partial \Omega \cap \mathbf{D}$, implying that the boundary values of h are zero on $\partial \Lambda$. Since h is bounded by 1, a comparison of boundary values shows that h is bounded above on Λ by the harmonic measure function $\omega(z, \Lambda, \partial \Lambda \cap \mathbf{T})$. Evaluation at $z = b$ gives $\omega(a, \Omega, \partial \Omega \cap \mathbf{T}) = h(b) \leq \omega(b, \Lambda, \partial \Lambda \cap \mathbf{T})$. Since this holds for $t > 0$, the conclusion of the lemma follows. $\square$

Proofs of the following two results are straighforward and are left to the interested reader.

LEMMA 3. *Given $\epsilon > 0$, there exists a constant $0 < \tau_\epsilon < 1$ so that for $t > 0$ and $\zeta \in \mathbf{T}$, $\omega(a, \Delta(\zeta, t), \partial \Delta(\zeta, t) \cap \mathbf{D}) < \epsilon$ for all $a \in \Delta(\zeta, t)$ satisfying $|a - \zeta| < \tau_\epsilon t$.*

LEMMA 4. *For $a \in \mathbf{D}$, $|a| > 1/2$, let ν_a denote harmonic measure with respect to a on $\partial \Delta(a/|a|, 2(1 - |a|))$. On I_a, Lebesgue measure and ν_a are mutually absolutely continuous, with $d\theta/d\nu_a \lesssim |I_a|$.*

LEMMA 5. $\|f + \mathrm{VMOA}(D)\|_* \lesssim \limsup\limits_{|a| \to 1} \dfrac{1}{|I_a|} \displaystyle\int_{I_a} |f - f(a)| \, d\theta.$

PROOF. Fix $f \in \mathrm{BMOA}$, $0 < \rho < 1$, and assume that

$$
\frac{1}{|I_a|} \int_{I_a} |f - f(a)| \, d\theta \leq M, \quad 1 - |a| \leq \rho.
$$

Then the above inequality also holds with f replaced by its real part u and $f(a)$ replaced by the average u_{I_a}. Now divide ∂D into n equal subarcs $I_1, \ldots, I_n$ where $\rho/4 \leq |I_1| \leq \rho/2$ and define the step function s_n by

$$
s_n = \sum_{j=1}^n u_{I_j} \chi_{I_j}.
$$

Let $I = I_1 \cup I_2$ and assume that these intervals are adjacent. It follows that

$$(4) \qquad |u_{I_1} - u_{I_2}| \leq \frac{2}{|I|} \int_I |u - u_I| \, dt \leq 2M$$

and hence the step function s_n has jump discontinuities of size no more that $2M$. It is now easily shown that

$$(5) \qquad \|u - s_n\|_* = \sup_I \frac{1}{|I|} \int_I |(u - s_n) - (u - s_n)_I| \lesssim M.$$

For intervals with $|I| < \rho/4$ this is quite clear and for longer intervals such as $I = I_1 \cup \cdots \cup I_k$ we compute

$$\frac{1}{|I|} \int_I |u - s_n| \, dt = \frac{1}{|I|} \sum_{j=1}^{k} \int_{I_j} |u - u_{I_j}| \, dt \leq M$$

and we see that (5) holds in general.

By (4) it is easily seen that there is a smooth function u_n satisfying $\|u_n - s_n\|_\infty \leq 2M$ and hence $\|u - u_n\|_* \lesssim M$. It now follows from Fefferman's inequality (see Garnett's book [G]) that the harmonic conjugates $\tilde{u}, \tilde{u}_n$, which are also smooth, satisfy $\|\tilde{u} - \tilde{u}_n\|_* \lesssim M$. Put $f_n = u_n + i\tilde{u}_n$. Then f_n is in the disk algebra and hence

$$\|f + \text{VMOA}\|_* \leq \|f - f_n\|_* \lesssim M. \qquad \square$$

PROOF OF THEOREM 2. Let $R = \limsup_{|a| \to 1} r_f(a, \delta)$. We may assume $R < \infty$, for otherwise, both sides of (3) are infinite by Theorem 1.

Given $\epsilon > 0$, choose $\eta > 0$ so that $r_f(a, \delta) < R + \epsilon$ if $1 - |a| < 2\eta$. Fix such an a and let Δ_a denote the set $\Delta(a/|a|, 2(1 - |a|))$. Let $\phi : \mathbf{D} \longrightarrow \Delta_a$ be a conformal mapping with $\phi(0) = a$ and note that in this case, ν_a (harmonic measure on $\partial \Delta_a$ with respect to a) corresponds under ϕ with Lebesgue measure on $\partial \mathbf{D}$. Applying Lemmas 2 and 4, Theorem 1, and (2), we have

$$\frac{1}{|I_a|} \int_{I_a} |f - f(a)| \, d\theta \lesssim \int_{I_a} |f - f(a)| \, d\nu_a$$

$$\leq \int_{\partial \Delta_a} |f - f(a)| \, d\nu_a = (1/2\pi) \int_{\mathbf{T}} |f \circ \phi - f \circ \phi(0)| \, d\theta$$

$$= \|(f \circ \phi)_0\|_1 \lesssim \|f \circ \phi\|_*$$

$$\lesssim \sup_{b \in \mathbf{D}} r_{f \circ \phi}(b, \delta) \leq \sup_{e \in \Delta_a} r_f(e, \delta)$$

$$\leq \sup\{r_f(e, \delta) : 1 - |e| < 2\eta\} < R + \epsilon.$$

We conclude from Lemma 5 that $\|f + \text{VMOA}\|_* \lesssim \limsup_{|a| \to 1} r_f(a, \delta)$.

For the opposite inequality, we use a method from [SS1]. Fix a sequence $\{a_j\}$, which we assume without loss of generality is converging to 1, with $r_f(a_j, \delta) \longrightarrow R$. Let h be an element of the disc algebra and consider $f + h$. Given ϵ, choose $\eta > 0$ so that $|h(z) - h(w)| < \epsilon$ for $z, w \in \Delta(1, \eta)$. Write Σ_j for the component of $\Omega_f(a_j, R - \epsilon) \cap \Delta(1, \eta)$ containing a_j. By the definition of R, for j large, $\omega(a_j, \Omega_f(a_j, R-\epsilon), \mathbf{T}) < \delta$, which by the maximum principle gives $\omega(a_j, \Sigma_j, \mathbf{T}) < \delta$. In conjunction with Lemma 3, this implies

$$\omega(a_j, \Sigma_j, \partial\Sigma_j \cap \Delta(1, \eta)) > 1 - \delta, \ j \text{ large}.$$

Since

$$|f(z) + h(z) - (f(a_j) + h(a_j))| > R - \epsilon - \epsilon, \ z \in \partial\Sigma_j \cap \Delta(1, \eta),$$

an argument identical to that establishing [SS1, (2.6)] proves that $\|(f+h)_{a_j}\|_1 > (R - 2\epsilon)(1 - \delta)$. By (2) we conclude that $R(1 - \delta) \lesssim \|f + h\|_*$. Since the disc algebra is dense in VMOA, $R = \limsup_{|a| \to 1} r_f(a, \delta) \lesssim \|f + \text{VMOA}\|_*$, and the proof of Theorem 2 is complete. $\square$

REMARK. Let $\phi : \mathbf{D} \to \Omega$ be a Riemann mapping function of the disk onto a simply connected domain Ω. Then, we can define the space $\text{BMOA}(\Omega)$ to consist of those f with $f \circ \phi \in \text{BMOA}(\mathbf{D})$ and appropriate invariant norm. Fix $a \in \mathbf{D}$ and let Δ_a be as in the proof of Theorem 2. The inequalities above show that

$$\frac{1}{|I_a|} \int_{I_a} |f - f(a)| \, d\theta \lesssim \int_{\partial\Delta_a} |f - f(a)| \, d\nu_a \le \|f\|_{\text{BMOA}(\Delta_a)}$$

and hence we have the following corollary:

COROLLARY 1. $\|f + \text{VMOA}(\mathbf{D})\|_* \lesssim \limsup_{|a| \to 1} \|f\|_{\text{BMOA}(\Delta_a)}$.

We may now apply geometric techniques in [Steg] and [HP] to obtain upper bounds for the $\text{BMOA}(\Delta_a)$ norm. For example, the fact (see [Steg] or [ATU]) that $\|f\|_*^2$ is dominated by the area of the range of f, $\text{area}(f(\mathbf{D}))$, implies

COROLLARY 2. *[Stan] If f is holomorphic on $\mathbf{D}$ and*

$$\limsup_{|a| \to 1} \text{area}(f(\Delta_a)) = 0$$

then $f \in \text{VMOA}$.

Note that there is no assumption here that f is bounded. Since there do exist unbounded functions whose cluster sets have area zero but which are not in VMOA, this corollary strengthens the Axler-Shapiro result mentioned in the Introduction.

3. Proof of Theorem B

Let $f \in H^\infty$ and fix $0 < \delta < 1$. For the inequality of Theorem B, it is enough by Theorem 2 to prove that for $\zeta \in \mathbf{T}$,

(6) $$\limsup_{a \in \mathbf{D}, a \to \zeta} r_f(a, \delta) \le 2\, r(\text{Cl}(f, \zeta), \delta).$$

Let $K_\zeta = \text{Cl}(f, \zeta)$ and write $R = r(K_\zeta, \delta)$. Given $\epsilon > 0$, there exists an open set U containing K_ζ such that $r(U, \delta) < R + \epsilon$. By definition of the cluster set, there exists ρ_0 so that $0 < \rho < \rho_0$ implies $\overline{f(\Delta(\zeta, \rho))} \subset U$. In particular, $r(f(\Delta(\zeta, \rho)), \delta) \leq r(U, \delta) < R + \epsilon$. Letting ϵ go to zero, we conclude that

$$\limsup_{\rho \to 0} \{r(f(\Delta(\zeta, \rho)), \delta)\} \leq R.$$

It remains to compare $r_f(a, \delta)$ and $2\, r(f(\Delta(\zeta, \rho)), \delta)$ for $|a - \zeta|$ small.

Let $\delta_0 = 1 - (1 - \delta)^2$, so $\delta < \delta_0 < 1$. By Lemma 1, $r(f(\Delta(\zeta, \rho)), \delta_0) < 2\, r(f(\Delta(\zeta, \rho)), \delta)$. Fix $R_0 > 2R$ and $\rho > 0$ so that $r(f(\Delta), \delta_0) < R_0$, where $\Delta = \Delta(\zeta, \rho)$. By Lemma 3, if a is chosen sufficiently close to ζ,

(7) $$\omega(a, \Delta, \partial\Delta \cap \mathbf{D}) < \delta_0 - \delta.$$

Fix such an a and write $\Omega = \Omega_f(a, R_0), \Gamma = \Gamma_f(a, R_0)$. If Λ denotes the component of $\Omega \cap \Delta$ containing a, then $\partial\Lambda$ is the disjoint union of A, B, C, where $A = \partial\Lambda \cap \partial\Delta \cap \mathbf{D}$, $B = \partial\Lambda \cap \Delta$, and $C = \partial\Lambda \cap \mathbf{T}$. We first argue that $\omega(a, \Lambda, B) < 1 - \delta_0$: Write $b = f(a)$. Define the harmonic function h on $U = f(\Delta) \cap D(b, t)$ by

$$h(z) = \omega(z, U, \partial D(b, t)).$$

Observe that $|f(z) - b| = t$ on B, so by a comparison of boundary values, $h(f(z)) \geq \omega(z, \Lambda, B)$, $z \in \Lambda$. In particular, $h(b) \geq \omega(a, \Lambda, B)$. But $h(b) < 1 - \delta_0$ by our choice of R_0. We conclude that $\omega(a, \Lambda, B) < 1 - \delta_0$.

Now note that $\omega(a, \Lambda, A \cup C) = 1 - \omega(a, \Lambda, B) > \delta_0$. By (7), $\omega(a, \Lambda, A) < \delta_0 - \delta$, and with the preceeding inequality, this implies $\omega(a, \Lambda, C) > \delta$. Since $C \subseteq \Gamma$ and $\Lambda \subseteq \Omega$, the maximum principle implies $\omega(a, \Omega, \Gamma) \geq \omega(a, \Lambda, C) > \delta$. Thus $r_f(a, \delta) \leq R_0$, and inequality (6) follows.

The second part of Theorem B requires a construction. A connected compact set K and δ are given. Write $R = r(K, \delta) > 0$. With routine arguments and normalizations we may assume that $0 \in K$ and that for any open set $U \supset K$,

(8) $$\omega(0, D(0, R), U^c) < \delta.$$

Let $\{U_j\}_{j=0}^{\infty}$ be a nested family of connected open sets converging down to K. From (8) it is clear that $K \cap \partial D(0, R) \neq \emptyset$, so we may modify the U_j slightly, if necessary, to ensure that each U_j shares a boundary point z_j with U_{j-1}, some $z_j \notin \overline{D(0, R)}$.

We obtain the function of interest *via* an application of the Riemann mapping theorem: In particular, we construct a simply connected Riemann surface $\mathcal{R}$ by stringing together the universal covering surfaces of the U_j using small cuts at the points z_j and simple branch points. This has a natural projection π to the plane and we obtain a function g as the mapping function from $\mathbf{D}$ onto $\mathcal{R}$ followed by π. Finally, g must be pared slightly to obtained the desired continuity behavior. For more details on this type of construction, see [Step] and [SS2].

Let $\widetilde{U}_j$ denote the universal covering surface of U_j, $j = 0, 1, \cdots$. The surfaces $\widetilde{U}_j$ may be viewed as surfaces spread over the plane, with natural projections π_j to U_j. For each $j > 0$ in succession, make a small cut on one sheet of $\widetilde{U}_j$, starting at an interior point p_j and ending at a boundary point over z_j; choose

p_j so that this cut projects to a straight line segment of length no greater than, say, $1/j$, and lying outside $D(0, R)$. The point z_j is also a boundary point of U_{j-1}, and since $U_j \subset U_{j-1}$, one can make the identical cut on one sheet of $\widetilde{U}_{j-1}$. Cross-connecting the two surfaces along these cuts forms a new surface with a simple branch point at p_j. Proceeding in this fashion for $j > 0$ (ensuring that the two cuts made on each $U_j, j > 0$, don't intersect) and taking the directed limit of the resulting nested surfaces yields a simply connected Riemann surface $\mathcal{R}$ and a natural projection π from $\mathcal{R}$ to the plane. Each $\widetilde{U}_j$ may be identified with a subsurface of $\mathcal{R}$ in a natural way; namely, so that $\pi|\widetilde{U}_j = \pi_j$. Since π provides a bounded analytic function on $\mathcal{R}$ (its range is U_0), $\mathcal{R}$ is conformally equivalent to $\mathbf{D}$. Designate a point $a \in \widetilde{U}_0 \subset \mathcal{R}$ lying over zero, let $\tilde{g} : \mathbf{D} \longrightarrow \mathcal{R}$ be a conformal mapping with $\tilde{g}(0) = a$, and define $g = \pi \circ \tilde{g}$.

We will need to modify g shortly, but let us observe some of its properties: g is a bounded analytic function on $\mathbf{D}$ with $g(0) = 0$ and having image surface $\mathcal{R}$. The cuts along which the various pieces of $\mathcal{R}$ were attached to one another correspond in $\mathbf{D}$ to nested cross-cuts $\{\gamma_j\}$; the surfaces $U_j \subset \mathcal{R}$ (minus the cuts) correspond in $\mathbf{D}$ with the sequence of open sets F_j between these cross-cuts. Because of the size and placement of cuts during the construction of $\mathcal{R}$, it is evident that $\omega(0, \mathbf{D}, \gamma_j) \downarrow 0$ as $n \to \infty$. In $\mathbf{D}$ this implies that the prime end defined by the sequence $\{\gamma_j\}$ corresponds with a single point of $\mathbf{D}$. Assuming without loss of generality that this point is 1 and recalling that $U_j \downarrow K$, we see that $\mathrm{Cl}(g, 1) \subseteq K$.

Within each F_j, designate a point w_j with $g(w_j) = 0$ and write Ω_j for $\Omega_g(w_j, R)$. Since the cuts used in constructing $\mathcal{R}$ were chosen to project outside $D(0, R)$, $\Omega_j \subset F_j, j = 0, 1, \cdots$. In particular, $g : \Omega_j \longrightarrow D(0, R) \cap U_j$ is a universal covering map, so (8) implies that

$$(9) \qquad\qquad \omega(w_j, \Omega_j, \mathbf{T}) < \delta.$$

Therefore, $\limsup\limits_{|a| \to 1} r_g(a, \delta) \geq R$.

The only difficulty with g is that it fails to be continuous on $\mathbf{T}\backslash\{1\}$. We will correct this by choosing an appropriate Jordan subdomain J of the disc and defining $f = g \circ \phi$, where $\phi : \mathbf{D} \longrightarrow J$ is a conformal mapping. We require the following of J:

(a) J is a Jordan region,

(b) $\overline{J} \cap \mathbf{T} = \{1\}$,

(c) $w_j \in J, j = 0, 1, \cdots$, and

(d) $\omega(w_j, \Omega_j, J^c) < \delta, j = 0, 1, \cdots$.

The first three conditions are easily met. As for (d), observe that for fixed j, (9) and the maximum principle imply that for $\rho_j < 1$ with $1 - \rho_j$ sufficiently small,

$$\omega(w_j, \Omega_j, \partial D(0, \rho_j)) < \delta.$$

Another application of the maximum principle implies that if $J \supseteq \Omega_j \cap D(0, \rho_j)$, then (d) holds. In light of this and since the sets F_j converge to 1 but are individually bounded away from 1, it is clear that one can construct the desired J.

If $\phi : \mathbf{D} \longrightarrow J$ is a conformal map with $\phi(0) = w_0$ and $\phi(1) = 1$, then $f = g \circ \phi$ has the properties we desire: It is bounded and analytic, (b) implies that it is continuous up to $\mathbf{T}$ except possibly at 1, and $\mathrm{Cl}(f, 1) \subseteq \mathrm{Cl}(g, 1) \subseteq K$. Writing $a_j = \phi^{-1}(w_j)$, note that $\Omega_f(a_j, r)$ is the component of $\phi^{-1}(\Omega_j \cap J)$ containing a_j. As ϕ is one-to-one, it preserves harmonic measure, so

$$\omega(a_j, \Omega_f(a_j, R), \mathbf{T}) = \omega(w_j, \Omega_j, J^c) < \delta,$$

implying $r_f(a_j, \delta) \geq R$. Since $a_j \to 1$,

$$\limsup_{|a| \to 1} r_f(a, \delta) \geq R = r(K, \delta).$$

An application of Theorem 2 completes the proof of Theorem B.

4. Harmonic estimates

In this section we prove Theorem A and the estimate (1). These rely on Theorem 2 and two results below, which involve the comparison of various measures of the size of a compact set K. The distance $d(\overline{z}, R(K))$ is known classically as the **analytic content** of K. It is well known that analytic content is zero if and only if $R(K) = C(K)$; moreover, a famous result of Vitushkin provides local necessary and sufficient conditions for determining equality (see Gamelin's book [Gam], Chapter VIII). In parallel with this, $d(|z|^2, h(K))$ is known as the **harmonic content** of K (see Khavinson [Kh].) We prove below that harmonic content is zero if and only if $h(K) = C(K)$, which is strictly stronger than $R(K) = C(K)$; see the comment in [Kh] on page 472, the example constructed in [McK] and page 192 of [Brow]. We also show that the quantity $r(K, \delta)$ provides local necessary and sufficient conditions for equality, and this forms the link to our earlier estimates.

THEOREM 3. *For K compact and $0 < \delta < 1$, there is a constant C_δ depending only on δ, so that*

$$r(K, \delta) \leq C_\delta \sqrt{d(|z|^2, h(K))} \ .$$

PROOF. Let U be an open neighborhood of K. Suppose $a \in K$ and $r > 0$ is such that $\omega(a, D(a, r), U^c) \leq \delta$. Write $b = d(|z|^2, h(K))$ and note that since $b = d(|z - a|^2, h(K))$, $\forall a$, we may assume without loss of generality that $a = 0$. Assume $r > \sqrt{b}$. Let $0 < \epsilon < b$ be given and choose $u \in h(K)$ so that $d(|z|^2, u) < b + \epsilon$.

Let Λ denote the component of $U \cap D(0, r)$ containing 0 and let ν denote harmonic measure on $\partial \Lambda$ with respect to 0. Write $\partial \Lambda$ as the disjoint union of the three sets

$$A = U \cap \partial D(0, r), \quad B = \partial U \cap D(0, \sqrt{b}), \quad C = (\partial U \cap D(0, r)) \backslash B.$$

We have

$$u(z) \begin{cases} \geq r^2 - b - \epsilon, & z \in A \\ \geq -b, & z \in B \\ \geq -\epsilon \geq -b, & z \in C. \end{cases}$$

Since $\nu(B \cup C) \leq \delta$, we have $\nu(A) \geq 1 - \delta$. Therefore

$$b + \epsilon \geq |u(0)| = |\int_{A \cup B \cup C} u \, d\omega| \geq \int_A u \, d\omega - |\int_{B \cup C} u \, d\omega|$$

$$\geq \int_A u \, d\omega - \int_{B \cup C} |u| \, d\omega \geq (r^2 - b - \epsilon)(1 - \delta) - b \delta.$$

That is, $(2b + 2\epsilon - \delta\epsilon)/(1 - \delta) \geq r^2$. Letting ϵ go to zero, we conclude that $r(K, \delta) \leq C_\delta \sqrt{b}$, where $C_\delta = \sqrt{2/(1 - \delta)}$. $\square$

The estimate (1) of the introduction follows from this theorem and Theorem 2. We do not know whether $r(K, \delta)$ and $\sqrt{d(|z|^2, h(K))}$ are comparable. If not, then the estimate of Theorem B is strictly stronger than (1).

THEOREM 4. *For K a compact set, the following are equivalent:*
(a). $h(K) = C(K)$,
(b). $d(|z|^2, h(K)) = 0$,
(c). $r(K, \delta) = 0$ for some $\delta \in (0, 1)$, and
(d). $r(K, \delta) = 0$ for any $\delta \in (0, 1)$.

PROOF. (a)$\Longrightarrow$(b) is immediate since $|z|^2$ is continuous. (b) $\Longrightarrow$(c) follows from Theorem 3 above. The quantity $r(K, \delta)$ is monotone increasing in δ, so the implication (c)$\Longrightarrow$(d) follows from Lemma 1.

We establish (a) from (d) via a contradiction: By the Hahn-Banach Theorem, if the harmonic functions are not dense in $C(K)$, then there exists a (Borel) measure $\mu \neq 0$ on K with $\int_K u \, d\mu = 0$ for all u harmonic on K. It suffices to assume that $\mu = \mu_1 - \mu_2$ where μ_1, μ_2 are positive measures on K. By the hypothesis on μ, the potentials

$$V_j(x) = \int_K \log(\frac{1}{|x - y|}) \, d\mu_j(y), \quad j = 1, 2,$$

satisfy $V_1 = V_2$ on K^c. To reach a contradiction, it suffices to prove that $V_1 \equiv V_2$ on K, for then $\mu = 0$.

Fix $x_0 \in K$ and suppose that

$$(10) \qquad\qquad V_1(x_0) > \alpha > V_2(x_0).$$

Since V_1 is lower semi-continuous, $V_1(x) > \alpha$ for $|x - x_0| < r$, some $0 < r < 1/2$. Let $0 < \epsilon < 1$ be given. Put $V_2(x) = v(x) + w(x)$, where

$$w(x) = \int_{D(x_0, r)} \log(\frac{1}{|x - y|}) \, d\mu_2(y).$$

Then $w \geq 0$ and v is harmonic on $D(x_0, r)$. Assume that r is small enough so that $|v - v(0)| < \epsilon$ on $D(x_0, r)$.

By condition (d), setting $\delta = 1 - \epsilon$, there is an open neighborhood U of K so that $\omega(x_0, D(a,r), U^c) > 1 - \epsilon$. Since $w \geq 0$ on $\partial D(x_0, r)$, $w = V_2 - v = V_1 - v > \alpha - v(x_0) - \epsilon$ on $\partial U \cap D(a,r)$, and w is superharmonic, we see that $w(x) \geq (\alpha - v(x_0) - \epsilon)\,\omega(x, D(x_0, r), U^c)$. Thus $w(x_0) \geq (\alpha - v(x_0) - \epsilon)\,(1 - \epsilon)$ for all small ϵ, implying $V_2(x_0) \geq \alpha$. This contradicts (10), proving that $V_1 = V_2$. $\square$

REMARK. The equivalence of (a) and (c) above gives a simpler description of the sets satisfying (a) then the classical conditions which can be found for example in Landkof's book. A compact set K satisfies (a) if and only if K has an empty interior and the complement of K has no irregular points, see Theorem 5.18 in [L]

References

[AS] S. Axler and J. H. Shapiro, *Putnam's theorem, Alexander's spectral area estimate, and VMO*, Math. Ann. **271** (1985), 161–183.

[ATU] H. Alexander, B. A. Taylor and J. Ullman, *Areas of projections of analytic sets*, Invent. Math. **16** (1972), 335 – 341.

[B] A. Baernstein II, *Analytic functions of bounded mean oscillation*, Proceedings of the NATO Advanced Study Institute at Durham, 1979, Ed. D.A. Brannan and J. G. Clunie, Academic Press, London, 1980.

[Brow] A. Browder, *Introduction to Function Algebras*, W. A. Benjamin, Inc., New York, 1969.

[CC] Juan José Carmona and Juliá Cufí, *On the distance of an analytic function to VMO*, J. London Math. Soc. (2) **34** (1986), 52–66.

[G] John B. Garnett, *Bounded Analytic Functions*, Academic Press, New York, 1981.

[Gam] T. Gamelin, *Uniform Algebras*, Prentice-Hall, Englewood Cliffs, N. J., 1969.

[HP] W. K. Hayman and Ch. Pommerenke, *On analytic functions of bounded mean oscillation*, Bull. London Math. Soc. **10** (1978), 219 –224.

[Kh] Dmitry Khavinson, *On uniform approximation by harmonic functions*, Michigan Math. J. **34** (1987), 465–473.

[L] N. S. Landkof, *Foundations of Modern Potential Theory*, Springer-Verlag, New York, 1972.

[McK] R. McKissick, *A nontrivial normal sup norm algebra*, Bull. Amer. Math. Soc. **69** (1963), 391 – 395.

[Stan] Charles S. Stanton, *Counting functions and majorization for Jensen Measures*, Pac. J. Math. **125 (2)** (1986), 459 –468.

[Steg] David A. Stegenga, *A geometric condition which implies BMOA*, Proceedings of Symposia in Pure Mathematics, vol. 35, American Mathematical Society, Providence, R. I., 1978, pp. 427–430.

[SS1] David A. Stegenga and Kenneth Stephenson, *A geometric characterization of analytic functions with bounded mean oscillation*, J. London Math. Soc. (2) **24** (1981), 243–254.

[SS2] David A. Stegenga and Kenneth Stephenson, *Generic covering properties for spaces of analytic functions*, Pacific J. Math. **119** (1985), 227–243.

[Step] Kenneth Stephenson, *The geometry of image surfaces of analytic functions*, Bounded Mean Oscillation in Complex Analysis, Ed. Ilpo Laine and Eero Posti, Univ. of Joensuu Publications in Sciences, No. 14, 1989, pp. 101–120.

DEPARTMENT OF MATHEMATICS, UNIVERSITY OF HAWAII, HONOLULU, HAWAII 96822

DEPARTMENT OF MATHEMATICS, UNIVERSITY OF TENNESSEE, KNOXVILLE, TENNESSEE 37996

Contemporary Mathematics
Volume **137**, 1992

On the One-Dimensional Extension Property

EDGAR LEE STOUT

Dedicated to Walter Rudin on the occasion of his retirement.

Introduction.

Recall that if D is a bounded domain in $\mathbb{C}^N$, the function $f \in \mathcal{C}(bD)$ is said to have the *one-dimensional extension property* if given a complex line λ in $\mathbb{C}^N$ that meets $\overline{D}$, there is a function F_λ continuous on $\lambda \cap \overline{D}$ and holomorphic on its interior with respect to λ such that F_λ and f agree on $\lambda \cap b\overline{D}$.

The general principle is that functions with the one-dimensional extension property are the boundary values of functions in $A(D)$, the space of all functions continuous on $\overline{D}$ and holomorphic on D. The first instance of this result seems to have been given by Agranovskiĭ and Valskiĭ [1], who proved the result in the case that the domain D is the unit ball $\mathbb{B}_N$ in $\mathbb{C}^N$. In [16] the result was reproved in the case of the ball by different methods, and it was shown that, in the ball case, certain smaller families of lines suffice. Let us say that the family $\mathcal{F}$ of complex lines in $\mathbb{C}^N$ is *sufficient for extension from bD* if given $f \in \mathcal{C}(bD)$ that has the one-dimensional extension property with respect to each $\lambda \in \mathcal{F}$, there is $F \in A(D)$ with boundary values f. The result of Nagel and Rudin is that the family of all lines at distance r, $r \in (0,1)$ from the origin, is sufficient for extension from $b\mathbb{B}_N$. (It remains an open problem to show that if $\Delta \Subset \mathbb{B}_N$ is a smoothly bounded, convex domain, then the family $\mathcal{T}(\Delta)$ of complex lines tangent to $b\Delta$ is sufficient for extension from $b\mathbb{B}_N$. The Nagel–Rudin result is the case that $\Delta = r\mathbb{B}_N$).

The result of Agranovskiĭ and Valskiĭ was generalized in a different direction in the paper [20] in which it was shown that if D is a bounded domain in $\mathbb{C}^N$ with bD of class $\mathcal{C}^2$, then an $f \in \mathcal{C}(bD)$ with the one-dimensional extension property extends through D as an element of $A(D)$. The proof in [20] was based on an

1991 *Mathematics Subject Classification*. Primary 32A40. ·

Research supported in part by NSF Grants DMS–8801932 and DMS–9001883 and in part by NSF Grant INT–8612981, U.S.-Sweden Several Complex Variables—A Cooperative Research Program at Institut Mittag-Leffler.

This paper is in final form and no version of it will be submitted for publication elsewhere.

application of the Radon transform. Subsequently, another proof was found by Kytmanov—see [2], p. 198, which is based on the Bochner-Martinelli integral. Recently a proof based on the Fourier transform has been given [7]. The question of finding small families of complex lines that are sufficient for extension from the boundary has been pursued in [6], in [7] and in [8]. In the latter paper it is shown, again by the method of the Radon transform, that if D is a bounded domain in $\mathbb{C}^3$ with bD of class $\mathcal{C}^2$, and if Γ is an algebraic curve in $\mathbb{C}^3$, then the family $\mathcal{I}_\Gamma$ of all complex lines that meet Γ is a family sufficient for extension from the boundary of D.

The results we have mentioned are not biholomorphically invariant—lines are typically not taken into lines by biholomorphic maps. It is our intention to give in this paper a version of the one-dimensional extension result that is invariant under biholomorphic maps. The result we obtain, when specialized to the case of a bounded domain in $\mathbb{C}^N$ is weaker than the results involving complex lines given above, because it involves a hypothesis about extension through slices by *all* nonsingular holomorphic curves. On the other hand, the result is biholomorphically invariant, in an evident sense, and it is valid for domains in a rather general class of manifolds.

I. Functions With The One-Dimensional Extension Property.

We shall prove the following result, which is closely related to the result obtained in [20].

THEOREM. *Let Ω be a bounded open set in $\mathbb{C}^N$, let $\mathcal{M}$ be a closed k-dimensional submanifold of Ω, and let D be a relatively compact domain in $\mathcal{M}$ with bD of class $\mathcal{C}^2$. If $f \in \mathcal{C}(bD)$ has the property that for every one-dimensional submanifold Σ of a neighborhood of $\overline{D}$ in $\mathcal{M}$ that meets bD transversally, $f|(bD \cap \Sigma)$ continues holomorphically into $D \cap \Sigma$, then f extends holomorphically into D.*

The analysis involved in the proof of this result follows the lines of that given in [20], but here we need to draw also on some analytic geometry. It would be interesting to know whether a proof of this result can be given that is based on integral formulas, a proof in spirit similar to that Kytmanov.

As the hypotheses of our theorem are invariant under biholomorphic mappings, it is worthwhile to note that there is an abstract characterization of the domains we are considering: Suppose that $\mathcal{M}'$ is a complex manifold of dimension k and that D' is a relatively compact domain in $\mathcal{M}'$. Assume that $\mathcal{O}(\mathcal{M}')$, the space of functions holomorphic on $\mathcal{M}'$ separates points on $\mathcal{M}'$ and that $\mathcal{O}(\mathcal{M}')$ provides local coordinates at each point of $\mathcal{M}'$ in that if $p_0 \in \mathcal{M}'$, there exist $f_1, \ldots, f_k \in \mathcal{O}(\mathcal{M}')$ such that $p \mapsto \mathbf{f}(p) = (f_1(p), \ldots, f_k(p))$ carries a neighborhood of p_0 in $\mathcal{M}'$ biholomorphically onto a neighborhood of $\mathbf{f}(p_0)$. Then every relatively compact neighborhood $\mathcal{M}''$ of $\overline{D}$ in $\mathcal{M}'$ embeds as a closed submanifold of a domain in $\mathbb{C}^N$. In particular, the result obtains for relatively compact domains in Stein manifolds.

PROOF OF THEOREM, FIRST STEP. The first step in our proof is to establish the orthogonality relation that if β is a $\overline{\partial}$-exact $(k, k-1)$-form on a neighborhood of $\overline{D}$ in $\mathcal{M}$, then

$$(1) \qquad \int_{bD} f\beta = 0.$$

We assume, as we may without loss of generality, that $\mathcal{M}$ lies in no proper affine subspace of $\mathbb{C}^N$. As β is $\overline{\partial}$-exact, we may write $\beta = \overline{\partial}\alpha$ with

$$\alpha = \sum_{\substack{|J|=k-2 \\ |K|=k}} A_{JK} d\overline{z}^J \wedge dz^K$$

in the usual multi-index notation. Initially, the functions A_{JK} are to be smooth on a neighborhood of $\overline{D}$ in $\mathcal{M}$, but we can suppose them to be defined and smooth on all of $\mathbb{C}^N$ and to have compact support. We have

$$\beta = \sum_{\substack{|J|=k-2 \\ |K|=k}} \overline{\partial} A_{JK} \wedge d\overline{z}^J \wedge dz^K.$$

We write A_{JK} as an inverse Radon transform as in [20], [5]:

$$A_{JK}(z) = \int_{C_{JK}} \psi_{JK}(\zeta, \zeta \cdot z)\omega'(\zeta) \wedge \omega'(\overline{\zeta})$$

with

$$\zeta \cdot z = \zeta_1 z_1 + \ldots + \zeta_N z_N$$

and with

$$\omega'(\zeta) = \sum_{j=1}^{N} (-1)^{j-1}\zeta_j d\zeta_1 \wedge \cdots \wedge \widehat{d\zeta_j} \wedge \cdots \wedge d\zeta_N$$

and similarly for $\omega'(\overline{\zeta})$. The path C_{JK} of integration is an appropriate compact path that does not pass through the origin, and the functions $\psi_{JK} : \mathbb{C}^N \times \mathbb{C} \to \mathbb{C}$ are smooth.

We have that

$$\int_{bD} f\beta = \sum_{\substack{|J|=k-2 \\ |K|=k}} \int_{C_{JK}} \left\{ \int_{bD} f\overline{\partial}_z \psi_{JK}(\zeta, \zeta \cdot z) d\overline{z}^J \wedge dz^K \right\} \omega'(\zeta) \wedge \omega'(\overline{\zeta}),$$

and we shall show that for fixed ζ, the integral in the braces is zero. The functional $z \mapsto \zeta \cdot z$ on $\mathbb{C}^N$ is not identically zero, so we can introduce new coordinates $w_1, \ldots, w_N$ on $\mathbb{C}^N$, the w's related linearly to the z's, so that $\zeta \cdot z = w_1$. Thus

$$d\overline{z}^J \wedge dz^K = \sum_{\substack{|R|=k-2 \\ |S|=k}} c_{RS} d\overline{w}^R \wedge dw^S$$

for some constants c_{RS}. Since

$$\overline{\partial}_z \psi_{JK}(\zeta, \zeta \cdot z) = \overline{\partial}_w \psi_{JK}(\zeta, w_1)$$
$$= \frac{\partial \psi_{JK}}{d\overline{w}_1} d\overline{w}_1,$$

the integral in braces is $\displaystyle\sum_{\substack{|R|=k-2 \\ |S|=k}} \mathcal{I}_{RS}$ if we denote by $\mathcal{I}_{RS}$ the integral

(2)
$$\mathcal{I}_{RS} = \int_{bD} f \frac{\partial \psi_{JK}}{\partial \overline{w}_1}(\zeta, w) d\overline{w}_1 \wedge d\overline{w}^R \wedge dw^S.$$

There are various cases to be dealt with.

First, we may have $1 \in R$, in which case $\mathcal{I}_{RS}$ vanishes because $d\overline{w}_1 \wedge d\overline{w}^R = 0$.

Assume now that $1 \notin R$ and denote by R' the multi-index obtained by adjoining 1 at the beginning of R.

The second case is that $R' \subset S$. (Recall that $|R'| = k - 1$ and $|S| = k$.) Suppose, for notational simplicity, that $R' = (1, \ldots, k-1)$ and $S = (1, \ldots, k)$. We are therefore considering

$$\mathcal{I}_{RS} = \int_{bD} f(w) \frac{\partial \psi_{JK}}{\partial \overline{w}_1}(\zeta, w_1) dw_k \wedge d\overline{w}^R \wedge dw^R.$$

Let $\pi : \mathbb{C}^N \to \mathbb{C}^{k-1}$ be the projection given by $\pi(w) = w' = (w_1, \ldots, w_{k-1})$. According to Fubini's theorem—see [21] for a version appropriate to our purposes—this integral can be written as

$$\mathcal{I}_{RS} = \int_{\mathbb{C}^{k-1}} \frac{\partial \psi_{JK}}{\partial \overline{w}_1}(\zeta, w_1) \left\{ \int_{\pi^{-1}(w') \cap bD} f(w) dw_k \right\} d\overline{w}^R \wedge dw^R.$$

For almost every $w' \in \mathbb{C}^{k-1}, \pi^{-1}(w) \cap \mathcal{M}$ is nonsingular and meets bD transversally. According to our hypothesis f extends holomorphically through the slice $\pi^{-1}(w) \cap D$, so Cauchy's theorem implies the vanishing of the last integral.

It remains for us to deal with the integral 2) in case that R' is not a subset of S. We again suppose that $R' = (1, \ldots, k-1)$, and we consider the projection π used above. Denote by E the compact set in $\mathbb{C}^{k-1}$ that consists of the critical values of $\pi|bD$ together with the set E' obtained as follows: The map $\pi|\mathcal{M}$ is holomorphic, and we may assume that it has (complex) rank $k - 1$ off a certain proper variety V in $\mathcal{M}$. Only finitely many global branches of V, say $V_1, \ldots, V_r$, meet $\overline{D}$. The set E' is defined to be the compact set $\pi\big((V_1 \cup \cdots \cup V_r) \cap \overline{D}\big)$, which has measure zero in $\mathbb{C}^{k-1}$. For every $w' \notin E$, $\pi^{-1}(w) \cap \mathcal{M}$ is an analytic variety that is nonsingular in a neighborhood of $\overline{D}$ and that meets bD transversally.

Consider a $w' \in \pi(\overline{D})\backslash E$. There is a neighborood $D'(w')$ of $\overline{D}$ in $\mathcal{M}$ in which $\pi^{-1}(w')$ is nonsingular. In $D'(w')$, there is a Stein neighborood, Ω, of the Stein manifold $\pi^{-1}(w') \cap D'(w')$ [19], [17]. It follows that a neighborhood $\Omega_1 \subset \Omega$ of $\pi^{-1}(w') \cap D'$ can be identified with a neighborhood of the zero section of the normal bundle $\mathcal{N}$ to the embedding $\pi^{-1}(w') \cap D' \hookrightarrow \mathcal{M}$ [11], pp. 256–257. Let $\eta : \mathcal{N} \to \pi^{-1}(w') \cap D'(w')$ be the bundle projection. As $\pi^{-1}(w') \cap D'(w')$ is a nonsingular curve, the bundle $\mathcal{N}$ is trivial [14], [9].

There is a zero-free holomorphic one-form, say $\vartheta_{w'}$, on the curve $\pi^{-1}(w') \cap D'(w')$, and $\eta^* \vartheta_{w'}$ is a holomorphic form on $\mathcal{N}$. If we identify Ω_1 with a neighborhood of the zero section of $\mathcal{N}$ as above and if we then consider $\eta^* \vartheta_{w'}$ as a form on Ω_1 by way of this identification, we recognize that

$$\eta^* \vartheta_{w'} \wedge \pi^*(dw_1 \wedge \ldots \wedge dw_{k-1})$$

is a nowhere vanishing holomorphic k-form on a neighborhood of $\pi^{-1}(w') \cap D'(w')$.

As the form dw^S is a holomorphic k-form on Ω_1, we may write

$$dw^S = F \eta^* \vartheta_{w'} \wedge dw_1 \wedge \ldots \wedge dw_{k-1}$$

for some function F *holomorphic* on Ω_1.

Thus, for some sufficiently small ball $B(w', \delta)$ centered at w' and of radius δ, we can write

$$\int\limits_{bD \cap \pi^{-1}(B(w',\delta))} f \frac{\partial \psi_{JK}}{\partial \overline{w}_1}(\zeta, w_1) d\overline{w}_1 \wedge d\overline{w}^R \wedge dw^S$$

$$= \int\limits_{B(w',\delta)} \frac{\partial \psi_{JK}}{\partial \overline{w}_1}(\zeta, w_1) \left\{ \int\limits_{\pi^{-1}(w) \cap bD} f F \eta^* \vartheta_{w'} \right\} d\overline{w}^{R'} \wedge dw^{R'}.$$

The inner integral again vanishes by Cauchy's theorem.

Almost all of the set $\pi(\overline{D}\backslash E)$ can be covered by a disjoint family of the balls $B(w', \delta)$ by the Vitali covering theorem, so we find that in the present case also the integral $\mathcal{I}_{RS}$ vanishes.

We have now proved that the integral 1) vanishes whenever β is a $\overline{\partial}$-exact $(k, k-1)$-form defined on a neighborhood of $\overline{D}$. In particular, f satisfies the tangential Cauchy-Riemann equations.

We will now prove the theorem in the two-dimensional case. Thus, suppose $k = 2$.

LEMMA 1. *Let Σ be a one-dimensional complex submanifold of a neighborhood of $\overline{D}$ that is transverse to bD. There is a neighborhood V of $\Sigma \cap \overline{D}$ in $\mathcal{M}$ such that f continues holomorpically into $V \cap D$.*

PROOF. The noncompact Riemann surface Σ is a Stein manifold, and so a neighborhood of it in M is biholomorphic for a neighborhood of the zero section of the normal bundle of the embedding of Σ in M. As holomorphic vector bundles on open Riemann surfaces are trivial, a neighborhood W of Σ is biholomorphically equivalent to $\Sigma \times U$, U the open unit disc in $\mathbb{C}$, under a map $p \mapsto (\pi(p), z(p))$ where π is a holomorphic retraction of W onto Σ and z is a holomorphic function on W with $z^{-1}(0) = \Sigma$. We have that $\Sigma = z^{-1}(0) = \Sigma \times \{0\}$ is transverse to bD, so if $\zeta \in \mathbb{C}$ is near the origin, then $\Sigma \times \{\zeta\} = z^{-1}(\zeta)$ is also transverse to bD; by rescaling if necessary, we may suppose that for every $\zeta \in U$, U the open unit disc, $z^{-1}(\zeta)$ is transverse to bD.

Fix an $r \in (0,1)$, and put

$$\Delta(r) = \{p \in W \cap D : |z(p)| < r\}.$$

We shall prove that $f|(b\Delta(r) \cap bD)$ continues holomorphically into $\Delta(r)$. Note, to begin with, that there is a natural way to continue $f|(b\Delta(r) \cap bD)$ into $\Delta(r)$ as a continuous function: $f|(z^{-1}(\zeta) \cap bD)$ continues holomorphically into the Riemann surface $D \cap z^{-1}(\zeta)$ for every $\zeta \in U$. This slice-wise continuation gives a continuous function on $\Delta(r)$.

We have to prove that this continuation is holomorphic. The question is entirely local. Fix a point $p_0 \in \Delta(r)$. The point $p_0' = \pi(p_0)$ lies in Σ, and there is a holomorphic function $\tilde{g}$ on Σ with $\tilde{g}(p_0') = 0$, $d\tilde{g}(p_0') \neq 0$, $\tilde{g}$ zero-free on $\Sigma \setminus \{p_0'\}$. Define $g : W \to \mathbb{C}$ by $g(p) = \tilde{g}(\pi(p))$.

We can define forms $\omega_{p'}'(\overline{g}, \overline{z})$ and $\omega(g,z)$ on W by

$$\omega_{p'}'(\overline{g}, \overline{z}) = (\overline{g}(p) - \overline{g}(p'))d\overline{z} - (\overline{z}(p) - \overline{z}(p'))d\overline{g}$$

and

$$\omega(g,z) = dg \wedge dz.$$

The kernel

$$\frac{\omega_{p'}'(\overline{g}, \overline{z}) \wedge \omega(g,z)}{\{|g(p) - g(p')|^2 + |z(p) - z(p')|^2\}^2}$$

is defined for p' near p_0 and $p \in b\Delta(r) \cap bD$.

Consequently, we can define $F(p')$ for p' near p_0 by

$$F(p') = \left(\frac{1}{2\pi i}\right)^2 \int_{b\Delta(r) \cap bD} f(p) \frac{\omega_{p'}'(\overline{g}, \overline{z}) \wedge \omega(g,z)}{\{|g(p) - g(p')|^2 + |z(p) - z(p')|^2\}^2}$$

(3)

$$+ \left(\frac{1}{2\pi i}\right)^2 \int_{\{|z|=r\} \cap bD} f(p) \left\{\frac{1}{z(p) - z(p')}\right\} \frac{\{\overline{g}(p) - \overline{g}(p')\}\omega(g,z)}{\{|g(p) - g(p')|^2 + |z(p) - z(p')|^2\}}.$$

In the first of these integrals, $b\Delta(r) \cap bD$ is given the orientation induced on it as part of $b(\mathcal{N} \cap D)$, and $\{|z| = r\} \cap bD$ is oriented with the orientation induced on it as $b(b\Delta(r) \cap bD)$.

The function F is holomorphic in p' when p' is near p_0. To see this, notice that

$$\bar\partial\left\{\left(\frac{1}{z(p)-z(p')}\right)\left(\frac{\overline{g}(p)-\overline{g}(p')}{|g(p)-g(p')|^2+|z(p)-z(p')|^2}\right)\right\}$$
$$=\frac{-\omega'_{p'}(\overline{g},\overline{z})}{\left(|g(p)-g(p')|^2+|z(p)-z(p')|^2\right)^2}$$

and

$$\bar\partial\left\{\left(\frac{1}{g(p)-g(p')}\right)\left(\frac{\overline{z}(p)-\overline{z}(p')}{|g(p)-g(p')|^2+|z(p)-z(p')|^2}\right)\right\}$$
$$=\frac{\omega'_{p'}(\overline{g},\overline{z})}{\left(|g(p)-g(p')|^2+|z(p)-z(p')|^2\right)^2}$$

(In these equations, p' is fixed, and the differentiations are with respect to the variable point p.)

Stokes's theorem lets us write that, for small $\varepsilon>0$, and for p' with $|z(p')-z(p_0)|$ small,

$$F(p')=\left(\frac{1}{2\pi i}\right)^2\int\limits_{\{|z(p)-z(p_0)|<\varepsilon\}\cap bD}+\left(\frac{1}{2\pi i}\right)^2\int\limits_{\{|z(p)-z(p_0)|=\varepsilon\}\cap bD}$$

(4)
$$=I_\varepsilon+J_\varepsilon$$

where the integrands that appear in the integrals in (4) are the same as those that appear in (3). That Stokes's theorem can be applied is a consequence of [15], Proposition I.9, because f satisfies the tangential Cauchy-Riemann equations. As ε is small, $g(p)-g(p')\neq0$ for p in $\{|z(p)-z(p_0)|<\varepsilon\}\cap bD$, and thus

$$I_\varepsilon=(\frac{1}{2\pi i})^2\int\limits_{\{|z(p)-z(p_0)|=\varepsilon\}\cap bD} f(p)\frac{1}{g(p)-g(p')}\left(\frac{\overline{z}(p)-\overline{z}(p')}{|g(p)-g(p')|^2+|z(p)-z(p')|^2}\right)\omega(g,z).$$

Thus,

$$F(p')=I_\varepsilon+J_\varepsilon$$
$$=\left(\frac{1}{2\pi i}\right)^2\int\limits_{\{|z(p)-z(p_0)|=\varepsilon\}\cap bD}\frac{f(p)}{|g(p)-g(p')|^2+|z(p)-z(p')|^2}$$
$$\times\left\{\frac{\overline{z}(p)-\overline{z}(p')}{g(p)-g(p')}+\frac{\overline{g}(p)-\overline{g}(p')}{z(p)-z(p')}\right\}\omega(g,z)$$
$$=\left(\frac{1}{2\pi i}\right)^2\int\limits_{\{|z(p)-z(p_0)|=\varepsilon\}\cap bD}\frac{f(p)\omega(g,z)}{\left(g(p)-g(p')\right)\left(z(p)-z(p')\right)}.$$

This integral depends holomorphically on p', so long as p' is near p_0. Thus, F is seen to be holomorphic.

We will now verify that $F(p_0)$ is the value obtained by continuing $f|\{z=z(p_0)\}\cap bD$ through the slice $\{z=z(p_0)\}\cap D$. To do this, we take $p'=p_0$ in (3) and let $\varepsilon \to 0$. As $\varepsilon \to 0$, the volume of the domain of integration in I_ε shrinks to zero, and the integrand stays bounded, so $I_\varepsilon \to 0$. This means that

$$F(p_0) = \lim_{\varepsilon \to 0} J_\varepsilon$$

$$= \lim_{\varepsilon \to 0} \left(\frac{1}{2\pi i}\right)^2 \int\limits_{\{|z(p)-z(p_0)|=\varepsilon\} \cap bD} \frac{1}{z(p)-z(p_0)} \left\{ \frac{f(p)\overline{g}(p)}{|g(p)|^2 + \varepsilon^2} \right\} dg \wedge dz.$$

We apply Fubini's theorem to find that

$$F(p_0) = \frac{1}{2\pi i} \int\limits_{\{z(p)=z(p_0)\} \cap bD} f(p) \frac{dg}{g(p)} .$$

By the Cauchy integral theorem and the choice of g, the last integral is the value at p_0 of the analytic continuation of $f|(\{z = z(p_0)\} \cap bD)$ through the slice $\{z = z(p_0)\}$.

To proceed, we shall need the following standard transversality fact.

LEMMA 2. *Let* Γ *be a closed submanifold of class* $\mathcal{C}^2$ *of an open subset of* $\mathbb{C}^N$, $\dim \Gamma = 2k - 1$. *If* $p_0 \in \Gamma$, *then almost every complex affine subspace of* $\mathbb{C}^N$ *that passes through* p_0 *and that has (complex) dimension* $N - k + 1$ *meets* Γ *transversely.*

The *almost every* is understood in the sense of the natural measure on the Grassmanian $\mathcal{G}_{N,N-k+1}(p_0)$ of affine subspaces of $\mathbb{C}^N$ of dimension $N - k + 1$ through p_0.

PROOF. As we may without loss of generality, we assume p_0 to be the origin. Denote by $\mathbb{C}^{N,N-k+1}$ the space of complex matrices of size N by $N - k + 1$, and define $F : \mathbb{C}^{N-k+1} \times \mathbb{C}^{N,N-k+1} \to \mathbb{C}^N$ by $F(z, A) = Az$. This map is holomorphic.

The differential of the map F at the point $(z_0, A_0) \in \mathbb{C}^{N-k+1} \times \mathbb{C}^{N,N-k+1}$ is the linear map $DF_{(z_0,A_0)} : \mathbb{C}^{N-k+1} \times \mathbb{C}^{N,N-k+1} \to \mathbb{C}^N$ given by

$$DF_{(z_0,A_0)}(z, A) = A_0 z + A z_0,$$

which is surjective whenever $z_0 \neq 0$: If $z_0 \neq 0$, then as A ranges through $\mathbb{C}^{N,N-k+1}$, the point $A z_0$ ranges over all of $\mathbb{C}^N$. Thus, $DF_{(z_0,A_0)}$ carries $\{0\} \times \mathbb{C}^{N,N-k+1}$ on to $\mathbb{C}^N$, provided $z_0 \neq 0$.

Thus, we can apply the result labelled *The Transversality Theorem* in [10], p. 68, to conclude that for almost every $A \in \mathbb{C}^{N,N-k+1}$ the partial map $z \mapsto F(z, A)$ is transversal to $\Gamma \setminus \{0\}$.

There is a point here that needs attention. The result quoted from [10] assumes that the manifold and maps in question are smooth, *i.e.*, of class $\mathcal{C}^\infty$, but in the case at hand, our manifold Γ is only of class $\mathcal{C}^2$. This is not a problem for us, however, for the following reason. We set $W = F^{-1}(\Gamma \setminus \{0\})$, a $\mathcal{C}^2$-submanifold of $\mathbb{C}^{N-k+1} \times \mathbb{C}^{N,N-k+1} \setminus F^{-1}(0)$. Let $\pi : \mathbb{C}^{N-k+1} \times \mathbb{C}^{N,N-k+1} \to \mathbb{C}^{N,N-k+1}$ be the natural projection. Then, $\pi|W$ is a $\mathcal{C}^2$-map from W into $\mathbb{C}^{N,N-k+1}$. The real dimension of $\mathbb{C}^{N,N-k+1}$ is $2N(N-k+1)$, and the real dimension of W is $2[(N=k+1)+N(N-k+1)-N]+2k-1 = 2N(N-k+1)+1$. Thus, the sharp version of Sard's Theorem given in [4], p. 316 yields that almost every point in $\mathbb{C}^{N,N-k+1}$ is a regular value for $\pi|W$.

Granted this observation, the proof of the transversality theorem given in [10] yields the result stated in the next to last paragraph. From it, we get immediately the conclusion of the lemma.

We continue to work in the two-dimensional case of the theorem. Let $p \in bD$. The transversality lemma just given implies the existence of a one-dimensional complex submanifold Σ in a neighborhood of $\overline{D}$ in $\mathcal{M}$ that is transverse to bD, and that passes through the point p. By Lemma 1, f extends holomorphically into a neighborhood of Σ, and in particular, there is a neighborhood W_p of p in M such that f continues holomorphically into $W_p \cap D$. Consequently, there is a neighborhood, say Ω, of bD in $\mathcal{M}$ such that f continues holomorphically into the open set $\Omega \cap D$.

We shall show that f continues through all of D.

Let us write

$$bD = \Gamma_1 \cup \ldots \cup \Gamma_s$$

where each Γ_j is a connected real submanifold of dimension three in $\mathcal{M}$, Γ_j of class $\mathcal{C}^2$. As D is supposed to be relatively compact, each Γ_j is compact.

Since Γ_j is a compact real hypersurface in the complex manifold $\mathcal{M}$, it is *maximally complex* in the sense of [12] and [13], and so there is an irreducible 2-dimensional variety X_j in $\mathbb{C}^N \backslash \Gamma_j$ such $\overline{X}_j = X_j \cup \Gamma_j$ is compact. The variety X_j has finite volume and satisfies the current equation $b[X_j] = \pm[\Gamma_j]$ where $[\overline{X}_j]$ denotes the current of integration over the variety X_j, and $[\Gamma_j]$ denotes the current of integration over the manifold Γ_j, which is endowed with the orientation induced on it as part of bD. Moreover, outside a thin set, the pair (X_j, Γ_j) is, near Γ_j, a $\mathcal{C}^2$ manifold with boundary.

Let us fix attention on Γ_1. Choose a point $p \in \Gamma_1$ and a $\delta > 0$ so small that $B(p,\delta) \cap \overline{X}_1$ is a $\mathcal{C}^2$ manifold with boundary. We may choose δ so small that $\Gamma_{1p} = B(p,\delta) \cap \Gamma_1$ is a 3-dimensional disk that is very close to a diameter of $B(p,\delta)$ and also so small that $(\mathcal{M}\backslash \Gamma_1) \cap B(p,\delta)$ consists of two components, Δ and Δ' with $\Delta \subset D$ and $\Delta' \subset \mathcal{M}\backslash\overline{D}$.

Denote by $Y_{1,p}$ the component of $X_1 \cap B(p,\delta)$ that abuts $B(p,\delta) \cap \Gamma_1$.

The two k-dimensional varieties $Y_{1,p}$ and Δ in $B(p,\delta)\backslash\Gamma_{1p}$ satisfy $\overline{Y}_{1p} \supset \Gamma_{1p}$ and $\overline{\Delta} \supset \Gamma_{1p}$, so by a theorem of Chirka [3], Th. 19.2, p. 193, only two cases are possible: 1) $\Delta = Y_{1p}$ or else 2) $\Delta \cup \Gamma_{1p} \cup Y_{1p}$ is a variety in $B(p,\delta)$.

In case 1) $\Delta \subset X_1$. This is so, for both Δ and $X_1' = X_1\backslash(\Gamma_1 \cup \ldots \cup \Gamma_s)$ are varieties in $\mathbb{C}^N\backslash(\Gamma_1 \cup \ldots \cup \Gamma_s)$, and Δ is irreducible. As Δ meets X_1' in an open set and is irreducible, it must be a branch of X_1', which implies $\Delta \subset X_1$ as claimed. In this case, $\Delta \cup \Gamma_{1p} \cup Y_{1p}$ is not a variety in $B(p,\delta)$.

Suppose now that Δ is not contained in X_1. Denote by E_1 the compact subset of Γ_1 that consists of the points at which the pair (X_1, Γ_1) is not a $\mathcal{C}^2$-

manifold with boundary. According to [13], the set E_1 has zero 3-dimensional measure. At each point $p \in \Gamma_1 \backslash E$, the set $\Delta \cup X \cup \Gamma_1$ is a variety, and the set $(D \cup X_1 \cup \Gamma_1) \backslash E$ is a purely 2-dimensional subvariety of $\mathbb{C}^N \backslash (\Gamma_2 \cup \ldots \cup \Gamma_s \cup E)$. By a removable singularity theorem of Shiffman [18], $D \cup X_1 \cup \Gamma_1$ is a subvariety of $\mathbb{C}^N \backslash (\Gamma_2 \cup \ldots \Gamma_s)$.

Next, we see that one of the X_j's contains D. If not, the set

$$Z = D \cup X_1 \cup \ldots \cup X_s \cup \Gamma_1 \cup \ldots \cup \Gamma_s$$

is compact and, by the analysis we have just given, it is a 2-dimensional variety. This is impossible: The only compact varieties in $\mathbb{C}^N$ are finite sets. Thus, some X_j, say X_1, contains the domain D.

It follows that $X_1 \supset \cup_{j=1}^s X_j$. To see this, consider X_2. We have that $X_1 \supset \Gamma_2 = \overline{X}_2 \backslash X_2$, so $X_2 \subset X_1$. In the same way, $X_1 \supset X_3, \ldots$.

The set X_1 is a bounded 2-dimensional subvariety of $\mathbb{C}^N \backslash \Gamma_1$ that contains D, and the function $f|\Gamma_1$ continues holomorphically into an open set $W \subset D$ with $bW = \Gamma_1 \cup S$ where $S \cap \Gamma_1 = \emptyset$. Call this continuation f_1. Thus, if we denote by Γ_1' a smooth hypersurface in D that is near Γ_1, then Γ_1' bounds a domain X_1' in X_1, and $f_1|\Gamma_1'$ is a smooth CR-function. It follows [12], Th. 4.11, that f_1 continues as a weakly holomorphic function into X_1'. Thus, f is seen to continue from Γ_1 into X_1 as a weakly holomorphic function. The restriction, F, of this function to D is holomorphic, and it agrees on Γ_1 with f.

We complete the proof in the two-dimensional case by showing that F coincides on $\Gamma_j, j = 2, \ldots, s$, with f.

To this end, fix a point $q_0 \in \Gamma_2$, and fix a surjective $\mathbb{C}$-affine transformation $T : \mathbb{C}^N \to \mathbb{C}^{k-1}$ with $T(q_0) = 0$. For $z \in \mathbb{C}^{k-1}$, the fiber $T^{-1}(z)$ is an affine subspace Π of $\mathbb{C}^N$ of dimension $N - k + 1$. If $z \in \mathbb{C}^{k-1}$ is chosen correctly, then $\Pi \cap \mathcal{M}$ will be a one-dimensional complex manifold that meets bD transversely and that contains a point $q_1 \in \Gamma_2$ near q_0. Connect q_1 to infinity (in $\Pi \cap \mathcal{M}$) with a smooth curve γ that lies in $\Gamma \cap \mathcal{M}$ and that emanates from q_1. The curve γ necessarily meets Γ_0. Let q_2 be the last point in γ that lies in $bD \backslash \Gamma_0$, and let Δ be the component of $\Pi \cap X_1$ that contains the component γ_1 of $\gamma \cap D$ that emanates from q_2. The other end point, q_3, of γ_1 lies in Γ_0. The functions F and f coincide on $\overline{\Delta} \cap \Gamma_0$, and $f|(\Pi \cap bD)$ continues holomorphically through the slice $\Pi \cap D$, say as f_Π. As F is holomorphic on D, it follows that f_Π and F must agree throughout Δ. Thus, they agree on the part of $b\Delta$ near q_2. Let Γ_j be the component of bD that contains q_2. Then, as our picture is stable under small perturbations, it follows that f and F must agree on an open subset of Γ_j and hence on all of it. We can repeat this process, mutatis mutandis, to find in a finite number of steps that f and F agree on the whole of bD, as we wished to prove.

At this point, Theorem 1 is proved in the particular case that $k = 2$.

However, granted the theorem in dimension two, the general case follows by induction on the dimension: Assume the theorem is correct for a given value of k, and consider Ω, $\mathcal{M}$ and D as in the statement of the theorem but with $\dim \mathcal{M} = k + 1, k \geq 2$. If we fix a point $p \in D$, then for almost all complex hyperplanes Π in the Grassmannian $\mathcal{G}_{N,N-1}(p)$, $\Pi \cap \mathcal{M}$ will be a complex submanifold of $\mathcal{M}$ of dimension k transverse to bD. The theorem, assumed valid in dimension k, implies that $f|b(\Pi \cap D)$ continues holomorphically through $\Pi \cap D$, say as $f_{p,\Pi}$. As $k \geq 2$, the value $f_{p,\Pi}(p)$ does not depend on the choice of Π: Two such Π's, say Π' and Π'' have the property that $\Pi' \cap \Pi'' \cap M$ contains at least an analytic curve, C, through p, and $f_{p,\Pi'}$ and $f_{p,\Pi''}$ agree on $C \cap b(\Pi \cap D)$. We define $F(p)$ to be the value this process yields. We have then a well-defined function $F : D \rightarrow \mathbb{C}$ that is bounded by $\sup\{|f(p)| : p \in bD\}$. Moreover, F is holomorphic on enough complex hypersurfaces in D that Hartogs's theorem applies to yield that F is holomorphic. As it assumes the desired boundary values f, the theorem is proved.

References.

1. M. L. Agranovskiĭ and R. E. Valskiĭ, *Maximality of invariant algebras of functions*, Sibirsk. Math. Zh. **12** (1971), 3–12; Sib. Math. J. **12** (1971), 1–7.

2. L. A. Aĭzenberg and A. P. Yuzhakov, *Integral Representations in Multidimensional Complex Analysis*, Translations of Mathematical Monographs, Vol 58, American Mathematical Society, Providence, 1983, pp. x + 283.

3. E. M. Čirka (Е. М. Чирка), Комплексные Аналитические Множества, Nauka, Moscow, 1985, 272 pp. "Complex Analytic Sets," Klumer, Dordrecht and Boston, 1989.

4. H. Federer, *Geometric Measure Theory*, Die Grundlehren der Mathematischen Wissen- schaften Bd. 153, Springer-Verlag, 1969, xiv + 676 pp.

5. I. M. Gel'fand, M. I. Graev, and N. Ya. Vilenkin, *Generalized Functions*, Vol. 5, Academic Press, New York and London, 1966, xvii + 449 pp.

6. J. Globevnik, *A family of lines for testing holomorphy*, Indiana Univ. Math. J. **36** (1987), 639–644.

7. J. Globevnik and E. L. Stout, *Boundary Morera theorems for holomorphic functions of several complex variables*, Duke Math. J. **64** (1991), 571–615.

8. E. Grinberg, *Boundary values of holomorphic functions, Radon transforms and the one- dimensional extension property*, preprint.

9. J. Guenot and R. Narasimhan, *Introduction à la théorie des surfaces de Riemann*, L'Enseignement Mathématique no. (II) XXI (1975), 123–328.

10. V. Guillemin and A. Pollack, *Differential Topology*, Prentice-Hall, Englewood Cliffs, 1974, xvi + 222 pp.

11. R. C. Gunning and H. Rossi, *Analytic Functions of Several Complex Variables*, Prentice-Hall, Englewood Cliffs, 1965, xii + 317 pp.

12. F. R. Harvey, *Holomorphic Chains and their boundaries*, Several Complex Variables, (Proc. Sympos. Pure Math., Vol. XXX, Part 1, Williams College, Williamstown, Mass., 1975), American Mathematical Society, Providence, 1977, pp. 309–352.

13. R. Harvey and B. Lawson, *On boundaries of complex analytic varieties. I,* Ann. Math (II) **102** (1975), 233–290.

14. A. Hurwitz and R. Courant, *Funktionentheorie*, Mit einen Anhang von H. Röhrl, 4. Auf., Springer-Verlag, Berlin, Göttingen, Heidelberg, New York, 1964, xiii + 706 pp.

15. G. Lupacciolu, *A theorem on holomorphic extension of CR-functions*, Pacific J. Math **124** (1986), 177–191.

16. A. Nagel and W. Rudin, *Moebius-invariant function spaces on balls and spheres*, Duke Math. J. **43** (1976), 841–865.

17. M. Schneider, *Tubenumgebungen Steinscher Räume*, Manuscripta Math. **18** (1976), 391–397.

18. B. Shiffman, *On the removal of singularities of analytic functions*, Mich. Math. J. **15** (1968), 111–120.

19. Y.-T. Siu, *Every Stein subvariety admits a Stein neighborhood*, Invent. Math. **38** (1976–77), 89–100.

20. E. L. Stout, *The boundary values of holomorphic functions of several complex variables*, Duke Math. J. **44** (1977), 105–108.

21. R. Sulanke and P. Wintgen, *Differentialgeometrie und Faserbündel*, Birkhäuser Verlag, Basel und Stuttgart, 1972.

DEPARTMENT OF MATHEMATICS, UNIVERSITY OF WASHINGTON, SEATTLE, WASHINGTON 98195

Contemporary Mathematics
Volume **137**, 1992

Analytic Γ-Almost-Periodic
Structures in Algebra Spectra

TOMA V. TONEV

Dedicated to Professor Walter Rudin.

1. Introduction

One of the central themes in uniform algebra theory is to locate some kind of analytic structure in algebra spectra such that the Gelfand extensions of algebra functions are analytic with respect to this analytic structure. This special interest is based on the fact that the existence of an analytic structure provides us with various regularity theorems and integral representations of algebra functions, with theorems for representing measures, derivations, approximations, etc.

Recall that a subset U in the spectrum $sp\,B$ of a commutative Banach algebra B has an *analytic*, resp *analytic Γ-almost-periodic structure* if there exists a nonconstant map ψ from a domain $\mathcal{W}$ in the complex plane $\mathbf{C}$ into U such that the composition $\widehat{f} \circ \psi$ is an analytic, resp. analytic Γ-almost-periodic function in $\mathcal{W}$ for each $f \in A$.

The problem of existence of one-dimensional and multi-dimensional analytic structures which live in algebra spectra has been given a thorough investigation (e.g. Bishop [6], Aupetit and Wermer [2], Basener [3, 4], Bear and Hile [5], Sibony [10], Tonev [18] etc.). Comparatively less is known about analytic almost periodic structures which live in uniform algebra spectra. Various conditions for the existence of analytic $\mathbf{Q}$-almost-periodic structures in spectra are given e.g. in [13,14,15,17,19].

In the present paper we remove some of the requirements for the results in [14] and expand them for a class of subgroups of $\mathbf{Q}$.

2. Γ-analytic functions

Recall that a continuous function f on the real line $\mathbf{R}$ is called *almost periodic* if it can be approximated uniformly on $\mathbf{R}$ by exponential polynomials of type $\sum_{k=1}^{n} a_k e^{i s_k x}$, where s_k are real numbers. If all *exponents* s_k belong to a fixed

---

1991 *Mathematics Subject Classification*. Primary 46J10.

This paper is in final form and no version of it will be submitted for publication elsewhere.

additive subgroup Γ of $\mathbf{R}$, then f is called a Γ-*almost-periodic function* on $\mathbf{R}$. A function f analytic in a domain $D \subset \mathbf{C}$ is said to be *analytic* Γ-*almost-periodic* in D if it can be approximated locally uniformly in D by exponential polynomials of type $\sum_{k=1}^{n} a_k e^{i s_k z}$, where s_k belong to the semigroup $\Gamma_o = [0, \infty) \cap \Gamma$. A continuous function f in $D \subset \mathbf{C}$ is Γ-*almost-periodic* in D if it can be approximated locally uniformly in D by functions of type $\sum_{k=0}^{n} a_k e^{i s_k z} + \sum_{l=1}^{m} a_l \overline{e^{i t_l z}}$, where s_k, t_l belong to Γ, $s_k \geq 0$, $t_l < 0$.

The algebra of bounded analytic Γ-almost-periodic functions in the upper half-plane $\mathbf{H} = \{z : \operatorname{Im} z > 0\}$ which are continuous on $\overline{\mathbf{H}}$ is isometrically isomorphic to the big disc algebra of so called *generalized analytic functions* in the sense of Arens and Singer [1]. This explains the tight relation between the class of analytic Γ-almost-periodic functions in domains of the complex plane and the class of Γ-analytic functions in domains of the big plane.

Let Γ be an additive subgroup of the real line $\mathbf{R}$. Γ_o will denote the "nonnegative" subsemigroup $[0, \infty) \cap \Gamma$ of Γ, and $G = \widehat{\Gamma}_d$ will denote the dual group of the group Γ equipped with the discrete topology. In fact G coincides with *all* homomorphisms from Γ into the unit circle S^1. G is a compact and connected group.

DEFINITION 1. *The big plane (or generalized plane) over a compact group G whose dual group is a subgroup of $\mathbf{R}$ is the cone $\mathbf{C}_G$ over G with generator $[0, \infty)$ provided with the factor topology, i.e.* $\mathbf{C}_G = [0, \infty) \times G / \{0\} \times G$.

The points in the big plane $\mathbf{C}_G$ will be denoted by $\mathbf{z}$, or $r \cdot g$, where $r \in [0, \infty)$ and $g \in G$ are the "polar coordinates" of $\mathbf{z}$. We regard all points of type $0 \cdot g$, $g \in G$, as identified to a point. In the sequel we denote this point by $\mathbf{0}$. Thus $\mathbf{0} = 0 \cdot g$ for every $g \in G$. The points of type $1 \cdot g$, where again $g \in G$, will be denoted in the sequel by g. The nonnegative number r, denoted also by $|\mathbf{z}|$, is called the *modulus* of $\mathbf{z}$.

DEFINITION 2. *We call the set* $\Delta_G(R) = [0, R) \times G / \{0\} \times G = \{\mathbf{z} \in \mathbf{C}_G : |\mathbf{z}| < R\}$ *the (open) big disc (or generalized disc) over G with radius R.*

Clearly $\Delta_G(1) = [0, 1) \times G / \{0\} \times G = \operatorname{Hom}(\Gamma, \overline{\Delta})$. The unit big disc $\Delta_G(1)$ will be denoted simply by Δ_G. Note that $G_R = \{R\} \times G$ is the topological boundary of the closed big disc $\overline{\Delta}_G(R) = [0, R] \times G / \{0\} \times G$. In particular, $G = G_1$ is the boundary of Δ_G. For a fixed $a \in \Gamma_o \setminus \{0\}$ by χ^a will be denoted the function

$$\chi^a(\mathbf{z}) = \chi^a(r \cdot g) = r^a g(a)$$

which maps G into S^1, and $\chi^o(g) \equiv 1$. We shall use also the following notations: $\mathbf{z}^a = (r \cdot g)^a = r^a g^a = \chi^a(\mathbf{z})$, where by g^a we mean $g(a)$. Observe that g is a character of the group Γ and the function $\chi^a(g) = g(a) = g^a$ is a character of G. Clearly, in the general case we have that $g \neq g^a = \chi^a(g) \in \mathbf{C}$ for any $a \in \Gamma_o$.

DEFINITION 3. *A polynomial (also Γ-polynomial, generalized polynomial) in $\mathbf{C}_G$ is any linear combination $P(\mathbf{z})$ of the functions $\mathbf{z}^a$, $a \in \Gamma_o$, with complex* *coefficients, i.e.*

$$P(\mathbf{z}) = \sum_{k=1}^{n} c_k \mathbf{z}^{a_k} = \sum_{k=1}^{n} c_k (r \cdot g)^{a_k} = \sum_{k=1}^{n} c_k \chi^{a_k}(r \cdot g),$$

where $a_k \in \Gamma_o$ and $\mathbf{z} = r \cdot g \in \mathbf{C}_G$. The space $A(\Delta_G)$ of all continuous functions on the closed unit big disc $\overline{\Delta}_G$ which can be approximated by polynomials on $\overline{\Delta}_G$ with respect to the uniform norm $\|f\| = \max\limits_{\mathbf{z} \in \overline{\Delta}_G} |f(\mathbf{z})|$ in $C(\overline{\Delta}_G)$ is called the big disc algebra. The functions in the algebra $A(\Delta_G)$ are called Γ-analytic functions (also generalized analytic functions in the sense of Arens-Singer) in the big disc $\overline{\Delta}_G$.

The algebra $A(\Delta_G)$ of Γ-analytic functions in the big disc is a direct generalization of the disc algebra $A(\Delta)$. Namely, as one can see, $A(\Delta)$ coinsides precisely with the big disc algebra $A(\Delta_G)$. Indeed, by choosing $\Gamma = \mathbf{Z}$ (and therefore $\Gamma_o = \mathbf{Z}_o = \mathbf{Z}_+$) we get: $G = S^1$, $\mathbf{C}_G = \mathbf{C}$ and $\chi^a(z) = \chi^a(r \cdot e^{i\theta}) = r^a e^{ia\theta} = z^a$ for every $a \in \mathbf{Z}_o$. Thus the big plane $\mathbf{C}_{S^1}$ coincides with the complex plane $\mathbf{C}$; $\mathbf{Z}$-analytic functions are classical analytic functions in Δ which are continuous up to the boundary S^1.

For any real number s define the character $\eta_s \in \widehat{\Gamma} = G$ as $\eta_s(a) = e^{ias}$, $a \in \Gamma$. The induced mapping $j_e : s + it \longmapsto (-\ln t) \cdot \eta_s$ is an isomorphism of the complex plane $\mathbf{C}$ onto a dense subset of $\mathbf{C}_G$ (see e.g. [8]). Recall that the composition $f \circ j_e$ is an analytic Γ-almost-periodic function in $\mathbf{C}$ for every Γ-analytic function f in $\mathbf{C}_G$.

Denote by Γ' the set $(0,1] \cap \Gamma$. In the sequel we require the group $\Gamma = \widehat{G}$ to possess the following property:

$(*)$

    (i) $\mathbf{Z} \subset \Gamma \subset \mathbf{Q}$, and

    (ii) *for any $a, b \in \Gamma'$ there exists a $c \in \Gamma'$ and integers k, l, such that $ck = a$ and $cl = b$.*

Clearly, the group $\mathbf{R}$ of reals satisfies neither one of these conditions. The group $\mathbf{Q}$ of rationals, the group $\mathbf{Z}$ of integers, the group of dyadic rationals are examples of groups Γ which possess property $(*)$.

Recall that the topology in $\mathbf{C}_G$ is generated by neighborhoods of type $Q = Q(\mathbf{z}_o; \chi^{a_1}, \ldots, \chi^{a_n}; \varepsilon) = \{\mathbf{z} \in \mathbf{C}_G : |\chi^{a_j}(\mathbf{z}) - \chi^{a_j}(\mathbf{z}_o)| < \varepsilon, \mathbf{z}_o \in \mathbf{C}_G, a_j \in \Gamma', \varepsilon > 0, j = 1, \ldots, n\}$. If Γ possesses property $(*)$, then the neighborhoods of type

(1) $\quad Q = Q(\mathbf{z}_o, a, \varepsilon) = Q(\mathbf{z}_o; \chi^a; \varepsilon) = \{\mathbf{z} \in \mathbf{C}_G : |\chi^a(\mathbf{z}) - \chi^a(\mathbf{z}_o)| < \varepsilon,$
$\quad\quad\quad \mathbf{z}_o \in \mathbf{C}_G, a \in \Gamma', \varepsilon > 0\},$

which depend on single functions χ^a only, form a basis of the topology in $\mathbf{C}_G$. Indeed, let $Q = Q(\mathbf{z}_o; \chi^{a_1}, \ldots, \chi^{a_n}; \varepsilon)$. By $(*)$ it follows that there exists a number $c \in \Gamma'$ and positive integers $k_1, \ldots, k_n$ such that $ck_j = a_j$, $j = 1, \ldots, n$. It is clear that the neighborhood $\{z \in \mathbf{C}_G : |\chi^c(z) - \chi^c(z_o)| < \delta\}$ is contained in the set $\{z \in \mathbf{C}_G : |\chi^{a_j}(z) - \chi^{a_j}(z_o)| < \varepsilon, \ z_o \in \mathbf{C}_G, \ a_j \in \Gamma', \ j = 1, \ldots, n, \ \varepsilon > 0\}$ for some $\delta > 0$ because for every $j = 1, \ldots, n$, one can find numbers $\delta_j > 0$ such that

$$|\chi^a(z) - \chi^a(z_o)| = |(\chi^c)^{k_j}(z) - (\chi^c)^{k_j}(z_o)| < \varepsilon$$

whenever $|\chi^c(z) - \chi^c(z_o)| < \delta_j$.

Let K be a compact set in the big plane $\mathbf{C}_G$. As usual $P(K)$ denotes the space of all continuous functions on K which can be approximated uniformly on K by Γ-polynomials. Similarly, $R(K)$ denotes the space of all continuous functions on K which are approximable uniformly on K by bounded Γ-rational functions (i.e. ratios of Γ-polynomials), and $A(K)$ denotes the space of all continuous functions on $\overline{K}$ which are locally approximable by Γ-polynomials.

Let Q be a neighborhood from the basis (1) of $\mathbf{C}_G$, which does not contain the origin $\mathbf{0}$ of $\mathbf{C}_G$.

Note that unlike the classical $\mathbf{Z}$-case, the neighborhoods Q from the basis (1) in general, are *not* homeomorphic to the unit big-disc. The next result is a version of maximality theorem for the algebras of type $P(\overline{Q})$, where Q is a basis neighborhood in $\mathbf{C}_G$.

THEOREM 1. *Let the group Γ possesses property (*) and let*

$$Q = Q(z_o, b, \varepsilon) = \{z \in \mathbf{C}_G : |\chi^b(z) - z_o^b| < \varepsilon, \ b \in \Gamma_o, \ \varepsilon > 0\}$$

be a neighborhood of $z_o \in \mathbf{C}_G$ of type (1) which does not contain $\mathbf{0}$. Then $P(\overline{Q})$ is a maximal algebra and a Dirichlet algebra.

PROOF. Since Q is a Γ-polynomially convex set in the big-plane $\mathbf{C}_G$, we have that $sp\, P(\overline{Q}) = \overline{Q}$ and $\partial P(\overline{Q}) = b\overline{Q}$ (see e.g. [13]), where $\partial P(\overline{Q})$ is the Shilov boundary of algebra $P(\overline{Q})$. Let B be a uniform algebra on bQ such that $P(\overline{Q})|_{bQ} \subset B \subset C(bQ)$. Assume that all functions of type $\chi^a(z) - z_o^a$, where $ak = b$ for some $a \in \Gamma'$ and some integer $k \in \mathbf{Z}_o$, are invertible in B. In this case B contains all functions of type $(\chi^a(z) - z_o^a)|_{bQ}^{-1}$ with $ak = b$, and consequently, $B \supset R(bQ) = C(bQ)$ (see [16, 20]). We conclude that now $B = C(bQ)$. The same conclusion is true in the case when

$$a' = \inf\{a \in \Gamma' : (\chi^a(z) - z_o^a)|_{bQ} \notin B^{-1}\} > 0.$$

Assume now that $a' = 0$. By (*) the function $(\chi^b(z) - z_o^b)|_{bQ}$ is noninvertible in B because $(\chi^a(z) - z_o^a)|_{bQ}$ is noninvertible for some $a \in \Gamma'$, where $ak = b$, $k \in \mathbf{Z}_o$. Since $B \subset C(bQ) = R(bQ)$, all functions in B are uniformly approximable on bQ by Γ-rational functions. Observe that every Γ-rational function r in B can be presented as $r = r_1 \circ \chi^d$ for some $d \in \Gamma'$ with $dk = b$, $k \in \mathbf{Z}_o$, where r_1 is a rational function in $\mathbf{C}$. Denote for a while by $R_d(bQ)$ and $P_d(bQ)$ the algebras $R_d(bQ) = [\{r \circ \chi^d|_{bQ} : r \in R(\chi^d(bQ))\}]$ and $P_d(bQ) = [\{p \circ \chi^d|_{bQ} : p \in P(\chi^d(bQ))\}]$ correspondingly. Consequently,

$$B \subset [\{\bigcup R_d(bQ) : dk = b, \ d \in \Gamma', \ k \in \mathbf{Z}_o\}].$$

We claim that $B \cap R_d(bQ) = P_d(bQ)$. Indeed, suppose that $g = h \circ \chi^d \in B \cap R_d(bQ) \setminus P_d(bQ)$. Observe that though $\chi^d(Q)$ might be disconnected, the set $\mathbf{C} \setminus \chi^d(Q)$ is always connected. Thus, there exists a component, say K_α, of $\chi^d(Q)$ such that $h \in R(bK_\alpha) \setminus P(bK_\alpha)$. Consider the algebra

$$B_{d,\alpha} = \{f \in C(bK_\alpha) : f \circ \chi^d \in B\}.$$

Obviously, $h \in B_{d,\alpha}$. Moreover, the algebra $B_{d,\alpha}$ contains $P(bK_\alpha)$ though it differs from $C(bK_\alpha)$ because $\chi^d(\mathbf{z}) - \chi^d(\mathbf{z}_o)$ is a noninvertible function in B and $\chi^d(\mathbf{z}_o) \notin \chi^d(bQ)$ (if we assume on the contrary that $\chi^d(\mathbf{z}_o) \in \chi^d(bQ)$ then we will have $\chi^b(\mathbf{z}_o) = (\chi^d(\mathbf{z}_o))^k \in (\chi^d(bQ))^k = \chi^b(bQ)$ which is absurd). By the maximality of the algebra $P(bK_\alpha)$ we obtain that $P(bK_\alpha) = B_{d,\alpha} = R(bK_\alpha)$ in contradiction with the choice of g. We conclude that $B \cap R_d(bQ) = P_d(bQ)$, and therefore,

$$B = [\{\bigcup P_d(bQ) : dk = b, \ d \in \Gamma', \ k \in \mathbf{Z}_o\}]$$
$$= P(bQ) = P(\overline{Q})|_{bQ}.$$

This proves the first part of the theorem. The second part follows from the fact that $P_d(bQ)$ is a Dirichlet algebra for each $d \in \Gamma'$.

In the sequel we shall assume, without mentioning it especially, that the group Γ possesses property (*). If $Q = Q(\mathbf{z}_o, a, \varepsilon)$ is a basis neighborhood in $\mathbf{C}_G$, we will assume also that $0 \notin Q$ whenever $\mathbf{z}_o \neq \mathbf{0}$.

COROLLARY 1. *Let U be an open subset of type $\{z \in \mathbf{C} : |e^{iaz} - e^{iaz_o}| < \varepsilon\}$ in some half-plane $\{z \in \mathbf{C} : \text{Im } z < \alpha\}$. Denote by $\overline{Q}$ the set $\overline{j_e(U)} \subset \mathbf{C}_G$, where $G = \widehat{\Gamma}_d$. Then the space of analytic Γ-almost-periodic functions on U is a Dirichlet and maximal sup-norm algebra on the set $j_e^{-1}(bQ)$.*

As an immediate consequence of Theorem 1 we get that $P(\overline{Q}) = R(\overline{Q}) = A(Q)$ for any neighborhood Q in $\mathbf{C}_G$. In particular, every function in $R(\overline{Q})$ and $A(Q)$ can be approximated uniformly on $\overline{Q} \subset \mathbf{C}_G$ by polynomials in $\mathbf{C}_G$.

DEFINITION 4. *A continuous function on an open subset U of the big-plane $\mathbf{C}_G$ is Γ-analytic (or generalized-analytic) in U if it can be locally approximated in U by polynomials in $\mathbf{C}_G$.*

The set of all analytic functions in $U \subset \mathbf{C}_G$ will be denoted in the sequel by $\text{Hol}(U)$.

The theory of analytic functions in one and several complex variables are closely connected with subsemigroups of commutative Banach algebras generated either by one or several of their elements. In a similar way the theory of analytic Γ-almost-periodic functions in $\mathbf{C}$ (and respectively — of Γ-analytic functions in the big-plane $\mathbf{C}_G$) is tightly related with commutative Banach algebras which are generated by specific multiplicative subsemigroups.

Let B be a commutative Banach algebra over $\mathbf{C}$ with unit. Let Γ_o be an additive subsemigroup of $\mathbf{R}$ which is dense in $[0, \infty)$ and which contains the semigroup $\mathbf{Z}_o$. Let $\Omega = \{b_{a_j}\}_{j=1}^{\infty}$ be a *multiplicative semigroup* of elements b_{a_j} in B with $\|b_{a_j}\| \leq 1$ which contains the unit of B and which is algebraically isomorphic to the semigroup Γ_o. The set $\Omega \cdot \Omega^{-1}$ is a subgroup of B which is algebraically isomorphic to the set $\Gamma = \Gamma_o + (-\Gamma_o)$. As it is not hard to be seen, Γ is a dense subgroup of $\mathbf{R}$ which contains $\mathbf{Z}$. Let $G = \widehat{\Gamma}_d$. Fix a linear multiplicative functional φ in $sp\,B$ such that $\varphi(b_a) \neq 0$ for some (and thus for any) element $b_a \in \Omega$. Consider the function

$$g_\varphi(a) = \begin{cases} \varphi(b_a)/|\varphi(b_a)| & \text{for } a \in \Gamma_o, \\ \overline{g_\varphi(-a)} = 1/g_\varphi(-a) & \text{otherwise.} \end{cases}$$

Clearly, g_φ is a well defined function on $\Gamma = \Gamma_o \cup (-\Gamma_o)$. In fact g_φ is a character of Γ (cf. [8]).

DEFINITION 5 [12]. *Let $\Omega = \{b_a\}_{a \in \Gamma_o}$ be a multiplicative subsemigroup of a commutative Banach algebra B. The continuous mapping $\tau_\Omega : sp\,B \longmapsto \mathbf{C}_G$ defined in the following way*

$$\tau_\Omega(\varphi) = \begin{cases} 0 & \text{if } \varphi(b_a) = 0 \text{ for some } a \in \Gamma_o \backslash \{0\}, \\ |\varphi(b_1)| \cdot g_\varphi & \text{if } \varphi(b_a) \neq 0 \text{ for each } a \in \Gamma_o \backslash \{0\}. \end{cases}$$

is the spectral mapping of Ω. The range $\sigma(\Omega) = \tau_\Omega(sp\,B) \subset \overline{\Delta}_G$ of τ_Ω is the spectrum of Ω.

Observe that

(2) $\chi^a(\tau_\Omega(\varphi)) = \widehat{b}_a(\varphi)$ for each $a \in \Gamma_o.$

Indeed, $\chi^o(\tau_\Omega(\varphi)) = \widehat{b}_o(\varphi) \equiv 1$ and

$$\chi^a(\tau_\Omega(\varphi)) = \chi^a(|\varphi(b_1)| \cdot g_\varphi) = |\varphi(b_1)|^a g_\varphi(a)$$

$$= |\varphi(b_a)|\frac{\varphi(b_a)}{|\varphi(b_a)|} = \varphi(b_a) = \widehat{b}_a(\varphi)$$

for every $a \neq 0$. Consequently, $\chi^a(\tau_\Omega(sp\,B)) = \widehat{b}_a(sp\,B)$.
    In other words (2) says that

(3) $\chi^a(\sigma(\Omega)) = \sigma(b_a),$

where $\sigma(b_a)$ is the spectrum of the element $b_a \in B$.
    The continuity of τ_Ω is an immediate consequence of (2). Obviously, $\sigma(\Omega)$ is a compact subset of the big-plane $\mathbf{C}_G$.

THEOREM 2. *Let $Q = Q(\mathbf{z}_o, a, \varepsilon) = \{\mathbf{z} \in \mathbf{C}_G : |\chi^a(\mathbf{z}) - \mathbf{z}_o^a| < \varepsilon\}$ be a basis neighborhood of $\mathbf{C}_G$ of type (1), and let A be a uniform algebra on bQ which contains the semigroup $\Omega = \{\chi^a\}_{a \in \Gamma_o}$. Then $sp\,A = bQ$ whenever $\sigma(\Omega) = bQ$, and $A \cong P(\overline{Q})$ whenever $\sigma(\Omega) = \tau_\Omega(sp\,A) = \overline{Q}$.*

PROOF. As mentioned before, $\partial P(\overline{Q}) = bQ$, $\overline{Q} = sp\,P(\overline{Q}) = sp\,P(bQ)$. Therefore, $P(\overline{Q}) = P(bQ) \subset A$ and $\tau_\Omega(sp\,A) \subset \tau_\Omega(\overline{Q}) = \overline{Q}$. Since, by Proposition 1, $P(bQ)$ is a Dirichlet algebra, for each $\mathbf{z}_o \in bQ$ the set $\tau_\Omega^{-1}(\mathbf{z}_o) \subset sp\,A$ contains only the point evaluation at $\mathbf{z}_o$. Hence τ_Ω is an one-to-one correspondence between the sets $\tau_\Omega^{-1}(bQ)$ and bQ. If in addition $\sigma(\Omega) = \tau_\Omega(sp\,A) = bQ$, then evidently $sp\,A$ coincides with bQ and this proves the first part of the theorem.

Let $\sigma(\Omega) = \overline{Q}$, $\mathbf{z}_1 \in Q$, and let φ_1, φ_2 be two functionals in the set $\tau_\Omega^{-1}(\mathbf{z}_1)$ with representing measures $d\mu_1$ and $d\mu_2$ on bQ respectively. Because of $\varphi_1(p) = \varphi_2(p) = p(\mathbf{z}_o)$ for each polynomial p and because $P(\overline{Q})$ is a Dirichlet algebra, we observe that both measures coincide. Thus $\varphi_1 = \varphi_2$; and therefore, the spectral mapping τ_Ω is injective on the set $\tau_\Omega^{-1}(\overline{Q})$. Since $\sigma(\Omega) = \overline{Q}$, we can identify $sp\,A = \tau_\Omega^{-1}(\sigma(\Omega)) = \tau_\Omega^{-1}(\overline{Q})$ with $\overline{Q}$. Hence $A \supset P(\overline{Q})$, $sp\,A = sp\,P(\overline{Q}) = \overline{Q}$ and $\partial A = \partial P(Q) = bQ$, so that $A \cong A|_{bQ}$. Because of $sp\,C(bQ) = bQ \neq \overline{Q}$, we see that $A \neq C(bQ)$. By the maximality of $P(bQ)$ (Theorem 1), we obtain finally that $A \cong A|_{bQ} \cong P(bQ)$.

Theorem 1 and Proposition 1 imply the following result for analytic Γ-almost-periodic functions in $\mathbf{C}$.

COROLLARY 2. *Let U be an open subset of type*

$$\{z \in \mathbf{C} : |e^{iaz} - e^{iaz_o}| < \varepsilon\}$$

in some half-plane $\{z \in \mathbf{C} : Im\,z < \alpha\}$. Denote by $\overline{Q}$ the set $\overline{j_e(U)} \subset \mathbf{C}_G$, where $G = \widehat{\Gamma_d}$ and let A be a sup-algebra of continuous Γ-almost-periodic functions on $\overline{U}$ which contains all functions e^{iaz}, $a \in \Gamma_o$. Then $sp\,A = bQ$ if $\sigma(\{e^{iaz}\}_{a \in \Gamma_o}) = bQ$, and A is isometrically isomorphic to the algebra of all bounded analytic Γ-almost-periodic functions on $\overline{U}$ whenever $\sigma(\{e^{iaz}\}_{a \in \Gamma_o}) = \overline{Q}$.

3. Analytic Γ-Almost-Periodic Structures

Throughout this section we suppose that the group Γ possesses property (*).

DEFINITION 6. *k-sheeted branched Γ-analytic big cover is said to be any triple (U, π, V), where:*

(1) *U is a locally compact Hausdorff space;*

(2) *V is a domain in $\mathbf{C}_G$;*

(3) *π is a proper continuous mapping of U onto V (i.e. $\pi^{-1}(K)$ is compact in U for every compact $K \subset V$) which is light on U;*

(4) *There exists a negligible set $\Lambda \subset V$ and an integer k such that π is a k-sheeted covering mapping of $U \setminus \pi^{-1}(\Lambda)$ onto $V \setminus \Lambda$;*

(5) *The set $U \setminus \pi^{-1}(\Lambda)$ is dense in U.*

Recall that π is *light* (or *isolated-to-one*) on U if the set $\pi^{-1}\big(\pi(y)\big)$ is discrete for any point $y \in U$. By a *negligible set* here we understand a nowhere dense in $D \subset \mathbf{C}_G$ set Λ such that for every subdomain $D_1 \subset D$ any function f which is Γ-holomorphic on $D_1 \setminus \Lambda$ and locally bounded in D_1 admits a Γ-holomorphic extension on D_1.

If $\Lambda = \emptyset$, then the Γ-analytic big cover (U, π, V) is called *nonbranched*, and if all conditions except (3) in Definition 6 are satisfied, then (U, π, V) is called *improper*.

The next result for the existence of Γ-analytic structure in algebra spectrum is proved in [16].

THEOREM 3. *Let Ω be a multiplicative subsemigroup of functions in a uniform algebra A which is isometrically isomorphic to Γ_o. If U is an open subset of $sp\,A \setminus \partial A$ such that the set $\tau_\Omega(U)$ is connected in $\mathbf{C}_G$, $G = \widehat{\Gamma_d}$, and if there exists a k, $1 \leq k \leq \infty$, such that $\#\tau_\Omega^{-1}\big(\tau_\Omega(m)\big) \leq k$ for each $m \in U$, then there exists a positive integer $k_1 \leq k$ such that U can be given the structure of a (possibly improper) branched k_1-sheeted Γ-analytic big cover such that the Gelfand extensions of all functions in A are Γ-holomorphic on U.*

The proof of this theorem is quite similar to its proof for the case $\Gamma = \mathbf{Q}$ from [14], but instead of the corresponding results for the big disc algebra it uses Theorems 1 and 2. Here we apply Theorem 3 to find other conditions for existence of embeddings ψ of open subsets $\mathcal{W} \subset \mathbf{C}$ in the uniform algebra spectrum $sp\,A$, giving rise to analytic Γ-almost-periodic structures in $sp\,A$, and in particular we prove Γ-versions for most of the results in [14] and [15].

THEOREM 4. *Let A be a uniform algebra on X and let τ_Ω maps an open subset U of $sp\,A$ homeomorphically onto some open connected set $\tau_\Omega(U)$ in the big-plane $\mathbf{C}_G$. If $\varphi \in U$ is a linear and multiplicative functional of A which possesses the property*

$$(4) \qquad \operatorname{Ker}\varphi = \Big[\bigcup_{a \in \Gamma_o} (f_a - \widehat{f}_a(\varphi))A \Big],$$

then there exists a continuous embedding j of some open subset $\mathcal{W}$ of the complex plane $\mathbf{C}$ into $sp\,A$ whose range is dense in U, such that $\widehat{f} \circ j$ is a bounded analytic Γ-almost-periodic function in $\mathcal{W}$ for each function f in A.

For the proof we shall need the following

LEMMA 1. *Under hypotheses of Theorem 4 the restriction $\tau_\Omega\big|_U$ of spectral mapping of Ω is a homeomorphism between U and $\tau_\Omega(U) \subset \mathbf{C}_G$, and $\widehat{h} \circ \tau_\Omega^{-1}$ is a Γ-analytic function on $\tau_\Omega(U) \subset \mathbf{C}_G$ for every function h in A.*

PROOF. Note first that (4) is equivalent to the identity

$$(5) \qquad \operatorname{Ker}\varphi = \Big[\bigcup_{a \in \Gamma'} [(f_a - \widehat{f}_a(\varphi))A] \Big],$$

where, as before, $\Gamma' = (0,1] \cap \Gamma$. By (4) one can easily observe that the functional φ is uniquely determined by its values on the functions in Ω. Clearly, by $\chi^a\big(\tau_\Omega(\varphi)\big) = \widehat{f}_a(\varphi)$, which holds for all $a \in \Gamma_o$, the point $\mathbf{z}_o = \tau_\Omega(\varphi_o)$ is also uniquely determined by the given data, from which we conclude that the full preimage $\tau_\Omega^{-1}(\mathbf{z}_o)$ of $\mathbf{z}_o$ consists of the element φ only. We claim that the restriction $\tau_\Omega|_U : U \to \tau_\Omega(U)$ is an one-to-one mapping. Note that the full preimage of each point $\mathbf{z} \in \tau_\Omega(U)$ consists of all maximal ideals in A which contain the ideal $J(\mathbf{z}) = \big[\bigcup_{a \in \Gamma'} (f_a - \mathbf{z}^a)A \big]$. We shall show that codim $J(\mathbf{z}) = \dim A/J(\mathbf{z}) = 1$ for each $\mathbf{z} \in \tau_\Omega(U)$. Under the ordering "$b \succ a$ if and only if $bk = a$ for some integer k" the set Γ' is a directed to the right system, i.e. for every $a, b \in \Gamma'$ there exists a $c \in \Gamma'$ which follows both a and b. Clearly, $\big[(f_a - \widehat{f}_a(\varphi))A\big] \subset \big[(f_b - \widehat{f}_b(\varphi))A\big]$ whenever $b \succ a$. Therefore, there arises an injective spectrum $\big\{ \big[(f_a - \widehat{f}_a)A + \mathbf{C}\big]; i_b^a; a, b \in \Gamma' \big\}$ of closed algebras $A_a(\mathbf{z}) = \big[(f_a - \widehat{f}_a(\varphi))A + \mathbf{C}\big]$, where i_b^a ($b \succ a$) is the natural inclusion. We have

$$A = \operatorname{Ker} \varphi + \mathbf{C} = \left[\bigcup_{a \in \Gamma'} \big[(f_a - \widehat{f}_a(\varphi))A\big] \right] + \mathbf{C}$$

$$= \left[\bigcup_{a \in \Gamma'} \big[(f_a - \widehat{f}_a(\varphi))A + \mathbf{C}\big] \right] = \left[\varinjlim_{\alpha \in \Gamma'} \big[(f_a - \mathbf{z}_o^a)A + \mathbf{C}\big] \right] = \big[\varinjlim_{\alpha \in \Gamma'} A_a(\mathbf{z}_o) \big].$$

Since the spectrum of an injective limit of algebras is the limit of the naturally arising projective spectrum of their spectra, $sp\,A$ is homeomorphic to $\varprojlim_{\alpha \in \Gamma'} \big\{ sp\,A_a(\mathbf{z}_o); (i_b^a)^*; a, b \in \Gamma' \big\}$, where all injective and projective spectra are considered with respect to the partial ordering "$\succ$" introduced above. The homeomorphic correspondence between $sp\,A$ and $\varprojlim_{\alpha \in \Gamma'} sp\,A_a(\mathbf{z}_o)$ can be expressed explicitly in the following way: To each maximal ideal $M = \big[\bigcup_{a \in \Gamma'} M_a \big] = \big[\varprojlim_{\alpha \in \Gamma'} M_a \big]$ of algebra A it assignes the element $\varprojlim_{\alpha \in \Gamma'} \{M_a; i_a^b\} \in \varprojlim_{\alpha \in \Gamma'} sp\,A_a(\mathbf{z}_o)$, where $M_a \in sp\,A_a(\mathbf{z}_o)$, $a \in \Gamma'$. Let now $\mathbf{z} \neq \mathbf{z}_o \in \tau_\Omega(U)$. Fix a number a in Γ' and consider the ideal

$$J_a(\mathbf{z}) = \big[(f_a - \mathbf{z}^a)A\big] \cap A_a(\mathbf{z}_o) \subset A_a(\mathbf{z}_o).$$

Denote by $S_a(\mathbf{z})$ the unit sphere of $J_a(\mathbf{z})$. We apply below Kato's theorem for perturbation of semi-Fredholm pairs of closed linear subspaces, which says: Given two closed linear subspaces M and N of a normed space Z such that codim $M = \dim Z/M = k < \infty$ and for which $\delta(M, N) = \sup_{u \in S(M)} \varrho(u, N) < 1$, where $S(M)$ is the unit sphere of M and $\varrho(a, b) = \|a - b\|$ is the natural distance in Z, then codim $N \leq$ codim M [9]. Take the algebra $A_a(\mathbf{z}_o)$ for Z, $J_a(\mathbf{z}_1)$ for M and $J_a(\mathbf{z}_2)$ for N, where $\mathbf{z}_1, \mathbf{z}_2 \in \tau_\Omega(U)$. By Kato's theorem codim $J_a(\mathbf{z}_2) \leq$ codim $J_a(\mathbf{z}_1)$ whenever $\delta\big(J_a(\mathbf{z}_1), J_a(\mathbf{z}_2)\big) < 1$. Let h be an arbitrary function in A such that the functions $u_j = (f_a - \mathbf{z}_j^a)h$, $j = 1, 2$, belong to $J_a(\mathbf{z}_j)$. We have:

$$\frac{u_1}{\|u_1\|} \in S_a(\mathbf{z}_1), \quad \frac{u_2}{\|u_1\|} \in J_a(\mathbf{z}_2) \quad \text{and} \quad \varrho\left(\frac{u_1}{\|u_1\|}, \frac{u_2}{\|u_1\|}\right) = \frac{|\mathbf{z}_1^a - \mathbf{z}_2^a|\,\|h\|}{\|u_1\|}.$$

Suppose that the point z_1 is contained in $\tau_\Omega(U)$ together with a basis neighborhood $Q(z_1, a, \varepsilon) \subset \mathbf{C}_G$, where $a \in \Gamma'$ and $0 < \varepsilon < 1$. Since

$$\|u_1\| = \max_{x \in X} \left| \widehat{f}_a(x) - z_1^a \right| |h(x)| = \max_{x \in \partial A} \left| \chi^a(\tau_\Omega(x)) - z_1^a \right| |h(x)|$$

$$\geq \varepsilon \max_{x \in \partial A} |h(x)| = \varepsilon \|h\|,$$

we have that

$$\varrho\left(\frac{u_1}{\|u_1\|}, J_a(z_2) \right) \leq \frac{1}{\varepsilon} |z_1^a - z_2^a|,$$

from which we conclude that

(6) $\displaystyle \sup_{u \in S_a(z_1)} \varrho\big(u, J_a(z_2)\big) < \frac{1}{2}$ whenever $|z_1^a - z_2^a| < \dfrac{\varepsilon}{2}.$

Hence by Kato's theorem we have that $\dim A_a(z_o)/J_a(z_2) \leq \dim A_a(z_o)/J_a(z_1)$ whenever $z_2 \in Q(z_1, a, \varepsilon/2)$. If $\dim A_a(z_o)/J_a(z_1) \leq 1$ for a $z_1 \in \tau_\Omega(U)$ and any $a \in \Gamma'$, then $\dim A_a(z_o)/J_a(z) \leq 1$ for every z in $Q(z_1, a, \varepsilon/2) \subset \tau_\Omega(U)$. Consequently the closed set $\{z \in \tau_\Omega(U) : \dim A_a(z_o)/ J_a(z_1) \leq 1\}$ is also open in $\tau_\Omega(U)$. It is nonempty because it contains at least the point z_o. By the connectedness of the set $\tau_\Omega(U)$, we see that $\dim A_a(z_o)/J_a(z) \leq 1$ on $\tau_\Omega(U)$. Actually, $\dim A_a(z_o)/J_a(z_o) = 1$ for all $z \in \tau_\Omega(U)$ because $J_a(z)$ is a proper ideal in the algebra $A_a(z_o)$. Recall that we can choose the number a arbitrarily close enough to 0. As an ideal of codimension one, $J_a(z)$ is a maximal ideal in $A_a(z_o)$. We can consider that the number a in Γ' from above is far enough with respect to the ordering "$\succ$". Consequently, the ideal $\varprojlim_{\alpha \in \Gamma'} J_a(z)$ belongs to the projective limit $\varprojlim_{\alpha \in \Gamma'} sp\, A_a(z)$ and by the above remarks its closure $\left[\bigcup_{a \in \Gamma'} J_a(z) \right] = \left[\bigcup_{a \in \Gamma'} (f_a - z^a)A \right]$ is a maximal ideal of A. We conclude that $J(z) = \left[\bigcup_{a \in \Gamma'} (f_a - z^a)A \right]$ is a maximal ideal of A for each $z \in \tau_\Omega(U)$. Since, according to the initial remark no other maximal ideal contains $J(z)$, we obtain that the full preimage of each point z in $\tau_\Omega(U)$ consists of $J(z)$ only. Consequently, the restriction $\tau_\Omega|_U : U \to \tau_\Omega(U)$ of the spectral mapping τ_Ω on U is one-to-one. Now the case $k = 1$ of Theorem 3 completes the proof of the lemma.

For the proof of Theorem 4 we take for j the composition $\tau_\Omega^{-1} \circ j_g$ of τ_Ω^{-1} with any standard embedding $j_g = j_e g$, $g \in G$, of $\mathbf{C}$ into $\mathbf{C}_G$; for $\mathcal{W}$ we take the set $j_g^{-1} \circ \tau_\Omega(U)$.

The next theorem is a generalized version of Theorem 4. The scheme of its proof is quite similar to that of the main result from [14], but we allow here the group Γ to be distinct from $\mathbf{Q}$ and the set $\tau_\Omega(U)$ from below not to contain $\mathbf{0}$.

THEOREM 5. *Let A be a uniform algebra on X and let U be an open subset of $sp\, A \setminus \partial A$ such that $\tau_\Omega(U)$ is an open and connected subset of the big plane $\mathbf{C}_G$. If the ideal*

(7) $$J(z_o) = \left[\bigcup_{a \in \Gamma'} (f_a - \widehat{f}_a(\varphi_o))A \right]$$

is of codimension $k < \infty$ for some linear and multiplicative functional φ_o which belongs to U, then the set U can be given the structure of a k_1-sheeted branched Γ-analytic big cover over $\tau_\Omega(U)$ for some $k_1 \leq k$ and for every open subset U_1 of U on which τ_Ω is a homeomorphism there exists a continuous mapping j of an open subset $\mathcal{W}$ of the complex plane $\mathbf{C}$ into $\mathbf{C}_G$ whose range is dense in U_1, such that the restrictions of functions $\hat{f} \circ j$ on $\mathcal{W}$ are analytic Γ-almost-periodic functions for every function f in A.

As before the proof is based on the following

LEMMA 2. *Under hypotheses of Theorem 5 the set U can be given the structure of a k_1-sheeted Γ-analytic big cover over $\tau_\Omega(U)$ for some $k_1 \leq k$ and $\hat{h} \circ \tau_\Omega^{-1}$ is a Γ-analytic function on this big cover for each function h in A. The cover is proper if and only if $\tau_\Omega(U)$ does not meet the set $\tau_\Omega(\partial A)$.*

PROOF. We claim that the number of elements in the preimages $\tau_\Omega^{-1}(\mathbf{z})$ of each $\mathbf{z} \in \tau_\Omega(U)$ is bounded from above by k. Proceeding in a similar way as in the proof of Theorem 4, we show first that $\dim A/J(\mathbf{z}) \leq k$ for every $\mathbf{z} \in \tau_\Omega(\varphi)$, $\varphi \in U$, where

$$J(\mathbf{z}) = \Big[\bigcup_{a \in \Gamma'} (f_a - \hat{f}_a(\varphi))A \Big].$$

Fix a linear basis of the space $A/J(\mathbf{z}_o)$, say $\{\mathbf{u}_1, \dots, \mathbf{u}_k\}$, where $\mathbf{u}_j = u_j + J(\mathbf{z}_o)$, $u_j \in A$, and denote by $B_a(\mathbf{z}_o)$ the Banach space $\mathbf{C}u_1 + \cdots + \mathbf{C}u_k + [(f_a - \chi^a(\mathbf{z}_o))A]$, where $\mathbf{z}_o = \tau_\Omega(\varphi_o)$. We have

$$A = \mathbf{C}u_1 + \cdots + \mathbf{C}u_k + J(\mathbf{z}_o) = \mathbf{C}u_1 + \cdots + \mathbf{C}u_k +$$

$$\Big[\bigcup_{a \in \Gamma'} [(f_a - \hat{f}_a(\varphi))A] \Big] = \Big[\mathbf{C}u_1 + \cdots + \mathbf{C}u_k + \bigcup_{a \in \Gamma'} [(f_a - \hat{f}_a(\varphi))A] \Big]$$

$$= \Big[\varinjlim_{\alpha \in \Gamma'} \{ \mathbf{C}u_1 + \cdots + \mathbf{C}u_k + [(f_a - \hat{f}_a(\varphi))A] \} \Big]$$

$$= \Big[\varinjlim_{\alpha \in \Gamma'} B_a(\mathbf{z}_o) \Big] = \Big[\bigcup_{a \in \Gamma'} B_a(\mathbf{z}_o) \Big],$$

where the inductive limit is taken with respect to the partial ordering "$\succ$" introduced in the course of proof of Lemma 1. Given a $\mathbf{z} \in \tau_\Omega(U)$, consider the space

$$E_a(\mathbf{z}) = B_a(\mathbf{z}_o) \cap \big[\mathbf{C}u_1 + \cdots + \mathbf{C}u_k + (f_a - \chi^a(\mathbf{z}))A \big] \subset B_a(\mathbf{z}_o).$$

By making use of Kato's perturbation theorem for semi-Fredholm pairs of closed linear subspaces, in a similar way as in Lemma 1 we get that $\dim B_a(\mathbf{z}_o)/E_a(\mathbf{z}) \leq \dim B_a(\mathbf{z}_o)/E_a(\mathbf{z}_o)$ for every $\mathbf{z} \in \tau_\Omega(U)$ and each $a \in \Gamma'$. Because of $E_a(\mathbf{z}) \subset B_a(\mathbf{z}_o)$ we have that $E_a(\mathbf{z}) = B_a(\mathbf{z}_o)$ and hence $A = \mathbf{C}u_1 + \cdots + \mathbf{C}u_k + J(\mathbf{z})$ for each $\mathbf{z} \in \tau_\Omega(U)$, wherefrom $\dim A/J(\mathbf{z}) \leq k$ for every $\mathbf{z} \in W$. Let $\mathbf{z}$ be a fixed point in $\tau_\Omega(U)$. Each element $\mathbf{g}$ in the factor-space $A/J(\mathbf{z})$ can be presented in a unique way as

$$\mathbf{g} = \sum_{\nu=1}^{k_1} \mathbf{a}_\nu(\mathbf{g})\mathbf{v}_\nu$$

for some $k_1 \leq k$, where v_ν, $\nu = 1,\ldots,k_1$, are functions in A whose cosets $\mathbf{v}_\nu = v_\nu + J(\mathbf{z})$ form a basis in the factor-space $A/J(\mathbf{z})$ and $\mathbf{a}_\nu = a_\nu + J(\mathbf{z})$, $a_\nu \in A$, are k_1 linearly independent elements in $\big(A/J(\mathbf{z})\big)^*$ which form a basis in the dual space $\big(A/J(\mathbf{z})\big)^* \cong \mathbf{C}^{k_1}$. Consequently, each element $g \in A$ possesses a representation of type

$$g = \sum_{\nu=1}^{k_1} a_\nu(g)v_\nu + h,$$

where $h \in J(\mathbf{z})$ and a_ν, $\nu = 1,\ldots,k_1$, are elements in A^* such that $a_\nu(f) = \mathbf{a}_\nu(\mathbf{f})$ for each $f \in A$. In particular $a_\nu\big|_{J(\mathbf{z})} \equiv 0$.

If q is a linear multiplicative functional of A such that $\tau_\Omega(q) = \mathbf{z}$, then $\widehat{f}_a = \chi^a(\mathbf{z}) = \mathbf{z}^a$ for every $a \in \Gamma_o$. Since the Gelfand extension of each function in the ideal $J(\mathbf{z}) = \big[\bigcup_{a\in\Gamma'} (f^a - \mathbf{z}_o^a)A \big]$ vanishes on every such q, we have that $\widehat{h}(q) = 0$ for the function h from above, and consequently

$$(8) \qquad\qquad q(g) = \widehat{g}(q) = \sum_{\nu=1}^{k_1} a_\nu(g)\widehat{v}_\nu(q)$$

for every function g in A and for each linear multiplicative functional q in $\tau_\Omega^{-1}(\mathbf{z})$. This means that considered as an element of A^*, any functional q in $\tau_\Omega^{-1}(\mathbf{z})$ coincides with the linear functional

$$(9) \qquad\qquad q = \sum_{\nu=1}^{k_1} \widehat{v}_\nu(q)a_\nu$$

on A. As linear multiplicative functionals, the points in $\tau_\Omega^{-1}(\mathbf{z})$ are linearly independent elements in the dual space A^*. Since a_ν are linear functionals over A^*, the identity (9) shows that the set $\tau_\Omega^{-1}(\mathbf{z})$ can not contain more than $k_1 \leq k$ elements, as claimed. Now we apply Theorem 3 to complete the proof.

By applying Wermer's technique from [21] one can strengthen Theorem 5 in the following way.

THEOREM 6. *Let A be a uniform algebra on a compact Hausdorff space X, let Γ be an additive subgroup of $\mathbf{R}$ which satisfies property $(^*)$, let $\Omega = \{f_a\}_{a\in\Gamma_o}$ be a multiplicative subsemigroup of functions in A isomorphic to Γ_o, and let W be a component of the set $\mathbf{C}_G \setminus \tau_\Omega(X)$, where $\tau_\Omega : sp\,A \to \mathbf{C}_G$ is the spectral mapping of Ω. If there exists a measurable subset N of W such that the Lebesgue measure of the set $\chi^a(N)$ is non-zero for some (and therefore for each) $a \in \Gamma_o$ and if τ_Ω is finite-to-one on $\tau_\Omega^{-1}(N)$, then the set W can be given the structure of a k_1-sheeted branched Γ-analytic big cover over $\tau_\Omega(W)$ for some $k_1 \leq k$, and* *for every open subset U of W on which τ_Ω is a homeomorphism, there exists a continuous mapping j of an open subset $\mathcal{W}$ of the complex plane $\mathbf{C}$ into $\mathbf{C}_G$ with dense range in U such that the restrictions of functions $\widehat{f} \circ j$ on $\mathcal{W}$ are analytic Γ-almost-periodic functions for every function f in A.*

In fact, the set $\tau_\Omega^{-1}(W)$ can be given the structure of a proper finite-sheeted branched Γ-analytic big manifold over W, and the Gelfand extensions of all functions in A are Γ-holomorphic on $\tau_\Omega^{-1}(W)$.

The case $\Gamma = \mathbf{Q}$ of Theorems 4 and 5 are proved in [13]. The classical case $\Gamma = \mathbf{Z}$ of Theorem 6 is proved by Bishop [6] (see also Wermer [21]). For the case when $\Gamma = \mathbf{Q}$ and $W = \Delta_G$ the corresponding result is proved in Tonev [15]. Note that by applying the technique of Aupetit and Wermer from [2], one can replace the requirement $dx\, dy\big(\chi^a(N)\big) > 0$ in Theorem 6 by the less restrictive one, namely the set $\chi^a(N)$ to be of positive logarithmic capacity (cf. [16, 20]).

The author is grateful to the referee for his or her helpful remarks.

References

1. R. Arens, I. Singer, *Generalized analytic functions*, Trans. Amer. Math. Soc. **81** (1956), 379-393.
2. B. Aupetit, J. Wermer, *Capacity and uniform algebras*, J. Funct. Anal. **28** (1978), 386–400.
3. R. Basener, *A condition for analytic structure*, Proc. Amer. Math. Soc. **36** (1972), 156–160.
4. _____, *A generalized Shilov boundary and analytic structure*, Proc. Amer. Math. Soc. **47** (1975), 98–104.
5. H. Bear, G. Hile, *Analytic structure in uniform algebras*, Houston J. Math. **5** (1979), 21–28.
6. E. Bishop, *Holomorphic completions, analytic continuation and the interpolation of seminorm*, Ann. Math. **78** (1963), 468–500.
7. C. Corduneanu, *Almost Periodic Functions*, Interscience, N.Y. (1968).
8. T. Gamelin, *Uniform Algebras*, Prentice-Hall Inc., Englewood Cliffs, N.J. (1969).
9. T. Kato, *Perturbation Theory for Linear Operators*, Springer Verlag (1966).
10. N. Sibony, *Multi-dimensional analytic structure in the spectrum of a uniform algebra*, Lect. Notes in Math., Springer Verlag **512** (1976), 139–175.
11. E. Stout, *The Theory of Uniform Algebras*, Bogden and Quigely, Tarrytown on Hudson (1971).
12. T. Tonev, *Algebras of generalized-analytic functions*, Banach Centre Publ. **8** (1982), 474-470.
13. _____, *Generalized-analytic coverings in the maximal ideal space*, Analytic functions, Lect. Notes in Math., Springer Verlag **1039** (1983), 436-442.
14. _____, *Some applications of perturbation theory for pairs of closed linear subspaces*, Complex Analysis'83, Sofia (1985), 218-234.
15. _____, *Generalized-analytic coverings in the spectrum of a uniform algebra*, Zeitschrift für Anal. and ihre Anwendungen **5** (1986), 179-184.
16. _____, *General Complex-Analytic Structures in Uniform Algebra Spectra*, preprint, Sofia (1987), 300 pp.

17. _____, *General complex-analytic structures in uniform algebra spectra − a survey*, Functional Analytic Methods in Complex Analysis and Applications to Partial Differential Equations, World Science (1990), 265-286.

18. _____, *Analytic manifolds in uniform algebras*, Houston J. Math. **17** (1991), 101-108.

19. _____, *Multi-dimensional analytic structures and uniform algebras*, Proc. of Symp. in Pure Math., v. **52**, part 2 (1991), 557-561.

20. _____, *Big-Planes, Boundaries and Function Algebras*, Elsevier − North Holland, (1992).

21. J. Wermer, *Banach Algebras and Several Complex Variables*, Springer Verlag (1976).

DEPARTMENT OF MATHEMATICS, UNIVERSITY OF MONTANA, MISSOULA, MONTANA 59812–1032

Contemporary Mathematics
Volume **137**, 1992

Recurrence for Lacunary Cosine Series

DAVID C. ULLRICH

Dedicated to Professor Walter Rudin on the occasion of his retirement.

0. Introduction

It seems that a few gaps still remain in the theory of lacunary series. For example, although such series have been appearing regularly in the work of various mathematicians for more than a century, it is not known whether they must be recurrent.

We shall say that the sequence (n_j) is *lacunary* if there exists a number $\lambda > 1$ such that $n_{j+1}/n_j \geq \lambda$ for all j. (Such sequences are sometimes called *Hadamard lacunary*.) A *lacunary series* is a trigonometric series of the form

$$(0.0) \qquad \sum_{j=1}^{\infty} a_j e^{i n_j t}$$

for some lacunary sequence (n_j) and complex coefficients (a_j); we will set

$$(0.1) \qquad s_N(t) = \sum_{j=1}^{N} a_j e^{i n_j t}.$$

Anderson and Pitt ([AP1], [AP2]) asked the following question: *Suppose that*

$$(0.2) \qquad\qquad |a_j| \leq 1 \qquad\qquad (j = 1, 2, \dots)$$

and

$$(0.3) \qquad \sum_{j=1}^{\infty} |a_j|^2 = \infty.$$

Does it follow that $\{s_N(t) : N = 1, 2, \dots\}$ *must be dense in the plane for almost every value of* t? (In such a case we will say that the series (0.0) is almost everywhere *recurrent*.) The author had been under the impression that this question originated with [AP1,2], but in fact Murai posed the same question (for

1991 *Mathematics Subject Classification.* Primary 42A55.

This paper is in final form and no version of it will be submitted for publication elsewhere.

the special case $a_j = 1$) somewhat earlier; see the highly recommended survey article [ME] (an English translation of [MJ]).

It seems likely that (0.2) and (0.3) should be sufficient to imply recurrence for a lacunary series, but this has not been proved. (Note that (0.3) is of course necessary for recurrence; in fact the proof of the almost everywhere convergence of a *lacunary* L^2-series is quite simple ([ZY] v.1, p.203). While one might conjecture that (0.3) together with a condition somewhat weaker than (0.2) should be sufficient [AN], it is clear that (0.2) cannot simply be omitted: If, for example, $a_j = 3^j$, then $s_N(t) \to \infty$ uniformly.)

Various partial results were obtained by Anderson and Pitt. In [AP2] it is shown that the series (0.0) is almost everywhere recurrent in the special case $a_j = 1$, $n_j = b^j$. This depends on a certain amount of algebra. The results proved in [AP1] are more "general", and the proofs are more "analytic", but they do not suffice to imply recurrence even in the special case considered in [AP2], much less to show that recurrence follows from (0.2) and (0.3).

In [AP1] it is shown that if (0.2) and (0.3) hold then the series (0.0) must be $\varepsilon - recurrent$ for some $\varepsilon = \varepsilon(\lambda) > 0$; this means that (for almost every t) the sequence $(s_N(t))$ must intersect every disc of radius ε infinitely often. The ε-recurrence is obtained by an application of Plessner's theorem ([CL] p. 147; [GA] p. 95) to the function $f(z) = \sum_{j=1}^{\infty} a_j z^{n_j}$; it follows from this result that the series (0.0) is actually recurrent if (0.3) holds and $a_j \to 0$.

The other main theorem in [AP1] states that if the sequence (n_j) actually satisfies $\lim_{j \to \infty} n_{j+1}/n_j = \infty$ then (0.2) and (0.3) are sufficient to imply recurrence of the series (0.0). This result may be understood as follows: Roughly speaking, the behavior of a lacunary series is very much like that of a sum of independent random variables uniformly distributed on the circle; now if $n_{j+1}/n_j \to \infty$ the individual terms should become arbitrarily close to independent as $j \to \infty$, and it is known that conditions somewhat weaker than (0.2) and (0.3) are sufficient to give recurrence in the independent case [BDC].

Indeed, many classical results concerning lacunary series simply show that they behave like sums of independent random variables in some sense or other; see [KH] or [ME] for various examples. A recent result due to Asmar and Montgomery-Smith [AMS] shows that in fact the *distribution* of a lacunary series is in a certain sense equivalent to the distribution of a sum of independent random variables; this theorem contains many of the classical inequalities as immediate corollaries. However, all of the results of which we are aware deal with the distribution of the series at a fairly large scale; the distribution at a very fine scale does not appear to be well enough understood to handle the question of recurrence. That is, the results of [AMS] give definitive information on the measure of the set where $|s_N(t)| \geq c \|s_N\|_2$, but for the present problem we need good lower bounds on the measure of the set where $|s_N(t)| < \varepsilon$, for arbitrarily small $\varepsilon > 0$. The "ε-recurrence" result of [AP1] contains such information for

$\varepsilon \geq \varepsilon(\lambda) > 0$, which is a step in the right direction, but not quite good enough. (It seems that information concerning the distribution at a fine scale might also be helpful in showing that the geometric mean of a lacunary series must be greater than a constant times the the L^2 mean; cf. [UL]).

The question of recurrence for lacunary series as above remains open; in this paper we will answer an easier question which was posed by Murai in [MJ] and [ME] (and again in [BH]): *Suppose (n_j) is lacunary and $\tau_j \in \mathbb{R}$, $j = 1, 2, \ldots$. Set*

$$(0.4) \qquad \sigma_N(t) = \sum_{j=1}^{N} \cos(n_j t + \tau_j).$$

Must $\{\sigma_N(t) : N = 1, 2, \ldots\}$ be dense in $\mathbb{R}$, for almost every $t \in \mathbb{R}$?

We shall prove that the partial sums of a lacunary "cosine series" as in (0.4) must indeed be recurrent. It seems extremely likely that one could use essentially the same method to prove that the sequence $(\mathbb{R}e(s_N(t)))$ must be recurrent (on $\mathbb{R}$) if s_N is as above and the coefficients satisfy (0.2) and (0.3); see the Remark at the end of the paper. (We have not worked out the details here, so that we could finish the paper in time for inclusion in the present volume; a paper on lacunary series seemed appropriate [RU].)

The question of whether (s_N) itself is recurrent remains open. This presumably requires something more than trivial modifications of the techniques in the present paper, because the proof we give below depends on the "intermediate value theorem" for real-valued continuous functions in an essential way. (One need only prove a version of Lemma 3 below with s_N in place of σ_N; the argument showing that the theorem follows from Lemma 3 remains valid in the complex case.)

1. A Special Case

We define $\sigma_N(t)$ as in (0.4) above, taking $\tau_j = 0$ merely to simplify notation (now (0.4) is in fact a "cosine series"). We shall prove the following:

THEOREM. *If $n_{j+1}/n_j \geq \lambda > 1$ for all j then the sequence of partial sums $(\sigma_N(t))$ is dense in $\mathbb{R}$, for almost every value of t.*

We begin with the following observation:

LEMMA 1. *For almost every t we have*

$$(1.1) \qquad \liminf_{N \to \infty} |a - \sigma_N(t)| \leq \frac{1}{2}$$

for every $a \in \mathbb{R}$.

PROOF. This is trivial in the present context. It is well known ([ZY] v.1, p. 205) that

$$(1.2) \qquad \sup_N \sigma_N(t) = \infty$$

and

(1.3) $$\inf_N \sigma_N(t) = -\infty$$

for almost every t; now any t satisfying (1.2) and (1.3) must also satisfy (1.1) for
every $a \in \mathbb{R}$, since $|\sigma_{N+1}(t) - \sigma_N(t)| \leq 1$.   □

NOTE. The proof of Lemma 1 uses already the fact that σ_N is real-valued; we
should point out that this was not really essential here, as Theorem 1 of [AP1]
shows that

(1.4) $$\sup_{a \in \mathbb{C}} \liminf_{N \to \infty} |a - s_N(t)| \leq c(\lambda) < \infty$$

for almost every t, if s_N is as in (0.1) (assuming (0.2) and (0.3)). This form of
Lemma 1 would suffice below (although the present version is both simpler and
more precise).

In the terminology of [AP1], Lemma 1 says that the sequence $(\sigma_N(t))$ is almost
everywhere "$\frac{1}{2}$-recurrent"; we shall see that this implies that the sequence is
almost everywhere recurrent (that is, ε-recurrent for every $\varepsilon > 0$) because of
the intermediate value theorem and the following proposition. Let us say that a
sequence of sets $E_N \subset [0, 2\pi]$ covers $[0, 2\pi]$ i.o.a.e. (for "infinitely often, almost
everywhere") if almost every point of $[0, 2\pi]$ lies in infinitely many E_N. The
notation $|E|$ will refer to the Lebesgue measure of the set $E \subset [0, 2\pi]$.

PROPOSITION 1. *Suppose that the two sequences of sets E_N, $F_N \subset [0, 2\pi]$
have the following property: There exist a constant $\gamma > 0$ and a sequence $\delta_N > 0$
tending to zero, such that for every $x \in E_N$ there exists an interval I of length
δ_N with $x \in I$ and $|F_N \bigcap I| \geq \gamma \delta_N$. If the E_N cover $[0, 2\pi]$ i.o.a.e. then the F_N
must cover $[0, 2\pi]$ i.o.a.e. as well.*

PROOF. If the F_N do not cover $[0, 2\pi]$ i.o.a.e. then we may find a set $A \subset
[0, 2\pi]$ with $|A| > 0$ and a number K such that $F_N \bigcap A = \phi$ for $N > K$. Now
almost every point of A must be a point of density; in particular there must exist
a point of density of A which is covered infinitely often by the sets E_N, which is
impossible. We leave the details to the reader.   □

The technical details in the proof of the theorem become somewhat simpler
if we suppose that n_{j+1}/n_j is bounded above; this seems curious, because it
appears that surely taking larger n_{j+1}/n_j should make the theorem easier. We
shall begin with a sketch of the proof for bounded n_{j+1}/n_j and then indicate
the modifications required for the general case. (The casual reader might in fact
take the truth of the theorem for bounded n_{j+1}/n_j as "proof" that the theorem
could not possibly be false in general, particularly in view of the fact that if
$n_{j+1}/n_j \to \infty$ then the a.e. recurrence of $(\sigma_N(t))$ follows from the results in
[AP1].)

Let us therefore suppose that $1 < \lambda \leq n_{j+1}/n_j \leq C < \infty$ for all j; now fix
$\varepsilon > 0$. We shall show that for almost every t we have $|\sigma_N(t)| < \varepsilon$ infinitely

often by applying Proposition 1; the same argument would show $|a - \sigma_N(t)| < \varepsilon$ i.o.a.e. for any $a \in \mathbb{R}$. We define

$$(1.5) \qquad E_N = \left\{ t \in [0, 2\pi] : |\sigma_N(t)| \le \frac{1}{2} \right\},$$

and we set

$$(1.6) \qquad F_N = \left\{ t \in [0, 2\pi] : |\sigma_{N+M}(t)| < \varepsilon \right\},$$

where M is some large positive integer to be determined; the value of M will depend only on λ. We shall see that if M is large enough then the sets E_N and F_N satisfy the hypotheses of Proposition 1. Now, Lemma 1 shows that the sets E_N cover $[0, 2\pi]$ i.o.a.e.; hence the sets F_N do so as well, so that $(\sigma_N(t))$ is almost everywhere recurrent.

Fix a point $t_o \in E_N$. Let I be the closed interval $[t_o - \alpha, t_o + \alpha]$, where $\alpha = 1/2n_N$. We shall see in a moment that if $M = M(\lambda)$ is large enough then σ_{N+M} must take both positive and negative values somewhere on I. It follows that σ_{N+M} must have at least one zero on I (this is the only part of the argument that depends in an essential way on the fact that σ_N is real-valued).

It is clear that

$$(1.7) \qquad \|\sigma'_{N+M}\|_\infty \le \sum_{j=1}^{N+M} n_j \le c\, n_{N+M}.$$

(Here and below the letter "c" will denote a finite positive constant depending only on λ, ε, and C (but independent of N), the value of which may change each time it appears. Constants the values of which remain fixed will be denoted c_1, c_2, etc.).

Now because we are supposing that n_{j+1}/n_j is bounded, inequality (1.7) implies

$$(1.8) \qquad \|\sigma'_{N+M}\|_\infty \le c n_N.$$

Since σ_{N+M} has a zero on I, this implies that $|\sigma_{N+M}(t)| < \varepsilon$ on a subinterval of I of length at least $c\varepsilon/n_N = c\varepsilon|I|$. In other words,

$$(1.9) \qquad \left| F_N \bigcap I \right| \ge c\varepsilon|I| = \gamma|I|,$$

as required.

This completes the proof of the theorem for sequences (n_j) such that n_{j+1}/n_j is bounded, modulo the assertion that σ_{N+M} must change sign on I if M is large enough. This will follow from a classical result of Kahane, Weiss and Weiss: It is well known that if $f(t) = \sum_{j=1}^{N} a_j \cos(n_j t + \tau_j)$ is a (real-valued) lacunary series then

$$(1.10) \qquad \sup_{0 \le t \le 2\pi} f(t) \ge c \sum_{j=1}^{N} |a_j|.$$

(Inequality (1.10) says that a lacunary sequence of integers is a "Sidon set"; see [RU].)

It seems quite clear that one should in fact obtain

$$(1.11) \qquad \sup_{t \in I} f(t) \ge c_2 \sum_{j=1}^{N} |a_j|$$

if I is a large enough interval; it is almost equally clear that $|I| \ge c/n_1$ should be "large enough". This is proved in [KWW]:

LEMMA 2. *([KWW], Thm. 1) If $\lambda > 1$ then there exist $c_1 = c_1(\lambda)$ and $c_2 = c_2(\lambda)$ such that (1.11) holds whenever I is an interval with $|I| \ge c_1/n_1$.*

We return to the proof of our theorem: We have $t_o \in I$ with $|\sigma_N(t_o)| \le \frac{1}{2}$ and $|I| = 1/n_N$. Choose L so that $\lambda^L \ge c_1$, where c_1 is as in Lemma 2. It follows that $|I| \ge c_1/n_{N+L+1}$, so that Lemma 2 may be applied, with $f = \sigma_{N+M} - \sigma_{N+L}$; we obtain

$$(1.12) \qquad \sup_{t \in I}(\sigma_{N+M}(t) - \sigma_{N+L}(t)) \ge c_2(M - L)$$

for every $M > L$.

Note now that

$$(1.13) \qquad \sup_{t \in I} \sigma_N(t) \le c = c(\lambda)$$

by an estimate like (1.7), with N in place of $N + M$, because $\sigma_N(t_o) \le \frac{1}{2}$. Since $|\sigma_{N+L}(t) - \sigma_N(t)| \le L$, it follows from (1.12) and (1.13) that

$$(1.14) \qquad \sup_{t \in I} \sigma_{N+M}(t) \ge c_2(M - L) - (c + L).$$

In particular, $\sup_{t \in I} \sigma_{N+M}(t) > 0$ if M is large enough, where M depends only on λ (since all the other constants in (1.14) depend only on λ). Similarly, $\inf_{t \in I} \sigma_{N+M}(t) < 0$. $\qquad \square$

2. The General Case

In the previous section we proved the theorem for sequences (n_j) such that n_{j+1}/n_j is bounded; in the present section we shall remove this restriction.

We take $\sigma_N = \sum_{j=1}^{N} \cos(n_j t)$ as before, and define E_N as in (1.5). Note however that "F_N" and "M" will not have precisely the same meaning as in the previous section: We will fix $\varepsilon > 0$ and set

$$(2.0) \qquad F_N = \left\{ t \in [0, 2\pi] : \min_{0 \le k \le M} (|\sigma_{N+k}(t)|) < \varepsilon \right\},$$

where the value of $M = M(\lambda)$ is to be determined later.

As before, we need only show that if $t_o \in E_N$ then $t_o \in I$, where I is an interval with $|I| = \delta_N \to 0$ and $|I \cap F_N| \ge \gamma |I|$, $\gamma = \gamma(\lambda, \varepsilon)$; this will show that the sets F_N cover $[0, 2\pi]$ i.o.a.e., by Proposition 1. In the previous section we were able to find a "large" interval contained in $I \cap F_N$; this is impossible here, but we will show that $I \cap F_N$ must contain a large number of somewhat smaller intervals.

We take $c_1 = c_1(\lambda)$ and $c_2 = c_2(\lambda)$ as in Lemma 2. We will need a name for a constant which remained anonymous above: the argument in the previous section proves the existence of $c_0 = c_0(\lambda)$ such that *if $|I| \ge c_1/n_1$ and $|a| \le c_2 N/2$ then there exists an interval $J \subset I$, with $|J| \ge c_0 \varepsilon / n_N$, such that*

$$(2.1) \qquad |a - \sigma_N(t)| < \varepsilon \qquad\qquad (t \in J).$$

(This was proved above without using the assumption that n_{j+1}/n_j was bounded; the boundedness of n_{j+1}/n_j was used only to deduce that $|J| \ge c\varepsilon/n_1$, with $c = c(\lambda, N)$.)

LEMMA 3. *For $\varepsilon > 0$ and $N = 1, 2, \ldots$ there exists $c_3 = c_3(\lambda, \varepsilon, N)$ such that if I is an interval with $|I| \ge c_1/n_1$ and $a \in \mathbb{R}$, $|a| \le c_2 N/2$, then*

$$(2.2) \qquad |\{t \in I : |a - \sigma_N(t)| < \varepsilon\}| \ge c_3 |I|.$$

PROOF. The proof is by induction on N. The case $N = 1$ is clear. Suppose we have proved the lemma (for all $\varepsilon > 0$) for $N < N_o$; suppose for notational convenience that $c_3(\lambda, \varepsilon, N)$ is a nonincreasing function of N, for $N < N_0$.

Let $C = 2c_1/(c_0 \varepsilon)$. If $n_{j+1}/n_j \le C$ for all $j < N_0$, then (2.2) (for $N = N_0$) follows from (2.1).

Now suppose on the other hand that $n_{j+1}/n_j > C$ for at least one $j < N_0$, and choose k so that $n_{k+1}/n_k > C$ but $n_{j+1}/n_j \le C$ for $j < k$. Suppose $|a| \le c_2 N_0/2$, and set $a = a_1 + a_2$, where $a_1 = (k/N_0)a$.

It follows that $|a_1| \le c_2 k/2$, so that we may find an interval $J \subset I$, with $|J| \ge c_0 \varepsilon / 2 n_k$, such that

$$(2.3) \qquad |a_1 - \sigma_k(t)| < \varepsilon/2 \qquad\qquad (t \in J),$$

as in (2.1). Note that $|J| \ge c(\lambda, \varepsilon, N)/n_1$, because $n_k \le C^{k-1} n_1$.

But $n_{k+1}/n_k > C$; this shows that $|J| \ge c_1/n_{k+1}$, so that we may find finitely many pairwise disjoint intervals $J_1, J_2, \ldots J_\nu \subset J$ such that $|\cup_{i=1}^\nu J_i| \ge |J|/2$ and $|J_i| = c_1/n_{k+1} (1 \le i \le \nu)$. We note that $|a_2| \le c_2(N_0 - k)/2$, so that we may apply the induction hypothesis, with $\sigma_{N_0} - \sigma_k$ in place of σ_N and J_i in place of I, to obtain

$$(2.4) \qquad |\{t \in J_i : |a_2 - (\sigma_{N_0} - \sigma_k)| < \varepsilon/2\}| \ge c_3(\lambda, \varepsilon/2, N_0 - 1)|J_i|$$

for $1 \leq i \leq \nu$. Now if we add the inequalities (2.4) for $1 \leq i \leq \nu$ we obtain (2.4) with J in place of J_i; together with (2.3) this shows

$$(2.5) \qquad |\{t \in J : |a - \sigma_{N_0}| < \varepsilon\}| \geq c(\lambda, \varepsilon, N)|J| \geq c(\lambda, \varepsilon, N)/n_1.$$

Since $J \subset I$, this gives (2.2) (with $N = N_0$), at least if $|I|$ is comparable to c_1/n_1. If $|I| \gg c_1/n_1$ we may partition I into finitely many subintervals of length comparable to c_1/n_1 and apply the above argument on each subinterval. $\square$

The proof of the theorem is similar: Define E_N as above, and suppose $t_o \in E_N$. Let $I = [t_o - \alpha, \, t_o + \alpha]$, where $\alpha = 1/n_N$. By now it is easy to see that there exists $C = C(\lambda, \varepsilon)$ such that whenever $k \geq N$ and $n_{k+1}/n_k > C$ then I is a union of essentially disjoint intervals $I_1, I_2, \ldots I_\nu$ such that $|I_i| \geq c_1/n_{k+1}$, while at the same time $|I_i|$ is small enough that $|\sigma_k(s) - \sigma_k(t)| < \varepsilon/2$ for all s, $t \in I_i$. Fix such a constant C.

Now, as in the previous section, there exists $M_1 = M_1(\lambda)$ such that if $n_{j+1}/n_j \leq C$ for $N \leq j < N + M_1$, then $|\sigma_{N+M_1}(t)| < \varepsilon$ on a subinterval of I of length at least $c\varepsilon|I|$, which shows that $|I \bigcap F_N| \geq \gamma|I|$ if F_N is as in (2.0) and $M \geq M_1$.

On the other hand, suppose that $N \leq k < N + M_1$ is such that $n_{k+1}/n_k > C$. Then (as above) $I = \bigcup_{i=1}^{\nu} I_i$ where the I_i are essentially disjoint intervals, $|I_i| \geq c_1/n_{k+1}$, but $|\sigma_k(s) - \sigma_k(t)| < \varepsilon/2$ for all s, $t \in I_i$.

Set $a_i = \sigma_k(t_i)$, where t_i is the center of I_i, $1 \leq i \leq v$. Recall that $t_o \in I$ and $|\sigma_N(t_o)| \leq \frac{1}{2}$. Since $|I| = 1/n_N$ this shows that $|\sigma_N(t)| \leq c(\lambda)$ for all $t \in I$, as in the previous section. Hence

$$(2.6) \qquad |a_i| \leq c(\lambda) + M_1 = c_4(\lambda) \qquad (1 \leq i \leq v),$$

since M_1 depends only on λ. Choose M so large that $c_4 \leq c_2(\lambda)(M - M_1)/2$.

Now $|I_i| \geq c_1 n_{k+1}$ and $|a_i| \leq c_2(N + M - k)$, so that we may apply Lemma 3 to show that

$$(2.7) \qquad |\{t \in I_i : |-a_i - (\sigma_{N+M}(t) - \sigma_k(t))| < \varepsilon/2\}| \geq c_3(\lambda, \varepsilon/2, M)|I_i|.$$

But $|\sigma_k(t) - a_i| \leq \varepsilon/2$ for all $t \in I_i$, so that (2.7) implies

$$(2.8) \qquad |\{t \in I_i : |\sigma_{N+M}(t)| < \varepsilon\}| \geq c_3|I_i|.$$

As before, this leads to (2.8) with I in place of I_i, which shows that $|I \bigcap F_N| \geq \gamma(\lambda, \varepsilon)|I|$, with $\gamma = c_3(\lambda, \varepsilon/2, M(\lambda)) = \gamma(\lambda, \varepsilon)$. $\square$

REMARK. As suggested in the Introduction, it seems likely that one could use the method above to prove that $(\sigma_N(t))$ is dense in $\mathbb{R}$ for almost every t, if (a_j) is a sequence of real numbers satisfying (0.2) and (0.3), and $\sigma_N(t) = \sum_{j=1}^{N} a_j \cos(n_j t + \tau_j)$. (One might try splitting the sum into a series of consecutive blocks, of essentially constant L^2 norm, and then dealing with these blocks as the individual terms were handled above. One would not obtain uniform estimates on derivatives as above, but one could obtain L^2 estimates (or even BMO estimates) leading to uniform estimates on relatively large sets.) As with regard to the question of bounded versus unbounded n_{j+1}/n_j, it is hard to imagine that this version of the theorem could be false, because the distribution of the σ_N must be smoother than in the case $a_j = 1$. (Further, [AP1] shows that s_N, hence σ_N, must be recurrent if $a_j \to 0$; the fact that σ_N is recurrent in this case is also an easy consequence of our theorem.)

References.

[AMS] N. Asmar and S. Montgomery-Smith, *On the Distribution of Sidon Series*, Preprint.

[AN] J. Anderson, *Le Comportment Radial des Séries Lacunaires*, C. R. Acad. Sci. Paris **291** (1980), 83-85.

[AP1,2] J. Anderson and L. Pitt, *On Recurrence Properties of Certain Lacunary Series. I. General Results. II. The Series $\sum_1^n \exp(ia^n\theta)$*, J. Reine angew. Math. **270** (1987), 65-82, 83-96.

[BDC] J. Bretagnolle and D. Dacunha-Castelle, *Théorèmes Limites à Distance Finie pour les Marches Aléatoires*, Ann. Inst. H. Poincaré **4** (1968), 25-73.

[BH] D. A. Brannan and W. K. Hayman, *Research Problems in Complex Analysis*, Bull. London Math. Soc. **21** (1989), 1-35.

[CL] E. F. Collingwood and A. J. Lohwater, *The Theory of Cluster Sets*, Cambridge, 1966.

[GA] J. Garnett, *Bounded Analytic Functions*, Academic Press, 1981.

[KH] J. P. Kahane, *Lacunary Taylor and Fourier Series*, Bull. Amer. Math. Soc. **70** (1964), 251-263.

[KWW] J. P. Kahane, G. Weiss and M. Weiss, *On Lacunary Power Series*, Ark. Mat. **5** (1962), 1-26.

[ME] T. Murai, *Gap Series*, Analytic Function Theory of One Complex Variable, (Y. Komatu, K. Niino and C. Yang, ed.) Pitman Research Notes in Mathematics 212, Longman Science and Technology, New York, 1989.

[MJ] T. Murai, *Gap Series (Japanese)*, Sugaku **35** (1983), 35-49.

[RU] W. Rudin, *Trigonometric Series with Gaps*, J. Math. and Mech. **9** (1960), 203-228.

[UL] D. Ullrich, *Khinchine's Inequality and the Zeroes of Bloch Functions*, Duke Math. J. **57** (1988), 519-535.

[ZY] A. Zygmund, *Trigonometric Series*, 2nd edition, Cambridge, 1959.

DEPARTMENT OF MATHEMATICS, OKLAHOMA STATE UNIVERSITY, STILLWATER, OKLAHOMA 74078-0613

Contemporary Mathematics
Volume **137**, 1992

Maximum Modulus Algebras

JOHN WERMER

Dedicated to Walter Rudin.

1. I am going to speak about a theorem of Walter's from 1953. The paper
is in the Duke Journal, vol. 20, and is entitled ANALYTICITY AND THE
MAXIMUM MODULUS PRINCIPLE.

This paper gave birth to a class of function spaces called *maximum modulus
algebras*, which are defined in the following way:

Let X be a locally compact Hausdorff space and let A be an algebra of continuous complex valued functions defined on X, which separates points on X and
contains the constants. Assume that for each compact set K contained in X, if
x^0 is a point of K, then for every g in A,

$$|g(x^0)| \leq \max_{\partial K} |g|$$

where ∂K denotes the boundary of K.

We then say that *A is a maximum modulus algebra (MMA) on X*. Walter's
result can be stated as follows:

RUDIN'S THEOREM. *Let Ω be a domain in the z-plane and let A be a MMA
on Ω. If A contains the identity function z, then every function in A is analytic
on Ω.*

Walter proved other results in this Duke Journal paper as well, but this is the
one I shall focus on.

Where do we find examples of MMA's? Let $\mathcal{A}$ be any commutative Banach
algebra with unit. We denote by $\mathcal{M}$ the maximal ideal space of $\mathcal{A}$ and by $\check{S}$ the
Šilov boundary of $\mathcal{A}$. We suppose that $\mathcal{A}$ is semisimple. By Gelfand's theory, $\mathcal{A}$
can be regarded as an algebra of continuous functions on $\mathcal{M}$. The Šilov boundary
behaves like the boundary of a domain in analytic function theory, so one could
expect that the locally compact space $\mathcal{M}\backslash\check{S}$ behaves like a domain, and hence
one could expect that $\mathcal{A}$, restricted to $\mathcal{M}\backslash\check{S}$, is a MMA on $\mathcal{M}\backslash\check{S}$.

1991 *Mathematics Subject Classification*. Primary 32E20. Secondary 46J10.

This paper is in final form and no version of it will be submitted for publication elsewhere.

In the 1950's quite a few of us tried to prove this statement, that the functions in $\mathcal{A}$ satisfy a local maximum modulus principle away from the Šilov boundary, but we tried without success. The reason for this lack of success appears to be that the techniques used in the study of Banach algebras at the time were mostly based on analytic function theory in one complex variable, and these techniques were inadequate. The problem was finally solved by Hugo Rossi in 1960; his proof made use of the solution of the Cousin problem in a domain of holomorphy in $\mathbf{C}^n$.

ROSSI'S THEOREM. *The restriction of $\mathcal{A}$ to $\mathcal{M}\backslash\check{S}$ is a MMA on $\mathcal{M}\backslash\check{S}$.*

A special case of Rossi's theorem is obtained by fixing a compact set $Y \subseteq \mathbf{C}^n$ and taking $\mathcal{A} = P(Y)$, the algebra which is the uniform closure on Y of polynomials in $z_1, \cdots, z_n$. With norm $\|f\| = \max_Y|f|$, $P(Y)$ is a Banach algebra and its maximal ideal space can be identified with the *polynomial hull* $\hat{Y}$ of Y. $\hat{Y}$ is defined by

$$\hat{Y} = \{z \in \mathbf{C}^n \mid |Q(z)| \le \max_Y |Q| \text{ for all polynomials } Q\}.$$

Rossi's theorem gives: *the algebra of polynomials restricted to $\hat{Y}\backslash Y$ is a MMA on $\hat{Y}\backslash Y$.*

It turned out that this result had, effectively, been proved by Oka in 1937, in [15]. (Stolzenberg pointed this out.)

A second source of MMA's comes from the study of singularity sets. We consider a product domain $G \times \mathbf{C}$ in $\mathbf{C}^2$, where G is a domain in the λ-plane. We consider a relatively closed subset X of $G \times \mathbf{C}$ which is a *singularity set* in the sense that there exists a function ϕ analytic on $G \times \mathbf{C}\backslash X$ and singular at each point of X.

PROPOSITION. *Let A be the algebra consisting of all restrictions to X of polynomials in the coordinates. Then A is a MMA on X.*

For a proof see Wermer, [28], and Slodkowsky, [22].

In 1909 Hartogs proved the following

HARTOGS' THEOREM. *Let X be a singularity set in $G \times \mathbf{C}$ which is a graph over G, i.e. X is given by*

$$w = f(\lambda), \qquad \lambda \in G,$$

where f is a continuous function on G. Then f is an analytic function on G.

The converse is clear: if X is the graph of the analytic function f, we may put $\phi(\lambda, w) = (w - f(\lambda))^{-1}$. Then ϕ has X as its singularity set.

Hartogs' theorem follows directly from Rudin's theorem. We take A to be the algebra of all functions on G of the form $P(\lambda, f(\lambda))$, where P is a polynomial in λ and w. Since polynomials in λ, w are an MMA on X, by the Proposition, A

is a MMA on G. Also A contains the identity function λ. By Rudin's theorem, every function in A is analytic on G. In particular, f is analytic.

Hartogs, in his 1909 paper [9], generalized his result as follows: let X be a singularity set in $G \times \mathbf{C}$ which lies n-sheeted over G, rather than 1-sheeted, as a graph. Then X can be defined by an equation:

$$w^n + A_1(\lambda)\, w^{n-1} + \cdots + A_n(\lambda) = 0, \lambda \in G,$$

where each A_j is an analytic function of λ. In particular, X is an analytic set.

In 1934 Oka wrote a short paper in the Journal of Science of the Hiroshima University, [14], pp. 93-98, in which he extended Hartogs' result:

OKA'S THEOREM. *Let E be a subset of G of positive logarithmic capacity and let X be a singularity set in $G \times \mathbf{C}$ which lies n-sheeted over E. Then the same conclusion holds as in Hartogs' theorem: X is an n-sheeted analytic set lying over G.*

Oka's paper contained no proofs and was not included in his collected works published in 1961. In 1962, T. Nishino published a paper SUR LES ENSEMBLES PSEUDOCONCAVES in the Journal of Mathematics of the Kyoto University, [13], in which he gave the proofs of the results of Oka's paper, as well as a general theory of pseudoconcave sets. A "pseudoconcave set" in $G \times \mathbf{C}$ is a set X whose complement $G \times \mathbf{C} \backslash X$ is pseudoconvex, and hence X is a singularity set.

Independently of the study of pseudoconcave sets, Errett Bishop in 1963 proved a theorem about the maximal ideal space of a Banach algebra, [6]. We fix a Banach algebra $\mathcal{A}$ as above and let $\mathcal{M}$ and $\check{S}$ have their earlier meanings. We then choose a function f in $\mathcal{A}$ and we draw the image set $f(\check{S})$ under the map f. This set is a compact subset of $\mathbf{C}$. We next choose a closed disk $\Delta \subset \mathbf{C} \backslash f(\check{S})$. We put

$$f^{-1}(\Delta) = \{x \in \mathcal{M} \mid f(x) \in \Delta\}, \text{ and}$$

$$f^{-1}(\lambda) = \{x \in \mathcal{M} \mid f(x) = \lambda\}, \lambda \text{ a } fixed\,point.$$

#T denotes the cardinality of the set T , and m denotes planar Lebesgue measure.

BISHOP'S THEOREM. *Assume $\exists E \subset \Delta$ such that $\#f^{-1}(\lambda) < \infty$ for all λ in E and $m(E) > 0$. Then*

(i) $\exists$ integer n such that $\#f^{-1}(\lambda) \leq n$ $\forall \lambda \in \Delta$ and

(ii) $f^{-1}(\Delta)$ can be given the structure of a 1-dimensional analytic set such that every h in $\mathcal{A}$ is analytic on $f^{-1}(\Delta)$.

In [3] Bernard Aupetit and I replaced the assumption that $m(E) > 0$ by the assumption that the logarithmic capacity of $E, \mathrm{cap}(E), > 0$ in Bishop's theorem and obtained the same conclusion.

Oka's theorem and Bishop's theorem have similar structure: one assumes finite-sheetedness over a small set E and concludes the existence of analytic structure. One suspects that one single fact lies behind the two situations.

The common background, of course, is that there is a MMA lurking about in each case. We consider a MMA on a locally compact space X and we fix a function f in A. We assume that f is a proper map of X onto a plane domain Ω.

HARTOGS-OKA-NISHINO-BISHOP ETC. THEOREM. *Let A be a MMA on X and choose f and Ω as above. Assume that $\exists E \subset \Omega$ such that $\#f^{-1}(\lambda) < \infty \forall \lambda \in E$ and $\mathrm{cap}(E) > 0$. Then X can be made into a 1-dimensional analytic space such that every h in A is analytic on X.*

For a proof. see [20] and [29]. We note that this result implies both Oka's theorem and Bishop's theorem.

Suppose now that we have a MMA on a space X and a function f as above. but we make no assumption about the finiteness of the fibers $f^{-1}(\lambda)$. What analytic structure must then exist in X? If we assume that the fibers are countably infinite, then all is well and analytic structure exists on most of X. This was shown by Oka and Nishino in the pseudoconcave case and by Basener in the Banach algebra case.

If we allow uncountably infinite fibers, analytic structure may cease to exist. I gave an example in [30], in Arkiv för matematik 20, (1982), which was as follows: Y is a compact set over the unit circle, lying in $\mathbf{C}^2$. The λ-projection of $\hat{Y}$ is the disk $|\lambda| \leq 1$, (so we may take $f = \lambda$), but $\hat{Y}$ contains no analytic structure, i.e. every analytic map of a disk into $\hat{Y}$ is constant.

The fibers of $\hat{Y}$ are totally disconnected sets here, in general uncountably infinite.

A faint shadow of analyticity does remain for an arbitrary MMA lying over a plane domain, with no assumption about the fibers. The following result is due to Oka, [15], for polynomial hulls, and Oka's method was used by myself to prove it in general, in [27] and [28], and it was also proved, independently, by Senichkin in [19]. See also Yamaguchi, (32).

SUBHARMONICITY THEOREM. *Let A be a MMA on X and let f in A be a proper map of X on the plane domain Ω. For each g in A, define $Z_g(\lambda) = \max |g|$ over the fiber $f^{-1}(\lambda)$, for $\lambda \in \Omega$. Then $\lambda \to \log Z_g(\lambda)$ is subharmonic on Ω.*

Note: If X is a Riemann surface lying over Ω and A consists of functions analytic on X, this subharmonicity property follows from the fact that the logarithm of the modulus of an analytic function is subharmonic.

2. In this paragraph we shall discuss a special class of polynomial hulls about which much has been found out in the last ten years, namely the polynomial hulls of sets in $\mathbf{C}^2$ which lie over the unit circle.

We fix a compact set Y in $\mathbf{C}^2$ such that the λ-projection of Y is the circle $\Gamma : |\lambda| = 1$. For each λ in Γ we denote by Y_λ the image in the w-plane of the fiber over λ. Thus

$$Y_\lambda = \{w | (\lambda, w) \in Y\}.$$

Y_λ is a compact set in the w-plane which varies with λ. With $\hat{Y}$ denoting, as earlier, the polynomial hull of Y, we seek conditions on Y under which $\hat{Y}\backslash Y$ has analytic structure. We assume that $0 \in \lambda(\hat{Y})$. Then $\lambda(\hat{Y})$ is the full disk $|\lambda| \leq 1$.

We fix a point $(0, w_0) \in \hat{Y}$, and choose a representing measure μ on Y, i.e. a probability measure μ on Y such that

$$P(0, w_0) = \int_Y P(\lambda, w)d\mu(\lambda, w) \quad \forall \, polynomials \, P.$$

Let μ^* be the projection of μ on Γ by the map λ. Thus

$$\int_Y \psi(\lambda)d\mu(\lambda, w) = \int_\Gamma \psi(\lambda)d\mu^*(\lambda)$$

for all $\psi \in C(\Gamma)$. We put $\psi(\lambda) = \lambda^n$. Then

$$0 = \int_Y \lambda^n d\mu = \int_\Gamma \lambda^n d\mu^*(\lambda), n = 1, 2, \cdots.$$

Since μ^* is a probability measure on Γ, it follows that $\mu^* = \frac{d\theta}{2\pi}$.

We form $L^2(Y, \mu)$ and inside this space we put $\mathcal{C} = \{F \in L^2(Y, \mu) \mid F$ is constant on each fiber $\}$. Each F in $\mathcal{C}$ has the form $F = \psi(\lambda)$ with ψ a function on Γ, and we see that $\|F\|_{L^2(Y,\mu)} = \|\psi\|_{L^2(\Gamma)}$, and so we can identify $\mathcal{C}$ with $L^2(\Gamma)$.

The coordinate function $w \in L^2(Y, \mu)$. We denote by W the orthogonal projection of w on $\mathcal{C}$. Then $w - W$ is orthogonal to $\mathcal{C}$, so $\int_Y (w - W)\overline{g}d\mu = 0 \, \forall g \in \mathcal{C}$, whence

$$(\dagger) \qquad \int_Y w\overline{g}d\mu = \int_Y W\overline{g}d\mu = \int_\Gamma W(\lambda)\overline{g(\lambda)}\frac{d\theta}{2\pi}.$$

With $g = \overline{\lambda}^n$, this gives

$$0 = \int_Y w\lambda^n d\mu = \int_\Gamma W(\lambda)\lambda^n \frac{d\theta}{2\pi}, n = 1, 2, \cdots.$$

Since $W \in L^2(\Gamma)$, it follows that W is in the Hardy space H^2.

Fix now $\lambda_0 = e^{i\theta_0} \in \Gamma$ and let α_ϵ be an arc on Γ of length 2ϵ centered at λ_0. Put

$$g_\epsilon(\lambda) = \frac{2\pi}{2\epsilon}\chi_\epsilon(\lambda),$$

where χ_ϵ is the characteristic function of α_ϵ. Then with $g = g_\epsilon, (\dagger)$ gives

$$\int_Y w\frac{2\pi}{2\epsilon}\chi_\epsilon(\lambda)d\mu = \int_\Gamma W(\lambda)\frac{2\pi}{2\epsilon}\chi_\epsilon(\lambda)\frac{d\theta}{2\pi} = \frac{1}{2\epsilon}\int_{\theta_0-\epsilon}^{\theta_0+\epsilon} W(\lambda)d\theta.$$

The measure $\frac{2\pi}{2\epsilon}\chi_\epsilon(\lambda)d\mu$ is a probability measure supported on $\lambda^{-1}(\alpha_\epsilon) \subseteq Y$. We denote by σ a weak-* limit of these measures, as $\epsilon \to 0$. Then

$$\int_{\lambda^{-1}(\lambda_0)} wd\sigma(\lambda) = W(\lambda_0), \, for \, a.a. \, \lambda_0 \in \Gamma.$$

We note that $\lambda^{-1}(\lambda_0) = \{\lambda_0\} \times Y_{\lambda_0}$. Since σ is a probability measure supported on $\lambda^{-1}(\lambda_0)$, it follows that $W(\lambda_0)$ lies in the convex hull of Y_{λ_0}. It follows that $W \in L^\infty(\Gamma)$, and so $W \in H^\infty$. We have shown

RESULT. $\exists W \in H^\infty$ such that $W(\lambda) \in co(Y_\lambda)$ for a.a. λ in Γ.

Moreover, by (†) with $g = 1$,

$$w_0 = \int_\Gamma W(\lambda)\frac{d\theta}{2\pi} = W(0).$$

Let us now assume that for each λ in Γ, the fiber Y_λ is convex. Then $W(\lambda) \in Y_\lambda$, a.a. $\lambda \in \Gamma$, and so $(\lambda, W(\lambda))$ belongs to Y for a.a $\lambda \in \Gamma$.

If P is a polynomial on $\mathbf{C}^2$ and a is a point in $|\lambda| < 1$, then the maximum principle gives

$$|P(a, W(a))| \leq \|P(\lambda, W(\lambda))\|_\Gamma \leq \max_Y |P|.$$

Thus the point $(a, W(a)) \in \hat{Y}$, and so the entire graph: $w = W(\lambda), |\lambda| < 1$, is contained in $\hat{Y}$. We thus have found:

If Y_λ is convex for each $\lambda \in \Gamma$, then through each point $(0, w_0)$ in $\hat{Y}$ there passes an analytic graph $w = W(\lambda)$ which lies over $|\lambda| < 1$ and is contained in $\hat{Y}$.

Using Möbius transformations we can draw a similar conclusion for each point (λ_0, w_0) in $\hat{Y}$ with $|\lambda_0| < 1$. We conclude

THEOREM 1. If Y_λ is convex $\forall \lambda \in \Gamma$, then $\hat{Y} \setminus Y$ is the union of all those bounded analytic graphs $w = \phi(\lambda)$ whose boundary over Γ is contained in Y.

Theorem 1 was proved in 1984 by Herbert Alexander and myself in [1] and independently by Slodkowski in [24]. Related results are due to Senichkin, [20].

If the fibers Y_λ are not merely convex, but actually are disks $|w - c(\lambda)| \leq r(\lambda), \lambda \in \Gamma$, with variable center and radius, more is true, [1]: Either $\hat{Y} \setminus Y$ is a single analytic disk over $|\lambda| < 1$, or $\hat{Y} \setminus Y$ is defined by an inequality

$$|\Phi(\lambda, w)| \leq 1, |\lambda| < 1,$$

where Φ is linear fractional in w with coefficients analytic in λ:

$$\Phi(\lambda, w) = \frac{A(\lambda)w + B(\lambda)}{C(\lambda)w + D(\lambda)}, A, B, C, D \, analytic \, on \, |\lambda| < 1.$$

Thus $\hat{Y} \setminus Y$ is a domain with boundary lying in $|\lambda| < 1$. The boundary consists of two pieces: the piece $|\Phi| = 1, |\lambda| < 1$, which is a Levi-flat hypersurface, and the piece Y, lying over Γ.

If each fiber Y is not a disk, and is not assumed convex, but is a Jordan domain which varies smoothly with $\lambda, \lambda \in \Gamma$, the problem of identifying $\hat{Y}$ is much harder. It was solved, with one restriction, by Franc Forstnerič in his paper [7] in the Indiana Univ. Math. J., vol. 37 (1988). He proved

THEOREM 2. *Assume that Y_λ is a Jordan domain varying smoothly with λ in Γ, such that $0 \in int Y_\lambda \, \forall \lambda \in \Gamma$. Then $\hat{Y} \backslash Y$ is a domain with boundary, whose boundary is the union of a Levi-flat hypersurface over $|\lambda| < 1$ and the set Y. Again $\hat{Y} \backslash Y$ is a union of analytic graphs over $|\lambda| < 1$. We note that the hypothesis $0 \in int Y_\lambda$ directly implies that the disk $w = 0, |\lambda| < 1$ is contained in $\hat{Y}$.*

This hypothesis was removed independently by Z. Slodowski in his paper *Polynomial Hulls in $\mathbf{C}^2$ and Quasicircles*, and by J. W. Helton and D. E. Marshall in their paper *Frequency Domain Design and Analytic Selections* in [10]. More generally, Slodkowski showed in his paper

THEOREM 3. *Assume that each Y_λ is simply connected and connected. Then $\hat{Y} \backslash Y$ is a union of bounded analytic graphs over $|\lambda| < 1$.*

See also [10], Theorem 2.

What can be said about $\hat{Y}$ if we make no assumption about the fibers? As we noted earlier, analytic structure may fail to exist in $\hat{Y} \backslash Y$, so Theorems 1, 2, 3 have no direct generalisation to arbitrary Y.

In the cases we have been considering where there is analytic structure, we found that through each point (λ_0, w_0) in $\hat{Y} \backslash Y$ there passes an analytic graph $\sum$ such that the boundary of $\sum$ is contained in Y in the sense that, if $\sum$ has the equation: $w = f(\lambda)$, then $f(\lambda) \in Y_\lambda$ for a.a. λ in Γ.

Let us fix such a graph: $w = f(\lambda)$ and let us put

$$\phi(\lambda, w) = (w - f(\lambda))^{-1}.$$

Then ϕ has the properties

(1) $\phi(\lambda, w)$ is analytic and bounded in $|w| > M$, where M is a constant, and

(2) For $\lambda \in \Gamma, w \to \phi(\lambda, w)$ is analytic outside Y_λ .

In fact,

$$\phi(\lambda, w) = \int_{Y_\lambda} \frac{d\sigma_\lambda(\zeta)}{w - \zeta}$$

where σ_λ is the unit point mass at $\zeta = f(\lambda)$.

Restricting ϕ to the complex line $\lambda = \lambda_0$ we have

$$\phi(\lambda_0, w) = (w - w_0)^{-1}, |w| > M.$$

Let now Y be an arbitrary compact set in $\mathbf{C}^2$ lying over Γ. Without loss of generality we assume that $Y \subset \{|w| < 1\}$. We generalize the preceding as follows:

We consider functions F of λ and w with the properties:

(3) F is analytic and bounded in $|\lambda| < 1, |w| > 1$. By Fatou's theorem, then, $F(\lambda, w)$ is defined a.e. on $|\lambda| = 1$ if $|w| > 1$.

(4) For a.a. $\lambda \in \Gamma, \exists$ probability measure σ_λ on Y_λ such that

$$F(\lambda, w) = \int_{Y_\lambda} \frac{d\sigma_\lambda(\zeta)}{w - \zeta}, |w| > 1.$$

We denote by $\mathcal{F}(Y)$ the space of all functions F satisfying (3) and (4). Fix a point (λ_0, w_0) in $|\lambda| < 1$, in $\mathbf{C}^2$.

CLAIM. *Suppose $\exists F \in \mathcal{F}(Y)$ such that $F(\lambda_0, w) = (w - w_0)^{-1}, |w| > 1$. Then $(\lambda_0, w_0) \in \hat{Y}$.*

PROOF. Fix a polynomial P in λ, w and write $\|P\|_Y = \max_Y |P|$. Define

$$\chi(\lambda) = \frac{1}{2\pi i} \int_{|w|=2} F(\lambda, w) P(\lambda, w) dw.$$

Because of (3), χ is bounded and analytic in $|\lambda| < 1$. From (4) we easily deduce that for a.a. $\lambda \in \Gamma$,

$$|\chi(\lambda)| \leq \max_{w \in Y_\lambda} |P(\lambda, w)| \leq \|P\|_Y.$$

Hence $|\chi(\lambda_0)| \leq \|P\|_Y$. Now

$$P(\lambda_0, w_0) = \frac{1}{2\pi i} \int_{|w|=2} \frac{P(\lambda_0, w)}{w - w_0} dw = \frac{1}{2\pi i} \int_{|w|=2} P(\lambda_0, w) F(\lambda_0, w) dw = \chi(\lambda_0).$$

Hence $|P(\lambda_0, w_0)| = |\chi(\lambda_0)| \leq \|P\|_Y$. Since P is arbitrary, we get our assertion.

The converse also holds. Alexander and I showed in [31] and [2]:

THEOREM 4. *Let Y be an arbitrary compact set in $\mathbf{C}^2$ over Γ with $Y \subset \{|w| < 1\}$. A point (λ_0, w_0) in $|\lambda| < 1$ in $\mathbf{C}^2$ belongs to $\hat{Y}$ if and only if $\exists F$ in $\mathcal{F}(Y)$ such that $F(\lambda_0, w) = (w - w_0)^{-1}$. Moreover, every F in $\mathcal{F}(Y)$ has a single valued analytic extension from $\{|\lambda| < 1\} \times \{|w| > 1\}$ to $\{|\lambda| < 1\} \backslash \hat{Y}$.*

Theorem 4 represents the polynomial hull of Y as the complement in $|\lambda| < 1$ of the "envelope of holomorphy " of the function space $\mathcal{F}(Y)$.

Note: Related results were given by Slodkowski in [23] and [22]. Analogous results concerning the analytic extension of CR-functions from the boundary of a given domain are due to Stout, [26], and Luppaciolu, [12].

3. Finally, we look at the possible generalization of Rudin's Theorem to the case when Ω is a domain in $\mathbf{C}^n$ with $n > 1$. Let A be an algebra of continuous functions on Ω which contains the coordinate functions $z_1, \cdots, z_n$ and contains the constants. We seek conditions on A which will force every function in A to be analytic in Ω.

Clearly the condition that A is a MMA on Ω is not adequate. We may, for instance, take Ω to be the open bidisk $|z| < 1, |w| < 1$ in $\mathbf{C}^2$ and let A be the algebra of all continuous functions $f(z, w)$ on Ω which are such that for each fixed w_0 the map $z \to f(z, w_0)$ is analytic on $|z| < 1$. Then A is a MMA on Ω, but A does not consist of functions analytic on Ω. One must assume more.

THEOREM. *Let A be an algebra of continuous functions on the domain Ω in $\mathbf{C}^n$ such that the coordinate functions and the constants belongs to A. Assume* *that for each complex line L such that $\Omega \cap L \neq \phi$ the restriction $A|_L$ of A to $\Omega \cap L$ is a MMA on $\Omega \cap L$. Then every function in A is analytic on Ω.*

PROOF. Fix a complex line L with $\Omega \cap L \neq \phi$. The restriction algebra $A|_L$ satisfies the hypothesis of Rudin's theorem and hence for each f in A, $f|_L$ is analytic on $L \cap \Omega$. By the classical theorem of Hartogs, it follows that f is analytic on Ω, and we are done.

Substantial results are known which contain the simple theorem just given as a very special case, and which also generalize the Hartogs-Oka-Nishino-Bishop etc. Theorem to situations of higher dimensional analytic structure. This work is discussed in Chapters 14, 15 and 16 of Helmut Goldmann's recent book, UNIFORM FRÉCHET ALGEBRAS, [8]. In particular, there are the papers of Basener, [5], Goldmann, (see[8]), Kumagai, [11], Rusek, [18], and Sibony, [21].

References.

1. H. Alexander and J. Wermer, *Polynomial hulls with convex fibers*, Math. Ann. **271** (1985), 99-109.

2. _____, *Envelopes of holomorphy and polynomials hulls*, Math. Ann. **281** (1988), 13-22.

3. B. Aupetit and J. Wermer, *Capacity and uniform algebras*, J. Funct. Anal. **28** (1978), 386-400.

4. R. Basener, *A condition for analytic structure*, Proc. Amer. Math. Soc. **36** (1972), 156-160.

5. _____, *A generalized Šilov boundary and analytic stucture*, Proc. Amer. Math. Soc. **47** (1975), 98-104.

6. E. Bishop, *Holomorphic completions, analytic continuations and the interpolation of semi-norms*, Ann. of Math. **78** (1963), 468-500.

7. F. Forstnerič, *Polynomial hulls of sets fibered over the circle*, Indiana Univ. Math. Jour. **37** (1988), 869-889.

8. H. Goldmann, *Uniform Fréchet Algebras*, North-Holland, 1990.

9. F. Hartogs, *Über die aus den singulären Stellen einer analytischen Funktion mehrerer Veränderlichen bestehenden Gebilde*, Acta. Math. **32** (1909), 57-79.

10. J. W. Helton and D. Marshall, *Frequency domain design and analytic selections*, Indiana Univ. Math. Jour. **39** (1990), 157-184.

11. D. Kumagai, *Maximum modulus algebras and multidimensional analytic structure*, Contemp. Math. **32** (1984), 163-168.

12. G. Luppaciolu, *A theorem on holomorphic extension of CR-functions*, Pac. Jour. of Math. **124** (1986), 177-191.

13. T. Nishino, *Sur les ensembles pseudoconcaves*, Jour. of Math. of Kyoto Univ. **1** (1962), 225-245.

14. K. Oka, *Note sur les familles de fonctions analytiques multiforms etc.*, Jour. of Sci. of the Hiroshima Univ. **4** (1934), 93-98.

15. _____, *Domaines d'holomorphie*, Jour. of Sci. of the Hiroshima Univ. **7** (1937), 115-130.

16. H. Rossi, *The local maximum modulus principle*, Ann. of Math. **72** (1960), 1-11.

17. W. Rudin, *Analyticity, and the maximum modulus principle*, Duke Math. Jour. **29** (1953), 449-457.

18. K. Rusek, *Analytic structure on locally compact spaces determined by algebras of continuous functions*, Ann. Polon. Math. **42** (1983), 301-307.

19. V. Senichkin, *The maximality of the algebra of functions etc.*, (English translation), Jour. Soviet Math. **2** (1974), 166-173.

20. _____, *Subharmonic functions and the analytic structure in the maximal ideal space of a uniform algebra*, Math. USSR-Sb **36** (1980), 111-126.

21. N. Sibony, *Multidimensional analytic structure in the spectrum of a uniform algebra*, Bekken, O.B. et al, eds., Spaces of analytic functions, Kristiansand, Norway (1975), Lecture Notes in Math. 512, Springer-Verlag, 1976, pp. 139-165.

22. Z. Slodkowski, *Analytic set valued functions and spectra*, Math. Ann. **256** (1981), 363-386.

23. _____, *Uniform algebras and analytic multifunctions*, Atti. Acad. Lincei, Rend. 75 (1983), pp. 9-18.

24. _____, *Polynomial hulls with convex sections and interpolating spaces*, Proc. Amer. Math. Soc. **96** (1986), 255-260.

25. _____, *Polynomial hulls in* $\mathbf{C}^2$ *and quasi-circles*, Ann. S. N. di Pisa **XVI** (1989), 367-391.

26. E. L. Stout, *Analytic continuation and boundary continuity of functions of several complex variables*, Proc. Royal Soc. Edinburgh **89** (1981), 63-74.

27. J. Wermer, *Subharmonicity and hulls*, Pac. Jour. of Math. **58** (1975), 283-290.

28. _____, *Maximum modulus algebras and singularity sets*, Proc. Royal Soc. Edinburgh **86A** (1980), 327-331.

29. _____, *Potential theory and function algebras*, Texas Tech Univ. Math. Series **14** (1981), 113-125.

30. _____, *Polynomially convex hulls and analyticity*, Arkiv för matem **20** (1982), 129-135.

31. _____, *Balayage and polynomial hulls*, Proc. of the Special Year in Complex Analysis, Univ. of Maryland (1987), Lecture Notes in Mathematics 1276, Springer-Verlag, pp. 303-312.

32. H. Yamaguchi, *Sur une uniformité des surfaces constantes d'une fonction entière de deux variables complexes*, Jour. Math. Kyoto Univ. **13** (1973), 417-433.

DEPARTMENT OF MATHEMATICS, BROWN UNIVERSITY, PROVIDENCE, RHODE ISLAND 02912